Trace Element Analytical Chemistry in Medicine and Biology

Trace Element Analytical Chemistry in Medicine and Biology

Volume 3

Proceedings of the Third International Workshop
Neuherberg, Federal Republic of Germany,
April 1984

Editors
Peter Brätter · Peter Schramel

W DE G

Walter de Gruyter · Berlin · New York 1984

Editors

Peter Brätter, Professor, Dr. rer. nat.
Hahn-Meitner-Institut für Kernforschung GmbH Berlin
Arbeitsgruppe Spurenelementforschung in der Biomedizin
Glienicker Straße 100
D-1000 Berlin 39

Peter Schramel, Dr. phil.
Gesellschaft für Strahlen- und Umweltforschung mbH
Physikalisch-Technische Abteilung
Ingolstädter Landstraße 1
D-8402 Neuherberg

CIP-Kurztitelaufnahme der Deutschen Bibliothek

Trace element analytical chemistry in medicine and biology :
proceedings of the . . . internat. workshop / . . . Internat. Workshop on Trace Element Analyt. Chemistry in Medicine and Biology. – Berlin ; New York : de Gruyter

NE: International Workshop on Trace Element Analytical Chemistry in Medicine and Biology

Vol. 3. Neuherberg, Federal Republic of Germany, April 1984. – 1985.
ISBN 3-11-009821-0

Library of Congress Cataloging in Publication Data

Trace element analytical chemistry in medicine and biology

Includes bibliographical references and indexes.
Contents: [v. 1.] Proceedings of the first international workshop, Neuherberg, Federal Republic of Germany, April 1980--
v. 2. Proceedings of the second international workshop, Neuherberg, Federal Republic of Germany. April 1982--v. 3. Proceedings of the third international workshop, Neuherberg, Federal Republic of Germany, April 1984.
1. Trace elements in the body--Congresses. 2. Trace elements--Analysis--Congresses. 3. Biological specimens--Analysis--Congresses. 4. Chemistry, Analytic--Congresses.
I. Brätter, Peter, 1935- . II. Schramel, Peter.
QP534.T7 616.07'56 80-26803
ISBN 3-11-008357-4 (v. 1)

 Printing: Gerike GmbH, Berlin. Binding: Lüderitz & Bauer Buchgewerbe, Berlin. Printed in Germany.

PREFACE

During the past two decades Trace Element Analytical Chemistry has become an indispensable part of biomedical research. Many of the recent advances in biochemistry, medicine, biology and nutrition research would have been impossible without the capacities of modern Trace Element Chemistry. New techniques, especially for speciation analysis, have been developed and have contributed to the understanding of the role and the function of trace elements in the living organism. Nevertheless we are just at the beginning and most of the problems are still unsolved. For example, recommended values for daily intake, as well as data for biovailability, distribution and interferences are not yet available. Therefore, the interaction between different disciplines and the mutual transfer of scientific knowledge must continue and should be intensified.

In an attempt to encourage interdisciplinary interaction the "International Workshop on Trace Element Analytical Chemistry in Medicine and Biology" is organized bianually by the GSF (Gesellschaft für Strahlen- und Umweltforschung mbH) and the Working Group "Trace Elements in the Life Sciences" of the AGF (Arbeitsgemeinschaft der Großforschungseinrichtungen).

The 3rd Workshop was specifically concerned with new developments in the field of the essential trace elements selenium, zinc and manganese as well as with current problems in analysis, nutrition and medicine. The actual state of knowledge about other recently recognized essential trace elements also played a dominant role. The increasing importance of selenium was taken into account by reserving one day for a detailed discussion on all relevant problems.

The workshop comprised 11 plenary lectures, 14 oral presentations and 46 posters. For the first time, an additional report was given and a discussion held on each poster in the auditorium, so that each author had the opportunity to point out the "highlights" of the presentation. All the contributions reflected the growing importance of interdisciplinary research.

These Proceedings contain the contributed lectures, papers and posters accepted by the Scientific Committee. In order to ensure rapid publication, the manuscripts were not extensively edited. The discussions after each session were compiled from tape recordings.

The editors would like to thank all participants for their contributions. Our thanks are also due to the chairmen of the individual sessions and to all who contributed to the discussions.

August 1984

P. Brätter
P. Schramel

CONTENTS

Session III: SELENIUM ANALYSIS
Plenary Lecture given by: G. Tölg

Session V: SPECIATION OF TRACE ELEMENTS
Plenary Lecture given by: U. Weser

Session VI: TRACE ELEMENTS IN CARDIOVASCULAR DISEASES

Session I

SELENIUM METABOLISM

Plenary Lecture given by: H.E. Ganther

Chairman:

M. Kirchgessner (FRG)

SELENIUM METABOLISM AND FUNCTION IN MAN AND ANIMALS

Howard E. Ganther
Department of Nutritional Sciences, University of Wisconsin, Madison, Wisconsin, 53706 USA

Introduction

Selenium is an essential trace element, a potent modifier of environmental hazards, and a naturally-occurring toxicant (for a comprehensive survey of research on this element, see reference 1). These biological effects occur at Se concentrations in the range of 0.05 to 5 μg/g of diet or tissue. In man and animals, the one well-defined function of Se is in glutathione peroxidase, catalyzing the destruction of peroxides by means of GSH. In this enzyme Se undergoes alternate oxidation and reduction, and an intermediate oxidized form containing glutathione covalently bound to Se can be isolated (2). There is evidence for other functions independent of glutathione peroxidase but no selenoenzyme other than glutathione peroxidase has been identified in animals (3). The most evident medical use for Se is to correct the deficiency that develops with long-term total parenteral nutrition (4). There is considerable interest in supplementing the human dietary intake to correct nutritional deficiencies, or with the objective of modifying the carcinogenic process. There is only limited knowledge of the chemical forms of Se in foods and the intermediary metabolism that converts ingested forms of Se to biologically active forms such as selenocysteine in glutathione peroxidase.

Trace Element - Analytical Chemistry in Medicine and Biology, Vol. 3

Naturally-occurring forms of selenium include selenoamino acids in proteins. Selenomethionine is a constituent of wheat (5), and selenocysteine occurs in animal tissues in the form of glutathione peroxidase (6,7). Plants take up inorganic forms of Se and make numerous selenium analogs of sulfur compounds (8); selenocystathionine occurs in nuts, and methylated derivatives such as Se-methylselenocysteine and Se-methylselenomethionine selenonium occur in species of *Astragalus* that accumulate high concentrations of Se. Little is known about the forms of Se in marine fish, but there is evidence for a low biological availability of Se in tuna (9) that may be related to the intrinsic chemical form of the element. The complex food chain for marine organisms can create novel chemical forms of trace elements, as shown for arsenic (10). Selenium may share with arsenic a general tendency to occur in methylated forms. Novel forms of Se occur in nucleic acids in the form of selenobases (11). There is little evidence for the occurrence of organoselenium compounds in lipids or carbohydrates, although Se is sometimes associated with these classes of substances (12,13).

Intermediary metabolism of selenium

In the absence of better information about the chemical nature of Se ingested in foods, the focus of any discussion on metabolic transformations must be the fate of inorganic selenium salts, and of selenoamino acids found in proteins. Sodium selenate and sodium selenite are converted to physiologically active forms of selenium, including selenocysteine in glutathione peroxidase. These inorganic forms are not present to any extent in foods, but are widely used in nutritional studies and as supplements for human consumption.

Unlike sulfur, selenium is reluctant to assume the +6 oxidation state (14), and selenate is not a major product of selenium intermediary metabolism. However, selenate given orally or by injection is rapidly excreted in urine, and its urinary excretion can be increased by sulfate administration (15). Although there is no pathway for sulfate reduction in monogastric animals, selenate reduction must occur because dimethyl selenide and other reduced forms of selenium are formed from selenate by animals. Direct information on selenate reduction in animal tissues is lacking, but a mechanism similar to one proposed for selenate reduction in yeast and plants may be involved (16). The first step would be activation of selenate by means of ATP and sulfurylase, forming adenosine-5'-selenophosphate. A non-enzymic thiolytic cleavage of the phosphoselenate anhydride by thiols such as GSH was postulated to give selenite:

$$SeO_4^{=} + ATP \xrightarrow[Mg^{++}]{\text{sulfurylase}} \text{adenosine-5'-selenophosphate} + \text{pyrophosphate}$$

$$\text{adenosine-5'-selenophosphate} \xrightarrow{GSH} \underset{\text{"thio-selenic acid"}}{GSSeO_3^-} \xrightarrow{GSH} HSeO_3^-$$

Selenite, in contrast to sulfite, is a fairly strong oxidizing agent, and reduction to red or gray elemental forms is frequently observed upon exposure to organic matter. This property can cause practical problems when selenite is incorporated into diets as a pre-mix with reducing sugars, or added to parenteral nutrition formulas containing ascorbic acid and glucose (and sometimes containing bisulfite added as a preservative). The tendency toward reductive metabolism of selenite reflects the differing chemical properties of sulfur and selenium at the +4 oxidation state (Figure 1). Oxidation of selenite by sulfite oxidase, the enzyme catalyzing the final step in sulfur oxidation, is not known to occur. Evidence is lacking for the formation of selenite from selenoamino acids, whereas sulfite can be formed by metabolism of cysteine to cysteine sulfinate and ß-sulfinyl pyruvate (17). In cells, reduction of selenite is catalyzed by glutathione reductase, following an initial non-enzymic reaction of selenite with glutathione (18) (Figure 2). The first isolatable product, GSSeSG, is very similar to the common substrate for the enzyme (GSSG). In yeast cells elemental Se is deposited and there is no evidence for further reduction. Elemental Se may arise by spontaneous decomposition of the rather labile selenopersulfide product, GSSeH, as demonstrated *in vitro* (18). However, fungi and animals are known to metabolize selenite to dimethyl selenide, indicating that some mechanism exists for reducing Se to the -2 oxidation state. Studies with crystalline yeast glutathione reductase (19) or glutathione reductase isolated from rat liver or kidney (20) showed that acid-volatile selenium having the properties of hydrogen selenide was formed by NADPH-linked reduction of GSSeH (the initial reduction product). The use of anaerobic conditions was essential for good activity in broken-cell preparations, because of the sensitivity of H_2Se to oxidation. These observations explain many of the characteristics of the cell-free liver system that converts selenite to dimethyl selenide, including the specific requirement for GSH and NADPH. Although non-enzymic reduction of selenite to H_2Se is possible using a large excess of a thiol, it is important to note that NADPH is required for optimal activity of the glutathione reductase system even in the presence of a high level of GSH. Evidence that hydrogen selenide is formed in whole cells was provided by a study showing that erythrocytes take up selenite and expel a product having the properties of H_2Se, which becomes tightly associated with plasma proteins (21).

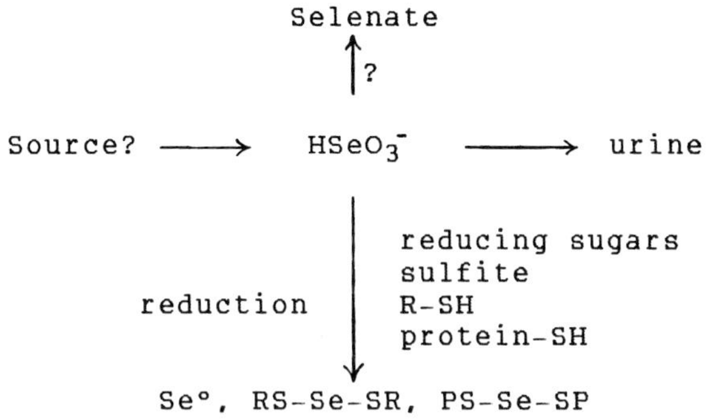

Fig. 1. Selenite Metabolism

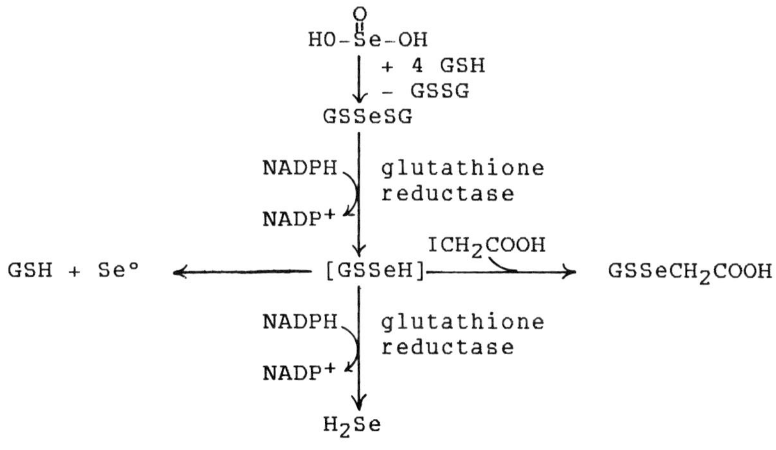

Fig. 2. Glutathione Reductase Pathway of Selenite Reduction

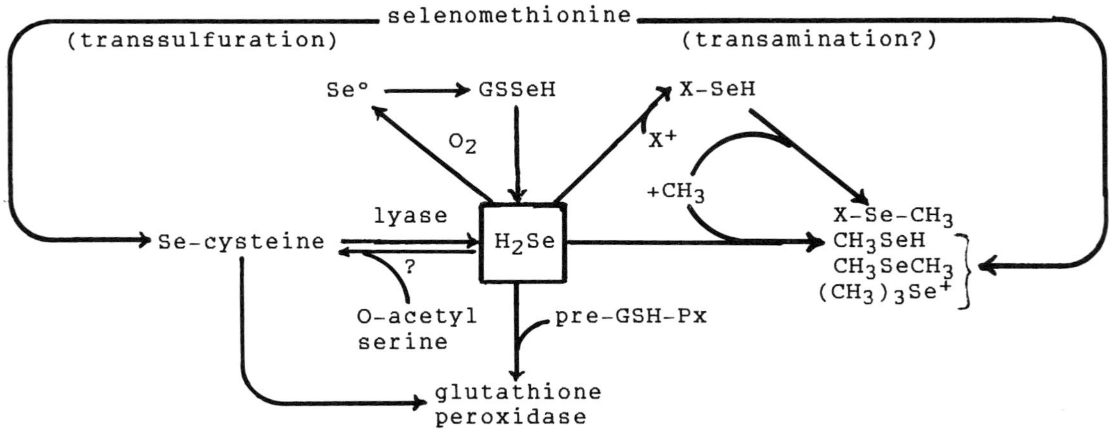

Fig. 3. Selenide metabolism

The key to verifying the course of the glutathione reductase pathway was to carry out the reaction in the presence of iodoacetate, thus trapping the labile selenopersulfide as the stable carboxymethyl derivative, GSSe-CH_2COOH (Figure 2). The lack of good methods for identifying H_2Se prompted the development of a method based on the reaction of H_2Se with Sanger's reagent (FDNB), which allows identification by mass spectrometry (see below).

Besides the formation of H_2Se by reduction of inorganic forms of Se, the enzyme-catalyzed release of H_2Se by a pyridoxal phosphate-dependent enzyme isolated from liver has been reported (22). Because of the relatively high K_M of the lyase compared to the low concentration of selenocysteine that is likely to be present in tissues, the significance of this reaction as a source of H_2Se is uncertain. The selenocysteine lyase may be a source of H_2Se when the cellular sources of selenium are selenoamino acids, whereas the glutathione reductase pathway may be more important when inorganic forms of Se are provided.

Hydrogen selenide, existing as HSe^- at physiological hydrogen ion concentrations, is highly reactive and is believed to have a central role in selenium metabolism (Figure 3). Hydrogen selenide is converted to dimethyl selenide by S-adenosylmethionine and a microsomal enzyme (20) that is similar to or identical with thiol-S-methyltransferase (23). It is possible that H_2Se undergoes reactions with electrophilic metabolites (X^+, Figure 3) to form selenol derivatives of lower toxicity. These might undergo further methylation or deselenomethylation, similar to sulfur analogs. H_2Se certainly can undergo reactions with metals to form insoluble complexes that reduce the biological availability of both the selenium and the metal. This interaction may play a role in reducing the toxicity of metals when large amounts of selenium relative to metal are given, but it cannot account for the ability of selenium at nutritional levels to modify metal toxicity where the metal is in large excess (24). Complexation of Se by a metal can explain the ability of silver and other metals to induce a conditioned deficiency of Se and block the biosynthesis of glutathione peroxidase in animals given an otherwise adequate level of Se (24). Utilization of Se for anabolic purposes via H_2Se is believed to be an important process, and biosynthesis of glutathione peroxidase by rat liver perfused with selenite has been demonstrated (25). Other selenoproteins (functional and nonfunctional) as well as other naturally occurring forms of Se such as selenobases in nucleic acids could also be derived from selenide metabolism.

There are two postulated mechanisms for incorporating Se into glutathione peroxidase - translational and posttranslational. The problem with the translational mechanism is how to make the process specific for selenocysteine in the presence of cysteine. Incorporation of inorganic selenium into selenocysteine of glutathione peroxidase by either mechanism is believed to involve H_2Se as an intermediate. Synthesis of selenocysteine by transfer of H_2Se to a suitable 3-carbon acceptor such as O-acetyl serine would provide the preformed amino acid for translational incorporation (Figure 3). The posttranslational hypothesis is favored by precedent, since there are numerous examples of proteins with unusual amino acid residues formed by modifying a polypeptide chain precursor. There is also direct experimental evidence with perfused livers showing that inorganic forms of Se are incorporated more readily into glutathione peroxidase compared to selenocystine (25). Thus, a pre-glutathione peroxidase having a serine or cysteine residue in the appropriate position might undergo modification by a specific selenocysteinyl synthetase. Hydrogen selenide could be inserted to form the selenocysteine residue by direct displacement of the functional group on the ß-carbon or by addition to a dehydroalamine residue following ß-elimination (3).

The translational hypothesis involves insertion of selenocysteine directly into the polypeptide chain, by means of a selenocysteine-specific transfer RNA and appropriate codon. Hawkes et al. (26) claim to have isolated a selenocysteine-specific transfer RNA from rat liver. It is not clear whether subsequent reports from the same laboratory demonstrating apparent preferential incorporation of ^{75}Se from labeled selenocysteine-tRNA reflect a tRNA-mediated process or whether some other effect is involved (for a recent review of this subject, see reference 3).

Methylation is an important aspect of selenium metabolism (14) (Figure 4). Dimethyl selenide and trimethylselenonium (TMSe) are known excretory products. A pathway for the synthesis of dimethyl selenide from hydrogen selenide (reactions 1 and 2) has been worked out (20), and in microsomes these reactions are probably catalyzed by the same thiol S-methyltransferase that methylates H_2S and CH_3SH (23). This enzyme has a relatively low K_M for hydrophobic substrates like H_2S and phenylsulfide compared to more polar thiols, and might catalyze further methylation of dimethyl selenide (reaction 3) to form TMSe, but direct evidence is lacking. If reaction 3 is catalyzed by an enzyme having a low activity for dimethyl selenide as substrate, saturation of this activity could explain why dimethyl selenide is exhaled when high doses of Se are given. Reactions (5) and (6) are hypothetical, but a microsomal oxygenase might bring about the formation of methylseleninic acid or dimethyl selenoxide. A Japanese group has reported that dimethyl selenoxide is a major

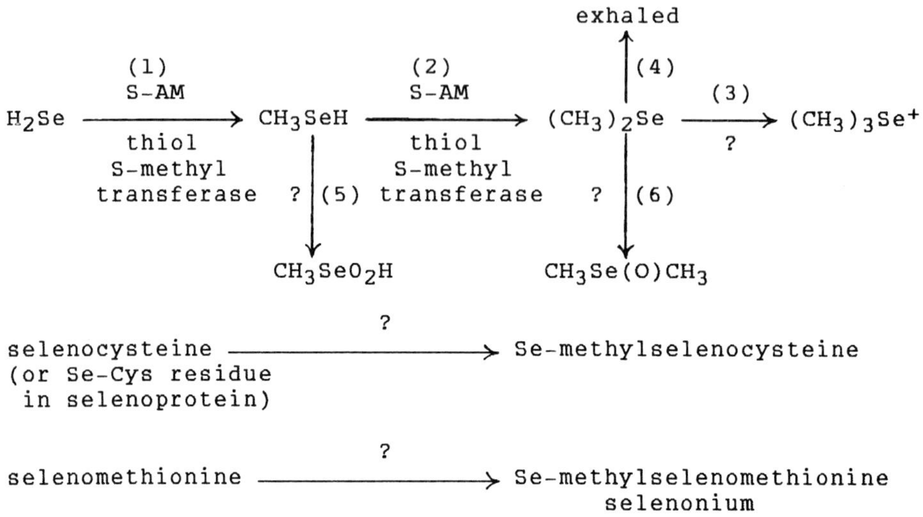

Fig. 4. Metabolism of Methylated Selenides

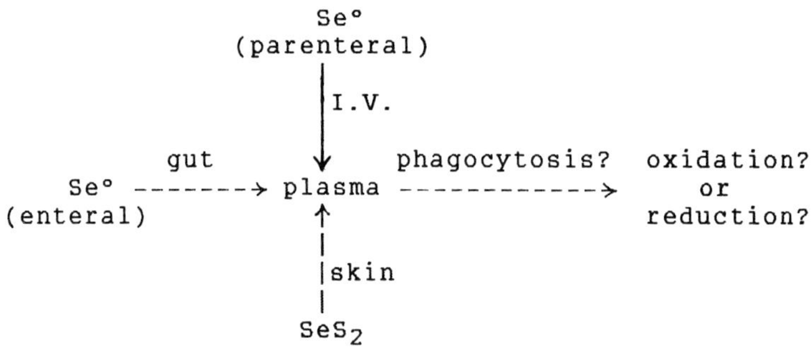

Fig. 5. Elemental Selenium Metabolism

metabolite (55% of total urinary Se) in rats given selenite or selenate (27). The basis for this unconfirmed identification apparently was paper chromatography. The proportion of selenium excreted as dimethyl selenide could depend on the activities of several different enzymes, and capacity for oxidative metabolism (reactions 5 and 6), as well as methylation (reaction 3) might determine the fate of the selenide.

Besides the formation of methylated selenides from hydrogen selenide, they may be formed independently by the metabolism of selenoamino acids. By analogy to sulfur metabolism (17), selenomethionine might be metabolized by a transamination pathway (28) rather than by transsulfuration, forming methylselenol (Figure 3). Se-methylselenocysteine, if formed from free or protein-bound selenocysteine (Figure 4) could be metabolized similarly. Both would yield the same triad of mono-, di-, and tri-methylated selenides that result from inorganic selenium metabolism. On the other hand, it is possible that methylation to form selenonium derivatives of the selenoamino acids might occur prior to separation of the Se from the amino acid, followed by the breakdown of the selenonium derivative to give dimethyl selenide or TMSe (see below). Preliminary studies of the fate of the Se-methyl selenoamino acids and their selenonium derivatives by Dr. S. Foster in my laboratory has shown that a high percentage of these compounds are excreted as TMSe or dimethyl selenide.

Elemental Se has a very low nutritional activity because its low solubility precludes absorption from the intestinal tract. This does not necessarily rule out biological activity, however, if the element is directly infused. Particles of Se° might be expected to undergo phagocytosis (Figure 5), and if subjected to oxidative attack by the peroxide-generating system of phagocytic cells, Se° could be rendered soluble and available. Alternatively, cleavage of Se-Se bonds by thiols might create biologically active selenide derivatives. The point is of some interest because of the likelihood that selenite in total parenteral nutrition formulas can become reduced to Se°; some particles may be large enough to be removed by the standard microfilters used with intravenous feeding (29), but some Se° might reach the plasma. My laboratory has shown that for selenite incubated at room temperature for 24 hours with TPN formulas containing glucose and 2 g/l of ascorbic acid, then subjected to thin layer electrophoresis at pH 5.3, up to 40% of the Se remains at the origin, which is consistent with a change in chemical form. Selenium sulfide (SeS_2) is used in certain anti-dandruff shampoos; a portion may be absorbed through skin and metabolized by pathways similar to those for other inorganic forms of Se.

Urinary forms of selenium and their metabolic origin

The fact that only one well-characterized urinary metabolite is known indicates the need for further research on the forms of Se in urine. Since it was first identified in rat urine in 1969 (30,31) TMSe has been regarded as a general and quantitatively important urinary metabolite (32). The extension of this concept to humans, however, has been a very uncertain matter, for several reasons: (1) almost all of the work has been done with rats; (2) the form of Se administered was often selenite, which is not likely to be a form of Se in foods consumed by humans; (3) the amount of Se administered was often much higher than what would be regarded as a physiological intake. The only reported identification of TMSe in human urine was for a patient given [^{75}Se]selenite (1 μg Se), in which case 21% of the ^{75}Se in urine collected after 5-10 hours co-chromatographed with TMSe (33). My laboratory has recently obtained evidence that TMSe is a normal metabolite of Se in humans fed a defined, meat-based diet as the source of Se. TMSe was measured using a well-controlled procedure involving cation exchange chromatography and Reineckete precipitation (Figure 6). About 12% of the urinary Se was accounted for by TMSe. Nearly half of the urinary Se was retained on a strong cation exchanger, but much of this was not TMSe. For rats, Palmer et al. (31) had reported that TMSe was a major metabolite (40% of urinary Se) of selenite, even at low, physiological doses. In contrast, Nahapetian et al. (34) have reported that at low doses (16 µg Se/kg body weight), TMSe was a minor metabolite of selenite and selenoamino acids, but was a major metabolite at high doses (1500 µg Se/kg).

Regarding other urinary metabolites, several laboratories (32,33) have reported the presence of a second metabolite (called U-2) as a rather diffuse spot near TMSe in paper chromatograms. U-2 is still unidentified. There are probably many more different forms of Se in urine. In theory, some of these might be Se analogs of urinary sulfur metabolites, which include many trace metabolites as well as the more abundant oxidized products of sulfur metabolism. The chemistry of Se would favor compounds of +4 oxidation state rather than the +6 state, in contrast to sulfur.

Regarding the origin of TMSe, the prevailing concept has been that it is formed by methylation of dimethyl selenide, as discussed earlier. However, I would like to offer an alternative hypothesis for consideration, whereby TMSe arises as a stable breakdown product of an unstable selenonium precursor of the type $(CH_3)_2Se^+$-R. This idea grew out of observations on the facile appearance of TMSe during the synthesis of certain selenonium compounds, and is supported by evidence in the chemical literature. Burness (35) has shown that dialkyl thetin salts (thetins are sulfonium compounds having a carboxylate moiety,

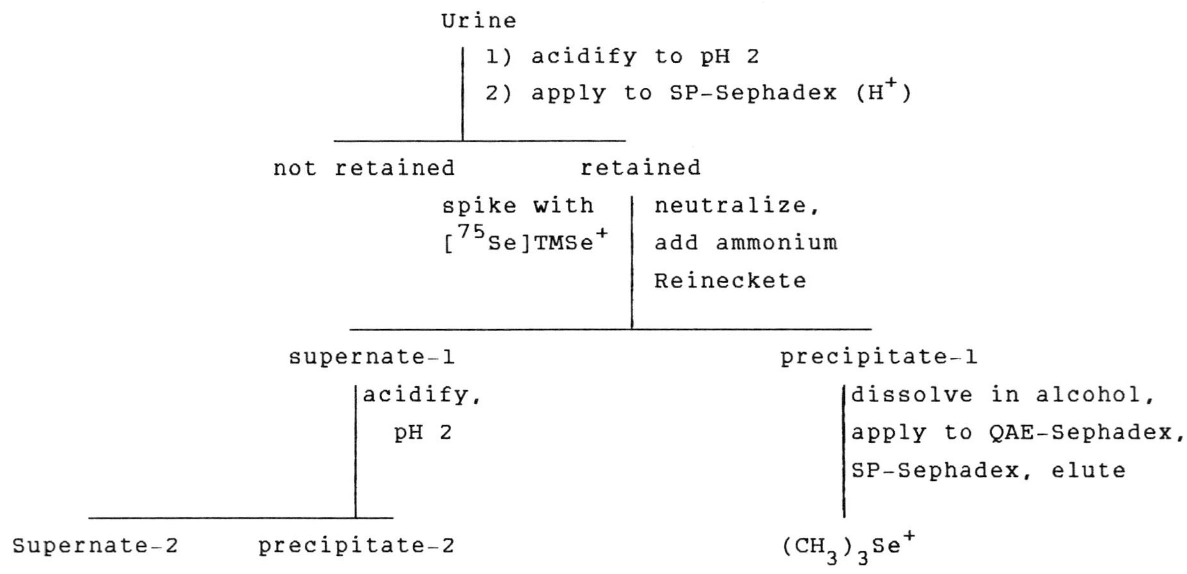

Fig. 6. Fractionation of Urinary Selenium

analogous to betaine) undergo facile decarboxylation, forming a methyl group out of the $-CH_2COOH$ moiety. This type of reaction would be especially favorable for the selenetine (selenonium analogue of dimethyl thetin):

$$CH_3-\overset{\overset{\displaystyle CH_3}{|}}{Se^+}-CH_2COOH \longrightarrow CH_3-\overset{\overset{\displaystyle CH_3}{|}}{Se^+}-CH_3 \quad + \quad CO_2$$

Both products have exceptional stability, thus favoring the reaction. In addition, expansion of the valence shell of the Se atom in an intermediate compound, analogous to the mechanism postulated for decarboxylation of thetins (35), probably is a favorable factor and can explain why the analogous reaction is virtually unknown in the betaine (ammonium) series. This postulated mechanism for the origin of TMSe predicts that the third methyl group is derived from a carbon chain, rather than by transfer of a methyl group. This hypothesis is being explored using ^{14}C- and ^{75}Se-labeled precursors.

Methodology for characterizing selenium compounds

This workshop provides an opportunity to describe some methods developed in my laboratory that permit unequivocal identification of certain forms of Se by mass spectrometry. The approach is based on derivatization of selenols with Sanger's reagent, 1-fluoro-2,4-dinitrobenzene (FDNB), forming the dinitrophenyl (DNP) derivative. The hydrophobic DNP moiety expedites isolation of the Se in a reasonably stable form. The DNP chromaphore is well suited for monitoring the isolation by means of HPLC or for locating compounds on thin layer plates (36). The DNP derivatives can either be identified by co-chromatography or can be converted to volatile derivatives for unequivocal identification by mass spectrometry, using the simple direct

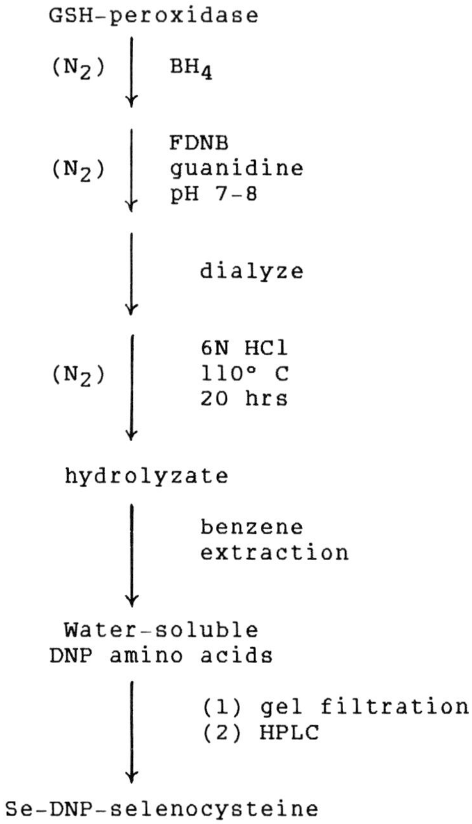

Fig.7. Derivatization of selenocysteine with Sanger's reagent

probe technique for sample introduction. The resulting mass spectrum usually includes the molecular ion as a major peak and the presence of several fragment ions containing the characteristic Se isotope pattern aids interpretation.

The procedure was successfully applied (7) to the identification of Se-cysteine in glutathione peroxidase (Figure 7). Treatment of reduced enzyme with FDNB under denaturing conditions was followed by acid hydrolysis under strictly anaerobic conditions, giving excellent recovery (over 90%) of the Se. Separation of Se-DNP-selenocysteine from other water soluble DNP derivatives was achieved by isocratic HPLC. Initial attempts at converting Se-DNP-selenocysteine into more volatile derivatives suitable for mass spectrometry using silylating agents and other methods were unsuccessful. The problem was traced to the facile migration (Se→N) of a DNP moiety (Smiles rearrangement), with release of the Se in a selenol form. By deliberately inducing this shift with weak base in the presence of methyl iodide, the selenol was trapped as a stable selenoether (Se-methyl-N-DNP-selenocysteine):

$$\text{DNP-Se-CH}_2\overset{\overset{\text{NH}_2}{|}}{\text{C}}\text{HCOO}^- \xrightarrow[\text{methyl iodide }(\text{N}_2)]{\text{B}^-\text{ (Smiles rearrangment)}} \text{CH}_3\text{-Se-CH}_2\overset{\overset{\text{HN-DNP}}{|}}{\text{C}}\text{HCOO}^-$$

After esterification of the carboxyl group, a well-behaved derivative is obtained that gives a clean mass spectrum. This is the first time the occurrence of selenocysteine in a selenoprotein has been unequivocally demonstrated by mass spectrometry. The limitation of co-chromatography as a means of identifying Se in proteins has been overcome, and anyone should be able to use this method for characterizing selenocysteine in proteins. Complete details can be found elsewhere (7,36).

The same technique is well suited for identifying hydrogen selenide or methyl selenol (37). Unlike earlier procedures for trapping H_2Se using metals, reaction of H_2Se with FDNB gives a benzene-soluble derivative suitable for direct identification by mass spectrometry, HPCL, or TLC (Figure 8). Any form of Se reducible to H_2Se can be identified by this procedure. One useful application of the technique is to differentiate inorganic forms of Se bound to proteins from Se that is incorporated into the polypeptide chain as a selenoamino acid. Se is notorious for interacting non-specifically with proteins (38), as by the reaction of selenite with sulfhydryl groups to form S-Se derivatives or elemental Se that is adsorbed hydrophobically.

This procedure should be applied to all putative "selenoproteins" (commonly, a ^{75}Se-labeled peak observed during fractionation of tissue extracts). Release of H_2Se will indicate that the Se is not a selenoamino acid and may simply be adsorbed non-specifically to the protein, and therefore of uncertain biological significance. Alternatively, a non-heme iron protein containing Se in place of S might be indicated. In either case the further characterization as a selenoprotein would require purification to homogeneity and evidence for the presence of Se in stoichiometric amounts.

It is also helpful to have ^{75}Se-labeled compounds for use as internal standards for co-chromatography and to validate isolation procedures. We have developed a convenient procedure for the synthesis of radiolabeled trimethylselenonium ion (39), starting with commercially-available [^{75}Se]selenocystine. The advantage of the procedure is that it eliminates the need to manipulate volatile intermediates. The availability of radioactive TMSe and related selenonium compounds will expedite further studies of their metabolism.

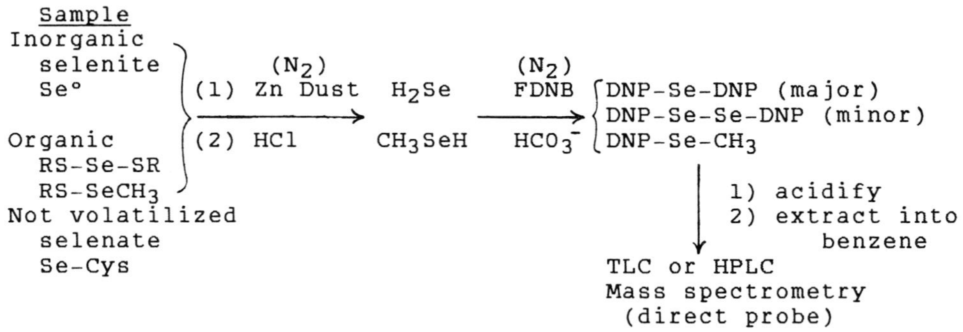

Fig. 8 Identification of Hydrogen Selenide and Related Compounds

References

1. Spallholz, J.E., Martin, J.L., Ganther, H.E. (eds.): Selenium in Biology and Medicine, AVI, Westport, Conneticut (1981).

2. Kraus, R.J., Prohaska, J.R., Ganther, H.E.: Biochim. Biophys. Acta 615, 19-26 (1980).

3. Sunde, R.A.: J. Am. Oil Chem. Soc. (in press).

4. Kein, C.L., Ganther, H.E.: Am. J. Clin. Nutr. 37, 319-328 (1983).

5. Olson, D.E., Novacek, E.J., Whitehead, E.I., Palmer, I.S. Phytochemistry 9, 1181-1188 (1970).

6. Forstrom, J.W. Zakowski, J.J., Tappel, A.L.: Biochemistry 17, 2639-2644 (1978).

7. Kraus, R.J., Foster, S.J., Ganther, H.E.: Biochemistry 22: 5853-5858 (1983).

8. Brown, T.A., Shrift, A.: Biol. Rev. 57, 59-84 (1982).

9. Douglas, J.S., Morris, V.C., Soares, J.H., Jr., Levander, D.A.: J. Nutr. 111, 2180-2187 (1981).

10. Edmonds, J.S., Francesconi, K.A.: Nature 289: 602-604 (1981).

11. Wittwer, A.J.: J. Biol. Chem. 258, 8637-8641 (1983).

12. Gennity, J.M., Bottino, N.R., Zingaro, R.A., Wheeler, A.E., Irgolic, K.J.: Biochem. Biophys. Res. Comm. 118, 176-182 (1984).

13. Zingaro, R.A.: Chemica Scripta 8A, 51-57 (1975).

14. Ganther, H.E.: Adv. Nutr. Res. 2, 107-128 (1979).

15. Ganther, H.E., Baumann, C.A.: J. Nutr. 77, 408-414 (1962).

16. Dilworth, G.L., Bandurski, R.S.: Biochem. J. 163, 521-529 (1977).

17. Cooper, A.J.L.: Ann. Rev. Biochem. 52, 187-222 (1983).

18. Ganther, H.E.: Biochemistry 10, 4089-4098 (1971).

19. Hsieh, H.S., Ganther, H.E.: Biochemistry 14, 1632-1636 (1975).

20. Hsieh, H.S., Ganther, H.E.: Biochim. Biophys. Acta 497. 205-217 (1977).

21. Gasiewicz, T.A., Smith, J.C.: Chem.-Biol. Interactions 21, 299-313 (1978).

22. Esaki, N., Nakamura, T., Tanaka, H., Soda, K. J. Biol. Chem. 257, 4386-4391 (1982).

23. Weisiger, R.A., Jakoby, W.B.: Arch. Biochem. Biophys., 196, 631-637 (1979).

24. Ganther, H.E.: Ann. N. Y. Acad. Sci. 355, 212-226 (1980).

25. Sunde, R.A., Hoekstra, W.G.: Biochem. Biophys. Res. Comm. 93, 1181-1188 (1980).

26. Hawkes, W.C., Lyons, D.E., Tappel, A.L.: Biochim. Biophys. Acta 699, 183-191.

27. Hasegwa, A., Urakubo, G.: J. Hygen. Chem. 33, 277-281 (1977) (in Japanese).

28. Case, G.L., Benevenga, N.J.: J. Nutr. 106, 1721-1736 (1976).

29. Shils, M.E., Levander, O.A.: Am. J. Clin. Nutr. 35, 829 (abstr) (1982).

30. Byard, J.L.: Arch. Biochem. Biophys. 130, 556-560 (1969).

31. Palmer, I.S., Fischer, D.D., Halverson, A.W., Olson, O.E.: Biochim. Biophys. Acta 177, 336-342 (1969).

32. Palmer, I.S., Gunsalus, R.P., Halverson, A.W., Olson, O.E.: Biochim. Biophys. Acta 208, 260-266 (1970).

33. Burk, R.F.: In: Trace Elements in Human Health and Disease, Vol. II (ed. Prasad, A.S.) pp. 105-133, Academic, New York (1976).

34. Nahapetian, A.T., Janghorbani, M., Young, V.R.: J. Nutr. 113, 401-411 (1983).

35. Burness, D.M.: J. Org. Chem. 24, 849-852 (1959).

36. Ganther, H.E., Kraus, R.J., Foster, S.J.: Methods Enzymol. 107, (1984) (in press).

37. Ganther, H.E., Kraus, R.J.: Anal. Biochem. (1984) (in press).

38. Ganther, H.E.: Chemica Scripta 8A, 79-84 (1975).

39. Foster, S.J., Ganther, H.E.: Anal. Biochem. 137, 205-209 (1984).

D i s c u s s i o n Session I
Plenary Lecture

Subjects of the discussion

- synthesis of selenocysteine
- mechanism of converting inorganic Se into selenocysteine
- Se and diseases
- Se and enzymes
- biochemical criteria for the determination of Se requirement
- glutathione perioxidase
- Se dependent - and independent glutathione perioxidase

Contributors: Gundermann, Kirchgessner, Oster, Weser

Session II

SELENIUM IN NUTRITION

Plenary Lectures given by: P. Brätter (FRG)

T. Westermark (Finland)

Chairman:

U. Weser (FRG)

EFFECTS OF SELENIUM INTAKE IN MAN AT HIGH DIETARY LEVELS OF SELENIFEROUS AREAS OF VENEZUELA

Peter Brätter, Virginia E. Negretti S., Ulrich Rösick
Hahn-Meitner-Institut for Nuclear Research Berlin, F. R. G.
Trace Element Research in Biomedicine

Werner G. Jaffé, Hernán Mendez C., Guillermo Tovar E.
Fundacredesa, Caracas, Venezuela

Introduction

Selenium takes a special place among the trace elements. Toxic action is known as well as deficiency syndroms and the optimal range of beneficial action seems to be smaller than for the other trace elements. Supported by epidemiological studies (1,2) there is an increasing interest in the deficiency situation of selenium in human nutrition and health. However, the treatment of a lowered selenium status of the organism is still a problem due to the small margin between the toxic and physiological dose response. In order to define the levels of a safe and adequate selenium dosage in prophylaxis and therapy information is needed regarding the influence of high selenium intakes and the physiological activity of other essential trace elements or their uptake rate. However, few data are presented in the literature on the consequences of longterm human exposure to elevated dietary selenium levels.

In order to study effects of high selenium ingestion in humans and the relationships between this element and other trace elements a research program was started in 1982 on a collaborative basis between the Hahn-Meitner-Institut Berlin and the national Venezuelan foundation for the investigation of the

Trace Element - Analytical Chemistry in Medicine and Biology, Vol. 3

health status of the population FUNDACREDESA.

A field investigation was carried out in the seleniferous area of Villa Bruzual in Venezuela. In this natural laboratory Jaffé (3) started 14 years ago a clinical and biochemical investigation of school children in order to study possible health risks. In the present study the examination was extended to adults, elderly persons and lactating mothers. A total of 134 persons was examined and samples of blood, hair, breast milk and aliquots of the families meal were collected.

Sample collection and preparation

Hair

Hair was cutted from every person at the left occipital region and sealed in plastic foils. Only the first two centimeters of the hair above the roots were used for the analysis. The samples were cleaned according to the recommendation of the IAEA (4) applying a 3 step procedure using ether, bidistilled water and acetone as rinsing fluids.

Breast milk

Breast milk was obtained by manual pressure of the breast and directly collected into precleaned plastic containers avoiding any contact with the breast. A good aid for the stimmulation of the milk output was the simultaneous feeding of the child with the other breast.

Blood, serum

The blood was taken after the person rests for some minutes in a lying position. All persons were advised to be empty till the blood collection in the morning. Monovettes (Fa. Sarstedt, Germany) were used as syringes, which have been checked to be free of contamination with the elements of interest.
After coagulation (about 4o minutes waiting time) the serum was separated by centrifugation and transfered into pre-cleaned plastic tubes.

From all families food samples from meal taken before the clinical investigation were collected and sealed in plastic bugs. All samples collected in the field study were stored at dry ice temperature in a Dewar container and transported to Berlin, where the trace element analysis and the determination of some serum components has been carried out at the Hahn-Meitner-Institute.

Sample analysis

Instrumental neutron activation analysis

All fluid samples were dryed at 50 ^{o}C and carefully sealed in vials of high purity quarz, each vial containing about 100 - 300 mg dry material. Food samples were homogenized and freeze dryed before sealing. The samples were irradiated for 10 d at a neutron flux density of about $5 \cdot 10^{13}$ n cm^{-2} s^{-1} in the Berlin research reactor (BER II). Bovine Liver and Orchard Leaves powder (National Bureau of Standards, Washington) were used as a standard reference material.

The trace elements Co, Cs, Fe, Rb, Se and Zn were determined via the decay of their long lived radio nuclides by means of high resolution γ- spectrometry. In hair mercury could be determined after the calculation of the Se-75 interference at E_{γ} = 279 KeV.

Atomic Absorption Spectrometry

The element copper was measured in serum by means of graphite furnance AAS, and to determine mercury in breast milk samples the hydrid-amalgame technique was applied.

Serum constituents

Special serum proteins were determined as follows: Albumine, α_2-macroglobuline and ceruloplasmine by Laser-nephelometry, C -reactive protein (CRP) and prealbumine by radial immunodiffusion, and ferritin via radioimmunoassay.

Results and Discussion

Serum-proteins - Health status

In general useful preliminary information on the nutritional and health status of the population group can be obtained from the levels of serum constituents. Albumine and prealbumine may serve as nutritional indicators of the long and short term protein supply, respectively. Further, C-reactive protein, α_2-macroglobuline and ceroluplasmine the so-called acute phase proteins are very often elevated in the course of infections. The measured ferritin level gives some information about the iron storage capacity.

		ALL		ADULTS		CHILDREN	
		N	%	N	%	N	%
ALBUMIN <3500 MG/DL	↓	46	33	18	23	28	65
FERRITIN <30 NG/ML	↓	34	24	16	20	18	42
CRP >0.5 MG/DL	↑	25	18	22	28	3	7
α_2-MACROGLOBULIN <200 MG/DL >400 MG/DL	 ↓ ↑	14 6	14	11 4	19	3 2	11
CEROLUPLASMIN <25 MG/DL >45 MG/DL	 ↓ ↑	12 14	18	6 13	24	6 1	16
PREALBUMIN <10 MG/DL	↓	4	3				

Tab 1: Deviation of Serum Protein Values from the Normal Range

In Table 1 the deviation of the measured values from the normal range are summarized. From the investigation of a total number of 134 persons including 46 children it is remarkable that about 65 % of the children have a low albumine serum level, indicating a certain protein deficit. Furthermore in about 42 % of the children the ferritin level is low, which might be due to an anaemic situation (latentor pre-latent).

Nearly a quarter of the adults showed a deviation in at least one of the characteristic serum proteins.

During the field investigation special attention was paid to the incidence of clinical signs which might be related to high selenium intake. Symptoms of dermatitis, pathological nails or loose hair were found only in few cases.

Analysis of food samples

From the analysis of samples of the meals of 45 families quantitative information of the mean selenium ingestion in Villa Bruzual was obtained (Fig. 1). A mean of 0.54 ppm Se was calculated on the dry weight basis. The two cases of Se-values greater than 2.2 ppm in Fig. 1 were found to be reflected also in corresponding higher Se blood levels of this families. From the mean Se-content of the meals rough estimation of the selenium intake of the population group yields a range of 300 - 400 μg Se per day without taking into account the intake of beverages. That is higher than the estimated human dietary requirement of 50 - 200 μg Se per day (5).

The measured frequency distribution of Fig. 1 corresponds to an occurance of different foods in the meals as given in Fig. 2A. From the determination of the selenium content of some of the locally produced foods (Fig. 2B) the highest value of 28 ppm dry matter was found in the egg white. Egg accumulates obviously selenium and it might be a good monitor to localize the occurence of this element in animal feed and the environment.

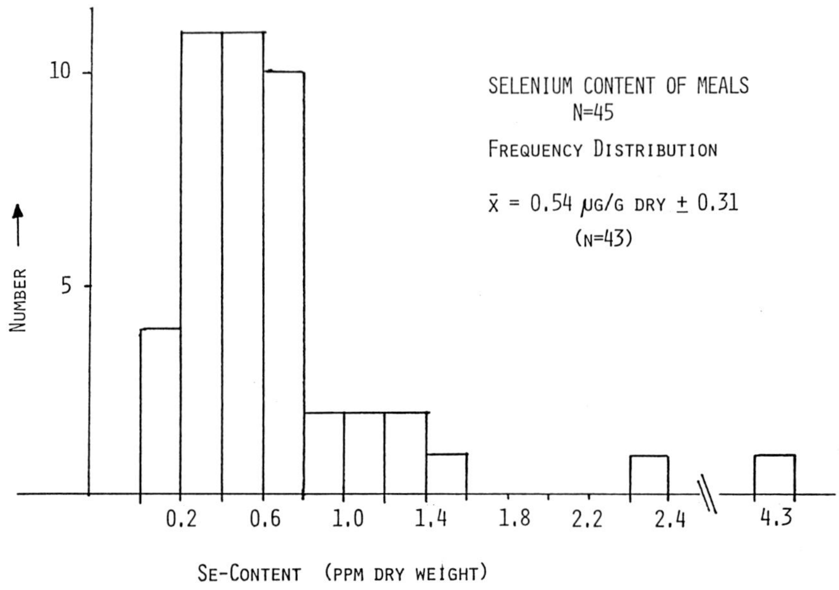

Fig. 1 Selenium content of meals

Se in serum and red blood cells

The population investigated was divided into age groups according to the classification used in the clinical examination. As shown in Fig. 3 an increase of the Se-content of serum and red blood cells (RBC) with age has been found. This increase amounts to a factor of 2 if the children and the adults aged 31 - 50 years are compared.

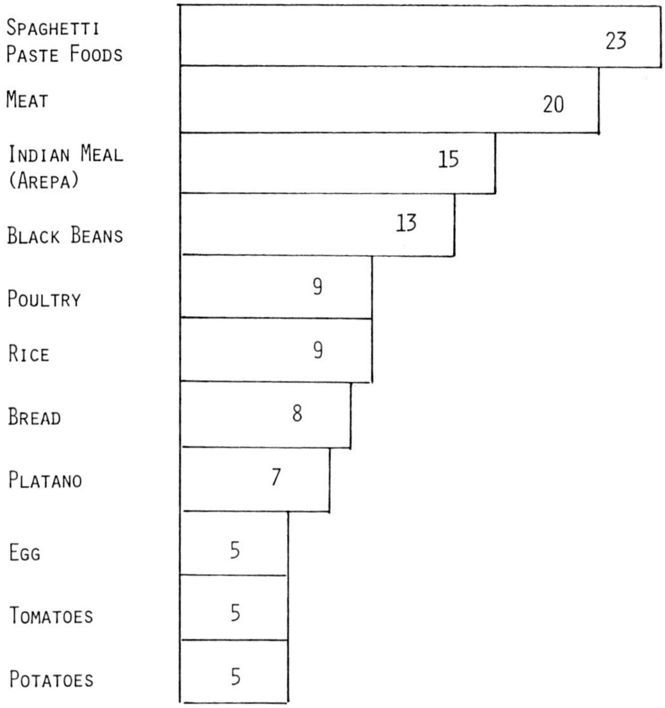

Fig. 2 A Occurence of different foods in 43 meals from Villa Bruzual

Sesame	2.8
Sorgo	0.76
Rice	0.96
Guayaba	1.82
Egg white	6.2 / 28.8
Egg yolk	2.6 / 6.2

Fig. 2 B Selenium content of different foods (µg/g dry)

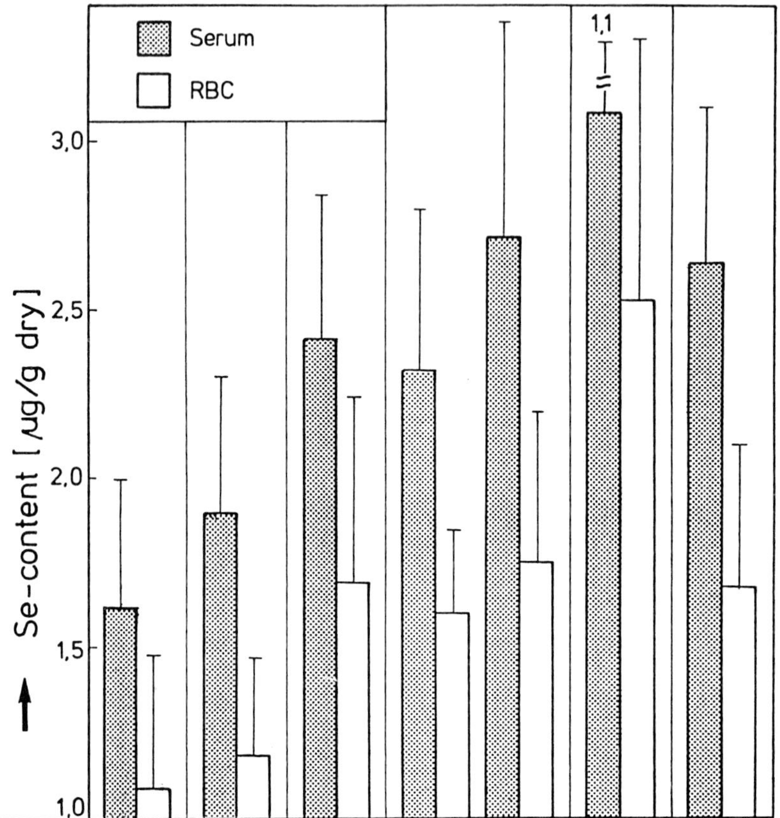

Age Group [a]	≤ 2	3 - 6	7-20 ♂ 7-15 ♀	Females 16 - 30	Mothers 16 - 30	31 - 50	> 50
Number	22	11	15	20	16	22	8

Fig. 3: Se-content of Serum and Red Blood Cells (RBC) Dependence of the age groups

The blood selenium values tend to decrease in the senior group and that might be due to a lower food consumption of elderly people. In general the RBC-selenium is always higher than the serum level if calculated on the wet weight basis. Another point of interest might be the RBC-serum ratio which seems to be nearly constant for all age groups.

This correlation has been proved including all persons as shown in Fig. 4. The correlation coefficient of 0.923 demonstrates a good relationship but at serum values greater than

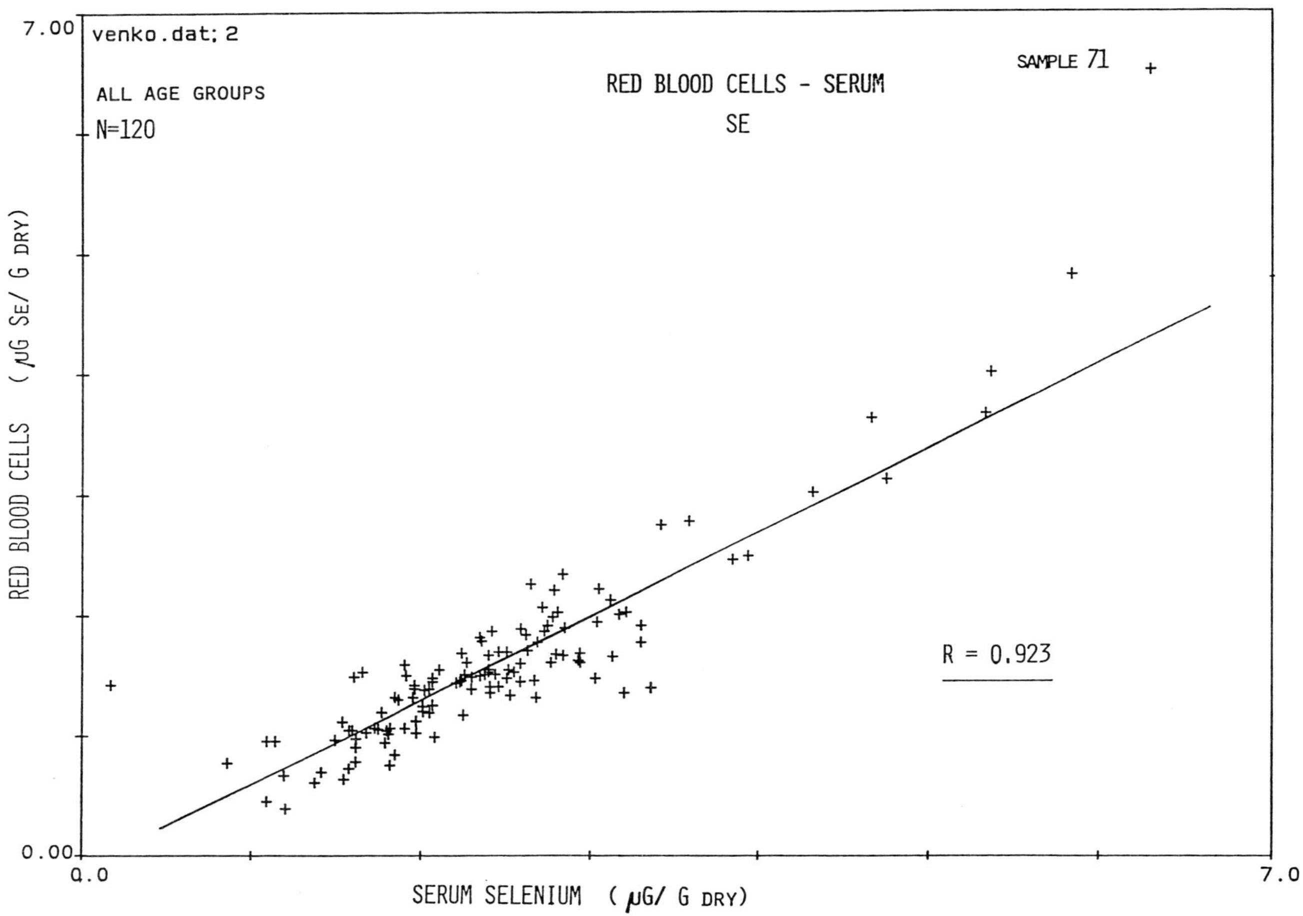

Fig. 4: Selenium in RBC as a function of the serum Se-content

VENEZUELA VILLA BRUZUAL	SE (NG/G) $\bar{X} \pm S$	FE (µG/G) $\bar{X} \pm S$	ZN (µG/G) $\bar{X} \pm S$	CU (µG/G) $\bar{X} \pm S$	RB (µG/G) $\bar{X} \pm S$	CO (NG/G) $\bar{X} \pm S$
MOTHERS 18 - 40A N = 17	265 ± 60	1.37 ± 0.58	0.83 ± 0.47	1.23 ± 0.20	0.22 ± 0.06	0.69 ± 0.52
ADULTS 16 - 30A N = 20	217 ± 60	1.45 ± 1.20	0.74 ± 0.12	1.11 ± 0.53	0.21 ± 0.07	0.55 ± 0.18
ADULTS 31 - 50A N = 28	314 ± 135	1.34 ± 0.40	0.84 ± 0.14	1.28 ± 0.33	0.19 ± 0.06	0.54 ± 0.23
GERMANY BERLIN N = 8	92 ± 6.3	1.96 ± 0.18	0.83 ± 0.55	N.D.	0.196 ± 0.024	0.52 ± 0.28

Tab. 2 Serum trace element content of adults

4 μg Se/g dry the relative increase of selenium in the red blood cells appears to be higher and the ratio between the both blood fractions tends to increase. With respect to this findings it is interesting to note, that for the low selenium intake situation of New Zealand residents also a constant ratio of Se-RBC to Se-serum has been found (6). The difference lies in the value of the ratio which yield about 1.54 for New Zealand and about 2.0 for the high Se-intake situation of Villa Bruzual. From our data it might be assumed that in the case of a longterm and nearly constant selenium intake of a population a balance could be established which yields a constant ratio for the distribution of Se between both blood fractions. The deviation at higher Se-serum values in our case needs further examination.

Another objective of the study was to investigate the influence of high dietary selenium intake on the concentration of other essential trace elements.

The results of the determination of Co, Cu, Fe, Se and Zn in blood serum are summarized in Tab. 2. For comparison, the corresponding values of a control group in Berlin are given. The mean serum selenium content was found to be 3 times higher than those of the Berlin group, but there was no significant difference in the values of the other trace elements. Copper values were also found within the known normal range. The data suggests that there is no direct effect of the high dietary selenium status of Villa Bruzual on the serum level of the other trace elements investigated.

Breast milk

From 17 mothers in Villa Bruzual samples of mature milk were analyzed. The results of trace element analysis in Tab. 3 are expressed on the wet and the dry weight basis. A comparison with 45 samples from Berlin shows that the higher dietary selenium level is reflected in the breast milk as well.

	Venezuela, Villa Bruzual				Germany, Berlin
	Wet weight			Dry weight	
	$\bar{x} \pm s$		Range	$\bar{x} \pm s$	$\bar{x} \pm s$ (n = 45)
Se (ng/g) n = 17	57.70	37.40	21 - 158	416 ± 236	121 ± 36
Co (ng/g) n = 17	0.53	0.24	0.31 - 1.27	4.1 ± 2.3	4.8 ± 3.5
Cr (ng/g) n = 17	4.20	2.25	1.0 - 10.3	33.1 ± 18.0	42 ± 22
Cs (ng/g) n = 17	2.00	1.20	0.9 - 5.3	15.7 ± 10.3	30 ± 13
Fe (µg/g) n = 16	0.66	0.34	0.21 - 1.30	5.1 ± 3.0	4.5 ± 2.5
Rb (µg/g) n = 17	0.60	0.13	0.35 - 0.90	4.5 ± 1.3	5.4 ± 1.3
Zn (µg/g) n = 16	1.14	0.60	0.27 - 2.18	8.6 ± 5.1	45.2 ± 18.0
Hg (ng/g) n = 6	1.20	0.20	1.0 - 1.5	9.9 ± 1.7	13.3 ± 11.5

Tab. 3 Trace element content of breast milk

The breast-fed infants of Villa Bruzual receive about 3 times more Se compared to the Berlin infants. On the other hand, we found that the mean zinc content of breast milk samples from Villa Bruzual amounts to 1.14 µg/g wet or 8.6 µg/g dry weight. These values are significantly lower than those of the Berlin group. In consequence, the corresponding zinc intake of the infants meets only the lower limit of the recommended values.

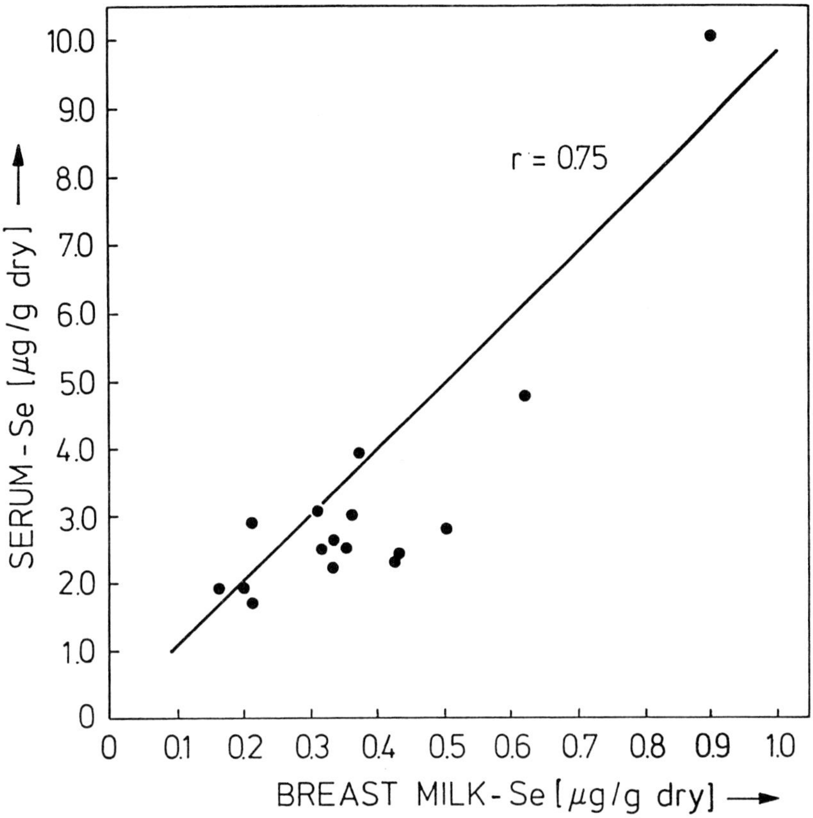

Fig 5: Serum Se-content as a function of Se in breast milk

A linear regression (Fig. 5) of serum selenium of the mothers versus their breast milk Se-content shows a positive correlation. However, from the point of view of statistical significance the number of cases has to be extended.

The serum zinc content of the mothers of Villa Bruzual (Tab.2) was found in the normal range suggesting that the low Zn-content of the breast milk is independent of the maternal Zn-status.

However, at this time, it remains an open question if the low Zn-content of the breast milk might be related to the high dietary selenium intake and the corresponding high Se-level of the breast milk.

It is known that high levels or chronic ingestion of mercury may lower the physiological activity of selenium due to the antagonistic relationships of both elements. Therefore, mercury was analyzed in the milk samples. Because of the low order of the values found, an influence of mercury was not taken into consideration. This was also supported by the Hg-level of the hair samples (Tab. 4), when they are used as monitor of the environmental burden.

(VILLA BRUZUAL, VENEZUELA)

ELEMENT	MEAN (µG/G)	RANGE
Co	0.074	0.008 - 0.390
Fe	54.2	11.3 - 270
Hg	0.86	0.09 - 7.0
Se	1,56	0.62 - 9.4
Zn	211	23 - 580

Tab. 4 Trace element content of hair (N=129)
2 cm segment close to the scalp

Hair

From all persons hair samples were taken in order to compare their trace element content with the Se-content of the blood fractions. Such kind of investigation is reasonable if the people do not wash the hair with sulfur containing shampoos

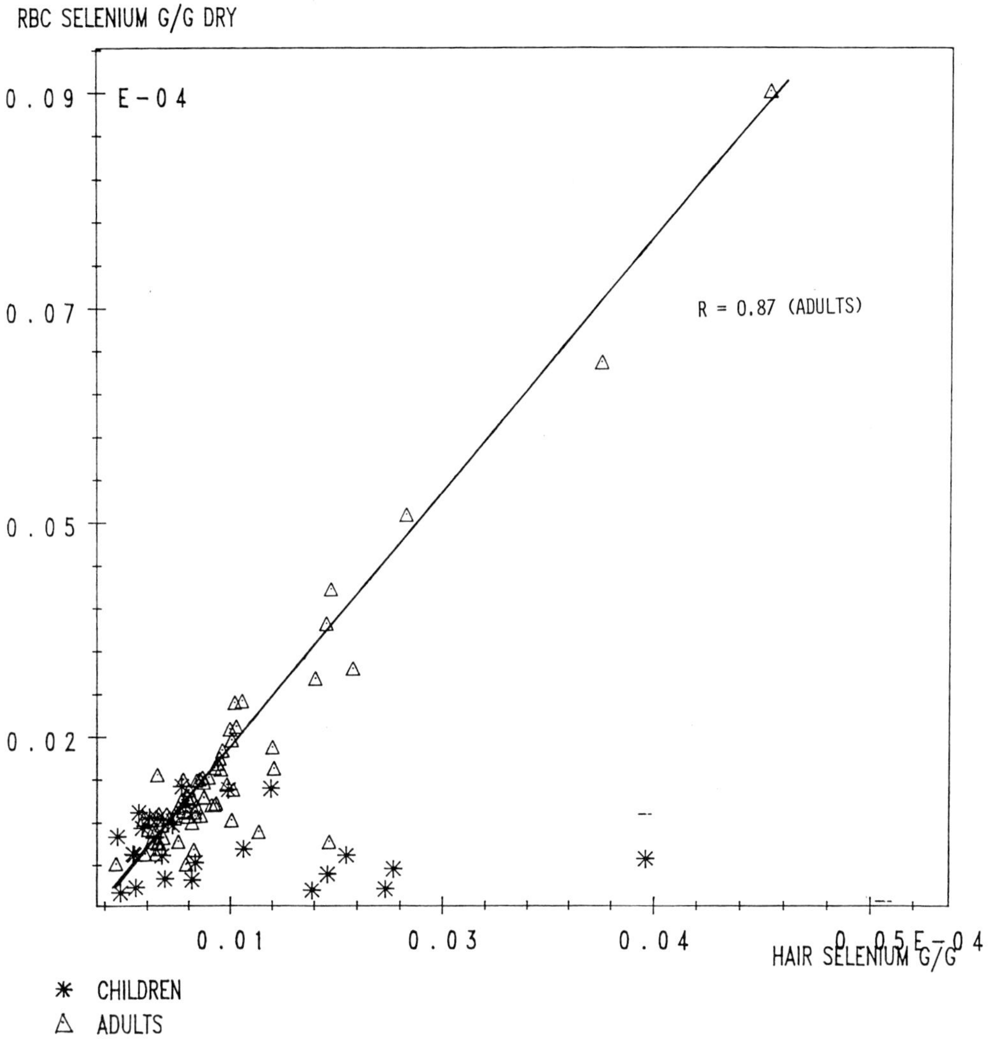

Fig. 6: Selenium content of red blood cells in relation to selenium in hair of children and adults

as it is the case in this region.

Hair receives its trace elements mostly from the blood and is able to uptake selenium in its matrix during keratinization. If the permanent dietary Se-level of a population group is reflected in the blood Se-level, a relation may also exist between hair concentration and diet. We regarded the red blood cells as a measure of the long term Se-status of the organism and tried to correlate their content with that of the hair, as shown in Fig. 6. Different results were obtained for children and adults. For adults a linear regression yielded a significant positive correlation but no correlation was found for the children.

To obtain some idea of the relation between Se in hair and the Se-intake the data of the seleniferous area of Villa Bruzual were compared with data from China (7) . The chinese values were obtained from regions of medium and low dietary Se-ingestion. The comparison in Tab. 5 shows clearly the connection between the blood- and hair Se-content at the different dietary levels.

	SELENIUM INTAKE		
	HIGH	MEDIUM	LOW
SE - LEVEL	VENEZUELA	CHINA	
BLOOD $\bar{x}$ (µg/ML)	0.420	0.095	0.021
HAIR $\bar{x}$ (µg/G)	1.56	0.359	0.074

Tab. 5 Dependence of the blood and hair selenium level on the selenium intake

The factor of about 3.5 between the blood and hair content in all 3 cases may not be a casual factor, but it describes an equilibrium between the Se-concentration in the organs and the compartments of the body.

Summary

From the study of high dietary Se-intake the following results were obtained:

1. The Se-levels in blood, breast milk and hair were influenced.
2. The Se-content of serum and RBC is age-dependent.
3. The level and the distribution of Se between serum and RBC are positively correlated. The ratio between the Se-content of serum and RBC was constant in the range of 1 to 4 µg Se/g dry serum, but above that value the RBC-Se increases much more than the corresponding serum-Se.
4. No deviation from the normal serum levels of other trace elements including Co, Cu, Fe, Rb and Zn was observed.
5. The Se-content of breast milk is positively correlated with that of the serum of the mothers. In contrast to the elevated Se-level ($\bar{x}$ = 416 ng/g dry) the Zn-content of the breast milk ($\bar{x}$ = 8.6 µg/g dry) was found to be below the normal range. If the high Se-intake may cause the low Zn-content of the breast milk remains an open question.
6. Hair might be a good monitor to investigate the dietary Se-ingestion of different population groups under certain conditions (permanent dietary Se-level, exclusion of the use of sulfur containing shampoos).
7. Clinical symptoms of high Se-intake as nausea, pathological nails or loss of hair occured only in a few cases.

References

1. Group of Environment and Endemic Disease, Institute of Geography, Academy of Sciences, Acta Geophica Sinica 34, 85 (1979).
2. Keshan Disease Research Group of the Chinese Academy of Medical Sciences, Chin.Med.G. 92, 471-477 (1979).
3. Jaffé, W.G., Raphael, M., Mondragón, M.C. and Cueras, M.A., Arch.Latinoamer.Nutr. 22, 570 (1972).
4. International Atomic Energy Agency, IAEA/RL/50 (1978).
5. Recommended Dietary Allowances, The National Council, Ninth Edition, Washington D.C., (1980)
6. Rea, H.M., Thomson, C.E., Campbell D.R., Robinson, M.F., Br. J. Nutr. 42, 201-208 (1979).
7. Zun, L., Proc. Symp. on Trace Element Metabolism in Man and Animal (TEMA-5), (ed. Howell, J.McC, Grawthorne J.M., White, C.L.), Australien Academy of Science, Canberra, 514-516 (1981).

Acknowledgement

The authors thank the following people for their valuable assistance in performing sampling, the trace element analysis or the clinical and dietary information:
M.L. Alvarez, F. Contreras, U. Fossi, W. Gatschke, I. Morales and G. Röhlicke.

The study was supported in part by the MAVESA, LAS LLAVES S.A.

D i s c u s s i o n Session II - 1
Plenary Lecture

Subjects of the discussion

- effects of high intakes of Se
- high content of Se in sesam meal in connection with animal experiments
- antidote for Se-toxicity
- expression of results on dry - or wet - weight basis
- differences in the health status between areas of low and high Se-intake

Contributors: Ganther, Gökmen, Oster, Piepponen

CONSEQUENCES OF LOW SELENIUM INTAKE FOR MAN

Tuomas W. Westermarck
Helsinki Central Institution for the Mentally Retarded
Killinmäki, SF-02400, Kirkkonummi, Finland

Introduction

Evidence that selenium is an essential trace element for man has recently become obvious (1, 2). A seleno-enzyme, glutathione peroxidase, plays an important role in preventing tissues from peroxidative cell damage by destroying the hydrogen peroxide generated by dismutation (3, 4), and hence, makes it possible for humans to live in an atmosphere containing oxygen. Studies of the effects of long-term selenium deficiency have shown a decrease in glutathione peroxidase activity in different tissues (5).

Oxidative tissue damage is caused by free radicals with unpaired electrons, such as superoxide ($O_2^{\cdot -}$) and hydroxyl ($OH^{\cdot}$) radicals. Pro-oxidants e.g. chemotherapeutic drugs, stress caused by forced exercise, exposure to radiation, high intakes of dietary fats, and infectious diseases are believed to increase the production of free radicals (6) resulting in a need of nutrients such as selenium with antioxidant influence.

Levander et al. (7) in the U.S.A. have estimated that a daily selenium intake of about 70 µg is needed to maintain the body selenium status of healthy young males; but in young women living in New Zealand the amount of selenium needed has been estimated to be only 20 µg/day (8). Thus, a precise adequate selenium requirement is not known, but an approximate safe selenium intake for adults is between 50 and 200 µg/day and correspondingly lower intakes for children (9).

Trace Element - Analytical Chemistry in Medicine and Biology, Vol. 3

In 1970 Mertz (10) defined a trace element deficiency not as the absence of a certain element but as its presence in sub-optimal levels resulting in disturebed function, or as a clinical condition responding to supplementation. The purpose of this paper is to review consequences of low selenium intake for man, and to report some clinical conditions responding to selenium supplementation.

Dietary intake of selenium

There are regional selenium intake differences in small as well as in large countries. This is reflected in the blood selenium levels of the population (Table 1). Thus, blood selenium

Table 1. Estimated dietary selenium intakes and blood selenium levels in human subjects.

Country	Dietary selenium intake (µg/day)	Blood selenium level (µg/L)
China (11, 12)		
Endemic selenosis area	4990	3180
Low selenium area (Keshan disease area)	11	5 - 10
Finland (13, 14)	30	
Southern Finland (Helsinki)		87 67 (serum)
Southeast of Finland (Lappeenranta)		56 42 (serum)
Northern Finland (Utsjoki)		108 (serum)
New Zealand (15, 16)		
North Island	56	83
South Island	28	59

concentrations are sensitive to the dietary intake of selenium; serum levels reflect recent and erythrocyte levels long-term intakes (17).

Selenium has appeared to be particularly important for infants and pregnant women. The fetus seems to receive a lot of selenium during intrauterine growth, because the serum selenium concentrations were significantly decreased in pregnant women during the gestation (18): The liver seems to serve as a depot of selenium. There are reports on an effect of age on liver selenium concentrations: the liver levels were highest in premature neonates, declined to the lowest levels in infants, then gradually rose to the adults´levels (19, 20). However, premature infants seemed to have a lower selenium concentration in serum than full-term infants and children (13, 21). There are also data showing that the selenium content of human milk decreases with lactating time (22).

Too low selenium intake has been found to be responsible for the decreased selenium status in residents of China, New Zealand and the Northern countries (Scandinavia) (5), in vegeterians (23), infants fed with comercially available milk formula and baby foods (24), and in children receiving semi-synthetic amino acid mixtures of protein hydrolyzates because of inborn errors in amino acids (25). The clear cases of clinical selenium deficiences are most likely to be found in these goups of people.

Parenteral alimentation (Long-term TPN)

Ricour et al. (26) have recommended 3 µg selenium per kg body weight (BW) a day given as sodium selenite in the total parenteral nutrition (TPN) of premature and deseased newborn babies. Van Rij et al. (27) described a woman in New Zealand who after a 30-day TPN developed a pain in her quadriceps

Table 2. Reports on total parenteral nutrition (Long-term TPN) without selenium supplements resulting in muscular disorders.

Muscular disorders recognized	Not observed
Van Rij et al. (27)	Kay and Knight (33)
Collip and Chen (28)	
Johnson et al. (29)	
Lane et al. (30)	
Fleming et al. (31)	
Kien and Ganther (32)	

and hamstring muscles aggrevated by walking. The plasma selenium dropped down to 9 µg/L. Selenium, 100 µg, as selenomethionine was infused daily with the TPN solution and the disorder resolved in a week. There are several other reports on apparent selenium deficiency during TPN (Table 2).

Kien and Ganther (32) described a boy who after receiving TPN for about 1,5 years developed intermittent leg muscle pain and tenderness, and white fingernail beds. His condition improved after intravenous selenium supplementation (42 µg selenium/day as sodium selenite). Cardiomyopathy due to selenium deficiency have also been observed in patients who have received parenteral nutrition (28, 29, 30, 31). However, Kay and Knight (33) found no symptoms of selenium deficiency in 43 adults from New Zealand during TPN, although the patients' selenium levels were very low.

Pathological conditions associated with selenium deficiency

It has been suggested that selenium deficiency could be

associated with such human pathological conditions as alcoholic cardiomyopathy (34), alcoholic cirrhosis (35), cardiac beriberi (36), cataract and other pathological eye conditions (5, 37), haemolytic diseases (5, 38), thromboasthenia (39), and mental retardation (19). Depressed blood glutathione peroxidase levels were found in patients with various dermatoses: psoriasis, eczema, atopic,dermatitis, vasculitis, mycosis fungoides, and dermatitis herpetiformis (40). The glutathione peroxidase levels were observed to increase in 50 patients after 6-8 weeks of treatment with tablets containing 0.2 mg selenium as sodium selenite and 10 mg tocopheryl succinate, which were taken twice a day (40). The clinical response in these patients were also encouraging indicating conditions with selenium deficiency.

The following diseases have also been associated with selenium deficiency: cardiovasculas diseases, cancer, cystic fibrosis (5, 41), congestive cardiomyopathy (42), Kashin-Beck's or Urov's disease (43,44), Legionnaires' disease (45), multiple sclerosis (46, 47), myotonic dystrophy (48), neuronal ceroid lipofuscinosis (19, 49, 50), progressive muscular dystrophy (19, 51, 52), and rheumatoid arthritis (53). The virtual eradication of the so called Keshan disease with selenium supplementation was the first demonstration that selenium is an essential trace element for man (11, 41, 54).

Keshan disease

Keshan disease is an endemic myocardio-necrosis disease with mitochondrial changes and myofibrillar atrophy of an unknown cause. More than ten million residents in a wide belt of China were affected by this disease, the patients being mainly children under 15 years of age and women of childbearing age (54). The selenium intake of the people affected was less than 30 µg a day (11).

Keshan disease is a congestive cardiomyopathy, and is characterized by gallop rythm, abnormal electrocardiogram changes, heart enlargement, heart insufficiency, cardiogenic shock, and hemiplegia risk (11). The average hair selenium level of the residents living in the Keshan disease area was 0.074 ± 0.050 µg/g and blood selenium content 21 ± 10 µg/L according to Yang et al. (12).

The disease was practically eradicated by supplementing the residents with 0.5 - 1 mg sodium selenite a week (0.15 - 0.3 mg selenium/week) orally. Chen et al. (11) reported that in 1974 and 1975 the incidence rate of the disease was 13.5 and 9.5/1000 in the control group, and 2.2 and 1.0/1000 in the group treated with selenium.

Muscular disorders

The heart is affected in almost all patients with progressive muscular dystrophy of Duchenne type, a congenital disease characterized by a progressive degeneration of the skeletal musculature and replacement of muscle tissue by collagen and fat. The histological and ultrastructural hallmarks of the cardiac myopathy of this most common form of progressive muscular dystrophy occuring predominantly in young males, are multifocal dystrophic areas characterized by fibrosis at the cellular level and a total loss of thick and thin myofilaments at the subcellular level. The histological changes in the skeletal muscles are similar to the changes observed in heart muscle. At the age af 20 years heart insufficiency and a sudden death occur in 80 per cent of patients (55).

The biological half-life of selenium-75 is significantly shorter in patients with progressive muscular dystrophy than in normal persons. Therefore, the capacity to retain selenium is lessened, although the intestinal absorption of the selenium-75, as selenite, seemed to be normal (51). These findings indicated an increased elimination of selenium and an elevated requirement of selenium supplements. This conclusion was also made by Ahlrot-Westerlund and Carlmark (56) after finding an increased urinary excretion of selenium in patients with Duchenne muscular dystrophy. The heart failure of a Duchenne patient was reported to improve after selenite nad vitamin E treatment (52). However, Burri et al. (57), Hunter et al. (58), and Robinson and Thomson (5) were unable to link low selenium status with muscular dystrophy; but Oster et al. (42) recently reported that a selenium deficiency may be present in a number of patients with congestive cardiomyopathy.

Örndahl et al. (48) found that in myotonic dystrophy the selenium level in serum decreased parallel with the progression of the illness. They also reported that selenium supplementation resulted in a positive clinical response in myotonic dystrophy patients (59).

Cardiovascular disease

In 1977 Westermarck (19) found in Finland that the blood selenium levels of patients with acute myocardial infarction were decreased compared with the levels of healthy adults. However, no difference was observed in the heart or liver selenium contents of patients who died from acute myocardial infarction compared with patients who died from other diseases.

The results of recent study, in which sera was prospectively collected, indicated that a very low serum selenium level, below 45 µg/L, might be associated with increased clinical manifestation of coronary heart disease (60). However, a longitudinal case-control study of 33 patients with one or more risk factors for coronary heart disease and 64 controls showed that serum selenium of 50 to 105 µg/L was not associated with

The development of the clinical manifestation of coronary heart disease during the follow up of 5 to 7 years (61). Furthermore, Dr. Sven Punsar et al., Finnish Heart Foundation, Helsinki, have observed that the inhibitants of the municipalities Mellillä and Pöytyä, in the southwest of Finland, had an exceptionally low selenium status, but nevertheless, the mortality from cardiovascular diseases in these Finns is very low compared with the other inhabitants of Finland.

Thus, the evidence linking selenium and cardiovascular disease is controversial. A number of studies have shown that low blood selenium level is connected with cardiovascular disease in humans (19, 60, 62, 63), whereas several research groups have found no link between low selenium level and human cardio-vascular disease (61, 64, 65, 66, 67).

Cancer

Shamberger and Frost (67), Shamberger and Willis (68) have claimed an inverse correlation between cancer mortality and the selenium level in human blood. During the period 1966 - 1970 the crude incidence rates of cancer were the highest among the Finns living in the southern and south-western parts of Finland. In these parts of the country the serum selenium concentrations were variable. The lowest incidence rates were found in the northern part of Finland; a region where the highest serum selenium levels were also encountered (13).

In New Zealand cancer patients did not differ with regard to their selenium status from elderly subjects and surgical patients without cancer (69). Similar observations have been made in Finland (9). However, the serum selenium concentration of 15 cancer patients admitted into a hospital in Eindhoven, the Netherlands, was found to be 35 per cent lower than the mean value of the controls (70). McConnell et al. (7) found that carcinoma patients had significantly lower selenium levels than non-carcinoma patients. Some studies have shown that the lower the level of selenium in cancer patients the greater the risk of metastases, multiple primaries, recurrence of the cancer, and early death (71, 72).

Knuutila et al. (73) recently found that dietary selenium may increase the radiation-induced chromosome and nuclear aberration in mouse bone marrow; oxygen,too, potentiates the radioation effects (74). This finding may explain the benefical influence observed when using various selenium treatments in combination with X-ray therapy to treat mammary metastases (75, 76).

Juvenile neuronal ceroid lipofuscinosis

Sodium selenite and vitamin E supplementation resulted in a positive clinical response in juvenile neuronal ceroid lipofuscinosis (NCL) patients (19, 49, 50). The juvenile or Spielmeyer-Sjögren- Vogt type of NCL is a slowly progressive fatal disease. It begins with a gradual loss of vision at the age of 4 - 8 yrs and is accompanied with progressive dementia and the impairment of speech and motor functions, and usually leads to profound mental retardation and to death by the age of 19 yrs.

Ceroid lipofuscin pigments accumulate in the body. Similar lipofuscin pigments are deposited in tissue during the normal ageing processes and are believed to arise during the ageing processes and are believed to arise during the destruction of cells and organelles by lipid peroxidation and during other forms of oxygen radical damage. Ceroid seems to be a potent binder of iron (77, 78). The rates of lipid peroxidation and superoxide dependent hydroxyl radical ($OH^{\bullet}$) formation may be accelerated by the presence of a suitable iron catalyst (79). The catalytic iron concentration in cerebrospinal fluid (CSF)

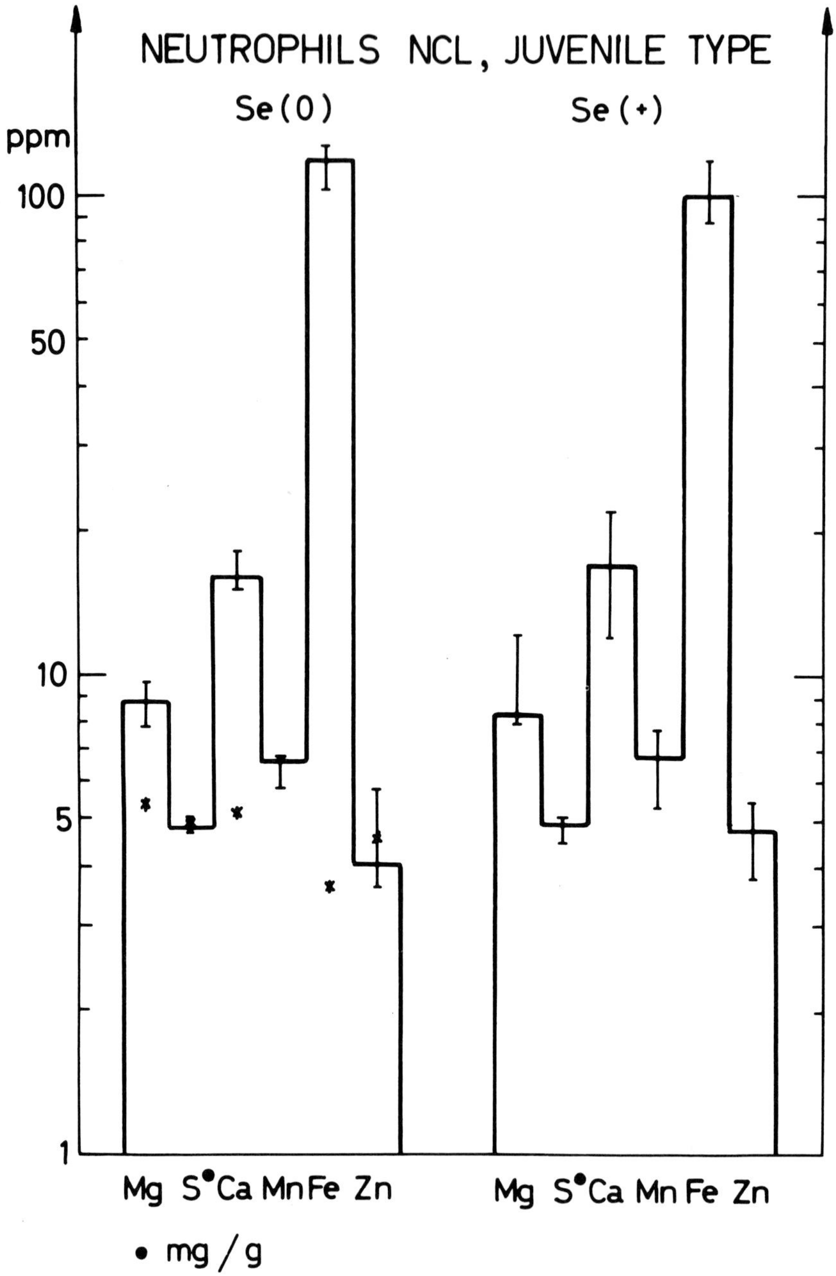

Fig. 1: The elemental profiles of the neutrophils of NCL patients, untreated (N=3) vs treated (N=5). The concentration is given as median values, μg per g dry matter (78). The asterisks represent a control group (N=26). (With the permission of the Medical Biology).

of patients with NCL seems to be elevated (80) and the intestinal absorption of iron increased (81).

Superoxide dismutase (SOD), catalase and glutathione peroxidase are the principal enzymes which protect against oxygen-radical damage in human cells (79). Although it has been shown that juvenile NCL patients have normal brain and serum levels of glutathione peroxidase, the selenium dependent glutathione peroxidase ativity has been demonstrated to be low in the erythrocytes (80, 81), lymphocytes and granulocytes (81) of these patients. This selenium dependent glutathione peroxidase activity was observed to be corrected after selenium supplementation (80). In a selenium-supplementation study (82), involving a healthy person, it was observed that sodium selenite reacted in an antagonistic way towards iron, copper, and manganese in the neutrophils; in juvenile NCL patients the administration of sodium selenite did not seem to decrease significantly the iron content of the neutrophils, because the iron level in treated patients did not differ significantly from that in untreated patients (Fig.1). Changed uptake patterns of iron and changed storing mechanisms of Fe might explain the unusual Fe metabolism and pathogenesis in NCL.

Even a long-term selenium supplementation seemed to keep the glutathione peroxidase activity elevated (49). The serum selenium concentration needed for that was between 2-5 µmol/L. The selenium was administered as sodium selenite. Vitamin E, B_6 and B_2 were later added to the selenium supplementation (89). At present the supplementation consists of sodium selenite (0.05-0.1 mg Se/kg BW), vitamin E (0.014-0.05 g/kg BW), vitamin B_6 (3.75-10.0 mg/kg BW) and vitamin B_2 (0.12-0.3 mg/kg BW) a day. About 50% of the NCL patients treated have shown a positive response, and in some patients the improvement has been considerable, e.g. EEG was normalied. Some patients whose condition has improved shown no deterioration during a follow-up time of eight years (50). However, the deterioration of vision has not been stopped by the therapy.

Kashin-Beck's disease

In Asia, an unusual disorder due to disturbed iron and collagen metabolism, known in China as Kashin-Beck's and in Russia as Urov's disease, was described to be the consequence of an excessive intake of iron in the drinking water (83). Blood and urine hydroxyproline and bone calcium concentrations of these patients were decresed compared with the values of healthy subjects living in the same region (84). The iron, lead, manganese, zinc, and silver levels (mean) in the femur of Kashin-Beck's disease patients were high compared with the values of the controls (table 3).

Children with Kashin-Beck's disease had knee-deformities, oedema of the forehead, decreased bone strenght, and a low average hair selenium level of 0.071 µg/g (44). Domaev (85) studied the EEG of 112 Kashin-Beck's patients, and the data

Table 3 Concentrations (Mean) of trace elements (mg%) in patients with Kashin-Beck's disease. From Rosin et al. (84).

Element	Epiphysis of femur		Diaphysis of femur	
	Controls (n=30)	Patients (n=50)	Controls (n=30)	Patients (n=50)
Iron	310	480	300	460
Lead	38,4	87.2	35.2	81.3
Manganese	10.0	16.3	9.2	15.6
Zinc	46.0	93.0	36.0	72.0
Silver	0.1	1.4	0.1	1.3

indicated a diffuse lesion of the brain. It was considered that a correlation between the findings of clinical, EEG and other studies pointed to a chronic disorder leading to an early aging of the brain. X-ray findings of the therapeutic effects of Na-selenite and vitamin E in patients with Kashin-Beck´s disease were also reported (43). The disease seems to be preventable by selenium supplementation (Personal comm. Prof. Guang-Lu Zu, Xi´an Medical College, P.R. China, 1983).

Iron-selenium interactions

In an oxygen-rich environment two active oxygen species $O_2^{\cdot -}$ and $OH^{\cdot}$ radicals are formed from O_2. The $OH^{\cdot}$ radical is extremely reactive and will interact indiscriminately with any molecule in its immediate surroundings. It is able to abstract hydrogen atoms from unsaturated lipid molecules and to initiate chain reactions. Hydroxyl radical formation reactions may be catalysed e.g. by iron and copper (79). The iron of non-protein bound iron, haem-iron, and protein-bound iron including transferrin-bound can catalyse organic oxygen radical reactions such as lipid peroxidation. Primary biochemical actions of selenium occur in man at the active site of glutathione peroxidase which, using glutathione as a cosubstrate, reduces intracellular formation of lipid hydroperoxides, hydrogen peroxides and free radicals (86, 87). Thus, selenium may detoxify or modulate the reactions catalysed by iron. Vitamin E acts synergistically with selenium dependent glutathione peroxidase in preventing the peroxidation of structural membrane components (87) stimulated by iron salts and haemoproteins.

The positive efficacy of selenium and vitamin E in preventing iron-induced epilepticform discharge in rats suggests that peroxidative injury may be critical in the development of seizure discharge (88), and that patients with epileptic seizures may have an extra high requirement of selenium. These findings were confirmed by Santavuori et al. (89) who found that the NCL patients´mean age at the onset of epilepsy was 11.1 years among the patients not given selenium and other antioxidants, and 13.6 among those having these treatments.

There are also evidences that high tissue iron levels play a crucial role in certain diseases such as Kashin-Beck´s diseases, NCL; and rheumatoid arthritis (80,84,90), and that in these diseases antioxidant treatment including selenium constitutes a promising therapeutic approach (43, 50, 91). Cardiac arrhythmias are seen in elderly NCL patients as in patients with end-stage haemochromatosis, a disorder af iron metabolism.

Discussion

As Verlinden ei al. (92) have pointed out several reaseons may result in a low selenium status during early infancy, low dietary selenium intakes are likely to play a major role. The liver in neonates may also play an important role as a selenium depot. The observations on selenium in blood in healthy adults and elderly people seemed to deny that age could bear a relationship on selenium paramenters (92).

The pathological conditions due to selenium deficiency found during long-term total parenteral nutrition, and the role of selenium as the eradicator of the Keshan disease have given us some most important clues of the consequences of low selenium intake for humans.

Although low selenium in the diet have been supposed to be a risk factor in human cancer (93), the low-selenium status of cancer patients is more likely a consequence of their illness rather than the cause of the cancer (41).

Depressed selenium level and inhibition of glutathione peroxidase activity by mercury and silver were experimentally found in rat tissues, whereas cadmium and lead were not observed to have any affect on this activity (94). Furthermore, increases in serum or plasma iron are brought about by decreased blood formation on lead poisoning (95). Increased plasma transferrin levels and transferrin to alb min ratios are found in Japanese quail given supplemental cadmium at the rate of 75 ppm of the diet (96). Thus the well known selenium antidote effect against heavy metals (5) may be influenced by iron; it is very likely that selenium does not act on its own.

Conclusion

1. Selenium intake prevents Keshan disease; it may also prevent Kashin-Back's disease.
2. Infants, women of childbearing age, and young TPN-children seem to be especially vulnerable to diseases connected with selenium deficiency.
3. Low selenium intake of patients with NCL, myotonic dystrophy, progressive muscular dystrophy may result in an acceleration of the development of the disease.
4. A high intake of heavy metals. iron, and polyunsaturated fatty acids combined with low intake of antioxidants may greatly increase the requirement of selenium, particularly in infants, women of childbearing age, patients treated with oxygen, and in patients predisposed to Keshan disease, Kashin-Beck's disease, cancer, cardiovascular disease and cardiomyopathy, and patients suffering from neuronal ceroid lipofuscinosis of juvenile type, and those from muscular disorders.

Acknowledgements

This work was supported by grants from the Research Department, Rinnekoti Institution for the Mentally Retarded, Espoo; Finska Läkaresällskapet, and Sigrid Jusélius Foundation, Helsinki Finland.

Key words

Low selenium intake; iron-selenium interactions; consequences for healthy and diseased humans.

References

1. Awasthi, Y.C., Beutler, E., Srivastava, S.K.: J. Biol. Chem. 250, 5144 - 5149 (1975).
2. McKeehan, W.L., Hamilton, W.G., Ham. R.G.: Proc. Natl. Acad. Sci. U.S.A. 73, 2023-2027 (1976).
3. Flohé, L., Günzler, W.A., Ladenstein, R.: Glutathione: Metabolism and Function, eds.: Arias. I.M., Jakoby, W.B., Raven Press. New York, 115 - 138 (1976).
4. Valenzuela, A., Rios, H., Neiman, G.: Experientia 33, 962-963 (1976).
5. Robinson, M.F., Thomson, C.D.: Nutrition Abstracts and Reviews in Clinical Nutrition, Series A, 53 3-26 (1983).
6. Del Maestro, R.F.: Acta Physiol. Scand. Suppl. 492,153-168 (1980).
7. Levander, O.A.: Proc. New Zealand Workshop on Trace Elements, ed. Dunckley, J.V., University of Otago, Dunedin, N.Z., 129-140 (1981).
8. Stewart, R.D.H., Griffiths, N.M., Thomson, D.D., Robinson, M.F.; Br. J. Nutr. 40, 45-54 (1978).
9. National Research Council. Recommended Dietary Allowances, 9th ed. National Academy Press, Washington, D.C. (1980).
10. Mertz, W.: Fed. Proc. 29, 1482-1488 (1970).
11. Chen, X., Yang, G., Chen, J., Chen, X., Wen, Z., Ge, K.: Biol. Trace Elem. Res. 2, 91-107 (1980).
12. Yang, G., Wang, S., Zhou, R., Sun, S.: Am. J. Clin. Nutr. 37, 872-881 (1983).
13. Westermarck, T., Raunu, P., Kirjarinta, M., Lappalainen, L.: Acta Pharmacol. Toxicol. 40, 465-475 (1977).
14. Varo, P, Koivistoinen, P.: Acta Agric. Scand. Suppl. 22, 165-171 (1980).
15. Watkinson, J.H.: N.Z. Med. J. 80, 202-205 (1974).
16. Thomson, C.D., Robinson, M.F.: Am. J. Clin. Nutr.: 33, 303-323 (1980).
17. Robinson, M.F., Rea, H.M., Friend, G.M., Stewart, R.D.H., Snow, P.G., Thomson, C.D.: Br. J. Nutr., 39, 589-600 (1978).
18. Hyvönen-Dabeck, M., Nikkinen-Vilkki, P., Dabeck, J.T.: Olin. Chem. 30, 529-533 (1984).
19. Westermarck, T.: Acta pharmacol. Toxicol. 41, 121-128(1977)
20. Casey, C.E., Guthrie, B.E., Friend, G.M., Robinson, M.F.: Arch. Environ. Hlth. 37, 133-135 (1982).

21. Amin, S., Chen, S.Y., Collip, P.J., Castro-Magna, M., Maddaiah, V.T., Klein, S.W.: Nutr.Metab.24, 331-340 (1980).

22. Lombeck, I., Kasperek, K., Bonnermann, B., Feinendegen, L.E., Bremer, H.J.: Eur. J. Pediatr. 129, 139-145 (1978).

23. Abdulla, M., Svensson, S., Nordeń, A., Öckerman, P-A.: Trace element metabolism in man and animals, eds.: Gawthorne, J.M., McHowell, C.J., White, C.L., Springer Verlag, Berlin, Heidelberg, New York, 14-17 (1981).

24. Lombeck, I., Kasperek, K., Feinendegen, L.E., Bremer, H.J.: Proc. New Zealand Workshop on Trace Elements, ed. Dunckley, J.V., University of Otago, Dunedin, N.Z., 141-149 (1981).

25. McKenzie, R.L., Rea, H.M., Thomson, C.D., Rodinson, M.F.: Am. J. Nutr. 31, 1413-1418 (1978).

26. Ricour, C., Duhamel, J.F., Nihoul-Fekete, C.: Arch. Fr. Pédiatr. 34, 161-162 (1975).

27. VanRij, A.M., Thomson, D.D., McKenzie, J.M. Robinson, M. F.: Am J. Clin. Nutr. 32, 2076-2085 (1979).

28. Collip. P.J., Chen, S.Y.: N. Engl. J. Med. 304, 1304-1305 (1981).

29. Johnson, R.A., Baker, S.A., Fallon, J.T., Maunard, E.P., Ruskin, J.N., Wen, Z., Ge, K., Cohen, H.J.: N. Engl. J. Med., 304, 1210-1212 (1981).

30. Lane, H.W., Dudrick, S., Warren, D,C.: Proc. Soc. Exper. Biol. Med. 167, 383-390 (1981).

31. Fleming, C.R., Lie, J.T., McCall, J.T., O'Brien, J.F., Baillie, E.E., Thitle, J.L.: Gastroenterology, 83, 689-693 (1982).

32. Kien, C.L., Ganther, H.E.; Am. J. Clin. Nutr. 37, 319-328 (1983).

33. Kay, R.G., Knight, G.S.: NZ Med. J. 90, 11-13 (1979).

34. Goldman, I.S., Kantrowitz, N.E.: Engl. J. Med. 305, 701.

35. Aaseth, J., Thomassen, Y., Alexander, J., Norheim, G.: N. Engl. J. Med. 303, 944-945 (1980).

36. Reasbeck, P.G.: N. Engl.. J. Med. 285, 294 (1982).

37. Ganther, H.E., Hafeman, D.G., Lawrence, R.A., Serfass, R. E., Hoekstra, W.G.: Trace Elements in Human Health and Disease, vol. II, ed.: Prasad. A.S., Academic Press, New York, San Francisco, London, 165-234 (1976).

38. Perona, G., Cellerino, R., Guidi, G., Casaril, M., Pizzolo, G.: Haematologica 62, 156-166 (1977).

39. Karpatkin, S., Weiss, H.J.: Engl. J. Med. 287, 1062-1066 (1972).

40. Juhlin, L., Edqvist, L.E., Ekman, L.G., Ljunghall, K., Olsson, M.: Acta Dermatovener. 62, 211-214 (1982).

41. National Research Council. Selenium in Nutrition Committee on Animal Nutrition, Subcommittee on Selenium.: Natl. Acad. Sci., Washington D.C. (1983).

42. Oster, O., Prellwitz, W., Kasper, W., Meinertz, T.: Clin. Chim. Acta 128, 125-132 (1983).

43. Li. C.Z.: Chung Hua Fang She Hsueh Tsa Chich 14, 6-8 (1980).

44. Schrauzer, G.N.: Selen, Publ. V.F.M., Heidelberg, 56-58 (1983).

45. Chen, J.R., Anderson, J.M.: Science 206, 1426-1427 (1979).

46. Wikström, J., Westermarck, T., Palo, J.: Acta Neurol. Scand. 54, 287-290 (1979).

47. Jensen, G.E., Gissel-Nielsen, G., Clausen, J.: J. Neurol. Sci. 48, 61-67 (1980).

48. Örndahl, G., Rindby, A., Selin, E.: Acta Med. Scand. 211, 493-499 (1982).

49. Westermarck, T., Santavuori, P., Marklund, S., Pohja, P., Salmi, A.: Caroid Lipofuscinosis (Batten's disease). Eds. Armstrong, D., Koppang, N., Rider, J.A., Elsevier Press, Amsterdam, 399-405 (1982).

50. Santavuori, P., Westermarck, T.; Medical Biology, in the press (1984).

51. Westermarck, T., Rahola, T., Kallio, A.-K., Suomela, M.: Klin. Pädiat. 194, 301-302 (1982).

52. Airenne, A.L., Westermarck, T.: Medical Biology, in the press (1984).

53. Aaseth, J., Munthe, E., Førre, Ø., Steinnes, E.: Scand. J. Rheumatol. 7, 237-240 (1978).

54. Keshan Disease Research Group of Chinese Academy of Medical Sciences (Ge, K.): Curr. Concepts Nutr. 11, 127-138 (1982).

55. Gilroy, J., Meyer, J.S.: Medical Neurology, third ed., Macmillan Publishing Co., Inc. New York. pp. 715-768 (1979).

56. Ahlrot-Westerlund, B., Carlmark, B.: International Symposium Health Effects and Intesctions of Essential and Toxic Elements, Lund, Sweden, 13-18 June 1983, Studentlitteratur, abstract p. 215 (1983).

57. Burri, B.J., Chan, S.G., Berry, A.J., Yarnell, S.K.: Clin. Chim. Acta 105, 249-255 (1980).

58. Hunter, M.I.S., de Vane, P.J.: Clin., Chim. Acta 115, 93-98 (1981).

59. Örndahl, G., Rindby, A., Selin, E.: Acta Med. Scand. 213, 237-239 (1983).

60. Salonen, J.T., Alfthan, G., Huttunen, J.K., Pikkarainen, J., Puska, P.: Lancet ii, 175-179 (1982).

61. Miettinen, T.A., Alfthan, G., Huttunen, J.K.,Pikkarainen, J., Naukkarinen, V., Mattila, S., Kumlin, T.: Br. Med. J. 287, 517-519 (1983).

62. Wang, Y.X., Böcker, K., Reuter.H., Kiem, J., Kasperek, K., Iyengar, G.V., Loogan, F., Gross, R., Feinendegen, L. E.: Klin. Wochenschr., 59, 817-818 (1981).

63. Oster, O., Meinertz, T., Geibel, A., Zehender, M.,Batke, E., Schmiedel, G., Prellwitz, W.: 3:rd Intermational Workshop Trace Elemnt Analytical Chemistry in Medicine and Biology, April 16-18, 1984, Neuherberg, FRG, abstract No 31 (1984).

64. Thomson, C.D., Rea, H.M., Robinson, M.F., Simpron, F.O.: Proc. Univ. Otago Med. Sch. 56, 1-3 (1978).

65. Nye, E.R., Simpson, F.O., Waal-Manning, H.J., Spears. G. F.S., Leek, G.M.: NZ Med. J. 90, 4-8 (1979).

66. Kiem, J., Feinendegen, L.E.: 3:rd Inteirnational Workshop Trace Element Analytical Chemistry in Medicine and Biology, April 16-18, 1984, Neuherberg, FRG, abstract No 29 (1984).

67. Shamberger, R.J., Frost, D.V.: Can. Med. Assoc. J. 100, 682 (1969).

68. Shamberger, R.J., Willis, C.E.: Crit. Rev. Clin. Lab. Sci. 2, 211-221 (1971).

69. Robinson, M.F., Godfrey, P.J.m Thomson, D.C., Rea, H.M., Van Rij, A.M.: A,. J. Clin. Nutr. 32, 1477-1485 (1978).

70. Kivits, H., Report Vakgroep Deeltjes Fysica, TH Eindhoven, NK-81-24, 90-92 (1981).

71. McConnell, K.P, Broghamer, W.L., Blotchky, A.J., Hurt,O, J.: J. Nutr. 105, 1026-1031 (1975).

72. Broghamer, W.L., McConnell, K.P., Blorcky, A.J.: Cancer 37, 1384-1388 (1976).

73. Knuutila, S., Erikss n, J., Juvonen, E., Kovanen, R., Luotola, J., Westermarck, T.: Submitted.

74. Chadwick, K.H., Leenhouts, H.P.: The Molecular Theory of Radiation Biology, Springer, Berlin (1981).

75. Todd, A.T., Scott, S.G., Coke, H.: Proc, Roy, Med. 28, 681 (1935); cited by Cowgill, U.M.: Biol. Trace Elem.Res. 5, 345-361 (1983).

76. Hernaman-Johnson, F-: Bull. Med. J.: 3880, 1052 (1935); cited by Cowgill, U.M.: Biol. Trace Elem. Res. 5, 345-361 (1983).

77. Siakotos, A.N., Goebel, H.H., Patel, V., Watanebe, I., Zeman, W.: Advances in Experimental Medicine and Biology, eds. Volk. B.W., Aronson, S.M., Plenum Press, New York, 53-61 (1972).

78. Johansson, E., Lindh, U., Alanen, T., Westermarck, T., Heiskala, H., Santavuori, P., Elovaara, I.: Medical Biology, in the press (1984).

79. Gutteridge, J.M.C., Westermarck, T., Santavuori, P.: Acta Neurol Scand. 68, 365-370 (1983).

80. Gutteridge, J.M.C., Rowley, D.A., Halliwell, B., Westermarck, T.: Lancet 2, 459-460 (1982).

81. Westermarck, T., Gutteridge, J.M.C., Heiskala, H., Santavuori, P.; Rahola, T., Suomela, M.: Life Chem. Rep., in the press (1984).

82. Johansson, E., Lindh, U.: Res. Suppl. in the press (1984).

83. Fairbanks, V.F., Fahey, J.L., Beutler, E.: Clinical Disorders of Iron Metabolism 2:nd ed., Grune and Stratton, New York, 1-486 (1971).

84. Rosin, I.V., Butko, V.S., Kalabukhov, E.P.: Ter. Arkh. 54, 80-82 (1982).

85. Domaev, Yu.A.: Zhnevropatpsikhiatkorsakov 76, 224-228 (1976).

86. Rotruck, J.T., Pope, A.L., Ganther, H.E., Swanson, A.B., Hafeman, D.G., Hoekstra, W.G.: Science 179, 588-590 (1973)

87. Diplock, A.T.: Phil. Trans. R. Soc. Lond. B.294, 105-117 (1981).

88. Willmore, L.J., Rubin, J.J.: Neurology, 31, 63-69 (1981).

89. Santavuori, P., Westermarck, T., Rapola, J., Pohja, P., Moreń, R., Lappi, M., Vuonnala, U.: Submitted.

90. Halliwell, B.: Medical Biology, in the press (1984).

91. Munthe E., Aaseth.: Xth European Congress of Rheumatology Abstract 673, (1983).

92. Verlinden, M., van Sprundel, M., Van der Auwera, J.C., Eylenbosch, W.J.: Biol. Trace Elem. Res. 5, 103-113 (1983).

93. Willet, W.C., Polk, B.F., Morris, J.S., Stampfer, M.J., Pressel, S., Rosner, B., Taylor, J.O., Schneider, K., Hames, C.G.: Lancet 2, 130-134 (1983).

94. Black, R.S., Whanger, P.D.: Biol. Trace Elem. Res. 1, 313-324 (1979),

95. Smith, J.L.: Trace Elements in Human Health and Disease II, ed. Prasad, A.S., Academic Press, New York, San Francisco, London, 443-452 (1976).

96. Underwood, E.J.: Trace Elements in Human and Animal Nutrition, 4:th ed., Academic Press, Nuw York, London 243-257 (1977).

D i s c u s s i o n Session II - 2
Plenary Lecture

Subjects of the discussion

- Se-deficiency and cardiomyopathy
- Se-deficiency and vitamin E status
- Se-deficiency and cystic fibrosis
- reaction of Se in low oxidation states with oxygen radicals
- changes in fingernails in connection with Se-deficiency

Contributors: Cochran, Ganther, Koh, Oster, Plantin, Weser

INTAKE OF Zn, Mn AND Se BY ADULT FEMALES

- A TOTAL DIET STUDY -

Rieder F.W. Schelenz

Bundesforschungsanstalt für Ernährung
Zentrallaboratorium für Isotopentechnik
Engesserstr. 20, D-7500 Karlsruhe 1

Introduction

The US National Research Council (NRC) has set up the "Recommended Dietary Allowances" (RDA) for zinc with 15 mg per day and person (1)(tab.1). Studies on the dietary zinc intake in various countries between 1970-1981 resulted in lower levels than recommended (2-18). Only a study from New Zealand reported a mean dietary zinc intake exceeding the RDA value (18,4 mg per day, determined on 4 female subjects) (9).

Table 1: NATIONAL RESEARCH COUNCIL (USA), 1980
RECOMMENDED DIETARY ALLOWANCES

ELEMENT	RECOMMENDED DIETARY INTAKE PER DAY mg
Zinc	15
Manganese	2,5 - 5,0*
Selenium	0,05 - 0,2*

* SAFE AND ADEQUATE DAILY DIETARY INTAKES

Trace Element - Analytical Chemistry in Medicine and Biology, Vol. 3

In most of the dietary studies on female adults between 1967-1982 known from the literature the Mn intake was adequate to the NRC recommendation of 2,5-5,0 mg per day and person (19-39).

The dietary intake of Se depends largely on the origin of the consumed foodstuffs. The selenium content of vegetable foods, e.g. cereals, depends mainly on the Se concentration of the soil on which the plants have been grown. Soils in Finland and New Zealand show comparably low Se concentrations (40,41). The Se content of wheat grown on these soils is lower than that of wheat from Canada or the USA, where Se concentrations of the soil are generally higher (42).

The average daily dietary Se intake in Finland was calculated to be 30 µg per person if 90 % of the consumed cereals where home grown (41). It has been calculated further that this would increase to 50-60 µg if 50 % of the home grown cereals would be replaced by imports from Canada or the USA (42). Total diet studies with women in New Zealand resulted in even lower Se intakes of 18-26 µg (40,43). These intakes are far below the "adequate and safe dietary intakes" of 50-200 µg Se per day recommended by the NRC (1).

Daily dietary Se intakes of 62 µg per day and person for US hospital diets and 60-150 µg Se per day for adults in the North-East of the USA have been reported (44). A study reported in 1970 resulted in an average daily dietary Se intake for adults in the USA of 132 µg (45). Based on the composition of the diet consumed by 100 women in Canada a Se intake of 131 $\pm$ 53 µg per day and person has recently been calculated using nutrition tables (46). A similar study from Japan resulted in a mean Se intake of 88 µg per day and person (47). Representative analyses of groups of ready-to-eat foods as well as of raw individual foodstuffs in the United Kingdom resulted in a calculated average dietary Se intake of 60 µg per day (48).

In the Federal Republic of Germany a study from 1972 reported an average intake of 11,8 mg Zn, 2,7 mg Mn and 59 µg Se per day for adult males. No analytical data are available for women concerning the dietary intake of Mn and Se in the Federal Republic of Germany (7). The study "Einkommens- und Verbrauchs-stichproben 1978" resulted in a dietary Zn intake of 9 mg per day and woman (19-50 years old), calculated from dietary records using nutrition tables.

One method to estimate the dietary trace element intake of man is to determine the element contents of numerous individual foodstuffs and based on the official food consumption statistics, to calculate the average element intake per capita. This procedure suffers from several disadvantages. In particular the possible effect of preparation of food on its element content is often disregarded. Therefore this paper is concerned with trace element determination of the daily total diet of selected persons with respect to their individual eating and drinking habits.

Materials and Methods

On 4 different week-days - randomly chosen - between March and June 1981 the total food intake of 26 women, 25-45 years old, living at Karlsruhe and Stuttgart (Federal Republic of Germany) was collected. The duplicate samples of all foods and beverages consumed by each woman during a day - corresponding to the individual eating and drinking habits - were weighed, homogenized, aliquots frozen at -18°C or freeze dried or dry ashed at 400°C corresponding to the nutrients to be determined. Dietary records were kept during the experimental days by each subject.

In all, 45 food constituents were determined in aliquots of the total daily diet (fig.1). The elements Zn, Mn and Se were determined applying purely instrumental neutron activation analysis (49). Homogeneity of the material was tested by repeated neutron activation analysis of Na, Cl, K and Mn in five different samples of the same batch (tab.2). The coefficients of variation - including the statistical counting error of radioactivity measurements - indicated that the elements were uniformly distributed in the sample down to 250 mg sample size.

TOTAL DIET

NUTRIENTS	MINERALS	TRACE ELEMENTS		RADIONUCLIDES	
		ESSENTIAL	TOXIC	β-, γ-	α-EMITTERS
ENERGY	CALCIUM	COBALT	ANTIMONY	CARBON-14	TOTAL-α
PROTEIN	CHLORINE	COPPER	ARSENIC	CESIUM-137	AMERICIUM
FAT	MAGNESIUM	IRON	CADMIUM	STRONTIUM-90	CURIUM
FATTY ACIDS (SAT., UNSAT.,)	PHOSPHOROUS	MANGANESE	LEAD	TRITIUM	PLUTONIUM
DIETARY FIBRE (CELLULOSE, LIGNIN, HEMICELLULOSE)	POTASSIUM	SELENIUM	MERCURY		THORIUM
THIAMINE	SODIUM	ZINC			URANIUM
NITRATE/NITRITE					
		OTHERS			
		BARIUM			
		CESIUM			
		GOLD			
		RUBIDIUM			
		SCANDIUM			

COMPONENTS ANALYZED: 45
DUPLICATE SAMPLES: 104
TEST PERSONS: 26 FEMALES (25-45 YRS.) WITH FAMILY AND 1 CHILD (MIN.) OLDER THAN 2 YRS.
COLLECTING DAYS EACH: 4
COLLECTING PERIOD: MARCH-JUNE 1981

Figure 1: FOOD CONSTITUENTS DETERMINED IN ALIQUOTS OF THE TOTAL DAILY DIET

Table 2: HOMOGENEITY OF ALIQUOTES OF A FREEZE DRIED TOTAL DIET SAMPLE

Element concentrations referred to wet mass

Sample No	Sample weight mg	Chlorine g/kg	Potassium g/kg	Manganese mg/kg	Sodium g/kg
1	231,4	2,40	-	1,7	1,67
2	228,9	2,39	1,92	1,5	1,62
3	257,3	2,42	2,13	1,4	1,66
4	282,7	2,49	1,77	1,8	1,94
5	267,7	2,36	1,93	1,7	1,86
Mean value:	253,6	2,41	1,94	1,6	1,75
Standard deviation:		0,05	0,15	0,16	0,14
Coefficient of variation (%):		2,0	7,6	10,0	8,1

Zinc

Concerning the 4 day average the 26 females ingested 10,6 ± 2,0 mg zinc per day, which is less than the recommended value of the RDA of 15 mg zinc per day (fig.2). It has to been pointed out that the bars drawn in figure 2 do not represent the standard deviation of the analytical method but indicate the variations of elemental intake in the 4 diets analyzed of each person. It is evident from figure 2 - based on a 4 day average - that none of the subjects exceeded the RDA level.

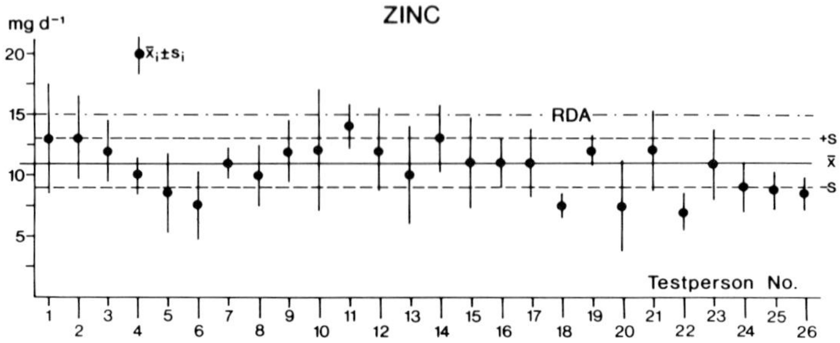

Figure 2: AVERAGE DAILY DIETARY Zn INTAKE PER PERSON

On 13 days, however, the RDA value was reached or exceeded by 11 women (fig.3). This is probably due to a higher consumption of beef and pork compared to the other persons. Animal protein is the main source of zinc in human nutrition. Additionally the total daily food intake of these persons was 20 % higher than the average consumption of the 26 women. The lowest zinc intakes were found when no meat was consumed and when additionally the total food intake was low compared to the mean food consumption of the other women.
The average daily dietary zinc intake of 10,6 mg found in this study is of the same order as reported from comparable studies between 1970-1981 in other countries (2-18).

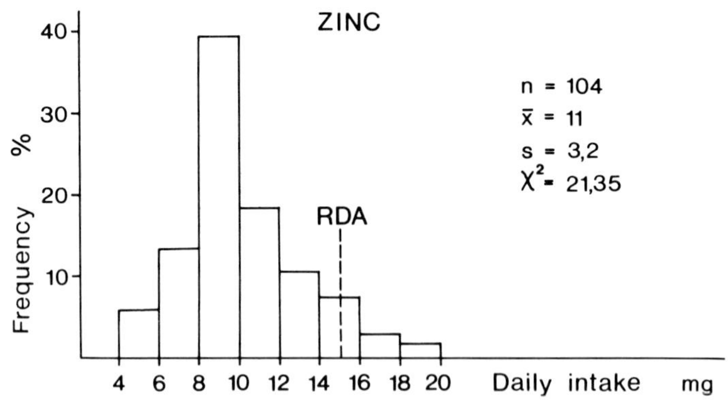

Figure 3: FREQUENCY DISTRIBUTION OF DAILY DIETARY Zn INTAKE

The correlation of zinc to protein, energy, Cu, Fe and Mn intake based on 104 daily total diets has been calculated from the individual analytical data (tab.3):

Table 3: CORRELATION COEFFICIENTS OF PAIRS OF NUTRIENTS

Pair of nutrients (Daily intake)		Coefficient of correlation n = 104 r	Significance
Zn/Protein	(mg/g)	0,71	+++
Zn/Energy	(mg/kJ)	0,65	+++
Mn/Protein	(mg/g)	0,37	+++
Mn/Energy	(mg/kJ)	0,39	+++
Cu/Zn	(mg/mg)	0,23	+
Fe/Zn	(mg/mg)	0,56	+++
Fe/Mn	(mg/mg)	0,47	+++
Mn/Zn	(mg/mg)	0,45	+++

+: $p < 0,05$; +++: $p < 0,001$
p: probability of error

The calculated coefficients of correlation indicate that a higher dietary zinc intake may also result in higher intakes of Cu, Fe and Mn. The uptake of the essential trace elements Cu, Fe and Mn also increases with protein or energy consumption.

There are some problems, however, to formulate a diet which contains adequate amounts of e.g. Zn without exceeding the recommended energy content of 9000 kJ per day. This is particularly the case when the consumption of offals is refused, which have a relatively high content of trace elements.

Manganese

In the 4 day average the 26 females ingested $3,56 \pm 1,23$ mg Mn per day. 21 (80 %) of the women had an intake within the RDA recommendations of "safe and adequate daily dietary intakes" of

2,5-5,0 mg Mn per day and person (fig.4). Based on the 4 day average only 2 women ingested more than 5 mg Mn per day.

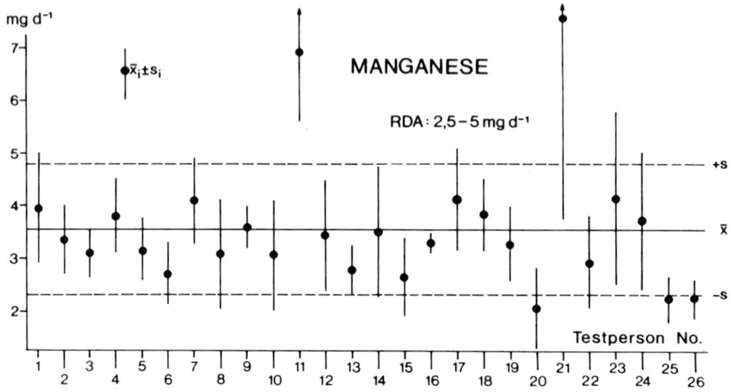

Figure 4: AVERAGE DAILY DIETARY Mn INTAKE PER PERSON

The frequency distribution of the Mn intake (fig.5) indicates that more than 80 % of the total diets contained manganese amounts to be "safe and adequate" (1). Twelve (11,5 %) daily total diets contained more Mn than corresponds to the recommended upper level of 5 mg per day. The highest dietary Mn intake of 12,9 mg per day was probably due to the consumption of 260 g whole wheat bread. Cereals, cereal products and whole meal bread are the main Mn sources for human nutrition. Test persons with Mn intakes above 5,0 mg per day (fig.5) consumed comparably more whole meal bread or "Müsli" with oaks or shredded wheat. The mean daily bread consumption (mainly consisting of whole meal bread) of 7 women with Mn intakes above 5 mg was 70 % above the mean bread consumption of 110 g of all subjects.

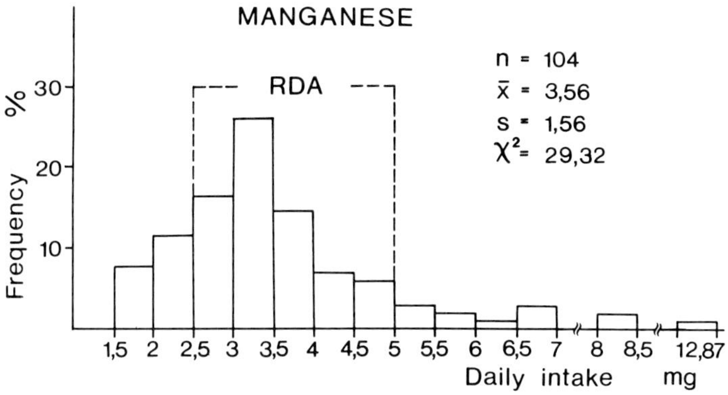

Figure 5: FREQUENCY DISTRIBUTION OF DAILY DIETARY Mn INTAKE

Studies from Great Britain indicate that 50 % of the Mn intake is due to black tea consumption (19,50). The Mn content of 1 cup of tea has been estimated to be 760 µg on average (19). Women of this study who consumed whole meal bread and also black tea had the highest Mn intakes. These women consumed 750-950 ml black tea daily.

On 18 (17,3 %) experimental days the dietary Mn intake was less than the lower RDA recommendation of 2,5 mg per day (fig.5). Looking at the 4 day average only 3 persons showed a Mn intake which was below the RDA value. None of these women consumed whole meal bread. The minimum Mn intake was 1,62 mg per day and person. This woman consumed in total nearly 40 % less than the other women and only half of the amount of bread. She ate no whole meal bread.

The average dietary Mn intake of 3,6 mg per day and person in this study is of the same order of magnitude as has been reported by other authors in recent years (19-39).

Selenium

The dietary Se intake of the 26 female adults - based on a 4 day average - was 55 ± 31 µg per day and person (fig.6). This value is near the lower margin of the recommended range of 50-200 µg Se per day considered to be "safe and adequate" (1).

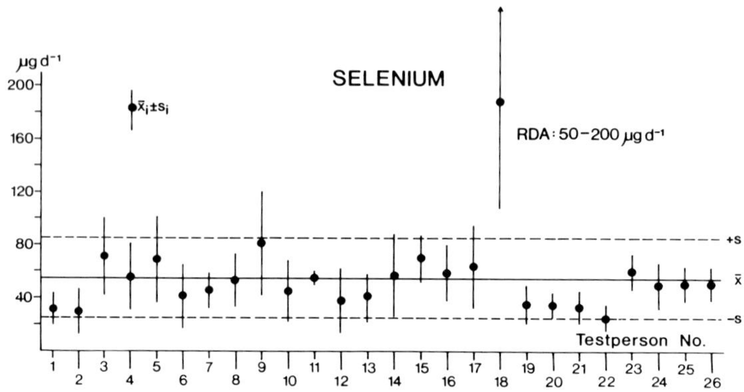

Figure 6: AVERAGE DAILY DIETARY Se INTAKE PER PERSON

Only 1 person (No. 18) exceeded the upper limit of the RDA value with dietary Se intakes of 234 and 276 µg per day on 2 experimental days . Similarly the 4 day average Se intake of 188 ± 80 µg was remarkably higher than that of the other persons. The higher Se intakes were probably due to the consumption of bread, whole meal bread, poultry meat, bacon, "Müsli" and farinaceous food which were commonly known to be comparably selenium rich foods. Furthermore this woman ate every day at breakfast a boiled egg. Eggs are also a selenium rich food (51). Fish was not consumed by this test person.

The Se intake of 7 women on 8 experimental days of 93-134 µg per day was significantly higher compared to the average intake of 55 µg per day of the other women (fig.7). These higher intakes were probably due to the consumption of selenium rich sea-fish, poultry meat and "Müsli".

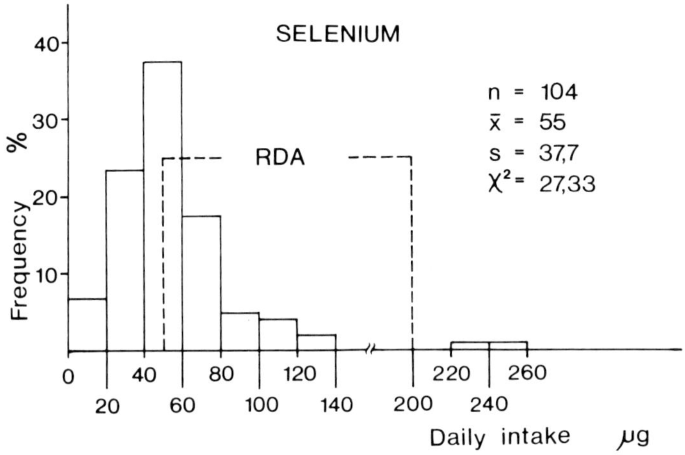

Figure 7: FREQUENCY DISTRIBUTION OF DAILY DIETARY Se INTAKE

The 10 days with consumption of sea-fish resulted in a higher average Se intake of 74 ± 31 µg per day and person.

Nearly half of the total diets contained Se concentrations which corresponded to an intake less than the lower RDA value of 50 µg per day and person.A Se intake of less than 20 µg per day was found on 6 experimental days for 6 different women (fig.7). The mean Se intake on these 6 days was 14 ± 3 µg with extreme values of 9,5-17 µg per day and person. None of these persons consumed fish and only 1 person consumed fried meat.

The average Se intake of 55 ± 31 µg per day found in this study is in the same order as that observed in the United Kingdom (60 µg) and recently in Finland (50-60 µg) (20,47). A similar study in the Federal Republic of Germany with male adults in 1972 resulted in a Se intake of 59 µg with extreme values of 9-170 µg per day and person (29). For the United Kingdom and Finland the relatively low Se intake was related to low Se concentrations in the soil.

Whereas the dietary intake of Se was found to be near the lower level of the RDA recommendation this does not a priori mean a marginal Se supply. A dietary Se intake of 20 µg per day and person is regarded by Gibson et al. to be sufficient for the maintenance of human health (39).

Besides the comparison of the daily intake of essential nutrients with the recommended values also the "Quality Index" represents a suitable standard. The Quality Index (QI) is the quotient of the analyzed or calculated nutrient density related to 1 MJ energy of the total diet to the energy related nutrient density which can be calculated from the recommended nutrient intakes (fig. 8). QI > 1 means excess supply and a QI < 1 marginal intake of an individual nutrient related to the energy of the diet.

Nutrient	Dimension	Recommended intake per day	Recommended intake per MJ and day	QI
Energy	MJ	8.8	1.00	1.00
Protein	G	52	5.91	1.32
Fat	G	68	7.73	1.42
Linoleic acid	G	10	1.14	1.23
Sodium	G	2-3	0.23-0.34	1.05-1.56
Chlorine	G	3-5	0.34-0.57	0.93-1.57
Potassium	G	2-3	0.23-0.34	0.91-1.35
Calcium	MG	700	79.5	1.23
Magnesium	MG	220	25.0	1.36
Iron	MG	18	2.04	0.59
Zinc	MG	15	1.70	0.65
Copper	MG	2-3	0.23-0.34	0.35-0.52
Phosphorous	MG	700	79.5	1.69
Manganese	MG	2.5-5.0	0.28-0.57	0.67-1.36
Selenium	/UG	50-200	5.7-22.7	0.27-1.09
Thiamine	/UG	1400	159.1	0.89

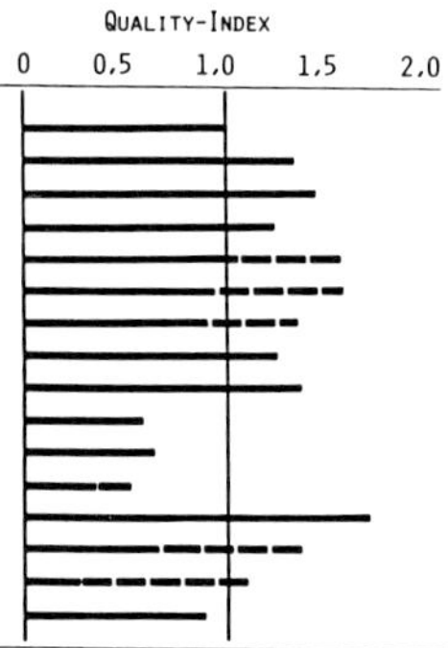

Figure 8: NUTRIENT DENSITY PROFILE: AVERAGE NUTRIENT CONSUMPTION OF 26 WOMEN (104 TOTAL DAILY DIETS) per 1 MJ

The right end of the dashed bar(fig.8) means: lower limit of the recommended intake as reference for QI. The left end of the dashed bar (fig.8) means: upper limit of the recommended intake as reference for QI.

The mean QI for Zn with 0,65 implies a low supply of this element related to energy. The energy related intakes of Mn and Se with QI values of 1,36 and 1,09, respectively, are adequate to the lower recommended limits but slightly deficient for Mn with a QI of 0,67 and marginal for Se with a QI of 0,27 to the upper recommended limits.These results are equivalent to those found using the analytical data for the dietary daily intake of these elements compared to the recommended values. As it can be seen also the energy related intakes of Fe and Cu are marginal.

Conclusion

Based on the 4 day mean values and compared to recommended intakes the dietary trace element intake of the 26 adult females showed no dietary deficiency for Mn, a barely sufficient intake of Se and considerably deficient Zn intakes (tab.4).

Table 4: AVERAGE DAILY DIETARY ELEMENT INTAKE OF 26 ADULT FEMALES BASED ON 104 TOTAL DIETS

Element	Average intake $\bar{x}^2 \pm$ s.d.[3] mg	Extreme values min. - max. mg	Recommended NRC value[1] mg
Zinc	10,6 ± 2,0	7,0 - 14	15
Manganese	3,56 ± 1,23	2,05 - 7,55	2,5 - 5,0[4]
Selenium	0,055 ± 0,031	0,025 - 0,188	0,05 - 0,2[4]

[1] NRC = National Research Council, USA, 1980

[2] $\bar{x}$ = arithmetic mean

[3] s.d. = standard deviation of the mean

[4] safe and adequate daily dietary intakes, NRC, 1980

Keywords

Total diet, female adults, duplicate portion method, neutron activation analysis, dietary recommendations, trace elements, trace element density, manganese, selenium, zinc.

References

1. Recommended Dietary Allowances
Food and Nutrition Board, National Academy of Sciences - National Research Council; 9th revised ed., Washington, D.C., USA, 1980

2. TRIBBLE, H., SCOULAR, F.I.: J.Nutr. 52 (1954) 209

3. GORMICAN, A.: J.Am.Diet.Ass. 56 (1970) 397

4. MURPHY, E.W., WATT, B.K., PAGE, L.: Regional variations of vitamin and trace element content of Type A school lunches In: Trace Elements in Environmental Health, IV., Columbia, Mo.: Univ. of Missouri, 1971
Hrsg.: Hemphill, D.

5. SCHLAGE, C, WORTBERG, B.: Acta Paediat.Scand. 61 (1972) 421, 648

6. OSIS, S., KRAMER, I., WIATROWSKI, E., SPENCER, H.: Am.J.Clin.Nutr. 25 (1972) 582

7. SCHELENZ, R.: J.Radioanal.Chem. 37 (1977) 539-548

8. BROWN, E.D., McGUCKIN, M.A., WILSON, M., SMITH, J.C.: J.Am.Diet.Ass. 69 (1976) 633

9. WHITE, H.S.: J.Am.Diet.Ass. 68 (1976) 243

10. GREGER, J.L., SCISCOE, B.S.: J.Am.Diet.Ass. 70 (1977) 37

11. GUTHRIE, B.E., ROBINSON, M.F.: Br.J.Nutr. 38 (1977) 55-63

12. WALKER, M.A., PAGE, L.: J.Am.Diet.Ass. 70 (1977) 260-266

13. Survey of Copper and Zinc in food. Food Surveillance Paper No.5, 1981. Ministry of Agriculture, Fisheries and Food. Her Majesty's Stationary Office, London

14. Einkommens- und Verbrauchsstichproben 1978 (EVS 78) cited in: "Material zum Ernährungsbericht 1980" Deutsche Gesellschaft für Ernährung, Frankfurt/Main 1980, S. 9-49

15. HOLDEN, J.E., WOLF, W.R., MERTZ, W.: J.Am.Diet.Ass. 75 (1979) 23-28

16. LYON, T.D.B., SMITH, H., SMITH, L.B.: Br.J.Nutr. 42 (1979) 413-416

17. GIBSON, R.S., SCYTHES, C.A.: Br.J.Nutr. 48 (1982) 241-248

18. ENGLISH, R.: J.Food and Nutrition 38 (1981) 63-65

19. CROSS, J.D., DALE, I.M., RAIE, R.M., SMITH, H., SMITH, L.B.: Radiochem.Radioanal.Letters 37, No.6 (1979) 365-376

20. WENLOCK, R.W., BUSS, D.H., DIXON, E.J.: Br.J.Nutr. 41 (1979) 253-261

21. KIRKPATRICK, D.C., COFFIN, D.E.: Can.J.Publ.Hlth. 68 (1977) 162

22. KIRKPATRICK, D.C., COFFIN, D.E.: J.Inst.Can.Sci.Technol. Aliment. 7 (1974) 56

23. SOMERS, E.: J.Fd.Sci. 39 (1974) 215

24. MERANGER, J.C., SMITH, D.C.: Can.J.Publ.Hlth. 63 (1972) 53

25. LANG, K.: Biochemie der Ernährung Steinkopff Verlag, Darmstadt, 1970, S. 366

26. SOMAN, S.D., PANDAY, V.K., JOSEPH, K.T., RAUT, S.J.: Health Physics 17 (1969) 35

27. NAKAGAWA, T.: Osaka Shiritsu Daigaku Igaku Zasshi 17 (1968) 401

28. GUTHRIE, B.E., ROBINSON, M.F.: Br.J.Nutr. 38 (1977) 55-63

29. GUTHRIE, B.E.: Daily dietary intakes of zinc, copper, manganese, chromium and cadmium by some New Zealand women Proc.Univ.Otago Med.School 51 (1973) 47

30. McLEOD, B.E., ROBINSON, M.F.: Br.J.Nutr. 27 (1972) 221

31. SCHLETTWEIN-GSELL, D., SEILER, H.: Mitt.Lebensm.Hyg. 63 (1972) 188

32. WHITE, H.S., GYNNE, T.N.: J.Am.Diet.Ass. 59 (1971) 27

33. GORMICAN, A.: J.Am.Diet.Ass. 56 (1970) 397

34. WHITE, H.S.: J.Am.Diet.Ass. 55 (1969) 38-43

35. TIPTON, I.H., STEWART, P.L., DICKSON, J.: Health Physics 16 (1969) 455

36. WALKER, M.A., PAGE, L.: J.Am.Diet.Ass. 70 (1977) 260-266

37. SRIVASTAVA, U.S., NADEAU, M.H., CARBONNEAU, N.: Nutr.Rep.Int. 18, No.3 (1978) 325-336

38. WOLF, W.R.: Assessment of Inorganic Nutrient Intake from Self-Selected Diets
Human Nutrition Research, BARC Symposium, Allanheld, Osmun, Totowa, ed. G.R. Beecher

39. GIBSON, R.S., SCYTHES, C.A.: Br.J.Nutr. 48 (1982) 241-248

40. STEWART, R.D.H., GRIFFITHS, N.M., THOMSON, C.D., ROBINSON, M.F.: Br.J.Nutr. 40 (1978) 45-54

41. Mineral Element Composition of Finnish Foods: N, K, Ca, Mg, P, S, Fe, Cu, Mn, Zn, Mo, Co, Ni, Sr, F, Se, Si, Rb, Al, B, Br, Hg, As, Cd, Pb and Ash
ed.: Koivistoinen, P.
Act.Agric.Scand.Suppl. 22 (1980) 1-171

42. VARO, P., KOIVISTOINEN, P.: Int.J.Vit.Nutr.Res. 51, No.1 (1981) 79-84

43. ROBINSON, M.F., McKENZIE, J.M., THOMSON, C.D., VanRIJ, A.I.: Br.J.Nutr. 30 (1973) 195

44. SCHROEDER, H.A., FROST, D.V., BALASSA, J.J.: J.Chronic Dis. 23 (1970) 227

45. MORRIS, V.C., LEVANDER, O.A.: J.Nutr. 100 (1970) 1383

46. GIBSON, R.S., SCYTHES, C.A.: Br.J.Nutr. 48 (1982) 241-248

47. SUZUKI, S., KAIZUMI, S., HARADA, H., ITO, K., TOTANI, T.: Ann.Rep. Tokyo Metropol.Res.Lab.Pub. Health 22 (1970) 153

48. THORN, J., ROBERTSON, J., BUSS, D.H.: Br.J.Nutr. 39 (1978) 391-396

49. SCHELENZ, R., FISCHER, E.: Neutronenaktivierungsanalyse von Spurenelementen in Lebensmitteln Berichte der Bundesforschungsanstalt für Ernährung, Karlsruhe, BFE-Bericht 1976/4, Mai 1976, 63 S.

50. MONIER-WILLIAMS, G.W.: Chapman and Hall, London, 1949

51. WEIGERT, P.: Z.Lebensm.Unters.Forsch. 171 (1980) 18-19

D i s c u s s i o n Session II - 3

Subjects of the discussion

- levels of transplutonium elements in the diet
- need for speciation studies
- interactions Mn - Cu and Fe - Mn
- trace elements in drinking water
- recommended daily dietary allowances in terms of species
- excretion of Mn
- bioavailability of trace elements in food

Contributors: Brätter, Gökman, Hurley, Koh, Machari

Session III

SELENIUM ANALYSIS

Plenary Lecture given by: G. Tölg (FRG)

Chairman:

G. Knapp (Austria)

Poster Reports given by: P. Brätter (FRG)

H.E. Ganther (USA)

G. Tölg (FRG)

SELENIUM-ANALYSIS IN BIOLOGICAL MATERIALS

Günther Tölg
Institut für Spektrochemie und angewandte Spektroskopie
and Max-Planck-Institut für Metallforschung Stuttgart,
Laboratorium für Reinststoffanalytik, Postfach 778,
4600 Dortmund 1, Federal Republic of Germany

Introduction

The scientific interest in trace analysis of selenium has been intensified since 1970, motivated by the distinctly ambivalent physiological activity of Se (1).
K. Schwarz and C.M. Foltz recognized as early as 1957 that not only would too high concentrations of selenium, taken up in foods or from work environment cause acute or chronic poisoning, but also that selenium deficiency within animal or human metabolism would cause disease or even fatality if the concentration fell below certain low threshold levels. Selenium deficiency was found to be connected with necrosis of the liver, muscular dystrophy, degeneration of the pancreas, the so-called mulberry disease, fertility insufficiency and further animal diseases. Additional efforts in the trace analysis of selenium were initiated by Shamberger's observations in 1979, that a diet supply of selenium compounds significantly reduced the number of dermic tumors of mice. A number of additional anti-cancerogeneous effects were observed in animal tests and were confirmed for humans by epidemiological studies (Shamberger et al.,1976; Schrauzer et al.,1977).

Trace Element - Analytical Chemistry in Medicine and Biology, Vol. 3

Owing to the affinity of selenium to a number of toxic heavy metals forming very stable selenides, the interest grows in the systematic investigation of its antagonistic effect on e.g. Hg, Cd, Pb. Finally, persistent motivations for the determination of selenium were again renewed by the occurrence of coronary diseases as well as a special form of congestive cardiac insufficiency of small children (Keshan-disease in China), occurring in cases of strong bio-deficiency of selenium.

For this workshop, I was given the task to comment critically on the most recent state of the analysis of selenium in biotic matrices. When contemplating on it, it was foremost in my mind that only the reliable and economic selenium determination will form a genuine basis for all further investigations of bio-effectiveness of selenium; and in this connection, geochemistry, many fields of bio-sciences, medical sciences and environmental disciplines are closely associated fields for analysis.
From all these different fields, widely different matrices are of interest in which selenium is determinable with practically every trace-analytical method available today. It is a hopeless attempt, therefore, to make fair references to the countless possible solutions of problems within the rather restricted space available. Therefore, the attempt is made to focus on the most powerful methods and to assess these in respect to their figures of merit, - power of detection, reliablity and economy. The starting point of my inquiry was a statistical survey (2) listing and evaluating from a methodical aspect, 2154 publications, i.e. all that were referred to on selenium and tellurium analytical work in the Chemical Abstracts during a period from 1971 to 1981. To improve clarity only the main methods are selected and their frequency of use is chronologically represented in Fig.1. The graphs are assumed to be representative for the analysis of selenium, because the number of papers containing

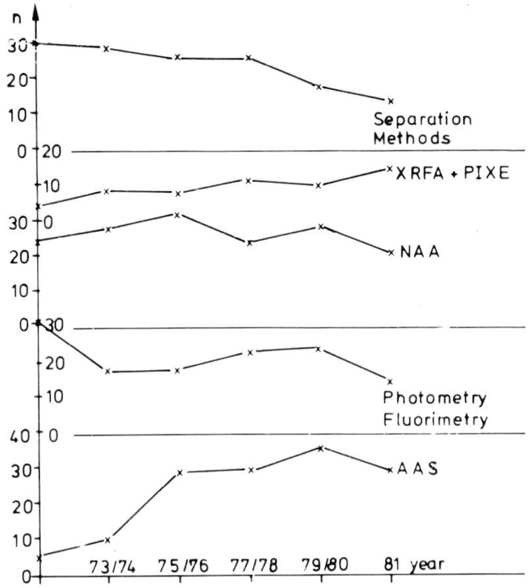

Fig. 1 Methods trends in selenium and tellurium analysis after L. Futekov et al. (2).

M a t r i x	Se-Contents to be determined $[ng \cdot g^{-1}]$
Earth crust (mean)	90
Rocks	≥ 50
Soils (normal)	≥ 100
Sea water	≥ 25
River and lake water	≥ 20
Ground water	≥ 1
Plants	≥ 10
Animal and human organs	≥ 10
Human blood serum	≥ 10

Table 1 Compilation of lowest selenium contents in environmental matrices from data of S.E. Raptis et al. (3) and H. Robberecht et al. (53).

work on tellurium is comparatively small and because the tellurium analysis is undergoing a similar trend. The decreasing tendency of the total activity as observed during the last years indicates a certain degree of establishment in method development. This tendency most likely refers only to the majority of papers on the determination of higher selenium concentrations (about $> 0.1\ \mu g/g$). A leading position in methods is taken up by atomic absorption spectroscopy (AAS). It has superceded chiefly photometry and fluorometry. While the number of references employing neutron activation analysis (NAA) displays a constant course, one can observe an ascending x-ray fluorescence analysis (XRFA)-graph during the course of time. This is probably due to the employment of energy dispersive XRFA, the rapid progress with the proton induced x-ray emission (PIXE)-method and the trend towards multielement analysis.

The number of papers dealing with separation methods decreased considerably in favour of those describing direct instrumental methods; on the other hand, the separation step may be included in the determination method, as is done in hydride generation-AAS (HG-AAS) and HG-optical emission spectrometry (HG-OES).

This statistical report agrees with the laboratory experience that the situation in routine analysis of selenium has improved significantly during the past years in respect of reliability of results. Since 1982, very informative review articles on the determination of selenium are available (3, 53), in which remaining points of weakness are critically indicated. This overall encouraging development permits me to restrict my contribution to mainly the future analytical chemistry of selenium. First of all, very small absolute amounts of selenium should be reliably determined. This is less mandatory for environment-relevant matrices usually containing levels of selenium > 10 ng/g (see Table 1), but more for use in trace analysis of micro-regions (i.e. tissue compartments, histological sections, protein fractions of blood serum).

For such tasks of picogramme analysis in micro-samples the determination methods available today are reduced to only a few methods with very low absolute limits of detection, compare Fig.2.

The electron capture detector gas chromatographic (ECD-GC) method holds the best power of detection, followed by the more easily performed hydride generation condensation-AAS (HGC-AAS) to which more detailed reference will be made at a later point. Any further determination methods of relatively high sensitivity can only briefly be referred to as regards their problems and limits.

Before going into these details, the general methodical trend and its effects on the analytical results are demonstrated by way of example of the determination of selenium content in blood serum that hitherto has not yet been solved satisfactorily.

In Table 2, a comparison of methods and results obtained with them are recorded in three periods of time. While in the first two periods until 1978, foremostly NAA and fluorometry were used for the determination of selenium in blood serum, the variety of methods greatly expanded during the period from 1979. Mainly HG-AAS, electrothermal atomization-AAS (ETA-AAS), XRFA and PIXE were employed in addition. Despite the considerable range of biological scatter of selenium contents of 'Normal-Blood-Serum' due to different geographical areas, habits of nutrition, age, sex etc., one can identify from this compilation a significant trend to lower selenium contents.

Period of publications	Methods	Number of papers	X $[ng \cdot mL^{-1}]$	$\bar{X}$ $[ng \cdot mL^{-1}]$	Reference
1967 - 1975	Fluor., NAA	9		111 ± 43	J. Versieck et al. [30]
1976 - 1979	Fluor., NAA	9		97 ± 34	J. Versieck et al. [30]
≥ 1979	NAA, AAS, XRFA, PIXE	11		93 ± 17	G. Tölg
1979	NAA		82 (74-89)		D. Behne et al. [31]
1979	ETA-AAS		114		K. Saeed et al. [32]
1982			62 (72)		G. Alfthan et al. [6]
1982	ETA-/HG-AAS		80		O. Oster et al. [33]
1982	HG-AAS		115		B. Lloyd et al. [34]
1982	HG-AAS		102		M. Verlinden [35]
1982/83	HG-AAS		89 (69)		B. Welz et al. [8, 36]
1983	XRFA, NAA, PIXE		106		H.J. Robberecht et al.[22]
1981	PIXE		80		J.R. Chen [37]
1983	PIXE		80		J. Pallon et al. [38]
1983	PIXE		110 (80-130)		G. Weber et al. [39]

Table 2 Methods trends of the determination of selenium in human blood serum and their effect on the analytical results.

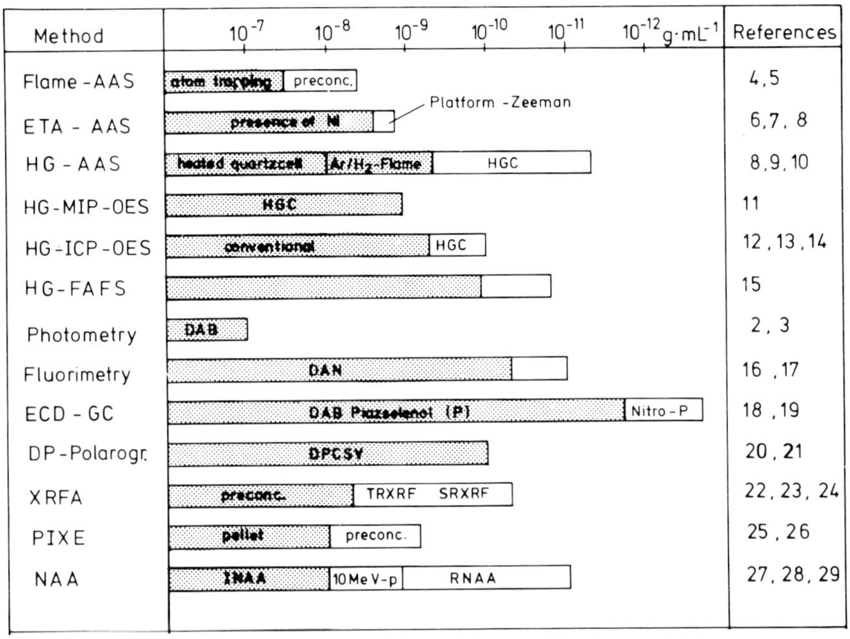

Fig. 2 Determination ranges of the most sensitive methods in selenium trace analysis.

Method	Reagents [mL,g]	Sample amount [g]	Decomp.-T [°C]	Se-amount [ng]	Remarks	Ref.
Dry ashing:	$Mg(NO_3)_2$ + MgO	≤ 5	≤ 600	≥ 10	quartz vessel, AAS	40, 41
Wet decomp.: open system	$HNO_3/HClO_4$ (≤5:1)	≤ 1	≤ 210	≥ 10	quartz vessel, AAS	35, 42
	$HNO_3/HClO_3,HClO_4$	≤ 0.2	≤ 240	≥ 1	quartz vessel, AAS	43
	$HNO_3/H_3PO_4/H_2O_2$	≤ 5		≥ 10	Fluorimetry	44
pressure bomb	HNO_3(≤ 4)	≤ 0.5	180	≥ 1	PTFE vessel, AAS	
			220	≥ 1	glassy carbon, DPCSV	46
					quartz vessel	47
Gaseous phase:						
O_2-stream	Oxygen	≤ 1	~1200	≥0,001	quartz tube with cooling trap (liq. N_2)	48
Trace-O-Mat®	Oxygen	≤1	≥1000	≥0,01	quartz device with cooling finger (liq. N_2).	49
HF-O-plasma	Oxygen	≤2	< 200	≥0,001	refluxing with HNO_3(<2)	50
Wickbold	H_2/O_2	≤2	≥ 1000	≥50	quartz dev., HGAAS	51

Table 3 Recommended decomposition methods of biotic materials in selenium trace analysis.

This and also the decreasing range of scatter of averages from different laboratories may be attributed in all probability to the improvement of the analytical techniques.

Blood serum is probably the matrix most suitable for such a critical comparison of the determination methods in respect to power of detection, reliability and economy - besides the few biotic standard reference materials with certified selenium contents - among all the numerous biotic matrices analyzed so far. In other matrices, however, elements may occur in excess or in different concentration ratios and thus cause interferences that alter significantly the order of merit of methods.

Recommended Decomposition Methods

As a rule, decomposition methods are required, which precede any determination method - except instrumental neutron activation analysis (INAA) - which may cause considerable systematic errors, chiefly by selenium losses and by insufficient mineralization of organic matrix. For brevity, I cannot discuss in detail the numerous decomposition methods proposed and the numerous experiments for their optimization, bringing to light the difficulty of this particular analytical step. Only the principal decomposition methods are selected in Table 3; the decomposition methods that are especially suitable for the determination of extremely low selenium concentrations are presented in some more detail.

1. <u>The decomposition using a quartz pressure bomb</u> (45), compare Fig. 3, permits sample (< 500 mg) digestion with HNO_3 at a temperature of up to 300° C with sufficiently high oxidation potential resulting in a 100% mineralization. The pressure development within the vessel during decomposition is automatically compensated by a counter pressure of up to 100 bar

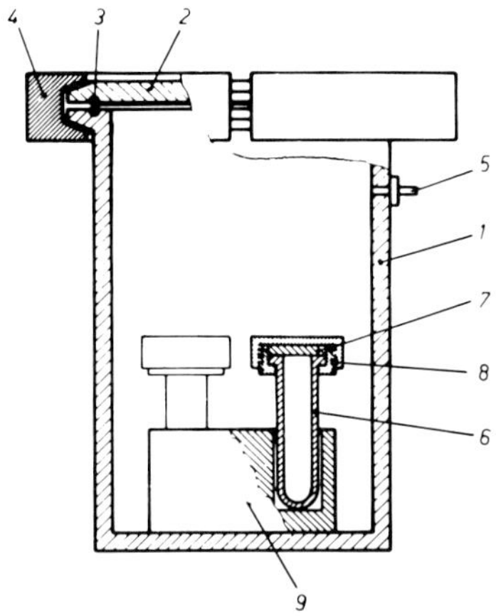

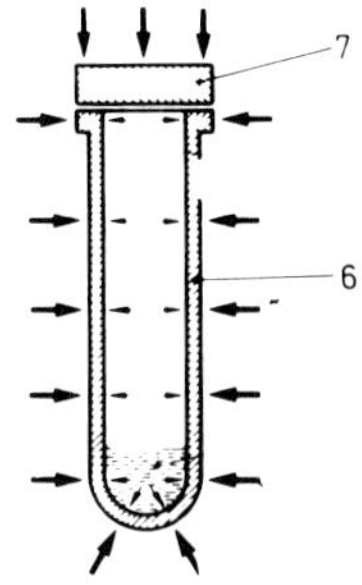

Fig. 3 Quartz pressure bomb device after G. Knapp (45)
1 pressure vessel; 2 cover; 3 graphite gasket; 4 sealing cap; 5 inlet for pressure gas; 6 quartz vessel; 7 quartz cover; 8 breech for quartz vessel; 9 heating block.

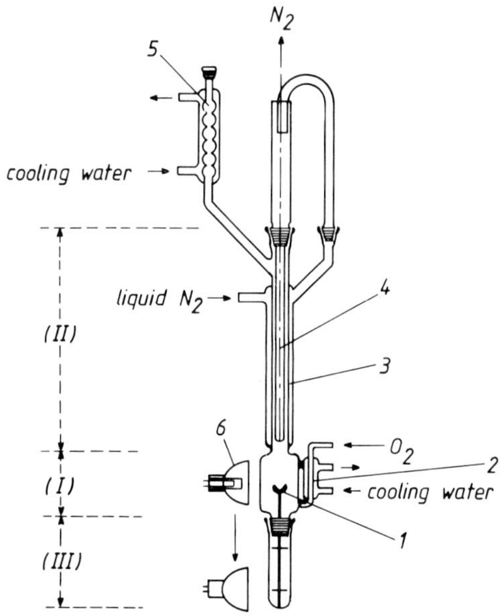

Fig. 4 Combustion unit (quartz) of the Trace-O-Mat(R)
(1) combustion chamber: (II) liquid nitrogen cooling system; (III) test tube (quartz).
1 sample holder (quartz); 2 capillary with cooler for oxygen supply; 3 cooling chamber; 4 cooling tube; 5 reflux condenser; 6 IR-heater lamp.

exerted from outside the quartz vessel, thus avoiding a bursting of the vessel.

2. Complete combustion of < 1 g of biotic matrix is obtained using pure oxygen in a quasi-static system in a quartz device (Trace-O-Mat(R)) (49), compare Fig. 4. The sample is oxidized under control with an adjustable supply of oxygen, after ignition with the aid of two heater lamps (upper position). At the combustion temperature, selenium is volatilized completely and condensed on a cooling finger (liquid N_2). In this manner, the selenium is separated from nonvolatile concomitant sample elements. The sample carrier and the combustion residues are removed out of the system. The selenium is dissolved, together with the other volatilized elements (such as Hg, Cd, Tl, Pb, Bi) by means of refluxing with a small quantity of extreme-purity acid (lower position of heater lamp). The selenium is collected in the quartz tube (lower part of Fig. 4) and submitted to the determination method.

3. The combustion of the sample (< 500 mg) may also be effected with h.f.-excited oxygen at a temperature of < 200°C (cold ashing) in a quartz apparatus equipped with a cooling finger, compare Fig. 5. This decomposition method has proved very useful even for substances very difficult to decompose (e.g. plastics, coal, graphite) (50).

Recently, also the acid digestion has been proved several times to work without any loss of selenium when using acid mixtures, e.g. $HNO_3/HClO_4$ or $HNO_3/HClO_3/HClO_4$, in quartz tubes (test tube form) at decomposition temperatures < 250 °C (35, 42). First and foremost simpler matrices (blood serum, proteins) can be mineralized without systematic errors, even in combination with voltammetric determination methods that are known to be susceptible to interferences if organic

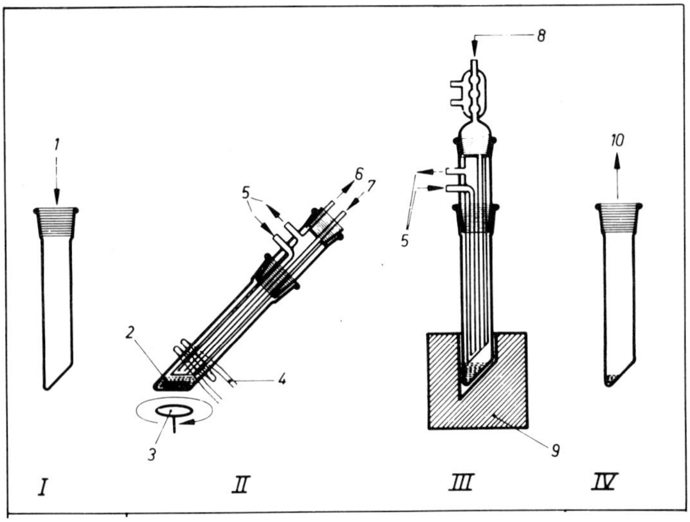

Fig. 5 Cold ashing by h.f.-excited oxygen after S.E. Raptis et al. (50). 1 sample; 2,3 magnetic stirrer; 4 h.f.-coil; 5 cooling water; 6 vacuum; 7 inlet for O_2; 8 inlet for acid; 9 heating block; 10 mineralized solution.

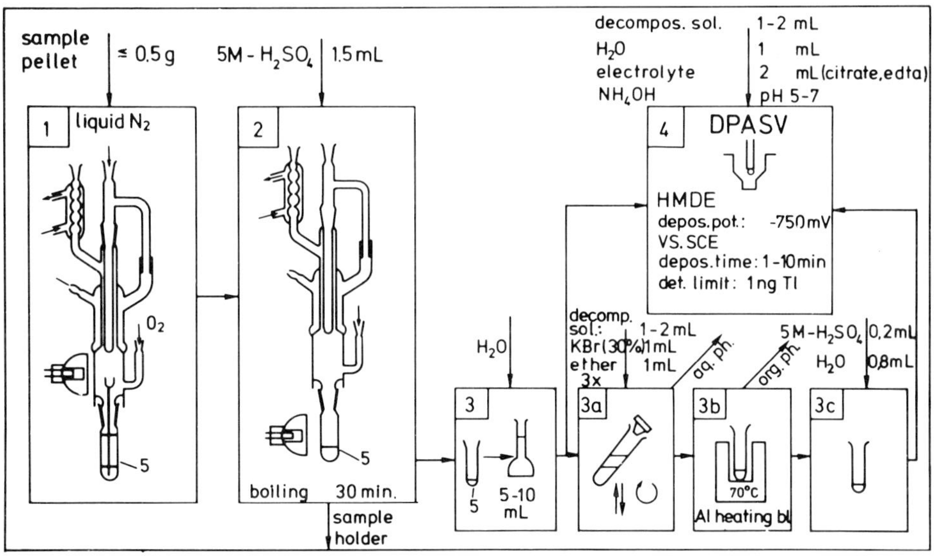

Fig. 6 Flow diagram of the determination of selenium in rocks, soils and biotic materials at the ng/g level after Trace-O-Mat combustion [1], [2] by selenium-hydride generation-AAS [5].

residues are still remaining in the decomposition solution (46).

No $HClO_4$ must be used during the course of a fluorometric selenium determination. It must be replaced by H_3PO_4 and H_2O_2 in the presence of HNO_3 (44).

Determination Methods with Highest Detection Abilities

HG-AAS is by far the most frequently used determination method; its numerous, initially overlooked interferences by concomitant elements are realized and could be eliminated in time. For details the reader is referred to the extensive respective literature. (52-61).

A proven method for most of the biotic matrices is a combination of HG-AAS with preceding decomposition in the Trace-O-Mat, compare Fig. 6.

The limits of detection of HG-AAS can be improved by two orders of magnitude, if the evolving H_2Se is condensed on a Chromosorb-column at approx. -150° C. By fast heating up the adsorption column, H_2Se is introduced into the cuvette of the AA-instrument in a single push, for measurement of the absorption (10), compare Fig. 7. The calibration curve in Fig.8 obtained with this apparatus demonstrates the operating efficiency of hydride generation condensation-AAS(HGC-AAS).

The only step that could still cause systematic errors within HG-AAS concerns the reduction of Se(VI) to Se(IV), because the selenium must be entirely present in the sample solution in this valency for the formation of H_2Se (53). In the majority of cases, the Se(VI) is reduced by boiling with HCl.As this reaction is concentration-, temperature- and time-dependent, there is still a requirement for thorough systematic investigation for optimizing the procedure.

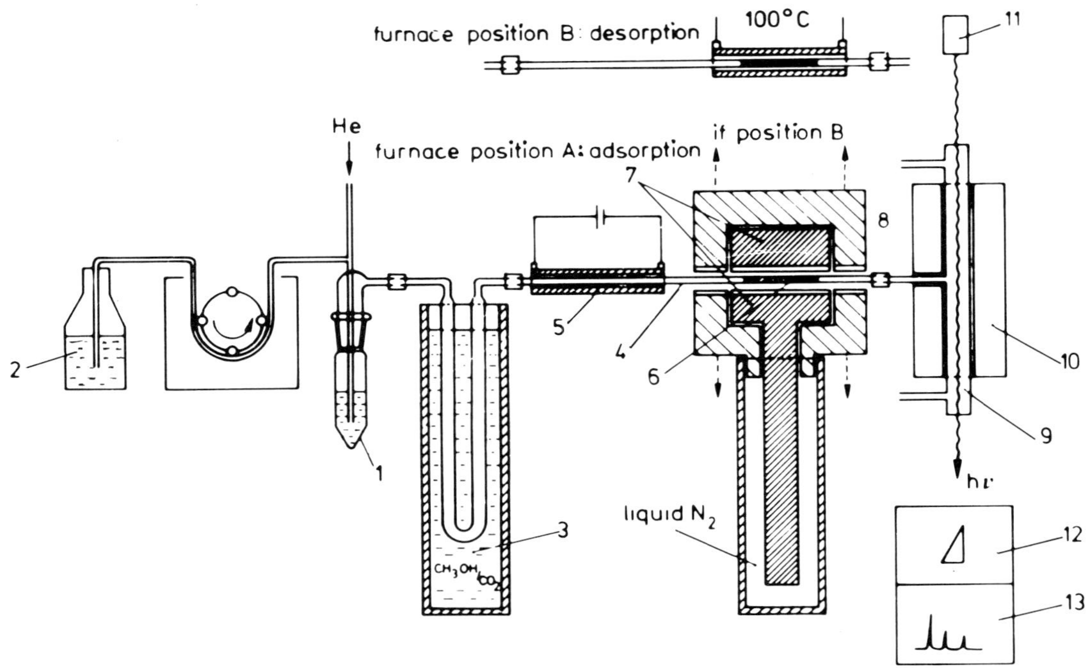

1	quartz vessel	5	heating system	9	quartz cuvette
2	reduction solution	6	adsorption region Chromosorb W 30/60	10	furnace
3	cold trap (-70°C)	7	liquid N_2 cooled Al-block	11	EDL
4	quartz adsorption tube	8	insulation	12	AA-spectrometer
				13	recorder

Fig. 7 HGC-AAS device for the determination of selenium in the pg/g level after J. Piwonka (10).

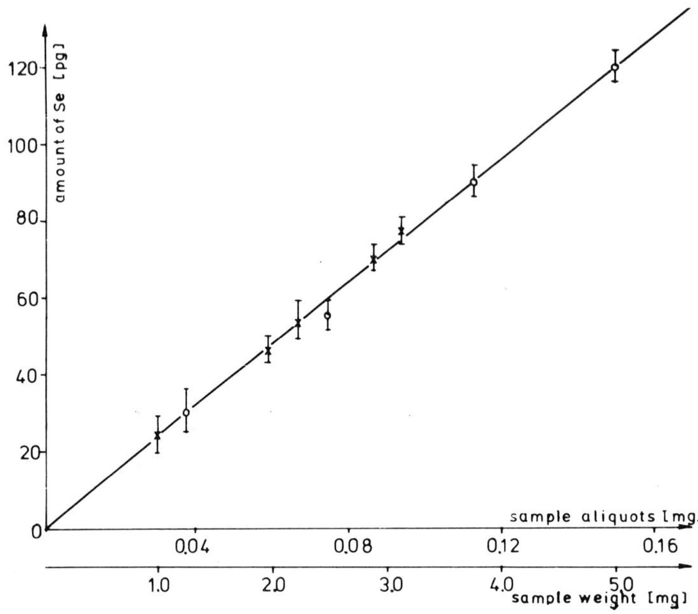

Fig. 8 Calibration graph for the determination of pg amounts of selenium by HGC-AAS (10) in human serum:
after decomposition with $HNO_3/HClO_4$: = x
after Trace-O-Mat decomposition: = o
aliquots from the decomposed sample weight were taken for the determination.

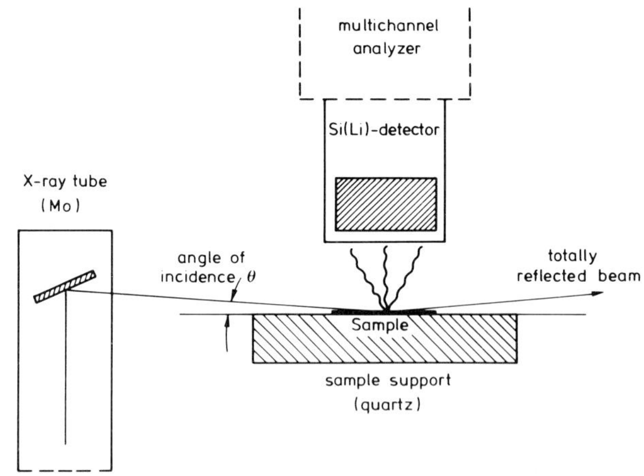

Fig. 9 XRFA with excitation by a totally reflected beam TR-XRFA (23).

A comparison of HG-AAS, hydride generation - inductively coupled plasma - OES (HG-ICP-OES) (12-14,62-66), and HG - microwave induced plasma - OES (HG-MIP-OES), (11, 67), always turns out in favour of HG-AAS due to its considerably higher power of detection at lower instrumental expenditure and comparable reliability, as long as no simultaneous determination with other hydride-forming elements is required. The HG-atomic fluorescence spectrometry (HG-AFS) (15, 68, 69) is even more powerful in detection but put to less use up to now.

The main source of error occurs in all HG methods to a comparable extent from the step of hydride formation. The conventional ETA-AAS using deuterium background compensation, platform technique and Ni, or Ag as matrix modifier (6, 70, 71) leads to negative interferences at the 196 nm selenium line, especially in the presence of e.g. higher concentrations of Fe(III)-, o-phosphate or calcium hydrogen-o-phosphate. These and further interferences by other elements of higher concentrations can be eliminated by preceding extractive separation of the selenium with e.g. 1,2-diamino-4-nitrobenzene (72) or by Zeeman background correction (7, 8).

To compare the powers of detection of the ETA-AAS and the HG-AAS techniques is difficult using data taken from literature at this stage of development as there is no case, where spectrometers of the same kind have been used under comparable working conditions.

The power of detection of ETA-AAS is claimed to be better by a factor of 2 (8) or even by a factor of $>$ 10 (68), compared with the conventional HG-AAS ; however, the detection power of the HGC-AAS technique as developed in our group has not been reached by far.

The fluorometry with diaminonaphthalene (DAN) (16, 17, 73-75) operates very reliably and is of superior detection power to all AAS techniques with the exception of HG-AAS. Here, indeed, special attention has to be paid to the decomposition of the sample (44).

Differential pulse anodic stripping voltammetry (DPASV) at a rotating gold disk electrode and differential pulse cathodic stripping voltammetry (DPCSV) at a hanging mercury drop electrode show limits of detection that are mutually comparable (~ 0.5 ng /mL, 60 s deposition time) and that can even be improved by longer deposition times, but DPCSV shows better reproducibility (21). In analytical application (76-81), the methods of stripping voltammetry are superior to the conventional AAS methods regarding the detection power, by one order of magnitude. As a prerequisite in all cases, sample decompositon must be performed with due care and 100% mineralization must be guaranteed. When operating with biotic samples, no interelement interferences are to be expected. The instrumental expenditure is considerably less than for all atomic spectrometric methods; however, the time consumption is somewhat higher. Highly experienced personnel are required for routine analysis.

When employing the ECD-GC-method, the absolute limit of detection arrived at is by far the lowest obtainable ($<10^{-12}$ g). The determination is made indirectly with an ECD, subsequent to gas chromatographical separation of 5-nitropiazselenol, which is formed quantitatively even from pg amounts of Se(IV) by conversion with 4-nitro-o-phenylenediamine in the sample solution at $1.0 < pH < 2.5$. The nitropiazselenol is quantitatively separated by extraction with toluene (18, 19, 82, 83) before introduction into the gas chromatograph. Extremely careful manipulations on micro-quantity scale are presupposed for success with this method, hence it is recommended only when its high power of detection is required.

A direct determination of selenium in biotic matrices by XRFA (energy- or wavelength-dispersive) using the pellet technique is only suitable for selenium contents > 0.1 µg/g, but it then offers the advantage of a multielement analysis (e.g. analysis of dust particles after filter sampling). The power of detection can be improved to approx. 10 ng/g by preconcentration of selenium in a thin-layer sample, e.g. on a Nuclepore membrane filter, by adsorption of colloidal precipitated selenium on activated carbon (22) or by co-precipitation of selenium with tellurium (26). The time spent for the analysis is considerably extended by sample decomposition and chemical separation preceding the actual determination in such cases; further, due to the preconcentration, one loses the advantage of a multielement analysis.

With total reflection-XRFA (TR-XRFA) the very thin sample is excited with the x-rays impinging under total reflecting angle of incidence (23), compare Fig. 9. This results in a power of detection for many elements which is better by orders of magnitude for TR-XRFA as compared with conventional XRFA. The calculated power of detection is particularly good for selenium (5 pg Se), when compared with other elements, see Fig. 10. A definite Se-line, compare Fig. 11, is obtained from an aliquot of the diluted decomposition solution (5 ng Se) in a determination of Se in blood (80 ng /mL Se). This excitation mode for micro-multielement analysis is very elegant in principle and very powerful in detection, but it always requires micro-techniques for the sample preparation. Very good prospects of TR-XRFA application are in the field of micro-distribution analysis of elements in biotic matrices with this very young method scarcely utilized so far.

PIXE, on the other hand, has already had its breakthrough in routine analysis. Its instrumentation is based on rather sophisticated accelerators, but these exists in sufficient

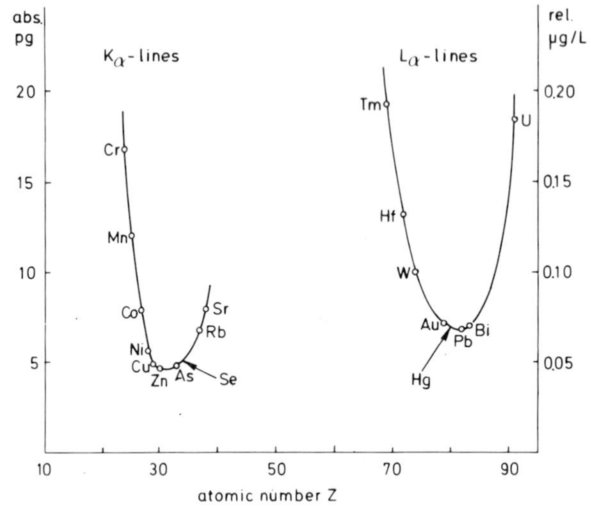

Fig. 10 Minimum detection limits of TR-XRFA (23) (aqueous solutions, 100 µL, Mo-tube, 15 mA, 60 kV, 1000 s).

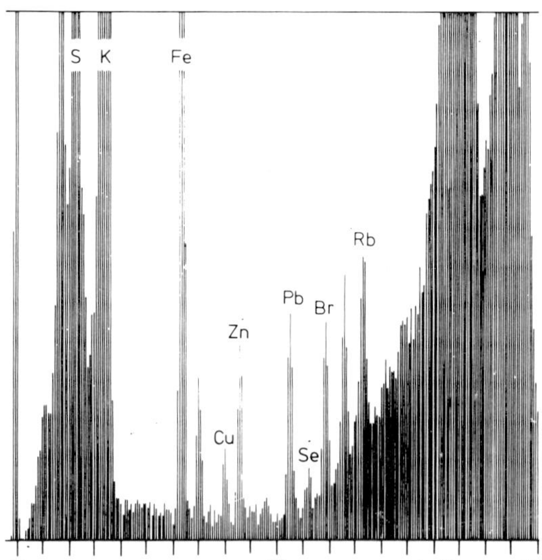

Fig. 11 TR-XRFA spectrum in arbitrary units including signal of 5 ng Se (aliquot of a solution of wet mineralized human blood with 80 ng/mL Se).

numbers in physical research institutes. In the Federal Republic alone, there are more than ten installations (11) suitable for PIXE analysis which could be made available for very intensive use in micro- and trace-analytical routine investigations on a wide scale. As a multielement determination method, its range of determinable elements is similar to that of XRFA; however, it possesses on average, a power of detection better by one order of magnitude, compare Fig.12. Just as in XRFA, the power of detection can be improved by including preconcentration steps at the expense of the multielement determination mode, (26) compare Fig. 13. The limit of detection for selenium at direct sample excitation (pellet) lies at > 100 ng/g (25). The far more promising application, however, points at the analysis of micro-regions because the exciting proton beam can be focussed to form a microprobe (85).

INAA has matured to a determination method and permits the determination of Se contents > 10 ng/g in all biotic matrices, but the results of this method are not necessarily always accurate. In NAA, too, considerable improvements of the power of detection are rendered possible by radiochemical separation (RNAA) (29, 86); also with this method, however, the additional separation steps are very time-consuming and hence uneconomical for routine investigations. Nevertheless, such methods are valuable for the certification of standard reference materials and for the control of the accuracy of other trace analytical determination methods (86).

Following this general methodical survey that should make clear today's basic problems and limits of trace analysis of selenium, I would like to briefly comment on the question of what sort of collaborative assistance the analyst should contribute on the subject of selenium from a methodical view-

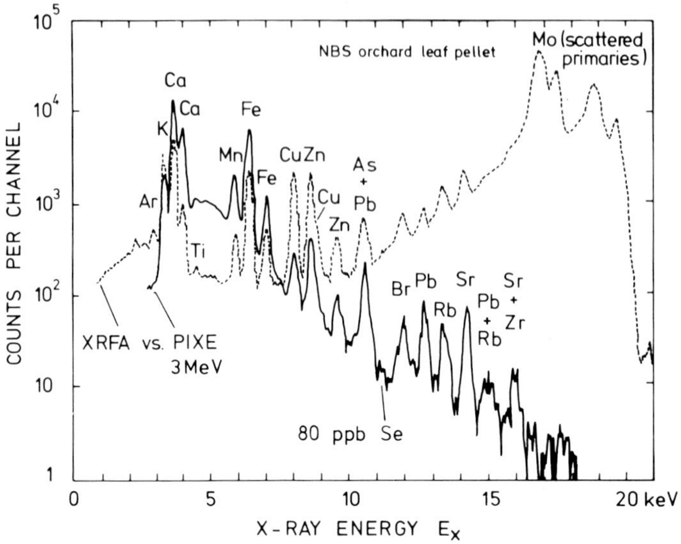

Fig. 12 Comparison of typical x-ray spectra of orchard leaves, (pressed pellets ....XRFA, ——— PIXE) after R. Garten 1984, in press (90).

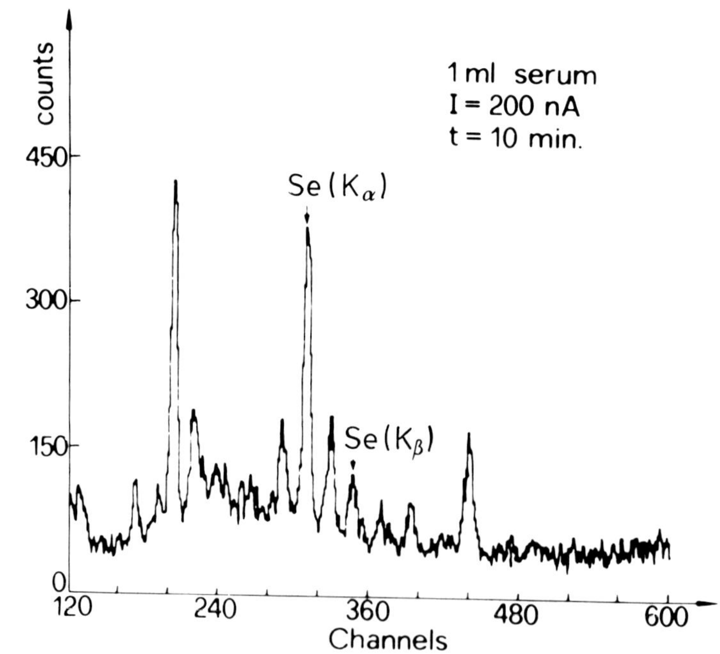

Fig. 13 PIXE determination of Se in a serum sample after coprecipitation using Tellurium (26).

point in the future. This question was initiated by my opinon that activity towards methodical developments of procedures will decrease soon for the determination of selenium contents in different biotic matrices, in view of power of detection, reliability and economy. It will certainly take a rather long period of learning, until all laboratories involved in the analysis of selenium will obtain accurate results in the most economical way. But even here, I feel quite optimistic. Compared to many other elements, trace analysis of selenium has proved to be relatively unproblematic. Its ubiquitous concentration level is relatively high, nevertheless, the often troublesome problem of blank values in trace analysis can be overcome without great difficulty in this case. The chemistry of selenium is relatively uncomplicated. The good volatility of its compounds permits smooth separation methods via the gas phase, effective even in the 100 fg-level. Finally, selenium can be determined independently using a number of determination methods available today in all levels of its natural occurrence with sufficient power of detection and now also reliably, as has been increasingly proved by inter-laboratory comparison studies.

Even so, we cannot relax in our efforts. As mentioned at the beginning, future work on selenium analysis should focus on micro-distribution analysis and speciation analysis. Further efforts should point on the development of methods clarifying its antagonistic activities towards other elements (e.g. Hg, Cd, Pb and other toxic heavy metals) and/or inter-actions with macro molecules (e.g. proteins, nucleic acid) within biological systems. The study of its reactivity with other elements requires economic methods for multielement determination of very high power of detection. However, it has been shown that the power of detection of those instrumental direct methods that are most economical, is not yet sufficient.

A particularly promising method would be mass spectrometry which has hardly been employed in the analysis of selenium so far.

Before discussing the selenium distribution in macro molecules in more detail, the toilsome path leading to the solution of still unsolved problems will be pointed out by way of a less spectacular, yet very instructive example. It has been attempted for years to draw conclusions from trace element contents in hair and fingernails on trace element content in blood, for evident reasons. In some cases, (e.g. Tl, Cd) a correlation is quite conceivable. In the case of selenium, however, new findings have confronted us with a new kind of contamination problem.

Using HGC-AAS we are now prepared to determine the selenium content not only in a single hair, but also in short segments of a hair and in a well reproducible and in all probability accurate manner (10), compare Table 4. During these investigations we found enormous fluctuations in Se-content occurring from one test person to the next, the cause of which was inexplicable at first, but later attributed to selenium contamination. These contaminations could not be completely removed even by very careful cleaning procedures. An explanation was only recently found from some PIXE-work, compare Fig. 14. From a gradient analysis across the cross-section of a hair using a proton microprobe, the authors found a strong enrichment of selenium on the hair surfaces clearly resulting from the application of Se-containing shampoos. It could be concluded from the thickness of the contamination zone that the selenium had diffused into the hair. For the determination of the selenium level in the hair bulk, however, the power of detection of PIXE is not sufficient. This example shows not only the necessity to avail oneself of the different methods, but emphatically also, that there is a great danger to become a victim of uncritical analytical

Matrix	Sample weight [mg]	$\bar{X}$ [ng/g]	S [ng/g]	S [%]	n
Finger nail	1,5 - 3,5	547	12,5	2,3	6
Single hair (person 1)	≦ 1				
Root		329	38	11,5	6
Tip		320	30	9,4	6
Single hair (person 2)	≦ 1				
Root		312	34	10,9	6
Tip		281	29	10,3	6

Table 4 Determination of selenium in human fingernails and single hairs in the root and tip part of the hair with HGC-AAS after $HNO_3/HClO_4$-decomposition (10).

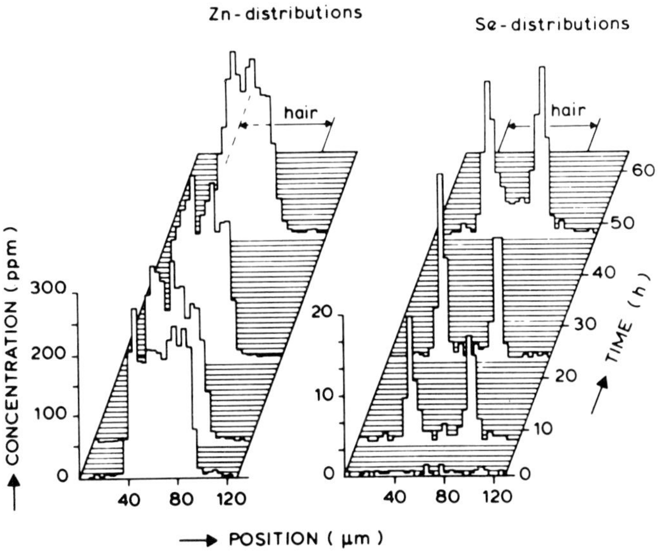

Fig. 14 Positional distribution of zinc and selenium across the diameter of hair segments soaked in a shampoo mixture with contact times as parameter (85).

conduct. In this particular instance, several authors who need not be named here, have been deluded by considerable blunder.

In my last example, a different working direction in our group is briefly referred to which has recently been reported on in more detail elsewhere (87). Several years ago, we started investigations into the transport mechanism of several trace elements in animal and human blood. Therewith, the distribution of beryllium in blood was successfully determined, even in the individual protein fractions of the serum (88). These promising results and the extremely interesting physiological conclusions drawn from the activity of beryllium in normal blood (1 ng/g Be) and blood contaminated with 10 ng/g Be encouraged us to extend such investigations on selenium utilizing the respective high power of detection offered by HGC-AAS (10).

The following basic requirements have to be met by the combined multistage analytical procedure: The separation of the proteins must keep physiological conditions, in order to avoid changes in linkage during separation procedures. The separation of proteins should be as complete as possible, entailing high demands on its separation power.

As there are only very low concentrations of selenium and other trace elements to be included in the investigations which are to be expected in the individual fractions, a separation is required on a preparative scale. The strategy to be employed is shown in Fig. 15.

As main separation method, we adopted the preparative isotachophoresis already approved in investigations on beryllium. In order to increase the serum amount to start with and to further improve the separation power, a preseparation by

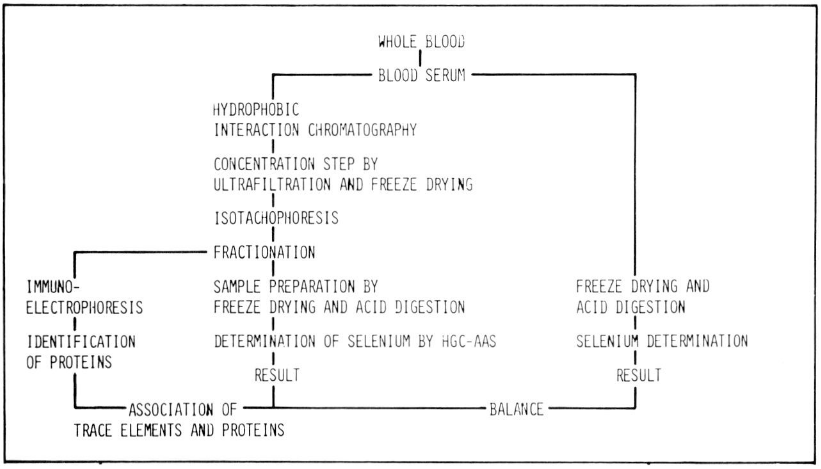

Fig. 15 Flow diagram for the separation and determination of selenium in protein fractions of human serum (87).

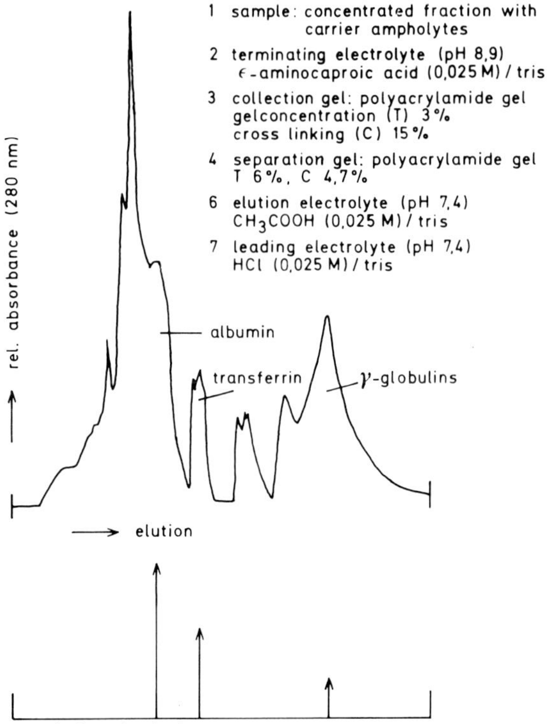

Fig. 16 Isotachophoretic separation of a human serum fraction (UV-diagram) (87).

hydrophobic interaction chromatography (89) was added. By appropriate modification of the hydrophilic properties of the buffer system, several eluate fractions of protein mixtures are separated according to their hydrophobic properties. These individual fractions are concentrated by way of ultrafiltration and subsequent freeze-drying, and then sequentially separated by isotachophoresis. The isolated proteins are continuously eluated, fractionated and assigned to a UV-diagram, registered simultaneously.

The proteins were identified by immunoelectrophoresis,while selenium was determined within the individual fractions by HGC-AAS. In Fig. 16, the results of the selenium distribution are presented exhibiting a preferential bonding of the selenium in the albumin- and γ-globulin fractions in normal blood.

With this start in the new direction of selenium analysis, it should be clear that the progress of analytical methods today is the imperative prerequisite for the selenium research of tomorrow; and this will be increasingly depend on interdisciplinary partnership and collaboration.

References

1. Shamberger, R.J.: in Spurenelemente (Ed. H. Zumkley), Georg Thieme Verlag, Stuttgart, New York, 181-190, (1983).

2. Futekov, L., Paritschkova, R., Specker, H.: Fresenius' Z. Anal. Chem. 315, 342-344 (1983).

3. Raptis, S.E., Kaiser, G., Tölg, G.: Fresenius' Z. Anal. Chem. 316, 105-123 (1983).

4. Lau, C.M., Ure, A.M., West, T.S.: Anal. Chim. Acta 141, 213 (1982).

5. Lund, W., Bye, R.: Anal. Chim. Acta 110, 279-284 (1979).

6. Alfthan, G., Kumpulainen, J.: Anal. Chim. Acta 140, 221-227 (1982).

7. Carrick, G.R., Manning, D.C., Slavin, W.: Analyst 108, 1297-1312 (1983).

8. Welz, B., Melcher, M., Schlemmer, G.: Fresenius' Z. Anal. Chem. 316, 271-276 (1983).

9. Ebdon, L., Wilkinson, J.R., Jackson, K.W.: Anal. Chim. Acta 136, 191-199 (1982).

10. Piwonka, J.: Doctoral Thesis, University of Stuttgart (1984).

11. Robbins, W.B., Caruso, A.A., Fricke, F.L.: Analyst 104, 35-40 (1979).

12. de Oliveira, E., McLaren, W.J., Berman, S.S.: Anal. Chem. 55, 2047-2050 (1983).

13. Hahn, M.H., Wolnik, K.A., Fricke, F.L., Caruso, J.A.: Anal. Chem. 54, 1048-1052 (1982).

14. Pruszkowska, E., Barrett, R., Ediger, R., Wallace, G.: At. Spectrosc. 4, 94-97 (1983).

15. Nakahara, T., Kobayashi, S., Wakisaka, T., Musha, S.: Appl. Spectrosc. 34, 194-200 (1980).

16. Watkinson, J.H.: Anal. Chim. Acta 134, 417-419 (1982).

17. Nazarenko, I.I., Kislova, I.V.: J. Anal. Chim. USSR 83, 1429 (1979).

18. Toei, K., Shimoishi, Y.: Talanta 28, 967-972 (1981).

19. Slu, K.W.M., Berman, S.S.: Anal. Chem. 55, 1603-1605 (1983).

20. Henze, G., Monks, P., Tölg, G., Umland, F., Weßling, E.: Fresenius' Z. Anal. Chem. 295, 1-6 (1979).

21. Adeloju, S.B., Bond, A.M., Briggs, M.H., Hughes, H.C.: Anal. Chem. 55, 2076- 2082 (1983).

22. Robberecht, H., Van Grieken, R.: Anal. Chim. Acta 147, 113-121 (1983).

23. Schwenke, H., Knoth, J.: in Trace Element Analytical Chemistry in Medicine and Biology (Eds. Brätter, P., Schramel, P.), Walther de Gruyter, Berlin, New York, 307-317 (1980).

24. Gilfrich, J.V., Skelton, E.F., Qadri, S.B., Kirkland, J.P., Nagel, D.J.: Anal. Chem. 55, 187-190 (1983).

25. Berti, K., Buso, G., Colautti, P., Moschini, G., Stievano, B.M., Tregnaghi, C.: Anal. Chem. 49, 1313-1315 (1977).

26. Buso, G.P., Colautti, P., Moschini,G., Xusheng, H., Stievano, B.M.: 3rd PIXE Conf., Heidelberg (1983) to be published in Nucl. Instr. Meth. 231 (1984).

27. Behne, D., Jürgensen, H.: J. Radioanal. Chem. 42, 447-453 (1978).

28. Debrun, J.L., Barrandon, J.N., Benaben, P.: Anal. Chem. 48, 167-172 (1976).

29. Massée, R., van de Sloot, H.A., Das, H.A.: J. Radioanal. Chem. 35, 157 (1977).

30. Versieck, J., Cornelis, R.: Anal. Chim. Acta 116, 217-254 (1980).

31. Behne, D., Wolters, W.: J. Clin. Chem. 17, 133-135 (1979).

32. Saeed, K., Thomassen, Y., Langmyhr, F.J.: Anal. Chim. Acta 110, 285-289 (1979).

33. Oster, O., Prellwitz, W.: Clin. Chim. Acta 124, 277-291 (1982).

34. Lloyd, B., Holt, P., Delves, H.T.: Analyst 107, 927-933 (1982).

35. Verlinden, M.: Talanta 29, 875-882 (1982).

36. Welz, B.: Personal Information, (1984).

37. Chen, J.R.: Nucl. Instr. Meth. 181, 337-343 (1981).

38. Pallon, J., Pakarinen, P., Akselsson, R.: 3rd PIXE Conf., Heidelberg (1983) to be published in Nucl. Instr. Meth. 231 (1984).

39. Weber, G., Robaye, G., Bartsch, P., Collignon, A., Beguin, Y., Roelandts, I., Delbrouk, J.M.: 3rd PIXE Conf., Heidelberg (1983) to be published in Nucl. Instr. Meth. 231 (1984).

40. Tam, G.K.H., Lacroix, G.: J. Assoc. Off. Anal. Chem. 65, 647-650 (1982).

41. May, T.W.: J. Assoc. Off. Anal. Chem. 65, 1140-1145 (1982).

42. Robberecht, H.J., Van Grieken, R.E., Van den Bosch, P.A., Deelstra, H., Vanden Berghe, D.: Talanta 29, 1025-1028 (1982).

43. Tölg, G.: Pure & Appl. Chem. 55, 1989-2006 (1983).

44. Reamer, D.C., Veillon, C.: Anal. Chem. 55, 1605-1606 (1983).

45. Knapp, G.: Fresenius' Z. Anal. Chem. 317, 213-219 (1984).

46. Kotz, L., Henze, G., Kaiser, G., Pahlke, S., Veber, M., Tölg, G.: Talanta 26, 681-691 (1979).

47. Uhrberg, R.: Anal. Chem. 54, 1906-1908 (1982).

48. Meyer, A., Hofer, Ch., Knapp, G., Tölg, G.: Fresenius' Z. Anal. Chem. 305, 1-10 (1981).

49. Han, H.-B., Kaiser, G., Tölg, G.: Anal. Chim. Acta 128, 9-21 (1981).

50. Raptis, S.E., Knapp, G., Schalk, A.P.: Fresenius' Z. Anal. Chem. 316, 482-487 (1983).

51. Seifert, G.: Landwirtsch. Forsch. 35, 214-219 (1983).

52. Welz, B.: Atomabsorptionsspektrometrie, 3rd Ed., Verlag Chemie, Weinheim; Deerfield Beach / Florida, Basel, 342-346 (1983).

53. Robberecht, H., Van Grieken, R.: Talanta 29, 823-844 (1982).

54. Nakahara, T.: Prog. Analyt. Atom. Spectrosc. 6, 163-223 (1983).

55. Meyer, A., Hofer, Ch., Tölg, G., Raptis, S., Knapp, G.: Fresenius' Z. Anal. Chem. 296, 337-344 (1979).

56. Dedina, J.: Anal. Chem. 55, 2097-2102 (1983).

57. Cutter, G.R.: Anal. Chim. Acta 149, 391-394 (1983).

58. Bye, R., Engvik, L., Lund, W.: Anal. Chem. 55, 2457-2458 (1983).

59. Mailer, R.J., Pratley, J.E.: Analyst 108, 1060-1066 (1983).

60. Welz, B., Melcher, M.: Analyst, in press (1984).

61. Welz, B., Melcher, M.: Analyst, in press (1984).

62. Thompson, M., Pahlavanpour, B., Walton, S.J., Kirkbright, G.F.: Analyst 103, 568-579 (1978).

63. Thompson, M., Pahlavanpour, B., Walton, S.J., Kirkbright, G.F.: Analyst 103, 705-713 (1978).

64. Nygaard, D.D., Lowry, J.H.: Anal. Chem. 54, 803-807 (1982).

65. Pruszkowska, E., Barrett, P., Ediger, R., Wallace, G.: Atomic Spectrosc. 4, 94-98 (1983).

66. McCarthy, J.P., Caruso, J.A., Fricke, F.L.: J. of Chromatogr. Sci. 21, 389-393 (1983).

67. Robbins, W.B., Caruso, J.A.: Anal. Chem. 51, 889A-899A (1979).

68. Brown, A.A., Ottaway, J.M., Fell, G.S.: Anal. Proc., 321-324 (1982).

69. Ebdon, L., Wilkinson, J.R.: Anal. Chim. Acta 136, 191-199 (1982).

70. Julshamn, K., Ringdal, O., Slinning, K.-E., Braekkan, O.R.: Spectrochim. Acta 37B, 473-482 (1982).

71. Welz, B., Schlemmer, G., Voellkopf, U.: Spectrochim. Acta, in press (1984).

72. Xiao-quan, S., Long-zhu, J., Zhe-ming, N.: At. Spectrosc. 3, 41-44 (1983).

73. Hasunuma, R., Ogawa, T., Kawanishi, Y.: Anal. Biochem. 126, 242-245 (1982).

74. Campbell, A.D.: Pure & Appl. Chem., in press (1984).

75. Stahr, H.M., Klinker, J., Nicholson, D., Hyde, W.: J. Liquid Chromatogr. 5, 1191-1200 (1982).

76. Ebhardt, K.B., Umland, F.: Fresenius' Z. Anal. Chem. 310, 406-409 (1982).

77. Jarzabek, G., Kublik, Z.: Anal. Chim. Acta 143, 121-130 (1982).

78. Lemly, A.D.: Environ. Technol. Lett. 3, 497-502 (1982).

79. Kavel, H., Umland, F.: Fresenius' Z. Anal. Chem. 316, 386-389 (1983).

80. Adeloju, S.B., Bond, A.M., Hughes, H.C.: Anal. Chim. Acta 148, 59-69 (1983).

81. Rahmalan bin Ahmad, Hill, J.O., Magee, R.J.: Analyst 108, 835-839 (1983).

82. Measures, C.I., Burton, J.D.: Anal. Chim. Acta 120, 177-186 (1980).

83. Jiang, S., De Jonghe, W., Adams, F.: Anal. Chim. Acta 136, 183-190 (1982).

84. Garten, R.P.H., Martin, B., Tölg, G.: Fresenius' Z. Anal. Chem. 317, 1-4 (1984).

85. Bos, A.J.J., van der Stap, C.C.A.H., Vis, R.D., Valkovic, V.: Spectrochim. Acta 38B, 1209-1215 (1983).

86. Byrne, A.R., Dermelj, M., Kosta, L., Tusek-Znidaric, M.: Mikrochim. Acta [Wien] 119-126 (1984).

87. Alt, F., Messerschmidt, J., Tölg. G., Kaiser, G., Piwonka, J.: Fresenius Z. Anal. Chem., in press.

88. Stiefel, Th., Schulze, K., Tölg, G.: in Trace Element Analytical Chemistry in Medicine and Biology, (Eds. P. Brätter, P. Schramel); Walter de Gruyter, Berlin New York, 427-438, (1980).

89. Rosengren, J., Pahlman, S., Glad, M. et al.: Biochem. Biophys. Acta 412 (1975).

90. Garten, R.P.H., Trends Anal.Chem. 3, in press (1984).

D i s c u s s i o n Session III - 1
Plenary Lecture

Subjects of the discussion

- PIXE for the analysis of hair segments
- handling of small sample amounts
- solid sample introduction into GFAAS

Contributors: Ganther, Knapp

THE SELENIUM CONTENT IN MILK AS AN INDICATOR OF THE SELENIUM STATUS OF COWS

W.T. Binnerts, J. Rijken, T.C. Viets

Department of Animal Physiology, Agricultural University College, Wageningen, The Netherlands

Introduction

Selenium, as well as iodine, is among the elements (1) of which milk analysis has been used to study the nutritional status of cows (2). There is no doubt that selenium is excreted into milk more or less in proportion to the quantity of intake, but the literature reports differ as to the shape of the dose-response curve. So, for instance, in recent studies only very small increases were noted upon feeding of selenium-containing compounds, whereas the effects on blood-Se content were large and source-related (3). Also in an other study (4) not much increase in milk selenium content was found upon feeding of supplements of selenium comparable to the normal daily intake. This contrasts with the results mentioned by Underwood (5) which lead him to believe that "the concentration of Se in cow's milk varies greatly with the Se intake of the animal". Furthermore the absolute amounts in the various reports are different. They range from some micrograms per liter to over 50 for natural milk, and 152 in feeding trials (6) to over 1000 in earlier reports (7). A substantial part of the differences is undoubtedly caused by analytical error. Also the source and quantity of selenium, and the duration of feeding may have had some effect. The purpose of the feeding experiment reported here has been to obtain some insight in the relative contributions of these effects. The results, apart from their physiological significance, could have some interest for humane

Trace Element - Analytical Chemistry in Medicine and Biology, Vol. 3

nutritional as well as toxicological purposes (8).

Analytical part

Se was determined with atomic spectrometry, using hydride generation, which followed an acid digestion with perchloric acid and reduction with HCl. The digestion procedure (9) leading to excellent recoveries of added selenite to milk samples, was later replaced by a more general one (10, 11) which also could be used for feed samples. The following details were found to be important for satisfactory work. In the acid digestion procedure the methods mentioned last use a small addition of sulfuric acid in order to evaporate the perchloric acid completely; without this the following reduction to Se (IV) by HCl which is needed for the hydride generation may not be complete. We found that for milk samples this effect was not large (90% recovery of selenite), but for feeds, liver, urine the recoveries without this precautions were very low (mostly below 50%) and the standard error was large, with low results. Then in the digestion also a fixed and rather large amount of nitric acid is used, which can only be done with HNO_3 of very low Se content. Methods with dropwise addition of as little HNO_3 as possible (9), just sufficient for maintenance of an oxidizing environment, according to our experience, should be distrusted because the blank may be too low, leading to a positive bias of the results. In the hydride generation direct injection of the borohydride reagent has advantages over slow pumping. This was revealed in the graphs made by an external recorder, which had a much more smooth appearance, though not regressing entirely to background immediately (presumably by some entrailed water vapor). Pure nitrogen was used as a carrier gas. The results of analysis were obtained by integrating over a period of 20 seconds, using the settings of the instrument (Perkin Elmer 2380 with MHS-10 hydride system and EDL lamp with power supply). The standard deviation of blanks (including background) was a relative 4% compared to the usual range of measurements with 5 ml of milk or approximately 0.5 g of milk ash. The results of at least four independent analyses were averaged for each sample; if the resulting standard error was larger than 10%, two additional determinations were performed, and the average of six was calculated. This always had a standard error lower than 10%. The absolute error of the determinations was estimated by comparison of the so obtained results for 10 different milk samples with similarly procured results of neutron activation analysis on the dried material. The latter analyses were performed at ECN, Petten, North-Holland, under the supervision of H.A, DAS & J.R.W. WOITTIEZ, with the short lived ^{77m}Se. The result of the comparison is presented in Fig. 1.

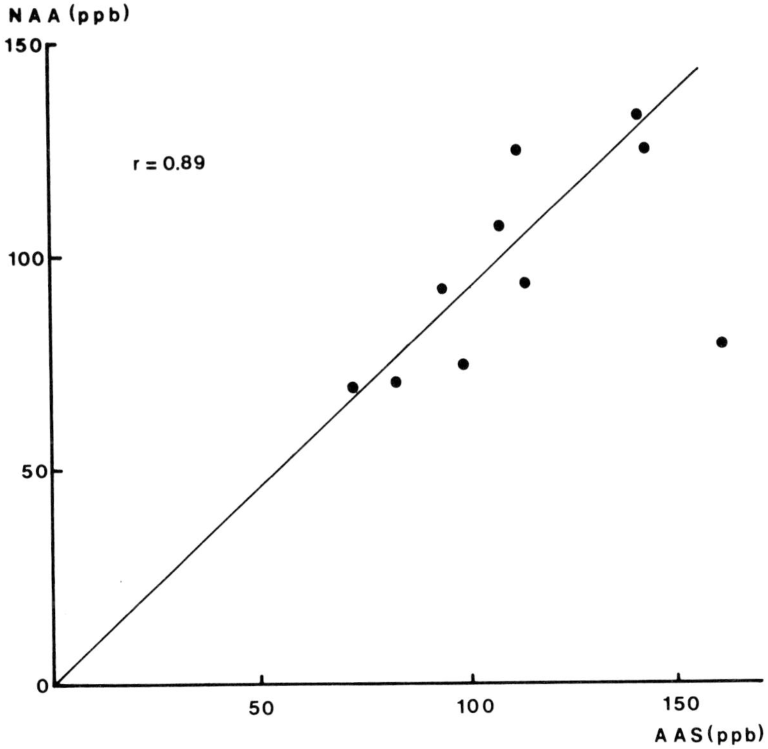

Fig. 1. Correlation of the results of atomic absorption (AAS) and neutron activation (NAA) analysis for 10 different milk samples. The outlier could not be explained and was not considered in the calculation of the correlation coefficient r.

This comparison was made after adjustment of the AAS method for correction of blanks due to nitric acid (see above). Further only nitric acid of low Se content was used. Without this adjustment AAS would have given too high results by 14 ng/g. In the NAA analysis the standard reference material Bovine Liver (N.B.S. Nr. 1577) was included with a result of 1.197 ± 0.018 ng/g, against 1.1 ± 0.1 for the specified value. The quantities used in the procedure were the following. A quantity of sample corresponding to 0.5 - 1 g of dry material was predigested overnight with 10 ml of (fuming) HNO_3 and some carborundum granules in a 50 ml Kjeldahl flask. After the addition of 2 ml perchloric acid the solution was boiled cautiously until the volume had been reduced to one-half. After cooling 2 ml of HNO_3 and 2 ml of H_2SO_4 were added. The solution was boiled again until white fumes appeared. Exceptionally a solution turned slightly brown, and then a small quantity (0.5 ml) of additional HNO_3 was added. To the cool solution 1 ml of HCl was added and it was boiled for 5 minutes. In the hydride system a quantity of digest was placed with 20 - 100 ng of Se, and diluted to 10 ml with double distilled water. A further 0.4 ml of HCl was added and the analysis was started according to the manufacturer's manual, with 2.5 ml of 3% sodiumborohydride solution in 1% NaOH. Standards of 100 ng Se, as selenite, in 1,5% HCl, were used.

Feeding experiment

Four cows with milk production of 15 to 25 kg daily were placed in metabolism stalls. Their feed intake of hay (12 kg/day) and concentrates (8 - 10 kg/day) had been calculated to provide sufficient energy and protein for production and maintenance. Urine and faeces were collected separately and quantitatively. After a preperiod for adaptation, 4.5 mg of Se, as selenite, and 40 mg of Kl was administered in gelatin capsules, twice daily. After 10 of the 12 days of administration, milk, urine, and faeces samples were taken and stored in polythene bottles at -20°C until analysis. For comparison similar samples were collected at the start and 20 days after conclusion of the experiment. It was remarkable, considering the low level of administration that after the experiment symptoms of Se-intoxication were noted in cow 5 and 8, which disappeared spontaneously. Some of the results are presented in Fig. 2 and 3.

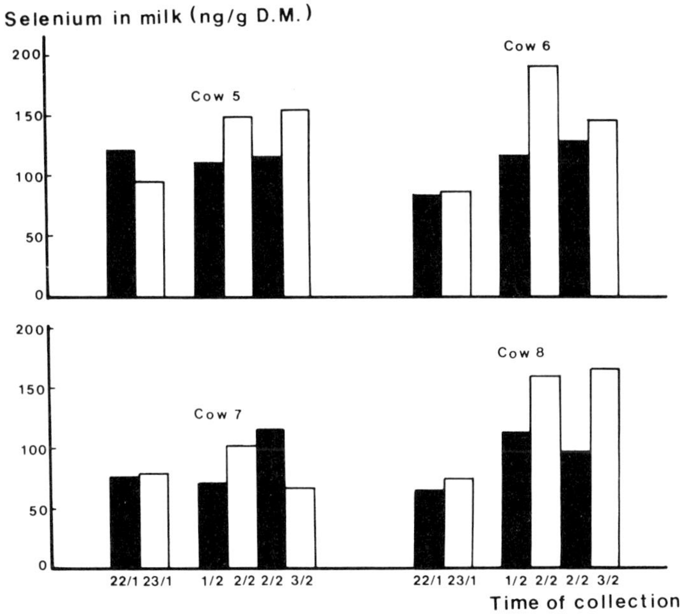

Fig. 2. Se excretion in milk after feeding for 10 days: the four last results compared to the starting values.

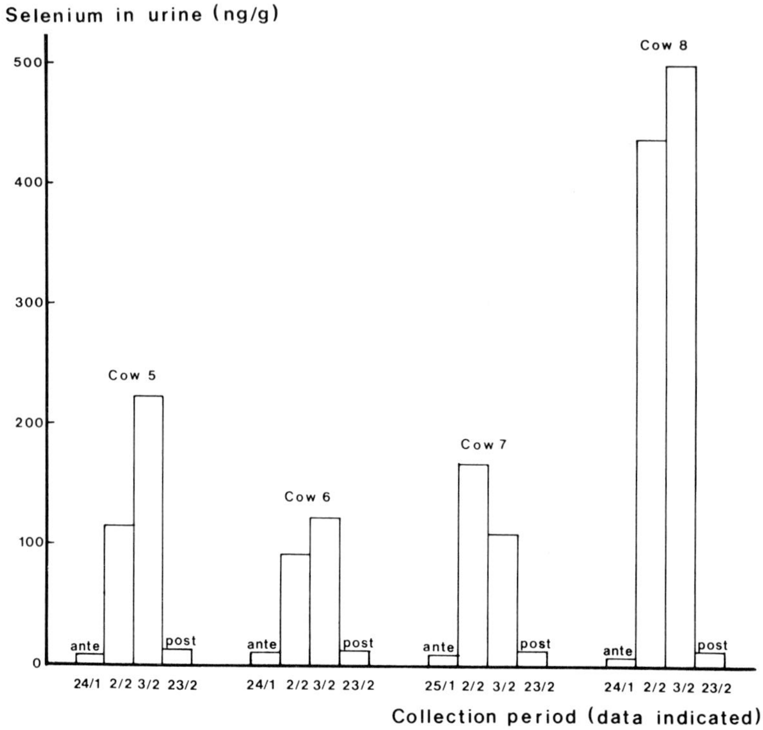

Fig. 3. Se excretion in urine after feeding for 10 days: the two last samples compared with samples taken before and after.

For the faeces a similar graph was obtained as for urine, with a somewhat smaller increase of Se content upon feeding. The balance calculations for the average of the last two days gave the following results (table 1).

	Milk	Urine	Faeces	Total
Cow 5	0.3	26	58	84
cow 6	1.5	14.3	75	91
cow 7	0.2	13.3	(35)	(49)
cow 8	0.7	39	52	92

Table 1. Recovery of fed selenium (in percentage of dose). The remaining part of the dose is supposed to have been incorporated into the tissues.

Discussion

The excretion in the milk of the fed selenite averaged only $0.53.10^{-3}$ per kg of milk, which is quite low compared to the 3.10^{-3} for the simultaneously fed iodide, and approximately 5.10^{-3} for the natural feed selenium. The low value for selenite was a disappointment, since it had been assumed that the rumen microflora would have converted whatsoever source of selenium to the same equilibrium mixture as selenium from natural sources. From the results it is further concluded that selenium from plant origin is ultimately very well transported into the milk, and conserved by the kidneys. Both of these effects can be ascribed to Se-methionine, which is the predominant form of plant selenium. Then again the rumen flora seems to have had little influence. It may be remembered that the comparable element, sulfur, undergoes extensive transformation in the rumen, whereby inorganic sulfur is an adequate source for the synthesis of aminoacids (12). In this experiment also the relatively large urinary excretion strengthens the argument that little conversion to Se-amino acids has taken place. Preliminary results of a feeding trial with lower amounts of selenite-Se (1.2 to 2.4 mg daily), are not much higher than the present ones in terms of the milk excretion ratio. Indeed, in this area of Se-levels a linear dose-response relation is obtained, but the slope is much lower than that of the natural selenium. Further trials with Se-methionine have still to be done.

A complicating factor in these considerations is the effect of the length of the feeding period. Although Table 1 leaves room for some uptake of selenium by the body tissues, 12 days is certainly not a long enough period to establish a near-steady state. The constant breakdown of body tissue will certainly have diluted the Se-pool(s) from which the milk selenium is obtained. This, however, is only a rather small part of the effect, and certainly not sufficient to explain the factor 10 difference between the expected and realized milk transfer ratio's.

Conclusions

1. Inorganic selenium was little excreted in milk and much in urine, and hence cannot be used for measuring the dose response curve for excretion in milk of the natural selenium of feed.
2. Se-analysis by atomic absorption spectrometry or (rapid) neutron activation analysis is reliable and sensitive enough to permit execution of analyses of the natural milk selenium content. Thus even exact definition of the increments due to variation of the natural Se-level of feed intake will be possible.
3. As long as it is difficult to simulate the composition of natural feeding stuff, in forms of selenium, feeding experiments should be limited to the variation mentioned under 2.
4. Additional information should be obtained from detailed studies, for instance with the rumen system, and with the milk gland, in order to define the exact behavior of the different forms of selenium.
5. In performing experiments as mentioned under 2. and eventually also 4. sufficient time should be allowed for adjustment to the environment. This has already been effectuated in the studies for assessment of the Se (and I) status of milk cows from milk analysis (2).
6. It is recommended to investigate the effect of feeding low protein levels on the incorporation of selenium in amino acids by the rumen micro flora. After all, it may be that the inactivity remarked in this study has to do with abundant supply of sulfur-containing protein.
7. For toxicity studies it is not so sure that the selenium level in milk will always be low (8) as suggested with feeding trials using selenite.

References

1. Binnerts, W.T. & Brouwer, E.: Nature 174, 974-975 (1954)
2. Binnerts, W.T.: Neth. Milk Dairy J. 33, 24-30 (1979).
3. Lein, D.H., Maylin, G.A., Braund, D.G., Gutenmann, W.H., Chase, L.E. & Lisk, D.J.: Cornell Veterinarian 70, 113-124 (1980).
4. Perry, T.W., Peterson, R.C. & Beeson, W.M.: J. Dairy Sci. 60, 1698-1700 (1977).
5. Underwood, E.J.: Trace Elements in Human and Animal Nutrition, fourth edition, Academic Press, New York. San Francisco. London, p. 308-309 (1977).
6. Grant, A.B. & Wilson, G.F.: N.Z.J. Agri. Res. 11, 733 (1968).
7. Smith, M.I.: J. Am. Med. Assoc. 116, 562-566 (1941).
8. Larson, B.L. & Smith, V.R.: Lactation, a comprehensive treatise, Vol. III, Academic Press, New York and London, 1974, p. 149-150.
9. Watkinson, J.H.: Analytical Chemistry 38, 92-97 (1966).
10. Ellen, G. & Douma, F.F.: De Ware(n) Chemicus 8, 154-161 (1978) (in Dutch).
11. Ihnat, M. & Miller, H.J.: Journal of the AOAC 60, 813-25 (1977).
12. Muth, O.H. & Oldfield, J.E. (Eds.): Symposium sulfur in nutrition, The Avi publishing company, inc., Westport, Connecticut, 1970, p. 136 and 169.

ASSESSMENT OF SELENIUM STATUS IN PATIENTS RECEIVING LONG-TERM TOTAL PARENTERAL NUTRITION

Jean Nève

Institute of Pharmacy, Free University of Brussels
Campus Plaine 205-1 - B-1050 Brussels, Belgium

Françoise Vertongen

Laboratory of Medical Chemistry, University Hospital
Saint-Pierre - B-1000 Brussels

Nadine Thonnart and Yvon A. Carpentier

Department of Gastro-Intestinal Surgery, University Hospital
Saint-Pierre - B-1000 Brussels

Introduction

Long-term total parenteral nutrition (TPN) is routinely used in the treatment of patients suffering from chronic pathologies of the digestive tract such as inflammatory or short bowel diseases. In these patients, it often represents the only way of providing an adequate nutritional intake. The precise composition of this intravenous diet is still rather controversial although a general consensus has been achieved for considering that excessive calorie intake could lead to severe side-effects (1,2). The importance of trace elements in human nutrition has been recognized for several years with the demonstration of various specific deficiency syndromes due to congenital abnormalities, inadequate supply or increased losses... The advent of TPN with solutions containing extremely low concentrations in essential trace elements has resulted in the characterization of some of these syndromes in the artificially-fed patients and provided an opportunity for studying the clinical and biochemical consequences of deficiency in man.

Trace Element - Analytical Chemistry in Medicine and Biology, Vol. 3

For example, muscle pain and tenderness in the thighs (3,4), white fingernail beds (4) and especially fatal congestive cardiomyopathy (5-7) have been described as a consequence of selenium deficiency in long-term TPN patients. These clinical manifestations have been associated with low blood and tissue selenium concentrations (3-11), elevated serum glutamic oxaloacetic and glutamic pyruvic transaminases and creatine kinase activities (4), and also with low blood and tissue activities of the selenoenzyme glutatione peroxidase (GPx), which plays an important role in the metabolisation of toxic oxygen derivatives in cells (3-11).
From all these valuable studies, it generally appears that whole blood selenium, plasma selenium and whole blood or erythrocyte GPx are liable to give a satisfactory assessment of selenium status and are therefore useful either in the characterization of selenium deficiency and/or in the monitoring of the response to selenium supplementation in the TPN patients. Nevertheless, further investigations are required for determining the sensitivity of these parameters to selenium depletion and repletion. These studies should provide a better knowledge of selenium requirements in normal and pathological conditions and of the appropriate route and form of supplementation. The purpose of the present work is to answer to some of these questions.

Subjects and Methods

In the first part of the study, ten adult patients (mean age 44 ± 20 years) with various gastro-intestinal diseases were regularly monitored for plasma and erythrocyte Se concentration as well as erythrocyte GPx activity before and during TPN providing less than 30 µg selenium/day (as sodium selenite in commercially available multi-element injectable forms - ampoules des Hôpitaux de Paris [R] -, mixed with the TPN solution).

TPN regimen consited in amino-acids solutions (1 g nitrogen/ 150 kCal) containing lipids and/or glucose as well as vitamins.

In the second part of the study, some of the patients, who were still on TPN or switched to semi-elemental enteral nutrition, were supplemented with higher selenium concentrations (sodium selenite from Merck, 130-150 μgSe/day and selenomethionine from Calbiochem-Behring, 50 to 100 μgSe/day. Injectable forms were manufactured by Mr. J. Timmermans, University Hospital Erasme, Brussels). Control subjects were 25 healthy adult volunteers (mean age 36 $\pm$ 9 years) living in the area of Brussels.

Blood samples were collected by venipuncture in acid-washed tubes containing heparin. Plasma was separated by centrifugation and the red cells washed twice with isotonic saline.

GPx activity in erythrocytes was directly assayed according to the method of Gunzler (12) using H_2O_2 as substrate, and the activity was expressed in μmol NADPH oxidized/min/gHb at 30°C (Hemoglobin was determined by the Drabkin's procedure).

Selenium was measured on digestates of plasma and erythrocytes by the previously described atomic-absorption spectrophotometric procedure (13,14) and the concentrations were expressed in μmol Se/ml and nmol Se/gHb, respectively. Student's t test was used for comparing the means.

Results and Discussion

1) Changes in three indicators of selenium status during selenium depletion.

Before the introduction of TPN, mean plasma selenium level and erythrocyte GPx activity of the patients were significantly lower as compared to normals ($p<0.05$ and $p<0.001$, respectively) although the majority of values remained in the normal range (Fig. 1 and 3).

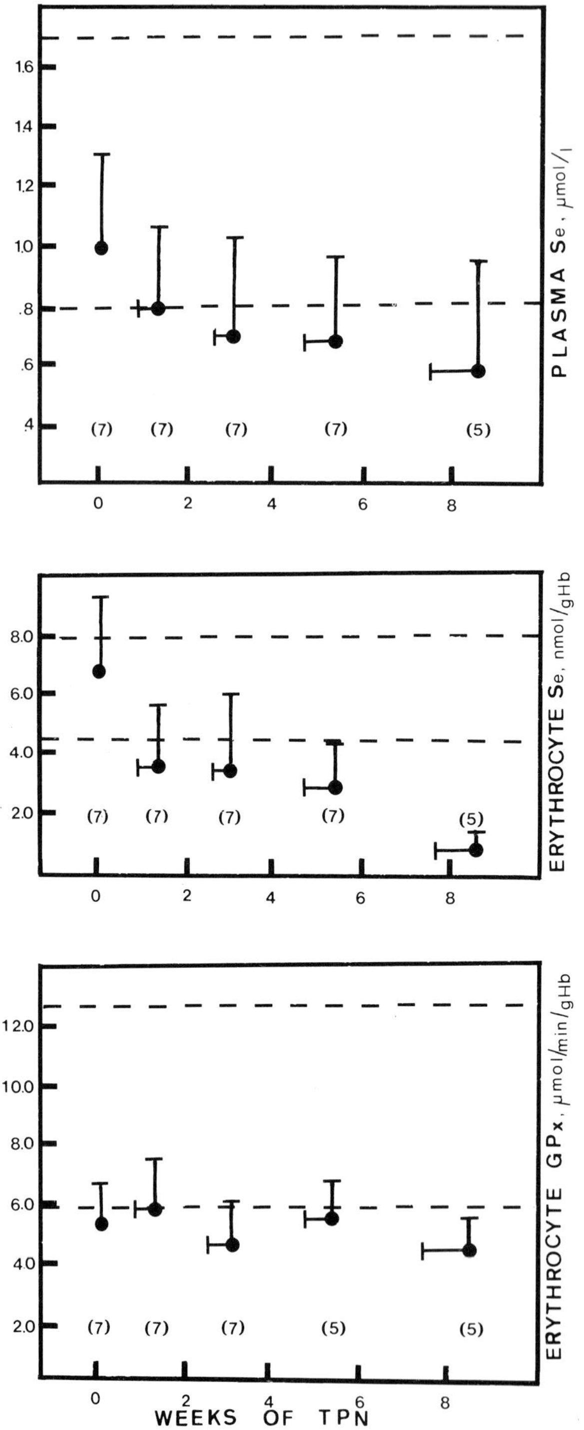

Fig. 1-3 : Mean plasma selenium levels, erythrocyte selenium levels and erythrocyte GPx activities in seven patients during TPN providing less than 30 µgSe/day (as sodium selenite).
Each point represents the mean $\pm$ SD; ( ) = number of determinations.
Dashed lines delimit normal levels ($\bar{x} \pm 2$ SD).

On the contrary, erythrocyte selenium was not statistically modified (Fig. 2). This situation may indicate a marginal selenium deficiency state that is in fact not unusual in patients suffering from gastro-intestinal diseases (3).
During administration of TPN with selenium intake lower than 30 µgSe/day, a rapid and significant decrease of plasma and erythrocyte selenium levels was observed (Fig. 1 and 2), whereas changes in erythrocyte GPx activity (Fig. 3) did not reach statistical significance. At the end of the investigation (6.7 $\pm$ 2.9 weeks TPN), mean plasma selenium levels reached 60 $\pm$ 19 % of initial levels ($p<0.025$), erythrocyte selenium levels 31 $\pm$ 17 % ($p<0.005$) and erythrocyte GPx activity 83 $\pm$ 21 % (NS).
The results confirm the value of plasma selenium in the assessment of short-term changes in selenium status (3,4,15-19) but also show the great sensitivity of erythrocyte selenium to these changes. Van Rij,etal.(3,18), in a similar study on patients receiving TPN, showed that red blood cell selenium decreased with time but admitted that their results may have been altered by blood transfusions. Levander,et al.(23) also indicated a significant decrease of erythrocyte selenium during a selenium depletion study in normal subjects. Until recently, the general opinion was that red blood cell selenium only reflects long-term changes in selenium status, probably because of the relatively long lifetime of the cells in the circulation (23). However, several experiments in animal and man suggest that a significant part of erythrocyte selenium is labile and rapidly exchangeable with plasma and other cells. For example, an important role for erythrocytes in selenium metabolism was demonstrated by the ability of these cells first to concentrate then to release selenium after oral ingestion or intravenous injection of the element (20-22).
The nature of the "labile" part of erythrocyte selenium is not entirely known but it does not seem to interfere with glutathione peroxidase.

Indeed, incorporation of selenium in the enzyme occurs during erythropoeisis and cell maturation (21) and in fact, only a small part of erythrocyte selenium - ± 10 % in the German adult (25) - accounts for the element as glutathione peroxidase. On the basis of our investigation, confirming the results of other authors (3), GPx activity in erythrocytes is not rapidly affected by selenium depletion and is therefore the less sensitive indicator to changes in selenium status.

2. Correlations between indicators of selenium status during depletion.

   Plasma selenium versus erythrocyte selenium.

   The correlation is significant in the patient's group ($r = 0.55$, $p < 0.001$, $n = 36$) and not significant in controls ($r = 0.39$, $p < 0.100$, $n = 22$). This finding support the existence of similar fluctuations for both parameters during depletion.

   Selenium versus glutatione peroxidase in erythrocytes.

   There was no significant correlation between the values measured before TPN ($r = - 0.03$, NS, $n = 9$) or when all results are examined together ($r = 0.29$, NS, $n = 34$). However, a significant correlation was observed between both parameters when, after ± 10 days TPN, erythrocyte selenium concentrations were decreased by 20% or more of the initial values (Fig. 4). This implies that GPx activity in erythrocytes becomes strongly selenodependent at low erythrocyte selenium concentrations (lower than 5.0 nmol Se/g Hb) (26-32).

   In the present study, four patients showed, after 6 weeks TPN, an average decrease of 25 % in GPx activity without any clinical consequence. Further decreases in GPx activity (75-90 %) have been associated with serious clinical side-effects (3,5).

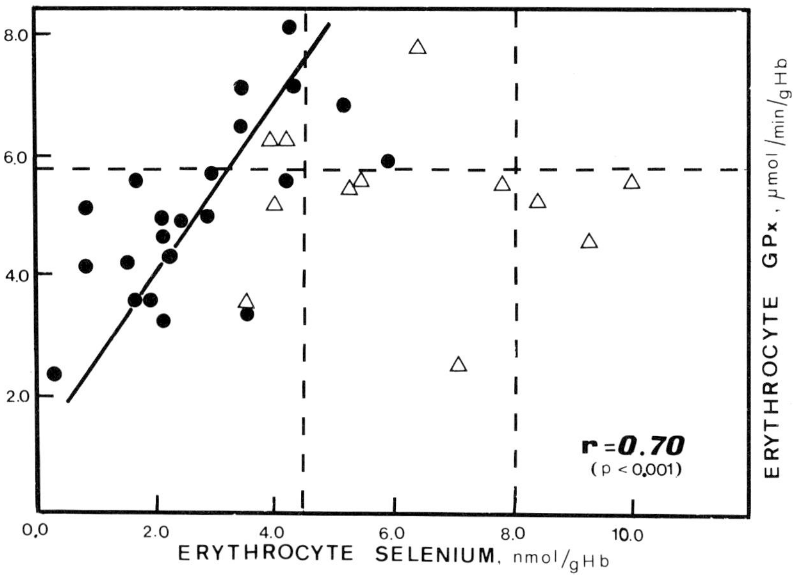

Fig. 4 : Correlation between Se level and GPx activity in erythrocytes of 10 patients receiving TPN providing less than 30 μgSe/day. Symbols ● are values corresponding to erythrocyte Se levels lower than 80 % of the initial levels (symbols △ are levels comprised between 80 and 100 %). Dashed lines delimit normal levels ($\bar{x} \pm 2$ SD).

3. Effects of selenium supplementation.

The effects of selenium supplementation was studied in those patients who had exhibited very low erythrocyte selenium concentrations. The element was given either intravenously (130-150 μg Se/day) as sodium selenite or orally (50-100 μg/Se day) as selenomethionine.

The effects of these various ways of supplementation on the three formentionned selenium indicators are summarized in table 1.

Table 1 : Selenium status for 10 patients on parenteral and enteral nutrition before Se supplementation and after 1-2 and 2-3 months of Se supplementation

Seleno-derivative	Se given/day (µg) and route		Before Suppl.			After Suppl. 1-2 months			After Suppl. 2-3 months		
			$\bar{x}$	SD	n	$\bar{x}$	SD	n	$\bar{x}$	SD	n
Na_2SeO_3	130-150 Parenteral	Plasma Se	0.64	0.43	4	1.03	0.17	4	0.95	0.16	2
		RBC Se	3.2	1.5	4	4.6	0.3	4	5.1	0.6	2
		RBC GPx	4.9	0.8	4	4.6	1.3	4	6.3	2.2	2
Se-Met	50 Oral	Plasma Se	0.52	0.01	2	1.09	0.01	2	1.3	-	1
		RBC Se	2.2	0.1	2	4.0	1.6	2	5.5	-	1
		RBC GPx	4.9	0.7	2	4.8	1.4	2	5.3	-	1
Se-Met	100 Oral	Plasma Se	0.80	0.23	4	1.39	0.23	4	1.62	0.56	3
		RBC Se	2.5	0.6	4	5.6	0.6	4	7.0	1.7	3
		RBC GPx	5.9	1.6	4	6.0	2.2	4	6.5	3.5	3

Units are expressed as indicated in the experimental part
Normal ranges are indicated in Fig. 1-3.

A rapid increase in both plasma and erythrocyte selenium concentrations was observed in response to selenium supplementation provided either intravenously or orally. This is in agreement with other observations (3,4,10,11,15-18,22,28,31, 33-35). In the conditions of our investigation, sodium selenite appeared to be as effective as selenomethionine in normalizing these two levels (after 1-2 months); however, plasma and erythrocyte selenium continued to rise in subjects receiving selenomethionine whereas selenite did not further modify these levels. Similar differences in the effects of various forms of selenium were reported by others (35,36), but this phenomenon has received no satisfactory explanation for the moment. GPx activity in erythrocytes showed a slight but not significant tendency to increase after more than two months selenium supplementation. A better response of GPx activity to selenium supplementation has however been described in more severe cases of selenium deficiency (10). Our results support the idea that GPx activity in erythrocytes is not a sensitive indicator of moderate changes in selenium status. This statement, which is shared by other authors (18,28,34,35), is probably related to the slow rate of erythropoeisis (28). A valuable work by Levander et al (35) has recently shown the great sensitivity of platelet GPx to these changes; unfortunately, their method cannot be applied in routine.

Conclusions

1. Erythrocyte Se as well as plasma Se are sensitive indicators of short-term changes in Se status. 2. On the contrary, erythrocyte GPx activity is not sensitive to these changes. 3. Se and GPx in erythrocytes are not correlated when erythrocyte Se levels remain in the normal range but become significantly correlated when this level is below 5.0 nmol Se/g Hb (± 0.06 µgSe/ml whole blood).

4. Se supplements of oral 50-100 μgSe as selenomethionine/day or intravenous 130-150 μg Se as sodium selenite/day are adequate for rapidly normalizing blood Se levels in Se-depleted patients but long-term effects of two forms are different.

Acknowledgements

Our thanks are expressed to Professor L. Molle for his constant assistance and to Mr. J. Timmermans, Professor A. Moës and Mr. Ph. Deleuze for help in preparing selenium supplements.

References

1. Homsy, F.N., Blackburn, G.L. : J. Am. Coll. Nutr. 2, 75-95 (1983).
2. Easton, L.B., Halata, M.S., Dweck, H.S. : Pediatr. Clin. North Amer. 29, 1171-1190 (1982).
3. Van Rij, A.M., Thomson, C.D., Mc Kenzie, J.M., Robinson, M.F. : Am. J. Clin. Nutr. 32, 2076-2085 (1979).
4. Kien, C.L., Ganther, H.E. : Am. J. Clin. Nutr. 37, 319-328 (1983).
5. Johnson, R.A., et al. : N. Engl. J. Med. 304, 1210-1212 (1981).
6. Collipp, P.J., Chen, S.Y. : N. Engl. J. Med. 304, 1304-1305 (1981).
7. Fleming, C.R., et al. : Gastroenterol. 83, 689-693 (1982).
8. Mc Kenzie, J.M., Rea, H., Van Rij, A.M., Robinson, M.F. : Proceed. Univ. Otago Med. School 54, 25-26 (1976).
9. Lane, H.W., Dudrick, S., Warren, D.C. : Proceed. Soc. Expl. Biol. Med. 167, 383-390 (1981).
10. Fell, G.S., et al. : Proc. Nutr. Soc. 39, 36 A (1980).
11. Baker, S.S., et al. : Am. J. Clin. Nutr. 38, 769-774 (1983).
12. Gunzler, W., Kremers, H., Flohé, L. : Z. Klin. Chem. Klin. Biochem. 12, 444 (1974).
13. Nève, J., Hanocq, M., Molle, L. : J. Pharm. Belg. 35, 345-350 (1980).

14. Nève, J., Hanocq, M., Molle, L., Lefèbvre, G. : The Analyst 107, 934-941 (1982).

15. Young, V.R., Nahapetian, A., Janghorbani, M. : Am. J. Clin. Nutr. 35, 1076-1088 (1982).

16. Levander, O.A., Sutherland, B., Morris, V.C., King, J.C.: Am. J. Clin. Nutr. 34, 2662-2669 (1981).

17. Smith, J.L., Goos, S.M. : J. Parent. Ent. Nutr. 4, 23-26 (1980).

18. Van Rij, A.M., Mc Kenzie, J.M., Thomson, C.D., Robinson, M.F. : J. Parent. Ent. Nutr. 5, 120-124 (1982).

19. Burk, R.F. : Trace Elements in Human Health and Disease, Academic Press, New York 1976.

20. Gasiewicz, T.A., Smith J.C. : Chem. Biol. Interactions 21, 299-313 (1978).

21. Mc Murray, C.H., Davidson, W.B. : Biochem. Biophys. Acta 583, 332-343 (1979).

22. Janghorbani, M., Christensen, M.J., Nahapetian, A., Young, V.R. : Am. J. Clin. Nutr. 35, 647-654 (1982).

23. Levander, O.A., Sutherland, B., Morus, V.C., King J.C.: Selenium in Biology and Medicine, Avi Publishing Co, Westport 1981.

24. Perona, G. et al : Brit. J. Haematol. 39, 399-408 (1978).

25. Schmidt, K., Heller, W. : Blut 33, 247-251 (1976).

26. Mc Kenzie, R.L., Rea, H.M., Thomson, C.D., Robinson, M.F. Am. J. Clin. Nutr. 31, 1413-1418 (1978).

27. Valentine, J.L., Kang, H.K., Dang, P., Schluchter, M. : J. Toxicol. Environ. Hlth. 6, 731-736 (1980).

28. Lombeck, I., Kasperek, K., Backmann, D., Feinendegem, L.E., Bremer, H.J. : Eur. J. Pediatr. 134, 65-68 (1980).

29. Robinson, M.F., Thomson, C.D. : Selenium in Biology and Medicine, Ari Publishing Co, Westport 1981.

30. Lane, H.W., Warren, D.C., Taylor, B.J., Stool, E. : Proceed. Soc. Expl. Biol. Med. 173, 87-95 (1983).

31. Lane, H.W., Warren, D.C., Martin, E., Mc Cowan, J. : Nutr. Res. 3, 805-817 (1983).

32. Thomson, C.D., Rea, H.M., Doesburg, V., Robinson, M.F. : Br. J. Nutr. 37, 457-460 (1977).

33. Levander, O.A. : Fed. Proceed. 42, 1721-1725 (1983).

34. Robinson, M.F. et al. : New Zealand Med. J. 683, 289-292 (1981).

35. Levander, O.A. et al. : Am. J. Clin. Nutr. 37, 887-897 (1983).

36. Thomson, C.D., Robinson, M.F., Campbell, D.R., Rea H.M. : Am. J. Clin. Nutr. 36, 24-31 (1982).

SELENIUM IN NUTRITION

Józef Kajfosz, Stanisław Szymczyk
Institute of Nuclear Physics, 31-342 Cracow, Poland

Introduction

Despite its toxicity at higher concentrations, selenium is known as essential for animal and human organisms. Following some reports of its role in the immunity of men against leukaemia /e.g. ref. 1/, the Commission on Public Health Protection of the Polish Academy of Sciences promoted the idea of increased selenium consumption in leukaemia prophylactics. In search for suitable selenium-containing food products the selenium concentrations of a considerable number of products were determined by the PIXE method in the Institute of Nuclear Physics in Cracow. The results presented here were obtained in cooperation with various medical and agricultural institutions in the course of the last few years.

Scope of Analyses

The first part of our studies of selenium concentrations in food products started in 1977 and was stimulated by the Clinic of Haematology of the Cracow Medical Academy, where the idea of leukaemia prophylactics was promoted. In these studies salt from different mining places, water, milk, meat, honey and other bee-products, corn grain and corn flakes, Antarctic fishes and Antarctic krill were analysed. Further, the Institute of Toxicology of the Silesian Medical Academy offered an opportunity to investigate different products of food industry and the Institute of Fruit-Growing in Skierniewice supplied us

Trace Element - Analytical Chemistry in Medicine and Biology, Vol. 3

with series of apple and tomato samples from different cultivations. For the last six years hundreds of samples /vegetables, fruits, cereals/ were investigated in cooperation with the Cracow Academy of Agriculture as a part of the research on traffic- and industrial pollutions. We have also determined selenium contents in various other food products. This report contains a brief account on selenium contents in food based on all these studies. The total number of analyses amounts to ca 4000.

Method of Analysis

All results were obtained by means of the PIXE method /Proton Induced X-ray Emission/. As a proton source the small C-48 cyclotron with proton energy of 2.6 MeV was used. More detailed data about the typical PIXE arrangement and its parameters were published elsewhere /refs 2,3/. Experience gained in the intercomparison studies organised by the International Atomic Energy Agency in Vienna /4,5/ indicate that the determination of Se content on a trace level is not an easy task since large dispersion of results in these studies shows many methods as not fully adequate.

In this respect at least two important advantages of PIXE are worth mentioning. First, no chemical pretreatment is required and thus the possible introduction of impurities from chemical reagents can be excluded. Second, the detection efficiency of the characteristic X-radiation is a smooth function of the atomic number of the elements determined. Since the nearby elements, Zn and Br, are commonly present in biological samples, the efficiency of measurement for Se can be conveniently interpolated from those of Zn and Br easily obtained from available standards, and thus good accuracy can be guaranteed. Moreover, the characteristic K_{α} X-radiation of selenium of the energy 11.2 keV is positioned in the high-sensitivity interval of ca 6 - 16 keV and the Se K_{α} line is well-separated from the neighbouring lines of Pb L_{α} and Br K_{α} , which assures a relatively low detection limit of 0.02 ppm and a relatively high precision in Se determinations.

Results

Salt from different deposits

Prevailing part of salt consumed in Poland comes from saline deposits. In the process of refinement all natural admixtures are removed. In order to increase the trace elements supply into human organisms the consumption of unpurified salt has been suggested. In a study stimulated by the salt-mining industry a series of salt samples from different deposits was analysed by PIXE. The results reveal a broad trace elements spectrum in which the extreme values found for the commonest elements were: K : 35 - 4200 ppm, Ca : 12 ppm - 2.5 %, Cu : 1 - 320 ppm, Fe : 9 ppm - 1.4 %, Zn : 1 - 62 ppm, Br : 16 - 310 ppm, Sr : 16 - 290 ppm, Pb : 0 - 2 ppm. Other elements observed in some samples were: Ti, Cr, Mn, Ni and Rb. The selenium content, however, rarely exceeded the detection limit and never reached the value of 0.1 ppm. Hence, salt cannot be considered a source of selenium.

Water

About 30 samples of tap water from water-pipes in Cracow and vicinity were analysed. The Se content was generally below the level of 0.05 ppm. A sample of mineral water from the health resort Busko-Zdrój was found to contain 0.04 ppm of Se. Investigations of mineral water from other water springs and other places are planned for the near future.

Cow-milk

Since milk constitutes an essential product in human nutrition and is consumed in various forms in relatively large amounts, the trace elements intake by milk consumption seems to be of special interest. However, our invastigations have proved that milk belongs to the products which are astoundingly pure despite the impurities in the environment. Even from regions strongly polluted by lead /detected in soil, animal nutrition and animal blood, hair and tissue/ no significant increase in lead content in milk was observed. The selenium content in the samples of milk and milk products was below the detection limit of 0.02 ppm. Thus milk and milk products do not contribute significantly to the selenium intake.

Honey and bee-products

Valuable nutrition and therapeutic properties of honey and other bee-products are well-known. Some honey producers and trade organisations stress the great role of trace elements in their products. However, our analyses do not confirm these claims. We found honey to be relatively pure, with selenium not above the detection limit. Consequently, those products cannot be regarded as a source of the trace elements. Neither can their therapeutic properties be related to those elements.

Corn /Zea mays/ grains and corn flakes

According to some reports corn grain was found to contain ca 4 - 5 ppm of Se /6/. This fact was given a broad publicity by some medical circles encouraging corn consumption for leukaemia prophylactics. Corn flakes were even advertized as a selenium--rich product. We analysed a series of corn grains grown in Poland, Africa, North and South Americas as well as a series of the flakes produced in Poland from imported corn. Our results did not confirm the earlier reports and claims. The Se content in these samples varied from 0.03 to 0.09 ppm and the mean value for the whole series found was 0.07 ppm.

Antarctic fishes and krill

Relatively large selenium concentrations of several ppm were determined in a series of samples of sea-born organisms, which included tissues and bones of several kinds of antarctic fishes and of the krill /Euphasia superba/. These large Se contents in marine species are somewhat strange in view of the very low Se concentrations /0.004 ppm/ in sea water. The concentrations obtained are listed in Table I.

Table I. The Se concentrations in Antarctic fishes and krill

Name of species	Kind of tissue	Se content in ppm
Notothenia gibberifrons	muscles	1.6
Notothenia rossi marmorata	muscles	1.5
Pseudochaenichthys georgianus	muscles	2.1
Chaenocephalus aceratus	muscles	1.1
Champsocephalus gunnari	muscles	2.2
Notothenia	milt	4.8
Chaenocephalus aceratus	milt	4.0
Chaenocephalus aceratus	hard roe	3.5
Krill /Euphasia superba/	muscles	6.2
Krill /Euphasia superba/	scales	6.4
Krill /Euphasia superba/	eyes	9.5

Fruits and vegetables

Various fruits and vegetables were analysed in the investigation of traffic- and industry pollution as well as in studies of natural trace elements contents. The Se content in vegetables rarely exceeds the value of 0.1 ppm and frequently is below the detection limit. Typical concentrations are thus of the order of 0.02 - 0.08 ppm. However, occasionally, among many samples of similar kind a few have increased Se contents to 0.2 - 0.5 ppm. Some examples are: rye grains, lettuce, pears, apples. This sudden increase in some individual cases may be connected either with increased local concentrations of Se in soil or with some inhomogeneous pollutions. The latter possibility is supported by the fact that the increased Se concentrations occur mainly in peripheral parts of the specimens /leaves, fruit-skin/. We found a general rule that the concentrations of pollutants in these parts are usually about 30 times higher than those in inner parts /inside of fruits, roots etc./

For some special purposes we have also examined Se content in the wine produced from dendilion /Taraxacum/ and in the roots of ginseng /Panax ginseng/, but no Se above the detection limit was found there. Both plants are of some importance in popular medicine. From the viewpoint of Se content the holophytes from the region of Camargues /France/ which contained up to 2 - 3 ppm of selenium were found to be most interesting.

Industrial food products

In the search for selenium-rich food products a number of products of food industry was investigated. Among others we analysed bread, different cakes, wheat and rye flour, cereals of different kinds, meats and meat products, ham, sausages, nuts, poppy, rice, peas, soybean, sugar, potatoes and potato meal. In all cases the selenium contents found were of the order of 0.05 ppm or smaller. A somewhat increased Se content of ca 0.15 ppm was found in wild strawberry juice.

Selenium-fertilized plants

An interesting possibility for obtaining selenium-rich food products for prophylactic or therapeutic purposes is the growing of plants under special conditions, e.g. on selenium rich soils or on selenium-rich fertilizers. Experiments of this kind were carried out in the Institute of Fruit-Growing in Skierniewice, where tomatoes were grown under such conditions. In a series of measurements we proved that the Se content increased from ca 0.03 - 0.05 ppm /for tomatoes grown in normal conditions/ to 2.3 - 2.6 ppm /for those fertilized with selenium compounds/.

Selenium in pollutions

A large part of industrial pollutions in Poland originates from coal combustion. In a series of samples from Silesian coal mines the mean concentration of Se was found to be 3.6 ppm. After combustion the ashes and fly-ashes were found to contain up to 33 ppm Se. This source of selenium is responsible for air pollution indicated both by means of sampling on air-filters and by means of biological indicators, the lichens /7/, in which we found selenium contents of 0.3 - 0.7 ppm. The selenium of industrial origin causes also continuous increase of Se content in soil of the exposed areas. However, since both the availability of selenium by biological organisms and its role /positive or negative/ in them strongly depends on the chemical form of its occurence, more detailed considerations would require the knowledge of data which cannot be obtained with the PIXE method.

Conclusions

The selenium concentrations in most of the investigated food products were on a low level, usually not exceeding the value of 0.1 ppm. Typical concentrations in agricultural products were ca 0.03 - 0.07 ppm. Occasionally, the Se content in individual samples increased to 0.5 - 0.7 ppm, probably because of increased local concentrations in soil. The believed abundance of selenium in maize, salt and bee-products has not been confirmed. The most promising sources of selenium for prophylactic diets are the products of marine origin, i.e. sea-water organisms, in which Se concentrations of up to 6 ppm were estimated. Another possibility is growing agricultural products on special, selenium-containing fertilizers.
In order to obtain more definite conclusions more elaborate and complex studies, including investigations of the chemical form of the selenium detected, would be necessary.
The PIXE method applied in these studies was found to be adequate and convenient for fast determinations of Se in food products.

References

1. Schrauzer, G.N.: Bioinorg. Chem. 5, 275-281 /1976/.
2. Szymczyk, S., Kajfosz, J., Hrynkiewicz, A.Z., Curzydło, J.: Nucl. Instr. and Meth. 181, 281-284 /1981/.
3. Kwiatek, W.M., Kajfosz, J., Szymczyk, S., Sznapka, Z., Walas, M.: Report /70/ from "Third International Workshop on Trace Element Analytical Chemistry in Medicine and Biology", Neuherberg, 1984.
4. M'Baku, S.B., Parr, R.M.: Proceedings of the 6th International Conference "Modern Trends in Activation Analysis", Toronto, 15-19 June, 1981.
5. Pszonicki, L.: Report /58/ from "Third International Workshop on Trace Element Analytical Chemistry in Medicine and Biology", Neuherberg, 1984.
6. Aleksandrowicz, J.: private communication.
7. Hrynkiewicz, A.Z., Szymczyk, S., Kajfosz, J., Olech, M.: Nucl. Instr. and Meth. 168, 517-521 /1980/.

THE SELENIUM OF MUSHROOMS

S. Piepponen, M.J. Pellinen and T. Hattula

Technical Research Centre, Food Research Laboratory
Biologinkuja 1, SF-02150 Espoo, Finland

Introduction

There has been considerable discussion in recent years concerning selenium deficiency in the diet of Finnish consumers and the possible effects of this deficiency on the health of the population. In this context mushrooms have frequently been cited as a significant source of selenium. In fact, however, only some mushroom species have a high selenium content. Many common species which are consumed in considerable quantities in Finland contain only small amounts of selenium (1).

Research has indicated that selenium contents may vary to a very great extent (0.01 - 37 mg/kg dry weight) between different mushroom species. The natural range of selenium content may also be very large for a single species. Different contents have even been reported in different parts of whole mushrooms (2,3,4). The growth location has not to date been reported to have a significant effect on selenium levels (1,2).

Naturally occurring champignon mushrooms contain on average three times more selenium than cultivated mushrooms of the same group. Cultivated *Agaricus bisporus* accounts for as much as 70 % of the world market in cultivated mushrooms. If the selenium content of cultivated *Agaricus bisporus* could be increased by a factor of 5 - 8, the significance of this mushroom as a source of dietary selenium would increase considerably.

Trace Element - Analytical Chemistry in Medicine and Biology, Vol. 3

It is generally held that selenium is located in biological material in the same groups of chemical compounds as sulphur, i.e. in proteins (5). In the *Astragalus*-species, well known as accumulators of selenium, incorporation may however take place in other materials as well as proteins (5). Because some mushrooms also share the capacity to accumulate selenium from the environment, there was a motivation to investigate in more detail the distribution of selenium in mushrooms.

The aim of this work was to survey the selenium contents of different mushroom species, to study the localization of selenium in mushrooms and to investigate the possibilities of increasing the selenium level in cultivated champignons.

Materials and methods

The wild mushroom samples were collected during the years 1979-83 from six sites in southern and central Finland. The survey of the selenium contents of edible mushrooms has been published earlier (1).

Cultivation studies in which selenium was added to the casing material were carried out in the Food Research Laboratory. In these studies, spawned horse manure-straw compost was obtained from a commercial mushroom farm. The compost was applied to plastic boxes (40 x 60 cm) as a layer about 18 cm thick. Each box thus contained about 13 kg of compost. The mushroom strain used was *Agaricus bisporus* Somycel 11. Samples were taken for selenium analysis from the first three flushes.

Selenium was added to the casing material in the form of sodium selenite (Na_2SeO_3) in the irrigation water. The doses used were 2, 5, 10, 30 and 60 mg of elemental selenium per cultivation box. Test cultivations were carried out using seven replicate boxes for each dosage, except in the experiment with radioactively labelled selenium, in which only two replicates were used. In the latter experiment the cultivation boxes were watered with 1.0 l of a solution containing 0.45 mCi ^{75}Se and 30 mg inactive selenium as carrier. Subsequent irrigation was with plain water.

The investigation of the localization of selenium in mushrooms was carried out using cultivated Agaricus bisporus Somycel 56 enriched with both inactive and radioactively labelled selenium and dried Boletus edulis.

The mushroom were homogenized using a Moulinette homogenizer, after which the fractionation was carried out according to the method of Sutherland and Wilkinson (6), except that stages III and V were omitted. The fractionation process is presented in Figure 2.

The selenium contents of the samples were assayed, after wet and dry ashing, by atomic absorption spectrometry using the hydride technique (1).

When radioactive ^{75}Se (main γ-energies: 136 KeV, 265 KeV and 280 KeV) was used in the cultivation and fractionation studies, the radioactivity of the samples was measured using a multichannel analyzer with a Ge-Li detector. The measurements were performed in standard geometry. The selenium content was calculated by comparing counts with equally made radioactive standards.

Results and discussion

The selenium contents of mushrooms vary considerably between different species. Of the 38 species investigated in our laboratory (1), by far the greatest selenium were measured in Boletus edulis (190 mg/100 g fresh weight in the mean). Other mushrooms having considerable selenium contents included Macrolepiota (47 mg/100 g f.w.), Gasteromycetes (27 mg/100 g f.w.), wild Agaricus spp. (21 mg/100 g f.w.), Marasmius oreades (16 mg/100 g f.w.) and Lactarius torminosus (17 mg/100 g f.w.). The contents in these mushrooms are sufficient to provide and amount of selenium which is nutritionally significant in relation to the total daily intake of selenium of the Finnish population. Other food mushrooms generally used in Finland, e.g. species belonging to Cantharellaceae, Russula, Boletaceae (other than B. edulis) and Lactarius (other than L. torminosus) contained only small amounts of selenium. The importance of these mushrooms as a source of selenium is therefore marginal.

Table 1. The selenium content of edible mushrooms (1).

Species		Number of samples	Dry weight (%)	Selenium (µg/100g fresh mushroom) Mean	Range
Boletus edulis	(Cep)	20	11	190	65 - 410
Other Boletaceae	(Boletes)	20	8.4	5	0.4- 18
Lactarius torminosus	(Woolly Milk Cap)	8	9.2	17	12 - 26
L. torminosus*	(" " ")	6	9.4	12	8 - 17
Other Lactarius	(Milk Caps)	17	8.9	2	0.5- 8
Russula	(Russulas)	13	11	2	0.6- 3
Cantharellaceae	(Chantarelles)	6	7,2	0,6	0.4- 0.8
Hydnum repandum	(Wood Hedgehog)	1	9,6	0,2	
Rozites caperata	(Gypsy Mushroom)	1	8,0	5	
Agaricus spp.	(Wild Champignons)	14	7.9	21	8 - 46
Agaricus bisporus**	(Cultivated Champignon)	8	8.5	7	4 - 10
Marasmius oreades	(Fairy Ring Champignon)	6	10	16	11 - 36
Coprinaceae	(Ink-Caps)	10	6.0	4	0.1- 13
Gasteromycetes	(Puffballs)	9	14	27	1 - 69
Lepiotaceae	(Parasol Mushrooms)	4	9.4	47	26 - 58
Albatrellus ovinus	(Schafporling (Ger.))	3	7,6	4	1 - 5
Lentinus edodes**	(Shiitake)	1	24	0,5	
Armillariella mellea	(Honey Fungus)	1	9,3	8	

* boiled sample
** cultivated sample

Wild Agaricus species contain about three times more selenium than cultivated A. bisporus. Addition of sodium selenite during watering increases the selenium contents of cultivated mushrooms.

The application of selenium in low concentrations did not have a detrimental effect on the harvest yield of Agaricus bisporus.

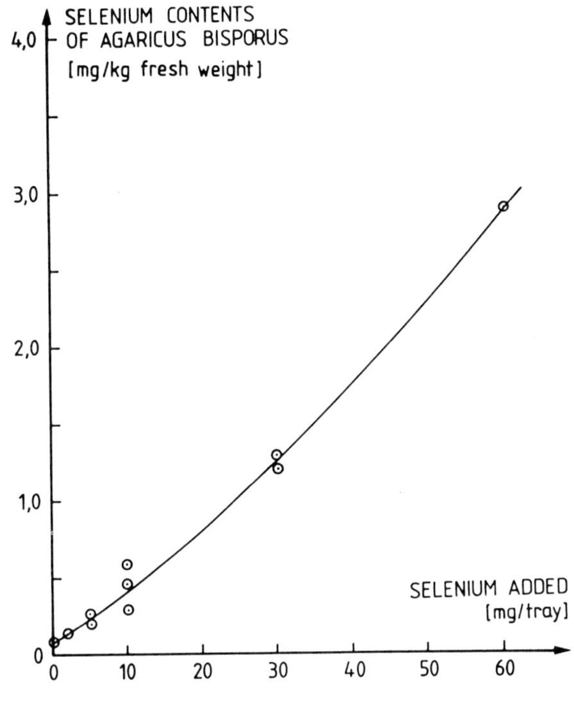

Figure 1. The incorporation of selenium added to the casing material of _Agaricus bisporus_.

On the basis of this investigation it appears that increase in the selenium concentration of cultivated champignon mushrooms can be accomplished rather simply in conjunction with watering of the casing material.

The distribution of selenium between different fractions of mushrooms is presented in Figure 2. Because the results obtained with _A. bisporus_ and _B. edulis_ in the fractionation experiments were very similar, the two species are grouped together in this discussion. A considerabel proportion of the selenium of mushrooms is located in low molecular weight organic or inorganic compounds. In contrast, Varimo <u>et al.</u> (7) reported that only one fifth of the total selenium content of yeast was located in low molecular weight compounds.

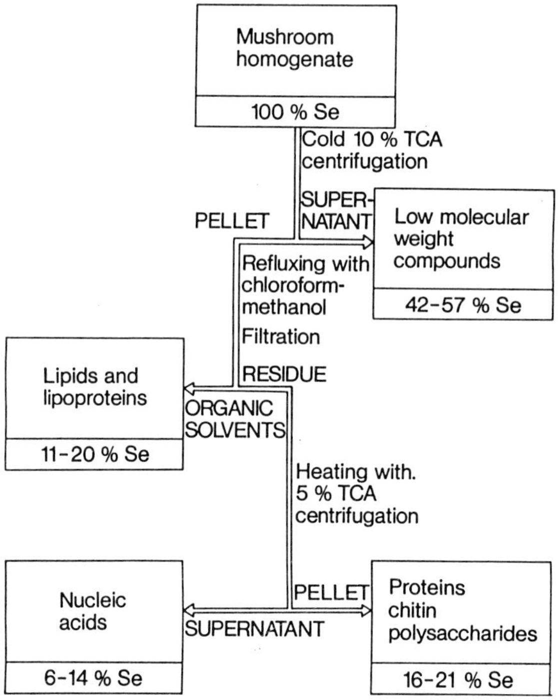

Figure 2. The localization of selenium in mushrooms.

About ten percent of the total selenium of mushrooms is located in lipids and a similar amount in nucleic acids. We have not yet investigated whether the selenium in lipids is actually localized in lipoproteins, as Lunde (8) reported in the case of fatty tissues of fish.

Only a rather small fraction of the selenium of mushrooms is located in the protein fraction, although selenium is usually associated mainly with proteins. According to Olson <u>et al.</u> (9) the selenium of wheat is mainly incorporated in the proteins, while Varimo et al. (7) found that 58 % of the selenium of yeast was located in soluble proteins or in nucleic acids. The best known selenium containing protein is glutathione peroxidase, which has been isolated from both animals and micro-organisms (10). To date, however, this protein has not been isolated from mushrooms.

The location of selenium in champignon mushrooms in three successive flushes is presented in Table 2. These results were obtained using mushrooms watered with radioactively labelled selenium. The distribution between different fractions is the same as that based on the analysis of unlabelled selenium and the observed distribution did not vary markedly between the different flushes. The amount of radioactivity incorporated per kilogram fresh weight of mushroom also showed little variation between the different flushes.

Table 2. The percentage distribution of selenium in different fractions obtained from three successive flushes of champignon mushrooms. The measurements were made on the basis of radioactive ^{75}Se concentrations.

	Flush I 27.12.83	Flush II 4.1.84	Flush III 13.1.84
Low molecular weight compounds	57 %	52 %	53 %
Lipids and lipoproteins	11 %	20 %	13 %
Nucleic acids	11 %	11 %	14 %
Proteins, chitin and polysaccharides	21 %	17 %	20 %
Specific total activity (nCi/g fresh weight)	5.75	6.10	5.50

A rather high proportion of the amino acids of mushrooms is contained in the fraction of low molecular weight compounds. According to our analyses 25 % of the nitrogen content of mushrooms is located in free amino acids and it is very likely that selenomethionine and selenocystein are partly present in the free state. On the other hand inorganic selenium may partly be adsorbed on cell organelles, as Varimo et al. (7) demonstrated in experiments with yeast.

References

1. Piepponen, S., Liukkonen-Lilja, H., Kuusi, T.: Z. Lebensm. Unters. Forsch. 177, 257 - 260 (1983).
2. Stijve, T., Diserens, J., Pletikosa, M.: Z. Lebensm. Unters. Forsch. 164, 201(1977).
3. Allen, R., Steinnes, E.: Chemosphere 4, 371(1978).
4. Byrne, A., Dermelj, M., Vakselj, T.: Chemosphere 8, 815(1979).
5. Shrift, A.: Metabolism of selenium by plants and micro-organism in organic selenium compounds: their chemistry and biology. Ed. Klayman, D. L. and Günther, W. H. H., John Wiley & Sons Inc., New York, pp. 763 - 813 (1973).
6. Sutherland, I. W., Wilkinson, J. F.: Chemical extraction methods of microbial cells in book "Methods in Microbiology", ed. Norris, J. R., Ribbons, D. W., vol 5B, Academic Press, London, pp. 378 - 383 (1971).
7. Varimo, K., Harju, K., Edelman, K., Korhola, M.: Kemia-Kemi 10, 1049 - 1050 (1983).
8. Lunde, G.: J. Sci. Fd. Agric. 23, 987 - 994 (1972).
9. Olson, O. E., Novacek, E. J., Whitehead, E. J., Palmer, J. S.: Phytochemistry 9, 1181 - 1188 (1970).
10. Ganther; H. E.: Chemica Scripta 8A, 79 - 84 (1975).

SELENIUM STATUS IN PATIENTS WITH JUVENILE CHRONIC ARTHRITIS

B. Åkesson, U. Johansson, P.-A. Öckerman
Department of Clinical Chemistry, University Hospital
Lund, Sweden

S. Portinsson, A. Åkesson, H. Svantesson
Department of Rheumatology, University Hospital, Lund, Sweden

Introduction

The joint involvement in juvenile chronic arthritis is often accompanied by anemia, impaired growth and muscle wasting (1). These phenomena may be influenced by the supply of dietary energy, protein and several micronutrients. Another reason for the growing interest in the relation between diet and rheumatoid arthritis is the increasing awareness that several nutrients are involved in immune reactions and in the defence against free radicals.

Selenium may be important in this respect since the seleno-enzyme glutathione peroxidase can reduce different peroxides formed during inflammatory processes. Low concentrations of plasma selenium has been observed inadults with rheumatoid arthritis (2), but so far no such studies in juvenile chronic arthritis have been carried out.

Material and Methods

Twentysix 12-17 year-old girls with juvenile chronic arthritis (JCA) according to the ARA critieria were studied.

Trace Element - Analytical Chemistry in Medicine and Biology, Vol. 3

Twentyeight healthy girls served as controls.

Blood samples were obtained after an overnight fast. Blood, plasma or serum was analyzed for the components shown in Tables 1 and 2. Plasma selenium was determined by gas chromatography, and glutathione peroxidase activity was assayed using tert-butylhydroperoxide as substrate.

Results and Discussion

Plasma selenium was significantly depressed in girls with JCA (Table 1). The selenium concentration in different subgroups of patients is shown in Figures 1-3. There was no apparent relation of plasma selenium to disease activity (Fig. 1), type of JCA (Fig. 2) or treatment with different drugs (Fig. 3).

Plasma glutathione peroxidase activity was unchanged among JCA-patients, but blood glutathione peroxidase was depressed (Table 1), which is compatible with a deranged selenium status.

Among the vitamins analyzed no differences between patients and controls were observed. Serum iron and plasma albumin were depressed in JCA but transferrin and ferritin levels were unchanged (Table 1).

The correlations between different indices of nutritional status among JCA patients is given in Table 2.

This study is the first showing that children with chronic arthritis have subnormal plasma selenium and blood glutathione preoxidase but normal plasma glutatione peroxidase activity. The low selenium levels agree with results from

Table 1. Indices of nutritional status in girls with juvenile chronic arthritis (JCA).
Mean values (SE) are given. Student's t-test: ^{xx}, $0.001<p<0.01$; ^{xxx}, $p<0.001$.

Parameter	JCA (n=26)	Controls (n=28)
Plasma selenium (umol/L)	0.82(0.04)^{xxx}	1.00(0.03)
Plasma glutathione peroxidase (umol/Lxmin)	434(19)	478(20)
Blood glutathione peroxidase (umol/gHbxmin)	36(1)^{xx}	42(2)
Blood folate (nmol/L	268(12)	278(12)
Serum cobalamine (pmol/L)	346(25)	335(22)
Serum α-tocopherol (umol/L)	19.3(0.8)	18.4(0.4)
Serum 25-hydroxy-vitamin D (nmol/L)	63(7)	76(5)
Plasma ascorbic acid (umol/L)	70(4)	75(4)
Serum iron (umol/L)	12(1)	18(1)
Serum transferrin (umol/L)	64(3)	66(2)
Serum ferritin (ug/L)	23(4)	26(3)
Plasma albumin (g/L)	36(1)^{xxx}	44(1)

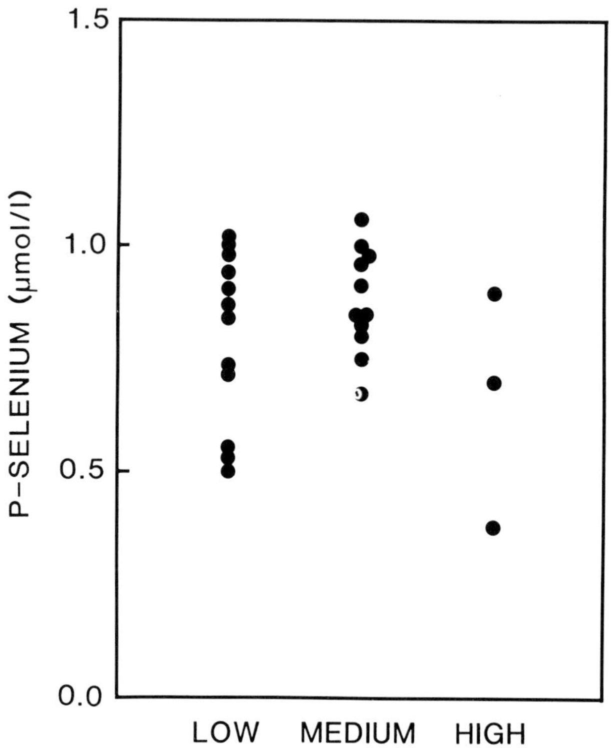

Figure 1. Relation of plasma selenium to disease activity in juvenile chronic arthritis. Disease activity was estimated from the number of swollen joints, erythrocyte sedimentation rate and plasma orosomucoid.

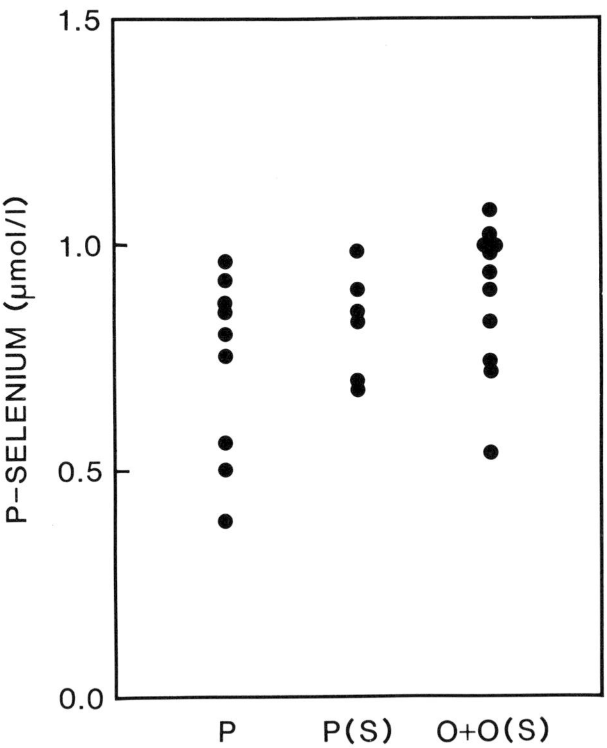

Figure 2. Relation of plasma selenium to the type of juvenile chronic arthritis. P, polyarticular type; P(S), polyarticular type with systemic manifestations; O + O(S), oligoarticular type with or without systemic manifestations.

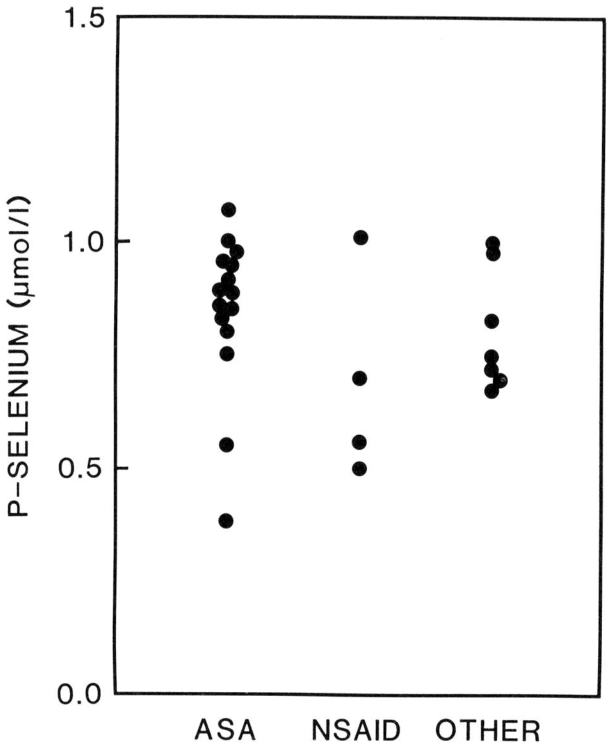

Figure 3. Relation of plasma selenium in juvenile chronic arthritis to antirheumatic drug treatment. ASA, acetylsalicylic acid; NSAID, other non-steroidal anti-inflammary drugs.

Table 2. Relation of selenium and glutathione peroxidase to blood indices of nutritional status and disease activity.
Linear correlation coefficients were calculated.
^{x}, $0.01<p<0.05$; ^{xx}, $0.001<p<0.01$

Parameter	Plasma selenium	Plasma glutathione peroxidase	Blood glutathione peroxidase
Plasma selenium	-	0.13	0.21
Plasma glutathione peroxidase	0.13	-	0.10
Blood glutathione peroxidase	0.21	-0.10	-
Blood folate	0.39^{x}	0.23	0.12
Serum cobalamine	0.10	0.09	0.00
Plasma ascorbic acid	0.49^{xx}	-0.24	0.17
Serum α-tocopherol	0.17	-0.36^{x}	0.12
Serum 25-hydroxy-vitamin D	0.03	0.30	0.04
Blood hemoglobin	0.00	-0.22	0.34
Serum iron	-0.23	-0.37^{x}	0.02
Serum transferrin	-0.03	0.02	0.25
Serum ferritin	-0.13	0.25	-0.02
Plasma albumin	-0.12	-0.32	0.10
Plasma orosomucoid	0.00	0.32	-0.23
Plasma ceruloplasmin	0.12	0.42^{x}	-0.15
Plasma C-reactive protein	0.16	0.10	-0.15
Plasma immunoglobulin G	0.03	0.19	-0.39^{x}
Plasma immunoglobulin A	-0.43^{x}	0.33	-0.18

adult rheumatics (2). In the Norwegian study (2) higher values for plasma selenium were reported, which may be explained by a higher selenium intake in Norway than in Sweden. The activity of glutathione peroxidase, has not previously been determined in patients with rheumatoid arthritis. The decrease in blood glutathione peroxidase is compatible with a deranced selenium turnover. No change in plasma glutathione peroxidase was observed indicating that the decrease in plasma selenium concerns other seleno-compounds.

This study was supported by the Swedish Medical Research Council (project 3968), the Swedish Council for Planning and Coordination of Research, A. Påhlsson's Foundation and Kock's Foundation. Ms B. Mårtensson and Ms I. Nilsson gave skilful assistance.

References

1. Robinson, W.D.: In Textbook of Rheumatology, p 337-349, [Kelley et al, editors] (1980).

2. Aaseth, J., Munthe, E., Førre, Ø., Steinnes, E.: Scand. J. Rheumatology 7, 237-240 (1978).

ZINC, COPPER, SELENIUM AND GLUTATHIONE PEROXIDASE IN PLASMA AND ERYTHROCYTES OF DOWN'S SYNDROME (TRISOMY 21) PATIENTS. INTERPRETATION OF SOME VARIATIONS

Françoise Vertongen*, Jean Nève**, Philippe Cauchie* and Léopold Molle**.
*Department of Clinical Chemistry, Hôpital Universitaire Saint-Pierre, B-1000 Bruxelles, and **Laboratory of Analytical Chemistry and Toxicology, Free University of Brussels, Campus Plaine 205-1, B-1050 Bruxelles - Belgium.

Introduction

Alterations in oxygen metabolism have been reported in trisomy 21 (Down syndrome, DS); particulaly relevant are the findings of abnormal activities of two erythrocyte enzymes involved in the scavenging of the deleterious oxygen derivatives, namely superoxide dismutase (SOD-1) and glutathione peroxidase (GPx). The elevated activity reported for the former (1 - 4) was attributed to a gene - dosage effect (5) and could account for some cell oxidative damages by generation of an excess of strong oxidants such as hydrogen peroxide (H_2O_2) or hydroxyl radicals ($OH^{\bullet}$) (6).

The level of erythrocyte GPx activity in DS patients however remains a matter of controversy : indeed, normal activities have been claimed by some authors (4,7), elevated activities by others (8,9). Increased activity could reflect a regulation mechanism secondary to accelerated peroxidative processes within DS cells.

Both enzymes contain trace elements in their structure. The erythrocyte SOD-1 molecule contains copper, indispensable for its catalytic activity and zinc more important for its stability but replaceable by other elements (10); selenium plays an essential role in the active center of GPx (11). In animal, a good correlation has been observed between dietary copper but not zinc, and SOD activity on the one hand and between selenium and GPx activity on the other hand (12-14). Both enzyme acti-

Trace Element - Analytical Chemistry in Medicine and Biology, Vol. 3

vities have therefore been studied as index of copper and selenium status in humans. In Down's syndrome, zinc and copper levels have been determined by many authors : plasma levels are in general normal but erythrocyte levels are significantly elevated probably as a result of increased SOD-1 activity (15-18). Selenium was only studied by two groups : Agar and Hingston in Australia found normal levels in whole blood (19) whereas two of us in a study in Paris (France) reported normal levels in erythrocytes but low levels in plasma (16). This finding and the fact that GPx activity in erythrocytes was significantly increased and did strongly correlate with selenium levels could suggest that selenium status of DS patients was inadequate (16).

The purpose of the present study was to assess copper, zinc and selenium status in Belgian DS patients by determining plasma and erythrocyte trace metal levels as well as erythrocyte GPx activities. In order to examine the influence of some factors liable to influence the results, the study was performed in two groups of patients living in different areas in Belgium and the results were compared with those obtained in healthy and non mongoloid mentally retarded subjects living in the same areas.

Patients

Ten DS patients aged 21-57 years living in an urban area (Brussels) formed a first group. Ten other DS patients, 23-57 years, living in a rural area (Meise), 20 km away from Brussels, composed a second study group.

Control subjects were 25 healthy adults, 23-57 years, for the first group and 7 healthy subjects, 24-49 years, and 10 non mongoloid mentally retarded patients, 23-55 years, for the second group. In the latter group, patients and controls took usually their meals together in the same community home.

Methods

Heparinized venous blood was collected in acid - washed tubes. Trace elements levels were determined as previously described by atomic absorption spectrophotometry (16). Results were expressed in µmoles/l for plasma concentrations and in nmoles/g haemoglobin (Hb) for red blood cell concentrations.

Erythrocyte GPx activity was measured as previously described with H_2O_2 as substrate (20). Results were expressed in units (U) (1U = 1 µmole of NADPH oxidized /min/g Hb). Standard haematological parameters were measured on an ELT 800 (Ortho); reticulocyte were counted by the cresyl blue method. Differences between group means were tested by the Student's t test or Cochran's test when the variances were statistically different.

Results

All patients and controls had normal haematological parameters (Hb level, haematocrit, red blood cell and reticulocyte counts).

Results of trace metals concentrations and erythrocyte GPx activities are shown in Table 1.

When compared with their respective controls, DS patients exhibited significantly lower selenium levels in plasma ($p < 0.010$) but not in erythrocytes ($p > 0.050$). Plasma zinc levels were significantly decreased in DS patients from Brussels ($p < 0.010$) but not in DS patients living in Meise. Plasma copper concentrations were similar in all investigated groups. Erythrocyte copper and zinc levels were increased in DS patients from both groups.

Erythrocyte GPx activity was not different in patients and in controls. No correlation between erythrocyte selenium level and corresponding GPx activity could be demonstrated, in any group.

Finally significant differences were detected between the two control groups : this hold true for plasma selenium ($p < 0.005$), erythrocyte selenium ($p < 0.001$), plasma zinc ($p < 0.005$), erythrocyte zinc ($p<0.001$) and erythrocyte GPx ($p < 0.010$).

PLASMA				
		Zn µmoles/l	Cu µmoles/l	Se µmoles/l
Brussels	Trisomy 21	11.9 ± 2.0	17.0 ± 4.4	1.06 ± 0.11
		*	NS	*
	Controls	15.0 ± 3.3	19.7 ± 3.6	1.25 ± 0.23
Meise	Trisomy 21	12.7 ± 1.5	17.1 ± 4.2	0.89 ± 0.09
		NS	NS	*
	All controls	12.2 ± 2.0	17.6 ± 4.4	1.05 ± 0.18

ERYTHROCYTES					
		Zn n moles/l	Cu n moles/l	Se n moles/l	GPx U
Brussels	Trisomy 21	691 ± 72	53.2 ± 4.7	5.76 ± 0.82	6.7 ± 1.6
		***	***	NS	NS
	Controls	441 ± 78	40.1 ± 4.6	6.28 ± 0.89	7.8 ± 1.8
Meise	Trisomy 21	687 ± 73	51.3 ± 10.4	4.48 ± 0.94	6.6 ± 1.9
		***	***	NS	NS
	Controls	538 ± 69	38.0 ± 4.2	5.26 ± 0.93	5.3 ± 1.5

Table 1 : Comparison of zinc, copper, selenium levels and GPx activity in plasma or erythrocytes of DS patients and controls. (NS : non significant ; * : $p < 0.010$; ** : $p < 0.005$; *** : $p < 0.001$

Discussion

Our results confirm normal plasma copper levels in DS patients (16). Decreased zinc levels found in some DS and normal subjects do probably not reflect a true zinc deficiency : it is indeed now well demonstrated that determination of plasma zinc alone is not sufficient to appreciate zinc status as this level is influenced by various non nutritional factors (21). In accordance with other studies, we found that copper and zinc concentrations were significantly increased in erythrocytes of both groups of DS patients.

As regard to selenium and GPx, it is firstly important to state that plasma selenium, erythrocyte selenium and erythrocyte GPx activity do significantly vary in both groups of controls, being higher in the urban area (Brussels) than in the rural one (Meise). The decrease in plasma selenium levels revealed in DS patients also display such a geographical variation : the level of the Meise group is lower than the other one but the relative decrease is similar in both groups. Moreover, in DS patients, erythrocyte selenium content and GPx activity are normal when compared to the controls of the same areas but geographical variations are again displayed.

Low plasma selenium levels in DS, still reported by two of us in the study performed in Paris (16), are not due to an insufficiency of selenium intake : indeed, controls who took the majority of their meals with the patients had higher plasma selenium levels. Moreover this decrease does not affect selenium bioavailability as GPx activity in DS erythrocytes remains normal.

In the study led in Paris, the enzyme activity was found significantly increased; this result was attributed to an enzymatic induction triggered by increased oxidative processes. The differences noted for the GPx activity in the two Belgian groups as well as the lack of significant correlation between selenium and GPx in erythrocytes could be due to a better protection against oxidative damages in our patients than in the French ones. Cell defences against oxygen toxicity are indeed rather complex, including enzymatic reductive processes and activity of membrane antioxidants. In these conditions, it seems that the increase of GPx activity in erythrocytes reported in some groups of DS patients is not characteristic of the disease but better reflects a greater exposure of some of these patients to peroxidative damages.

Conclusions

This study confirms the existence of high erythrocyte copper and zinc as well as low plasma selenium levels in Down's syndrome. The origin of this last decrease remains unclear but seems not due to an insufficient selenium intake; moreover, erythrocyte selenium and glutathione peroxidase activity are not affected. In these conditions and because there is no correlation between selenium and glutathione peroxidase in erythrocytes, it can be concluded that selenium status of the two Belgian DS patients groups remains adequate. Nevertheless, the origin of the modification of erythrocyte GPx activity in other groups of DS patients deserves further examination.

References

1. Sinet, P.M., Allard, D., Lejeune, J., Jerome, H.: C.R. Acad. Sci. Paris 278, 3267-3270 (1974).
2. Sinet, P.M., Michelson, A.M., Bazin, A., Lejeune, J., Jerome, H.: Biochem. Biophys. Res. Commun. 67, 904-909 (1975).
3. Feaster, W.W., Kwok, L.W., Epstein, C.J.: Am.J.Hum. Genet., 29, 563-570 (1977)
4. Brooksbank, B.W.L., Balazs, R.: Lancet i, 881-882 (1983).
5. Sinet, P.M., Couturier, J., Dutrillaux, B., Poissonnier, M., Raoul, M.O., Allard, D., Lejeune, J., Jerome, H.: Exp. Cell. Res. 97, 47-55 (1976).
6. Sinet, P.M.: Ann. N.Y. Acad. Sci. 396, 1-190 (1982).
7. Zakharov, A.F.: In : Theorical Aspects of Medical Genetics, Moscow, Meditsina (1979).

8. Sinet, P.M., Michelson, A.M., Bazin, A., Lejeune, J., Jerome, H.: Biochem. Biophys. Res.Commun., 67, 910-915 (1975).

9. Kedziora, J., Lukaszewics, R., Koter, M., Bartosz, G., Pawlowska, B., Aitkin, D.: Experientia, 38, 543-544 (1982).

10. Fridovich, I.: Adv. Enzymol. 41, 36-97 (1974).

11. Rotruck, J.T., Pope, A.L., Ganther, H.E., Swanson, A.B., Hafeman, D.G., Hoekstra, W.G.: Science 179, 588-590 (1973).

12. Bohnenkamp, W., Weser, U.: Biochim. Biophys. 444, 396-406 (1976).

13. Bettger, W.J., Savage, J.E., O'Dell, B.L.: Nutr. Rep. Int. 19, 893-900 (1979).

14. Bettger, W.J., Fish, T.J., O'Dell, B.L.: Proc. Soc. Exp. Biol. Med. 158, 279-282 (1978).

15. Mc Bean, L.D., Smith, J.C., Berne, B.H., Halsted, J.A.: Clin. Chim. Acta, 50, 43-51 (1974).

16. Neve, J., Sinet, P.M., Molle, L., Nicole, A.: Clin. Chim. Acta, 133, 209-214 (1983).

17. Halsted, J.A., Smith, J.C.: Lancet, i, 322-324 (1970).

18. Bjorksten, B., Back, O., Gustavson, K.H., Hallmans, G., Hagglof, B., Tarnvik, A.: Acta Paediatr. Scand. 69, 183-187 (1980)

19. Agar, N.S., Hingston, J.: Med.J.Australia, 556, (1980).

20. Vertongen, F., Heyder-Bruckner, C., Fondu, P., Mandelbaum, I.M.: Clin. Chim. Acta 116, 217-222 (1981).

21. Helgeland, K., Haider, T., Jonsen, J.: Scand. J. Clin. Lab. Invest., 42, 35-41 (1982).

TRACE ELEMENTS IN HUMAN PANCREAS

Minoru Terai

Department of Chemistry, Tokyo Metropolitan University, Hukasawa, Setagaya, Tokyo, 158, JAPAN

Michio Matsumoto, Hujihika Suzuki

School of Medicine, Juntendo University, Bunkyo, Tokyo, 113,

Yoshiyuki Tanizaki

Tokyo Metropolitan Radioisotope Researh Center, Hukasawa, Tokyo, 158, JAPAN

Masae Yukawa

National Institute of Radiological Science, Chiba, 226, JAPAN

Introduction

A pancreas is one of the most important organ because it functions to keep a normal state of organisms. In recent year, there are many reports about enzymes and hormones secreted by the pancreas, but it is hardly found reports(1, 2, 3, 4) on trace elements in the pancreas that is expected some important physiological functions. In this investigation, the trace elements in the pancreas were determined and a distribution profile of the trace elements was examined for three parts such as a head, a body and a tail of normal(it means not disease) pancreas and pancreatic cancer, and pancreatic calculi.

Trace Element - Analytical Chemistry in Medicine and Biology, Vol. 3

Experimental

Sample. Pancreatic samples were cut out of three parts, such as the head, the body and the tail, of the normal pancreas without the pancreatic disease, two parts such as carcinoma of the pancreatic head or body, and other part, and pancreatic calculi with chronic pancreatitis at 55 cases of pathological dissection. All samples were dipped in diluted formalin before drying. Each sample was dried up in a disiccator until a constant weight was reached.
Method. The sample pulverized was analyzed by neutron activa-analysis. The sample was irradiated in a rotary specimen rack in the TRIGA MARK II reactor (Rikkyo University, Yokosuka) with a thermal neutron flux of $5 \times 10^{-11} n \cdot cm^{-2} \cdot sec^{-1}$. The irradiation periods were 3 min for short-lived nuclides and 24 hours(intermittent irradiation of 6 hr a day for 6 days) for long-lived nuclides. The determination of γ-ray spectra was carried by Ge(Li)γ-ray detector with 4096 channel pulse height analyzer(Type 8100:Camberra Industries Inc. USA) 3 min for short-lived nuclides and 50 min for long-lived nuclides.

Results and discussion

The data have been analyzed by evaluating the mean values and the standard deviations for 4 elemenrs in each tissue, after rejection of the widely different values according to the method of Smirnoff, the results are shown in Table 1-3. The high content of zinc is the characteristic of the pancreas. The contents of zinc in the other tissues such as the brain, the kidney, the liver, the lung, and the heart are in the range of about 80 to 140 ppm/dry wt.(5). The content of zinc in the pancreas is higher than that of the other tissues on condition that all tissue is not pancreatic disorder. The content of zinc,however, is remarkably decreased in the

Table 1. Contents of elements in normal pancreas.

Elements	Cr	Co	Zn	Se
No.	55	55	55	55
Head	2.3±2.0	0.36±0.27	190±120	3.7±2.3
Body	1.8±0.7	0.35±0.19	210± 91	3.6±2.3
Tail	1.6±1.2	0.36±0.24	190±110	3.6±2.1

(ppm/dry wt.)

Table 2. Contents of elements in carcinoma of pancreatic head and around normal pancreas.

Elements	Cr	Co	Zn	Se
No.	2	2	2	2
Normal	0.96±0.38	0.18±0.03	23± 8	1.1±0.5
Cancer	0.49±0.30	0.33±0.19	20± 5	2.1±0.9

(ppm/dry wt.)

Table 3. Contents of elements in carcinoma of pancreatic body and tail and around normal pancreas.

Elements	Cr	Co	Zn	Se
No.	4	4	4	4
Normal	2.8±2.7	0.37±0.25	76±23	2.4±1.3
Cancer	1.4±0.6	0.32+0.09	74±28	2.9±2.3

(ppm/dry wt.)

carcinoma of the pancreatic head and the normal pancreas around the cancer. The contents of chromium and selenium are decreased in the carcinoma of the pancreatic head, too. The content of zinc in the carcinoma of the pancreatic body or tail is slightly decreased not so much as the case of the carcinoma of the pancreatic head. The contents of chromium and selenium is also decresed in the carcinoma of the pancreatic body or tail, but it is not so remarkable as that in the carcinoma of the pancreatic head. At the carcinoma of the pancreatic head, the notable atorophy and fibrosis are often found in the residual tissue of the pancreatic tail. This result agreed with the result of Griffith's report(6). In consequence, the content of zinc is decreased in the pancrea tic body and tail. It is supposed that the decrease of zinc in the carcinoma of pancreas is not a cause of the carcinogenesis but the result after getting cancer. One of author wrote that the concentration of some elements is changed at the boundary of the cancer and the tissue(7). Such a result can be anticipated on the physiological grounds. However, the quantitative experiment did not examine on the condition of good control.

The concentration of the elements in the pancreatic calculi are shown in Table 4. The data of calcium did not write in this pater, because these concentrations were very large. That is to say, the quantity of calcium is beyond the limits of the determination by neutron activation analysis. It did not calculate the mean value and the standard deviation of the elemental concentration in the pancreatic calculi. The concentrations of magnesium, manganese, iron, cobalt, cupper and selenium are distributed in a wide range. The concentration of zinc and cobalt in the pancreatic calculi are almost same as that of pancreas. The concentration of the chromium is changed to higher or lower values by the individual sample. Though the concentrations of the some elements are revealed for the pancreatic calculi, the relation between the concen-

Table 4. Concentrations of elements in pancreatic calculi.

Case No.	Mg	Cr	Mn	Fe	Co	Zn	Se
1	940	0.20	2.4	96	0.40	180	0.39
2	540	0.31	0.77	78	0.11	240	0.22
3	2000	2.1	4.6	180	1.2	480	0.50
4	1600	4.8	2.1	78	0.39	170	0.27
5	110	0.62	——	280	0.16	170	0.27
6	11000	4.7	7.5	660	1.3	23	3.0
7	90	2.6	——	80	0.30	310	——
8	140	1.6	7.6	——	0.38	300	0.18

(ppm/dry wt.)

tration of the elements and the chronic panreatitis still does not clarify. This research has just been started, now. We have less data about the elemental concentration and the physiological functions, yet. We have to make efforts for a further investigation.

References

1. Parr, R.M., Taylor, D.M.: Biochem. J, 91, 424(1964).
2. Ferr, G.S., Smith, H., Howie, R.A.: J. Clin. Pathol. 21, 8(1968).
3. Liebscher, K., Smith, H.: Arch. Environ. Health, 17, 882 (1968).
4. Koch, H.J, Smith, E.R., McNeely, M.D.: Cancer, 10, 151 (1957).
5. Terai, M., Akabane, A., Matsumoto, M., Hashimoto, K.: Bull. Bhem. Soc. Jpn. 52, 2892(1979).
6. Griffith, K., Wright, E.B, Dormandy, T.L.: Nature, 241 (1973).
7. Terai, M., Akabane, A., Matsumoto, M., Hashimoto, K: Proceedings of the Fourth International Conference of Nuclear Methods in Environmental and Energy Research, 1980, 427-436.

DETERMINATION OF SELENIUM IN HUMAN TISSUE SAMPLES USING GRAPHITE FURNACE ATOMIC ABSORPTION SPECTROMETRY BASED ON ZEEMAN EFFECT BACKGROUND CORRECTION.

Ole Ringdal, Kaare Julshamn

Institute of Nutrition, Directorate of Fisheries
N-5013 Bergen, Norway

Knut-Jan Andersen

Section for Clinical Research and Moleculare Medicine
Medical Department A, University of Bergen, Bergen, Norway

Einar Svendsen

Department of Pathology, The Gade Institute,
University of Bergen, Bergen, Norway

Introduction

Selenium is an essential trace element, but is toxic when present in high amounts in the diet (1). Great variations in the geophysical occurrence of the element is well established and has also been reflected in age-spesific death rates for cardiovascular, cerebrovascular, coronary, ishemic and hypertensive heart diseases (2, 3, 4).
The interest in selenium as a biological active element has increased rapidly due to recent developments in advanced analytical techniques (5). Among these techniques the graphite furnace atomic absorption (GFAA) offers speed and convenience when many samples are to be analyzed. On the other hand the analysis may suffer from different kinds of interferences (6,7). The discussion in the literature has emphasized the efficiency of deuterium background correction and matrix modifiers to remove effects of spectral interference. The Zeeman background corrector seems to reduce the interference appearing in the

Trace Element - Analytical Chemistry in Medicine and Biology, Vol. 3

atomizer (8). The novel background correction measures the background at the exact or very close to the analytical wavelength of the element and possible spectral interferences thus are eliminated. In addition, matrix modifiers, L'vovs platform and maximum power heating (MPH) tend to stabilize selenium in the graphite furnace thereby reducing loss and possible chemical interferences (9).

The purpose of this study was therefore twofold, firstly to optimize the determination of selenium using Zeeman Atomic Absorption Spectrometry (ZAA) and secondly to determine selenium content in several human tissues.

Experimental

Apparatus

All tests were performed on a Perkin-Elmer Zeeman/5000 Atomic Absorption Spectrophotometer equipped with a HGA-400 Graphite Furnace System and an AS-40 autosampler. The lamp, a Selenium EDL, was run in the modulated mode at 5W. All experiments were performed at 196.0 nm with 2 nm spectral slit width. Peak absorbance and integrated absorbance were calculated and displayed by a Perkin-Elmer Data System 10. Conventional graphite tubes and pyrolytically coated tubes with solid pyrolytic graphite platforms were compared. Argon with 99.99% purity was used. The optimal Zeeman furnace conditions are summarized in Table 1.

Reagents

Distilled, deionized water was used for the preparation of reagents and standard solutions. Selenium (IV) stock solution was prepared from Titrisol-solution (Merck) containing 1.000 g selenium to 1.000 l of water. The acids were of suprapure quality. Nickel matrix modifier was prepared from analytical reagent grade nickel(II)nitrate. Solutions of iron, calcium and phosphate were prepared from analytical reagent grade chemicals.

Biological material

Tissue samples (10-20 g) from human (n=12) liver, kidney, brain, heart, pancreas and spleen were obtained at autopsy within 24 hr and stored at -20°C. The frozen samples were then freeze-dried to constant weight, homogenized and stored in plastic bottles until analyzed. Tissues were carefully sampled from the same region of each organ examined.
The validity of the method was tested by using certified NBS-material, Bovine Liver (SRM 1577). A biological material relative high in selenium content (Crustacean) was used in the Charring and Atomization studies.

Experimental procedures

Freeze-dried samples (100 mg) were wet digested in a closed system using 2 ml conc. $HNO_3/HClO_4$ (9:1) as described by Julshamn et al. (7). After digestion the samples were made up to 10 ml with water.
Aliquots (0.9 ml) of digested samples or selenite standards (40 μgL^{-1}) in 1% (v/v) HNO_3-solutions were added to sample-cups together with 0.1 ml of a nickel-solution of appropriate concentration. Tissue digests were dispensed in triplicate to allow the use of the standard addition method. The total volume injected for each determination was adjusted to 20 µl.

Studies of matrix effects

The effect on the absorbance of a 0.8 ng selenite standard of increasing amounts of nitric and perchloric acid residues, and such ions as iron, calcium and ortophosphate were studied. The temperature program used is given in Table 1.

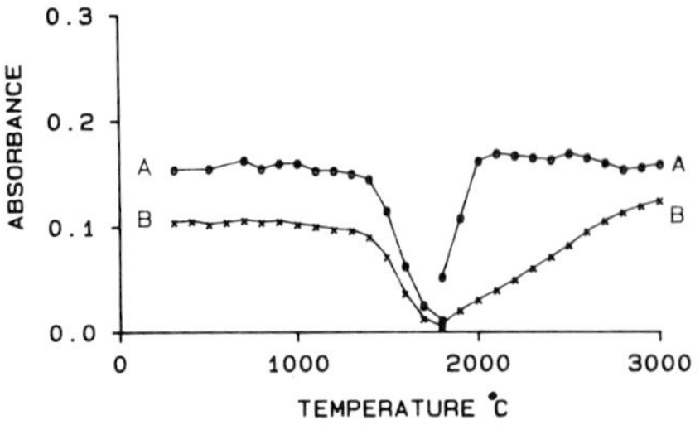

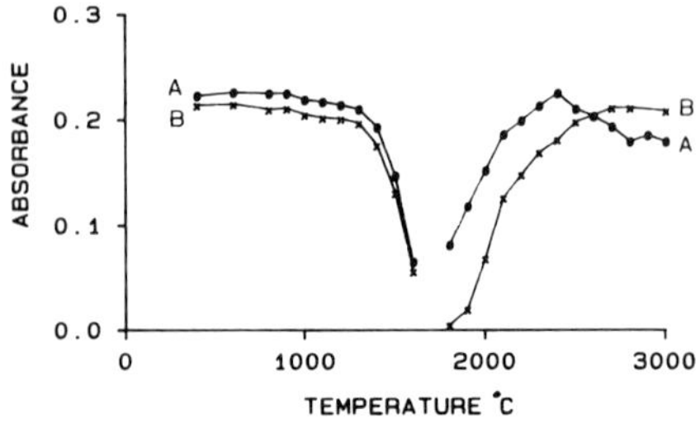

Figure 1. Figure 2.

Char and atomization curves for a selenite standard (40 μL^{-1}) in a conventional graphite tube (Fig. 1) and a pyrolytical coated graphite tube with a L'vov platform (Fig. 2). Curve A using Maximum Power Heating mode and curve B using Ramp Atomization of 1 sec.

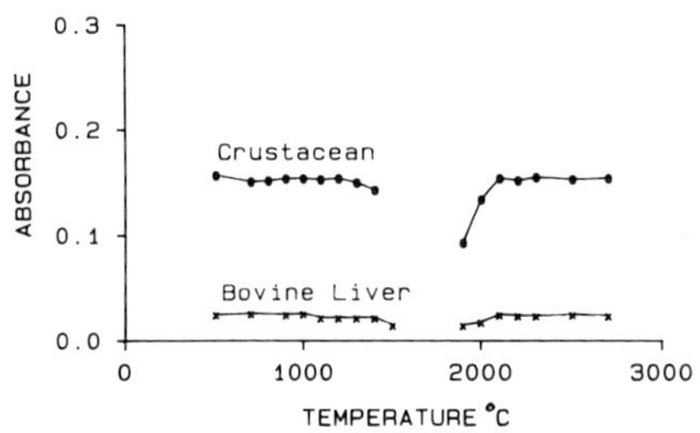

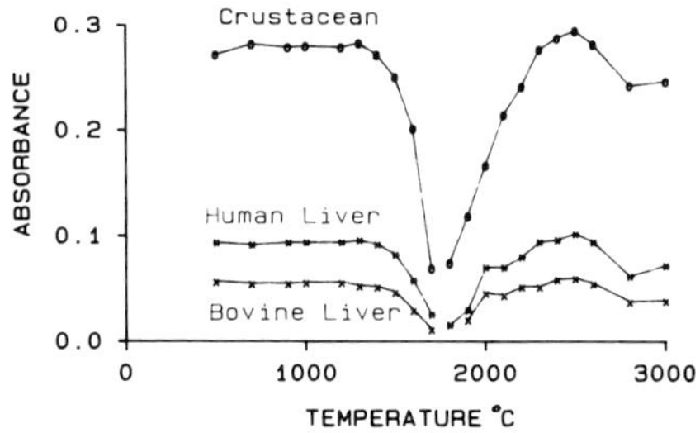

Figure 3. Figure 4.

Char and atomization curves for different biological matrices in a conventional graphite tube (Fig. 3) and a pyrolytical coated graphite tube with a L'vov platform (Fig. 4). MPH was used.

Results and Discussion

Selenium is among the most volatile elements and is typically lost at temperatures above 300°C unless special precautions are taken. Ediger (10) proposed that addition of nickel could stabilize selenium up to thermal pretemperatures of 1200°C. The stabilizing effect of adding increasing amounts of nickel, as a matrix modifier, on the signal of selenium from a standard and a sample solution has been studied. From these experiments a nickel concentration of 0.25% (w/v) was chosen in all analyses.

Charring and atomization study

The results of a charring and an atomization study comparing conventional graphite tubes and pyrolytical graphite tubes with the L'vovs platform are shown in Fig. 1-4. Absorbance profiles obtained using a selenite standard (40 μgL^{-1}) are shown in Fig. 1 and 2, while the corresponding profiles obtained for biological matrices are shown in Fig. 3 and 4. The results shown in Fig. 1 and 3 were obtained using conventional graphite tubes, while pyrolytical graphite tubes with platform were used for the studies shown in Fig. 2 and 4.

Little selenium was lost, when 50 µg nickel (20 µl of 0.25% w/v) was added to the standard solution until the char temperature exceeded 1000°C for either temperature settings (Fig. 1 and 2). Maximum power heating increased the absorbance by approx. 50% as compared to ramp atomization when conventional graphite tubes were used (Fig. 1). No such difference could be obtained with the use of a platform. As shown in Fig. 1, a stabilized selenium signal was obtained already at an atomization temperature of 2100°C (Curve A). Ramp atomization (Curve B) showed no such stabilizing effect on selenium. Fig. 2 shows the stabilizing effect on the selenium signals at an atomization temperature of 2400°C using MPH and 2600°C using ramp atomization. A decrease of approximately 15% in the absorbance was found when the atomization temperature increased above 2500°C using MPH.

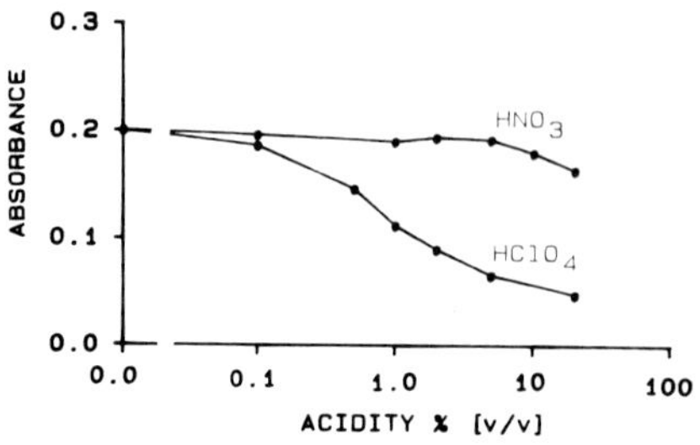

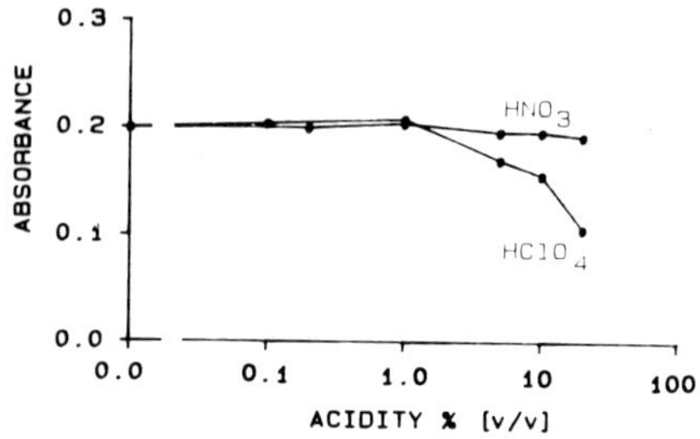

Figure 5. Figure 6.

Interference on the absorbance from a selenite standard ($40\ \mu g L^{-1}$) from nitric and perchloric acid using a conventional graphite tube (Fig. 5) and a pyrolytical coated graphite tube with a L'vov platform (Fig. 6). MPH was used.

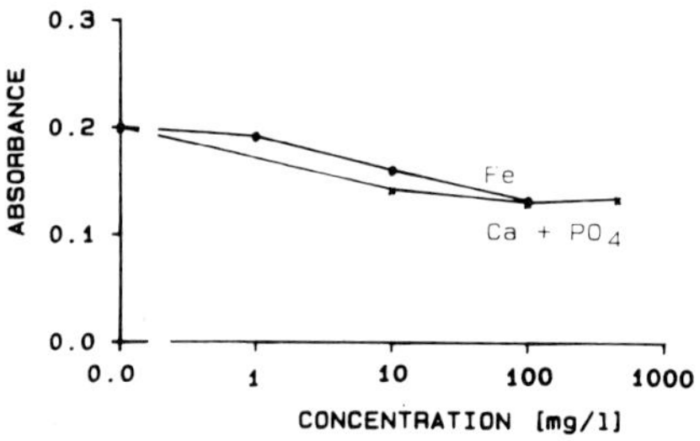

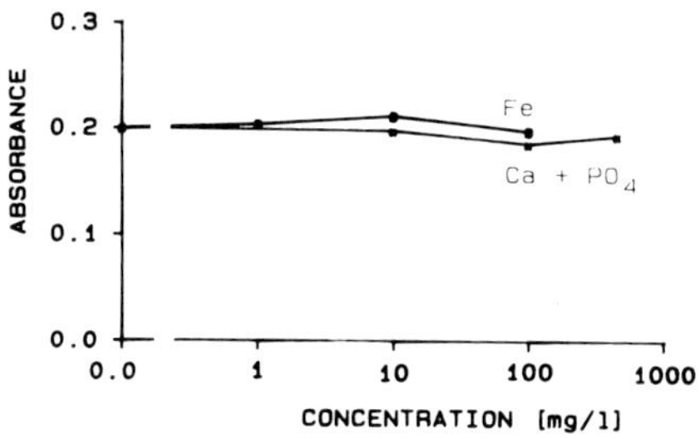

Figure 7. Figure 8.

Interference on the absorbance from a selenite standard ($40\ \mu g L^{-1}$) from iron and calcium-phosphate using a conventional graphite tube (Fig. 7) and a pyrolytical coated graphite tube with a L'vov platform (Fig. 8). MPH was used.

A profile similar to that in Fig. 2 curve A was found for the biological samples (Fig. 4).
Selenium loss during charring as reported by Welch et al. (11) using selenite standards was not observed in the present study. Temperature settings, chosen as optimal from the absorbance profiles displayed in Fig. 1-4 are summerized in Table 1.

Table 1. Zeeman Graphite Furnace Conditions. Programme 1 shows atomization from the graphite wall, whereas Programme 2 shows atomization from the L'vov platform.

	DRY		CHAR		ATOMIZE		CLEAN OUT
	Pr.1	Pr.2	Pr.1	Pr.2	Pr.1	Pr.2	
Temp., C^{o}	120	120	1000	1200	2200	2450	2700
Ramp, sec.	15	15	15	15	0	0	1
Hold, sec.	15	15	15	15	3	3	1
Int.gas flow (mL min^{-1})	50	50	50	50	0	0	50
Record, sec.	-	-	29	29	+	+	-
Read, sec.	-	-	-	-	+	+	-

Matrix effects

The suppressive effect of perchloric acid on the absorbance of trace elements in graphite furnace AAS is well established (7, 12). Organic samples are often digested and oxidized in a mixture of nitric and perchloric acid. It is difficult to ascertain the exact acid concentration in the final digests after the long boiling times used in the digestion procedure. An attempt was made in this study to determine the levels of nitric and perchloric acid which could be tolerated using Zeeman background correction. The interference from nitric and perchloric acid on the absorbance of a selenite standard using Zeeman background correction are shown in Fig. 5 and 6.
Little interference from nitric acid was found up to 5% (v/v) using conventional tubes (Fig. 5) and 10% (v/v) using pyro-

lytical tubes with platform (Fig. 6). The suppressive effect of perchloric acid on the signal was substantial even at 0.1% when conventional tubes were used (Fig. 5), while no interference was observed for concentrations below 1% using pyrolytical tubes with platform (Fig. 6). One may therefore conclude that although pyrolytic graphite tubes with platform tolerates more perchloric acid, the interference is still severe and therefore deserves further attention.
It is known that phosphate and iron provides a spectral interference when determining selenium at its primary resonance line at 196.0 nm (6, 13). Minor amounts of these substances do not seem to produce a negative signal, even with conventional deuterium background correction, when nickel is used as a matrix modifier (14). The interference from iron and calcium ortophosphate on the absorbance of a selenite standard using Zeeman background correction is shown in Fig. 7 and 8. While both iron and calcium ortophosphate were found to give a drop in sensitivity, even at concentrations below 5 mgL^{-1}, when conventional tubes were used (Fig. 7), no interference could be detected even at concentrations of 100 mgL^{-L}, when pyrolytic platform tubes were used (Fig. 8). This effect was surprising and needs further studies.

Quality of the method

Table 2 shows the reproducibility, detection limit and the recovery of the method used. The recovery of the digestion procedure was as reported by Julshamn et al. (7). The accuracy obtained on a NBS Bovine Liver sample was in good agreement with the certified NBS-value. The reproducibility was tested with an internal biological standard (Crustacean) with a selenium content of approximately 5 $mgkg^{-1}$ dry weight.

Table 2. Reproducibility, detection limit and recovery of the method.

Parameter	Conv. tubes	Pyr.coat./ platform
Reproducibility (10 injections)	2.5%	1.1%
" (10 replicates), RSD	7.0%	6.2%
[a]Detection limit, µg/l digest	0.12	0.18
" " , µg/kg dry matter	13	20
Recovery (7)	100%	100%
Referance material, Bovine liver NBS SRM1577, 1.1±0.1 µg/g	1.11±0.09	1.16±0.09

a) Sum of the variance of a blank and a sample with low content

Table 3. Selenium content in different human tissues.

TISSUE		DRY WEIGHT		WET WEIGHT	
	(N)	mg/kg	RSD (%)	mg/kg	RSD(%)
KIDNEY, cortex	10	3.77±0.50	13.3	0.53±0.13	24.5
" , medulla	12	1.67±0.28	16.8	0.23±0.08	34.8
LIVER	11	1.94±0.55	28.4	0.41±0.07	17.1
SPLEEN	12	1.36±0.22	16.2	0.24±0.07	29.2
PANCREAS	11	0.54±0.14	25.9	0.17±0.06	35.3
HEART	12	1.20±0.18	15.0	0.20±0.05	25.0
BRAIN, gray matter	3	1.15±0.11	9.6	0.13±0.05	38.5
" , white matter	4	0.53±0.08	15.1	0.11±0.06	54.5

Determination of autopsy material

Table 3 shows pilot data from our institutes on the selenium content in different human tissues. The data given were averaged without considering the clinical history of the patients and therefore should not be compared with controls or normal values reported by others. Differences in sample preparation and the analytical methods used must also be considered. Nevertheless, the tissue distribution shown in table 3 agrees with previous studies demonstrating that maximum levels occur in kidney and liver, lower ones being found in spleen, myocardium and brain (1).

Acknowledgments

The authors thank the Medical Research Center, University of Bergen, for providing the analytical apparatus used in this study.

References

1. Tsalev, D.L., Zaprianov, Z.K.: Atomic Absorption Spectrometry in Occupational and Environmental Health Practice. Vol. 1, CRC Press, Inc., Boca Raton, Florida, 182-187 (1983).
2. Shamberger, R.J., Tytko, S.A., Willis, C.E.: Trace Substances in Environmental Health, IX, Hemphill D.D., ed., University of Missouri, Columbia, Mo., 15-22 (1975).
3. Shamberger, R.J., Gunsch, M.S., Willis, C.E., McCormack, L.J.: Trace Substances in Environmental Health, XII, Hemphill D.D., ed., University of Missouri, Columbia, Mo., 48-52 (1978).
4. Salonen, J.T., Alfthan, G., Pikkarainen, J., Huttunen, J.K., Puska, P.: Lancet ii, 175-179 (1982).
5. Raptis, S.E., Kaiser, G., Tölg, G.: Fresenius Z. Anal. Chem. 316, 105-123 (1983).
6. Manning, D.C.: At. Absorpt. Newsl. 17, 107-108 (1978).
7. Julshamn, K., Ringdal, O., Slinning, K-E., Braekkan, O.R.: Spectrochim. Acta 37B, 473-482 (1982).
8. Fernandez, F.J., Myers, S.A., Slavin, W.: Anal. Chem. 52, 741-746 (1980).
9. Slavin, W., Manning, D.C., Carnrick, G.R.: At. Spectrosc. 2, 137-145 (1981).
10. Ediger, R.D.: At. Absorpt. Newsl. 14, 127-130 (1975).
11. Welz, B., Schlemmer, G., Voellkopf, U.: Spectrochim. Acta 39B, 501-510 (1984).

12. Julshamn, K.: At. Absorpt. Newsl. 16, 149-150 (1977).

13. Saeed, K., Thomassen, Y.: Anal. Chim. Acta 130, 281-287 (1981).

14. Carnrick, G.R., Manning, D.C., Slavin, W.: Analyst 108, 1297-1312 (1983).

AN INTERLABORATORY SURVEY OF THE DETERMINATION OF SELENIUM IN BLOOD: A PRELIMINARY REPORT

Tee-Siaw Koh,

Veterinary Sciences Division,
South Australian Department of Agriculture,
c/- I. M. V. S.,
Box 14, Rundle Mall Post Office,
Adelaide, S A 5000,
Australia.

Introduction

In the past decade, an upsurge in the interest in selenium (Se) as a trace element has led to a proliferation of analytical methods for Se determination. This situation raises concern about the comparability of Se results obtained by different methods and by different laboratories. Recently Kumpulainen and Koivistoinen (1981) undertook an interlaboratory comparision of Se determination in serum among 11 laboratories. These authors found that further interlaboratory comparisions were necessary to achieve accuracy in the determination of Se in serum.

This paper reports the preliminary findings of an international interlaboratory survey involving 40 laboratories from 13 countries on the determination of Se in blood.

Trace Element - Analytical Chemistry in Medicine and Biology, Vol. 3

Methods

Preparation of Samples

Eight blood samples, with EDTA as an anticoagulant, were obtained from cows of different Se status. The Se content of the 8 samples ranged from about 0.2 umol/L (0.016 ug/mL) to 14 umol/L (1.1 ug/mL). To ensure that the participants would be determining metabolically incorporated Se, no Se salts were added to the blood samples. The 8 samples were filtered through a 125 micron nylon screen, homogenised and subsampled (ca. 2.8 mL) into 3 mL polypropylene vials. To minimise loss of Se by adsorption onto the walls of the vials, the blood samples in the vials were frozen and thawed 3 times over a 3-week period. Thereafter the blood was again pooled, filtered and re-dispensed into the conditioned vials. All blood samples were sterilised by gamma irradiation with 2.5 Mrad of ^{60}Co.

Homogeneity tests for Se on each of the 8 blood samples indicated coefficients of variation (CV) of less than 2% (the number of determinations for each blood sample was 20).

Invitation to Participate

Invitations to participate in the survey were sent to laboratories involved in Se research. Since the object of the survey was not an evaluation of a method but of performance, no discrimination in the selection of laboratories was exercised. The blood samples, together with instructions and a report form, were sent to 60 laboratories which agreed to participate.

Results and Discussion

Of the 60 laboratories that agreed to participate, 40 returned results, viz., Australia (6 laboratories), Belgium (2), Canada (1), China (5), Deutsche Demokratische Republik (1), Federal Republic of Germany (1), Finland (7), New Zealand (5), Norway (1), Poland (1), Sweden (1), United Kingdom (2) and United States (7).

The findings and analytical methods of the participants are summarised in Table 1.

Six methods were used by the participants. Of the 40 participants 24 used fluorometry, 11 used hydride generation, 2 used furnace AAS, 2 used gas chromatography, 1 used X-ray fluorescence and 1 used neutron activation analysis.

For all samples, except sample RM1-Se-C, the interlaboratory coefficients of variation were greater than 13% (Table 1). The interlaboratory performance was of particular concern at lower Se levels, viz. RM1-Se-A (CV=32.4%) and Adelaide-2 (CV=57.3%). One laboratory using a hydride generation method reported an indefinite value of <0.10 umol Se/L for sample Adelaide-2 while another using an X-ray fluorescence method reported that Se was undetectable in sample RM1-Se-A. These two results were not included in the statistical analysis of the overall results.

When the criterion of excluding values outside 3 standard deviations (SD) as outliers was employed, no outliers were detected in the results for samples Adelaide-3, RM1-Se-A, and RM1-Se-B. However, for the other samples, a number of outliers were detected. Excluding the outliers had little effect on

Table 1: Determination of Selenium in Whole Blood (umol/L)*: Analytical Methods and Results

Method		Adelaide-1	Adelaide-2	Adelaide-3	RM1-Se-A	RM1-Se-B	RM1-Se-C	RM1-Se-D	RM1-Se-E
		Blood Sample							
All Methods:	Mean	1.08	0.20	14.25	0.23	0.54	1.46	2.67	7.08
	Range	0.86-1.65	0.09-0.58	9.18-18.00	0.10-0.46	0.39-0.86	0.95-1.71	0.61-3.33	2.91-9.32
	CV, %	16.6	57.3	13.4	32.4	16.9	9.5	17.7	17.3
	No. of Labs.	40	36	40	40	40	40	40	40
Fluorometry:	Mean	1.05	0.20	14.34	0.24	0.55	1.47	2.73	7.12
	Range	0.86-1.39	0.09-0.58	10.73-18.00	0.13-0.46	0.42-0.76	1.27-1.71	2.03-3.29	5.03-9.23
	CV, %	12.7	56.3	13.4	30.8	15.6	7.4	12.0	13.8
	No. of Labs.	24	21	24	24	24	24	24	24
Hydride Generation:	Mean	1.07	0.16	14.82	0.19	0.52	1.49	2.78	7.63
	Range	0.89-1.29	<0.10-0.25	12.67-17.10	0.10-0.25	0.46-0.60	1.37-1.68	1.90-3.33	5.59-9.32
	CV, %	12.7	31.8	10.8	21.0	8.4	8.0	14.6	12.6
	No. of Labs.	11	9	11	10	11	11	11	11
Furnace A A S#		1.65/0.89	0.38/0.30	12.66/15.58	0.18/0.39	0.43/0.54	1.39/1.65	2.41/3.04	6.46/6.97
Gas Chromatography		0.99/1.17	0.20/0.10	9.18/15.32	0.32/0.15	0.42/0.48	0.95/1.39	0.61/2.79	2.91/7.85
X-Ray Fluorescence		1.58	0.57	11.78	undetected	0.86	1.37	2.37	4.94
Neutron A A		0.95	0.14	13.30	0.19	0.46	1.32	2.48	6.34

* To convert umol Se/L to ug Se/L, multiply values by 0.07896

\# Owing to the small number of laboratories using Furnace AAS, Gas Chromatography, X-Ray Fluorescence and Neutron Activation Analysis, only individual results are given.

the mean values but the reduction in the interlaboratory CV was marked for Adelaide-2, from 57% to 28%.

Although the hydride generation methods appear to give lower CV values than the fluorometric methods, the results from both methods are comparable (Table 1). The number of laboratories using the other 4 methods was too small to permit statistical analysis.

Variable detection limits were reported by the participants, even among those using similar techniques. Some laboratories reported the values of the detection limit in concentration units, others in absolute values. This is probably due to participants' preference in their way of quantitating Se standards for calibration. The range of detection limits for each method is shown in Table 2.

Table 2: Range of detection limits*

		Concentration (umol Se/L)	Absolute (umole Se)
Fluorometry	:	0.0004-0.13	0.004-0.25
Hydride Generation	:	0.0008-0.19	0.03-0.06
Furnace AAS	:	0.006/0.002	-
Gas Chromatog.	:	0.013	0.00006
X-Ray Fluoro.	:	4.0	-
Neutron AA	:	-	0.003

* To convert values to ug Se/mL or ug Se, multiply by 0.07896

In response to questions on quality control, 50% of the laboratories indicated not having included internal quality control materials for Se determination while 47.5% of the laboratories reported not using any certified standard reference materials for external quality control. The SRM Bovine Liver produced by the National Bureau of Standards, U.S.A. was the most common standard reference material used by the participants.

In trace-element research it is desirable that analytical results be accurate. This survey raises concern about the discrepancies in the Se results obtained by the participants. It appears that the most likely cause of the unsatisfactory performance is due to the failure to implement effective quality control schemes. To improve analytical performance it is essential that reference blood samples of known Se content be available.

Acknowledgements

I appreciate the co-operation of the 40 participants for their contribution towards this report. I am grateful to the staff of the Biochemistry Section, Veterinary Sciences Division for discussion and technical assistance.

Reference

Kumpulainen, J and Koivistoinen, P : Kemia-Kemi, 8, 372-373 (1981).

ACCURACY OF THE SELENIUM DETERMINATION IN HUMAN BODY FLUIDS USING ATOMIC ABSORPTION SPECTROMETRY

Bernhard Welz, Marianne Melcher, Gerhard Schlemmer

Department of Applied Research, Bodenseewerk Perkin-Elmer & Co GmbH, D-7770 Überlingen, Federal Republic of Germany

Introduction

Selenium has gained substantial interest as a possibly essential trace element; its accurate determination, however, is still a major challenge for the analyst. Recently we participated in an IUPAC interlaboratory trial on the determination of selenium in lyophilized human blood serum. The average values that we obtained were 55 µg/L using graphite furnace AAS, and 71 µg/L using hydride generation AAS, respectively (1). Both results are significantly lower than the accepted mean value of 91 $\pm$ 7 µg/L (2). This has stimulated further work to investigate the sources for possible systematic errors.

For graphite furnace AAS, a strong dependence of pre-atomization losses was found on the valency state of selenium and on the matrix modifier used (3). A copper nitrate/magnesium nitrate mixed modifier turned out to give best results in the presence of various inorganic matrices. In the present work we have investigated the extent to which these findings can be applied to the analysis of human body fluids.

For the hydride-generation technique, acid decomposition is of significant importance for the accuracy of the determination (4). We therefore investigated several digestion procedures and their influence on the selenium concentration found, as well as on the recovery of selenium from acid-resistant organoselenium compounds.

Trace Element - Analytical Chemistry in Medicine and Biology, Vol. 3

Experimental

A Perkin-Elmer Model Zeeman 3030 Atomic Absorption Spectrometer, equipped with an HGA-600 graphite furnace programmer and an AS-60 autosampler was used for all graphite furnace work. Pyrolytically coated graphite tubes with inserted L'vov Platform were used exclusively. A Perkin-Elmer Model 4000 Atomic Absorption Spectrometer, equipped with an MHS-20 hydride system was used for determinations with the hydride-generation technique. A selenium electrodeless discharge lamp, operated with a power of 6 Watts, was used for both instruments. Spectral slit widths of 2 nm and 0.7 nm were used for hydride-generation and graphite furnace AAS, respectively, to separate the 196.0 nm line.

All reagents were of analytical reagent grade or of higher purity, nitric acid was further purified by subboiling distillation. Mixed matrix modifiers were prepared to contain 0.5 g/L copper and 1.0 g/L magnesium in 0.2 % (m/V) or in 10 % (m/V) nitric acid. For the direct analysis of whole blood or serum, the matrix modifier contained in addition 0.15 % (m/V) Triton-X-100. A 10 µL aliquot of this solution corresponds to 5 µg Cu and 10 µg Mg. The Trace-O-Mat used for oxygen combustion as well as the Autoclave-3 for pressure decomposition and the heating block with temperature control for the perchloric acid digestion are described in detail elsewhere (5).

Results and Discussion

1. Graphite Furnace AAS.

In the presence of inorganic salt matrices it could be demonstrated that the addition of a copper nitrate/magnesium nitrate mixed modifier prevents selenium losses during thermal pretreatment (3). A characteristic mass of around 20 pg/0.0044 A.s is obtained under these conditions. A considerably lower sensitivity, however, results when serum, diluted with Triton-X-100, is dispensed directly onto the L'vov Platform in the graphite tube, and the mixed modifier is added on top of the sample.

The characteristic mass is typically around 27 pg/0.0044 A.s under these conditions and the analyte addition technique must be applied for the determination of serum. The average content found for selenium in Seronorm, Batch 103, using this direct analysis was 80 ± 5 µg/L (n = 9) whereas a value of 91 ± 7 µg/L has been accepted for this standard reference material (2). This indicates that the results are still somewhat low for the direct determination under these conditions and we suspect that the organic matrix has an influence on the determination of selenium.

A combustion of lyophilized blood serum in a stream of oxygen where the volatile trace elements are condensed in a cooling system filled with liquid nitrogen was therefore investigated (5, 6). After this combustion, no organic compounds are left and no carbon residues can be expected to remain in the graphite furnace. Using this decomposition technique, a characteristic mass of 27 pg/0.0044 A.s was obtained again, which is the same as for the undigested serum. This indicates that the signal depression is not due to organic but to inorganic constituents of the blood serum. The value found for selenium using this combustion technique, and applying analyte addition, was 93 ± 3 µg/L (n = 6) for Seronorm, Batch 103, which is in good agreement with the accepted value of 91 ± 7 µg/L (2). This leads to the conclusion that after complete destruction of the organic matter there are some residual influences on the atomization efficiency (expressed by the characteristic mass) of selenium by the inorganic constituents of the blood serum. Accurate results, however, can be obtained when the analyte addition technique is applied.

The combustion of lyophilized blood serum in a stream of oxygen cannot be applied for routine purposes. We therefore investigated if the introduction of oxygen into the graphite tube during thermal pretreatment would have the same effect. Serum or whole blood samples are diluted 1+3 with a Triton-X-100/copper nitrate/magnesium nitrate modifier solution. The time temperature program used for heating the graphite furnace is given in table 1. Again, as with the previous procedures, the characteristic mass for sample and references is considerably different, so that the analyte addition technique has to be applied. The results, however, are

Table 1: Time temperature program for the determination of selenium in blood and serum using graphite furnace AAS.

Step	1	2	3	4	5	6	7
Temp °C	90	500	500	1000	2100	2600	20
Ramp (s)	1	10	1	1	0	1	1
Hold (s)	50	30	20	10	4	5	5
Read					X		
Baseline					-1		
Int. Flow mL/min	300		300	300	0	300	300
Int. Alt. O_2 mL/min		300					

Table 2: Selenium in human whole blood using graphite furnace AAS (GF AAS) and hydride-generation AAS (HG AAS). Values in µg/L.

Sample	GF AAS	HG AAS
B.W.1	100	102
B.W.2	98	104
M.M.1	92	95
M.M.2	98	98
S.A.1	92	88
S.A.2	84	82
G.S.1	89	92
G.S.2	84	92

accurate and in good agreement with those obtained by other techniques as shown in table 2 for several whole blood samples. This leads to the conclusion that, to our present knowledge, complete destruction of the organic matter as well as the application of the analyte addition technique is required to obtain accurate results for selenium in human body fluids with graphite furnace AAS. Dilution with Triton-X-100/copper nitrate/ magnesium nitrate and thermal pretreatment with oxygen in the graphite tube were found to be the optimum procedure.

2. Hydride-Generation AAS.

In the hydride-generation technique, by the addition of sodium tetrahydroborate to the acid sample solution, gaseous hydrogen selenide (H_2Se) is formed, stripped from the solution and atomized in a heated quartz cell. For this reaction, it is essential that selenium is present in its tetravalent, ionic form, the selenite. Organic selenium compounds must be decomposed completely for the successful application of this technique.

Many selenium compounds are volatile and can be lost during a not well controlled destruction procedure. On the other hand, very acid-resistant organoselenium compounds, such as selenomethionine, selenocysteine and the trimethylselenonium ion are often not fully decomposed, hence causing low results (4). A digestion with nitric acid is apparently insufficient, and responsible for low selenium values found in earlier work (1).

We therefore investigated several decomposition procedures for their effectiveness to recover selenium from biological materials. Among the procedures were the combustion in a stream of oxygen already described for the graphite furnace, an acid digestion with nitric, sulfuric and perchloric acids at temperatures up to 310 °C in special, long-necked flasks (5), and a nitric acid digestion used in earlier work (1). The results found for six independent decompositions each of Seronorm, Batch 103, are summarized in table 3.

Table 3: Determination of selenium in lyophilized human blood serum, Seronorm, Batch 103. Accepted value: 91 ± 7 µg/L Se (2).

Sample No.	Digestion procedure		
	nitric acid	O_2 combustion	perchloric acid
1	64	79	91
2	64	75	89
3	60	88	95
4	64	73	88
5	68	90	86
6	64	82	86
x	64	81	89
s	2	7	3
RSD	3.7	8.5	3.9

As expected, the nitric acid digestion resulted in the lowest values for selenium due to incomplete destruction of the organic matter and of acid-resistant selenium compounds. The combustion in oxygen resulted in somewhat low values as well, which are, in addition, not very reproducible. This may be due to difficulties associated with the handling of lyophilized serum in the combustion apparatus and/or incomplete combustion. The highest values and a good repeatability were obtained after perchloric acid digestion. Our value of 89 ± 3 µg/L for selenium in Seronorm, Batch 103, is in good agreement with the accepted value of 91 ± 7 µg/L.

A sample of heparinized whole blood, without and with added selenomethionine, was investigated in addition to serum to check the efficiency of the proposed acid digestion procedure (table 4). The decomposition with nitric, sulfuric and perchloric acids in a special, long-necked flask

Table 4: Selenium in heparinized whole blood and recovery of selenomethionine (Se-Me) after different acid digestion procedures. All values in µg/L Se.

Added Se-Me	$HClO_4$ 310 °C direct		$HClO_4$ 310 °C HCl reduction		$HClO_4$ 260 °C direct		$HClO_4$ 260 °C HCl reduction		HNO_3 1+1 pressure, 160 °C		HNO_3 1+1 6 h, 90 °	
	µg/L Se	% recov	µg/L Se	% recov	µg/L Se	% recov	µg/L Se	% recov	µg/L Se	% recov	µg/L Se	% recov
0	103				65		106		92		94	
0	102				62		106		96		88	
0	102				60		104		94		90	
0			102				105		84		92	
0			104				107		96		90	
0			104				105		82		86	
$\bar{x}$	**102**		**103**		**62**		**105**		**91**		**90**	
79	184	104			106	56	179	94	146	70	120	38
79	174	91			108	58	184	100	144	67	120	38
79	180	99			104	53	187	104	156	82	112	28
79			182	100			180	95	160	87	120	38
79			184	102			184	100			114	30
79			186	105			180	95			124	43
$\bar{x}$	179	**97**	184	**102**	106	**56**	182	**97**	152	**77**	118	**35**

(5) was carried out with two different maximum temperatures, 260 °C and 310 °C respectively. An additional heating step to 90 °C for 20 min with 5 M HCl was included after this decomposition to reduce hexavalent selenium, which might have formed under the oxidizing action of perchloric acid, to the tetravalent state.

When a maximum decomposition temperature of 260 °C is applied, a substantial amount of selenium is apparently oxidized to the hexavalent state so that a reduction step is essential for an accurate determination. After a maximum decomposition temperature of 310 °C, however, all selenium is apparently present in the tetravalent state. At this temperature most of the perchloric acid is volatilized, and selenates decompose to selenites. The recovery of added selenomethionine is between 91 and 105 % after a digestion at 310 °C, and also at 260 °C, when followed by a reduction step with hydrochloric acid.

The decomposition with nitric acid was carried out in a boiling water bath (duration 6 hours) as well as under pressure at 160 °C (duration 1 hour). No reduction step is required after this digestion because selenium is not oxidized to its hexavalent state under these conditions. The value found for selenium in the whole blood sample using nitric acid digestion is 10 to 15 % lower than that obtained after perchloric acid digestion. The average recovery of added selenomethionine, however, is 77 % and 35 %, respectively, for the two nitric acid digestion procedures. Nitric acid alone can therefore not be recommended for digestion of blood samples for the determination of selenium with hydride-generation AAS.

Conclusion

The accurate determination of selenium in human body fluids like serum or whole blood apparently requires a destruction of the organic matter for hydride generation as well as for graphite furnace AAS. A nitric, sulfuric and perchloric acids digestion with a maximum temperature of 310 °C was found optimum for the hydride generation technique. Under these conditions, the selenium content found in a lyophilized serum sample agrees well with the results obtained by other workers using different techniques. Selenomethionine added to heparinized whole blood is completely recovered after the proposed digestion procedure.

Analyzing blood serum directly in a graphite furnace with the addition of a copper nitrate/magnesium nitrate matrix modifier results in a substantial loss of sensitivity. In addition, the values are 10 - 15 % low even when the analyte addition technique is used. Accurate values are obtained when the organic matter is destroyed completely. The sensitivity loss, however, persists so that the analyte addition technique has to be applied or calibration must be made against serum or blood standards. The procedure most suitable for routine purposes includes a dilution with Triton-X-100/copper nitrate/magnesium nitrate, and a thermal pretreatment in the graphite tube in the presence of oxygen.

References

1. Welz, B., Melcher, M., Schlemmer, G.: Fresenius Z. Anal. Chem. 316, 271 - 276 (1983).
2. Ihnat, M., Wolynetz, M., Thomassen, Y., Verlinden, M.: Pure Appl. Chem., in preparation (1984).
3. Welz, B., Schlemmer, G., Voellkopf, U.: Spectrochim. Acta 39 B, 501 - 510 (1984).
4. Verlinden, M.: Talanta 29, 875 - 882 (1982).
5. Welz, B., Melcher, M.: Anal. Chem., accepted for publication (1984)
6. Knapp, G., Raptis, S., Kaiser, G., Toelg, G., Schramel, P., Schreiber, B.: Fresenius Z. Anal. Chem. 308, 97 - 103 (1981).
7. Sinemus, H.W., Melcher, M., Welz, B.: At. Spectrosc. 2, 81 - 86 (1981).

D i s c u s s i o n Session III - 2

Subject of discussion

- Se analysis, sample preparation and -pretreatment
- Se pool in the organism
- Se-speciation
- Se exection
- indicators for the Se status
- problems of parenteral nutrition
- changes of Se-bounding in the organismus
- Se-uptake from the soil by mushrooms
- Se in relationship to different diseases

Contributors: Brätter, Ganther, Heilmaier, Machari,
Oster, Plantin, Schelenz, Tölg

CHROMIUM REVISITED

Rita Cornelis*, Bart Wallaeys

Laboratory Analytical Chemistry, Institute for Nuclear Sciences, Rijksuniversiteit Gent, B-9000 Gent, Belgium

* Senior Research Associate, National Fund for Scientific Research

Introduction

Over the past three decades, more than a thousand papers and several books have been published on the biological aspects of chromium. To date such studies must be considered a very mixed bag, as some show a marked impact of Cr while others find none at all. The ignorance of the possible link between Cr and clinical parameters is surprising. The problem is partly that many studies rely upon data without providing criteria satisfactory to ensure their accuracy. Inevitably all this probing will turn up answers to the questions that have been considered, but in the meantime controversy will prevail. "Chromium revisited" is a modest attempt to shed some light on the Cr-issue, unless it only adds to the confusion.

An excellent critical review on the nutritional role of Cr published in 1982 by Dr. Guthrie (1), supports the above notions. More recent publications strengthen the idea that a lot of caution is required when drawing clinical conclusions, based on what seem to be inaccurate analytical data.

Cr in Human Serum

The normal content of Cr in human serum continues to be a bone of contention (2), but a consensus grows between a widening group of researchers to accept a level in the order of 0.1-0.2 $ng.ml^{-1}$. Such values were obtained by two different analytical techniques in four research centres located in Belgium, Finland and U.S.A. They are summarized in table 1. It may be suggested that the contribution of all higher reported serum Cr-values be restricted to their historical impact in the search for the "true" content.

Trace Element - Analytical Chemistry in Medicine and Biology, Vol. 3

TABLE 1 : Normal values of Cr in human serum, at present accepted to be accurate

Reference	Method	$ng.ml^{-1}$	S.D. $ng.ml^{-1}$	range $ng.ml^{-1}$	n° subjects
Vanderlinde et al.(1979)(3)	GFAA	0.075	0.069		14
Kumpulainen et al.(1983)(4)	GFAA	0.12	0.05	0.08-0.20	6
Kayne et al. (1978)(5)	GFAA	0.14			8
Versieck et al. (1978)(6)	NAA	0.160	0.083	0.0382-0.351	20

GFAA : graphite furnace atomic absorption spectrometry
NAA : neutron activation analysis

These apparently inaccurate data should no longer be used as a reference to support clinical observations. Not only contamination during sample collecting and handling were at stake, but also analytical techniques per se lacking the necessary sensitivity and accuracy for those low levels. Criticism that the values of 0.1-0.2 ng $Cr.ml^{-1}$ serum may suffer from losses due to volatilization during sample drying or dry ashing can no longer be sustained. Careful studies by several investigators have proved that no perceptible losses occur (7).
A search through the literature reveals how different research groups try to define the specific role of Cr in disease and how they claim clinical meanings for their observed deviations in serum Cr. As accuracy and precision of the Cr-determinations are fundamental prerequisites, the reported value of the normal serum Cr-content may serve as a gauge for a prudent acceptance of a possible correlation. On this premiss, and considering 0.1-0.2 ng $Cr.ml^{-1}$ serum as the normal value for the time being, a selection of recent publications is briefly reviewed. No retrospection is made as all pioneering contributions refer to normal serum values several orders of magnitude higher than acceptable nowadays.

Serum Cr, glucose metabolism and diabetes

Many researchers try to define a role for Cr in relation to glucose metabolism, insulin action and diabetes.

Table 2 gives a general view of the various Cr-data, which are commented hereafter.

Vanderlinde et al.(3) claim increased serum Cr-levels in diabetics versus healthy subjects. Their 14 insulin requiring diabetics showed a mean serum levels ± S.D. of 0.24±0.17 ng $Cr.ml^{-1}$, which according to the authors was significantly higher (at $p < 0.0025$) than the 0.075±0.069 ng $Cr.ml^{-1}$ of their 14 controls. Taking into account the large S.D. values, it may be desirable to increase the size of both groups sampled as to support this clinical statement with an even better statistical level of significance.

Far from convincing are similar and opposite findings, each based on much higher serum Cr-values for the control group.

Rabinowitz et al.(8) assessed a mean ± S.D. of 1.7±1.4 ng $Cr.ml^{-1}$ plasma for the total of their diabetic group, versus 1.1±0.9 ng $Cr.ml^{-1}$ for their healthy group. The data for the ketosis prone, daily insulin dependent diabetics (n=21), are reported to be significantly different (p=0.02) from the normals and from the other diabetics. The same research group also designed a clinical trial of Cr- and yeast-supplements to study the possible influence on carbohydrate and lipid metabolism in diabetic men (9). No measurable change in fasting plasma Cr-levels were observed, as the mean ± S.D. of 1.7±0.65 ng $Cr.ml^{-1}$ (n=43) remained unaltered during 4 months Cr-supplementation. This consisted either of 150 μg Cr daily as $CrCl_3$ or of 18 μg $Cr.day^{-1}$ present in an extract of autolyzed brewer's yeast, or of 6 μg $Cr.day^{-1}$ as a different brewer's yeast.

Thompson (10) on the contrary reports for his normals a mean ± S.D. of 2.00 ±1.68 ng $Cr.ml^{-1}$ plasma, which decreased to 1.65±0.85 ng $Cr.ml^{-1}$ for his insulin treated diabetics, and describes these figures as statistically different at $p < 0.05$. His non-insulin treated diabetics exhibited no difference to the normals.

Liu and Morris (11) observed statistically increased serum Cr-levels for their hyperglycemics. They also expressed the opinion that the mean first dropped after a glucose load, both in a group of 15 normals and 12 hyperglycemics, but the drop was only significant for the 12 hyperglycemics. The outcome changed, however, as to show a similar increase in the mean serum Cr-level in both groups after 3 months daily supplementation with 5g brewer's

TABLE 2 : Summary of recently published serum Cr-levels for healthy individuals and for diabetics

Reference	ng $Cr.ml^{-1}$ serum or plasma mean ± S.D. healthy subjects	n°	diabetics or hyperglycemics	n°	comments
Vanderlinde et al.,1979 (3)	0.075±0.069	14	0.24±0.17	14	insulin dep. diabetics
Rabinowitz et al.,1980 (8)	1.1±0.9[b]	20	1.7±1.4[b]	46	tot. group diabetics
			2.0±1.4[b]	21	ket.prone, insulin dep.
			1.2±1.0[b]	25	ket.resist., non-obese,+ non ket., obese
Thompson,1980(10)	2.00±1.68[b]	144	1.65±0.85[b]	32	insulin dep.
			2.05±1.16[b]	53	non-insulin dependent
Liu and Morris, 1978 (11)	1.67±1.74[b]	15	1.89±1.25[b]	12	hypergly-cemics
Liu and Abernathy, 1982 (12)	1.02±0.89[b]	20			

[b] S.D. recalculated from SEM and n° of subjects

yeast, containing 4 µg Cr. Considering their actual Cr-data, a suspicion arises that the observed variations are mere artefacts. The women with normal glucose tolerance (n=15) (age 40-75 yr) exhibited 1.67±1.74 ng $Cr.ml^{-1}$ (mean ± S.D.), data which in the light of present knowledge are not acceptable as normal any more.

Lateron, Liu and Abernathy (12) studied serum Cr-levels in 20 young subjects (age 17-22 yr), with normal glucose tolerance, during a 3 h oral glucose tolerance test and blood collections at 30 min intervals. They reported fasting serum levels of 1.02±0.89 ng $Cr.ml^{-1}$ (mean ± S.D.), which did not significantly change throughout the experiment. Half of the group tested appeared to the low insulin secretors with a fasting mean ± S.D. serum Cr-level of 0.80±0.38 $ng.ml^{-1}$ (n=10). These values apparently were lower than the fasting value for the other half of the group, being high insulin secretors, with 1.23 ± 1.17 ng $Cr.ml^{-1}$. This difference appears not to be significant at the $p=0.05$ level. It is highly probable that also the Cr-values quoted in this study carry the burden of inaccuracy. Therefore the claimed correlations between fasting serum Cr-levels with fasting serum insulin levels ($r=0.512$), and further with the percentage of ideal body weight ($r = 0.637$) and with triceps skinfold thickness ($r=0.525$) carry little credibility.

Serum Cr and atherosclerosis

Chromium deficiency has been implicated as an important etiological factor in the pathogenesis of atherosclerotic disease.

Serum Cr in patients with recent and old myocardial infarction have been described by Abraham et al. (13). In their control group they found a mean $\pm$ S.D. of 1.71 ± 0.94 ng $Cr.ml^{-1}$ serum (n=45), which apparently remained unaltered in patients with previously documented infarction as these showed 1.84 ± 1.05 ng $Cr.ml^{-1}$ serum (n=34). In 37 patients with acute myocardial infarction, however, serum levels rose to a mean $\pm$ S.D. of 6.36 ± 3.10 $ng.ml^{-1}$ (n=37) during the first 5 days following the infarct, returning to normal over the next 5 days. It may be advisable to check again this hypothesis in conditions which produce more plausible normal levels.

Serum Cr and renal insufficiency

Patients with terminal renal insufficiency treated with continuous ambulatory peritoneal dialysis (CAPD), have been reported to possess abnormally high Cr-levels. Wallaeys et al. (14) investigated a mean value $\pm$ S.D. of 4.20 ± 1.73 ng $Cr.ml^{-1}$ in the sera of 5 CAPD patients, i.e. circa 26 times higher than their normal mean $\pm$ S.D. of 0.160 ± 0.083 ng $Cr.ml^{-1}$ (6). Additional research in CAPD patients, in hemodialysis patients as well as in uraemic patients on conservative treatment is most desirable.

Serum Cr and age

Abraham et al. (15) studied serum Cr and ageing in a healthy group (n=81).

The mean $\pm$ S.D. serum Cr-level in those under the age of 60 years was 1.6 ± 0.90 ng.ml^{-1} (n=52) whereas in those over the age 60 years it was 1.9 ± 1.05 ng.ml^{-1} (n=29). There was no statistical evidence of any change in serum Cr-levels with increasing age when all the subjects were considered.

Thompson (10) on the contrary claims that there is a statistical fall ($p < 0.01$) in the mean plasma Cr-levels of healthy men with age. They investigated a mean $\pm$ S.D. of 2.30 ± 0.30 ng Cr.ml^{-1} (n=4) for the youngest group, age 10-19 years old, whereas the group aged 80-89 years exhibited 1.10 ± 0.45 ng Cr.ml^{-1} (n=17).

Little evidence flows out of both these publications, as the quoted levels appear abnormally high.

Serum Cr and prothesis

The use of Cr-containing metallic implants is known to induce in-vivo alterations in the Cr-concentrations of the surrounding tissue and may possibly increase the serum Cr-content.

Manthey and Kübler (16) detected a mean $\pm$ S.D. of 2.54 ± 1.62 ng Cr.ml^{-1} serum in 66 patients who underwent cardiac valve replacement. This value widely exceeds their normal limits of 0.43 ± 0.19 ng Cr.ml^{-1} in 40 controls ($p < 0.001$).

Black et al. (17) analysed the serum Cr-content of 15 patients before undergoing conventional total hip replacement and continued their screening for periods up to 6 months postoperatively. Serum Cr-levels were found to rise from a pre-operative value of 0.55 ng Cr.ml^{-1} to a 7.20 ng Cr.ml^{-1} 15 days post-operatively and then to diminish, although still remaining 50% more elevated than the normal mean, by six months.

Serum Cr-level as an indicator for Cr-nutritional status

One of the main questions concerning serum Cr-levels, is to know if they are in equilibrium with body stores and hence constitute a sensitive indicator of Cr-nutritional status. A positive answer, putting aside previous theorems is to be read in an interesting study on the kinetics of $^{51}Cr(III)$ in the human body by Lim et al. (18). Their methodology using intravenous injections of carrier free ^{51}Cr gets around the difficulty of producing accurate Cr-analysis data and is, as such, liable to shed light on the metabolism and nutritional requirements of Cr(III).

It appears as if only a few of the above cited references may contribute to the outcome of serum Cr-levels in certain pathological conditions. Serum chromium is abnormally high in the small group tested of patients undergoing continuous ambulatory peritoneal dialysis, as well as in patients who received Cr-alloy prothesis. Neither the links between serum Cr, glucose metabolism and diabetes nor those between serum atherosclerosis and age can be put into evidence through lack of acceptable analytical data for Cr in serum.

Cr in Urine and the Search for Related Parameters

Most investigators seem to agree that normal urinary Cr-concentrations are ≤ 1 ng.ml^{-1}, which corresponds to daily excretions of approximately 0.3-2.0 µg (1).

Extensive research is going on with a view to seize the biological relevance of urinary Cr-determinations. The interest focusses on the effect of diabetes, Cr-supplementation, glucose loading, exercise ... on Cr-excretion. A selection of publications of the last 3 years is expected to reflect the actual knowledge on this matter.

Urinary Cr, Cr-supplementation and diabetes

An attempt was made to schematize the various findings in 2 tables : table 3 summarizes urinary Cr in diabetes and table 4 urinary Cr after Cr-supplementation. The data are commented with respect to their analytical relevancy.

Vanderlinde et al.(3) report a fasting normal $\pm$ S.D. urinary Cr-excretion of 0.39 ± 0.30 ng Cr/mg creatinine, or, supposing 2.14 g creatinine and 1.5 l urine per day, the recalculated data become 0.56 ± 0.43 ng Cr.ml^{-1} (n=14). Their insulin dependent diabetics excreted 0.85 ± 0.57 ng Cr/mg creatinine or 1.21 ± 0.81 ng Cr.ml^{-1} (n=24), which means a circa 2-fold increase.

Anderson et al.(19) systematically searched for possible correlations of clinical parameters and urinary Cr-excretion. In their double-blind cross-

TABLE 3 : Summary of urinary Cr-values in health and diabetes

Reference	mean Cr ± S.D. $ng.ml^{-1}$ healthy subjects	n°	diabetics	n°	remarks
Vanderlinde et al., 1979 (3)	0.56±0.43[a]	14	1.21±0.81[a]	24	insulin dependent
Kumpulainen et al., 1983 (4)	0.11±0.05	10	0.12±0.04[a]	16	
Ping et al., 1983 (24)	0.72±0.31	23	1.0±0.9	23	age depend. for
	0.39±0.19	19			healthy
Rabinowitz et al., 1983 (8)	2.9±1.8[a]	20	4.6±5.5[a]	21	ketosis prone
			8.0±10.5[a]	7	non ket., non obese
			2.4±1.9[a]	18	non ket., obese
Saner et al., 1983 (25)	0.37±0.20[a]	12	8.8±3.5[a]	14	Turner's syndrome

[a] recalculated data, see text

over study with 76 healthy subjects, the mean ± S.D. Cr-urine content was 0.20±0.09 $ng.ml^{-1}$. After 2 months daily supplementation with 200 µg Cr as $CrCl_3$, the mean ± S.D. urinary Cr-value increased to 1.02±1.05 $ng.ml^{-1}$, which remained the same after 3 months (i.e. 1.13±0.70 $ng.ml^{-1}$). These supplemented subjects did not exhibit an increase in urinary Cr-excretion in response to a glucose load. On the other hand approximately 80% of the non-supplemented subjects showed an increase in urinary Cr-excretion in response to a glucose load. The authors'general conclusion was, however, that urinary Cr-excretion after a glucose challenge was not predictable and dit not depend on Cr-status. The same group of 76 healthy individuals also served to test the effect of Cr-supplementation on serum glucose, insulin and lipid variables (20). When the data of all the subjects were combined, there was no sustained significant effect on fasting serum glucose levels, neither on the levels 90 min after consuming 1 g glucose per kg body weight. Insulin concentration, high density lipoproteins, cholesterol and body weight did not change significantly during Cr-supplementation. These findings plainly fail to support previous allegations on the physiological role of Cr.

TABLE 4 : Urinary Cr after daily Cr-supplementation

Reference	mean Cr ± S.D. $ng.ml^{-1}$	n°	µg Cr supplem.	chemical form
Anderson et al., 1982 (19)	0.20±0.09	76	-	
	1.02±1.05	76	200	$CrCl_3$
Kumpulainen et al., 1983 (4)	0.12±0.04*a	16	-	
	0.19±0.07*a	9	20	$CrCl_3$
	0.89±0.25*a	8	200	$CrCl_3$
Uusitupa et al., 1983 (22)	0.093±0.042*a	10	-	
	0.89±0.38*a	10	200	$CrCl_3$
Rabinowitz et al., 1983 (9)	4.7±3.9*a	43	-	
	5.5±4.5*a	28	150	$CrCl_3$
	6.2±5.5*a	28	18	yeast
	4.6±7.3*a	30	6	yeast
Saner et al., 1981 (25)	8.8±3.5*b	6	-	
	3.4±1.7*b	6	50	yeast

* diabetics or related

[a] recalculated data, supposing 1.5 l urine per day

[b] recalculated data, see text.

Anderson et al. (21) also assessed that the minimum Cr-absorption, calculated from urinary excretion, was about 0.4%. Increasing Cr-intake fivefold by Cr-supplementation led to a nearly fivefold increase in Cr-excretion. The authors concluded that their results suggest that urinary Cr-excretion does not appear to be a meaningful indication of Cr-status, but is a meaningful indicator of Cr-intake.

Kumpulainen et al. (4) assessed a mean ± S.D. Cr-urine value in 10 healthy individuals of 0.11±0.05 ng $Cr.ml^{-1}$. The figures for 16 maturity onset diabetics were 0.17±0.07 µg $Cr.day^{-1}$ (or supposing a 1.5 l urine per day, 0.12±0.04 $ng.ml^{-1}$) i.e. the same concentration as their normals. This research group confirmed the findings by Anderson et al. (21) that 24 h urinary Cr-excretion accurately indicates the dietary Cr-intake, be it in this case an evaluation with samples of maturity onset diabetics. The absorption of

Cr from food was estimated at approximately 0.5%. Uusitupa et al.(22) in their double-blind crossover study of Cr-supplementation of 10 noninsulin dependent diabetics found no significant improvement in glucose tolerance, the insulin response remained unchanged and there was no change in serum lipid levels. This study confirms the observations made by Sherman et al. (23) anno 1968, whereby 3 months Cr-supplementation failed to exert any effect on blood glucose levels in 10 diabetics. At that time, however, this was a lone finding.

Ping et al.(24) claim an age dependancy for Cr in urine as their mean $\pm$ S.D. content was 0.72 ± 0.31 $ng.ml^{-1}$ for 23 healthy men aged 23 to 35 y which decreased to 0.39 ± 0.19 $ng.ml^{-1}$ for 19 healthy men aged 47-69 y. Their 23 diabetic patients, aged 13-79 y, showed a mean $\pm$ S.D. value of 1.0 ± 0.9 $ng.ml^{-1}$ urine. The authors remark how the range covered by their diabetics extends from 0.1-3.2 ng $Cr.ml^{-1}$ and contains many values in excess of 1.3 $ng.ml^{-1}$, never observed for their healthy men.

Rabinowitz et al.(8) also report Cr-urine levels for normals and for diabetics. The mean $\pm$ S.D. for their 20 normals are 4.3 ± 2.7 µg $Cr.day^{-1}$ or supposing 1.5 l urine per day 2.9 ± 1.8 ng $Cr.ml^{-1}$. A significantly higher output is reported for their 7 non-obese diabetics with 11.3 ± 7.9 µg $Cr.day^{-1}$, but not for their 21 ketosis-prone, insulin dependent diabetics, neither for their 18 obese diabetics. In a more recent study (9), the same research group adds information on Cr-supplementation studies, as already described in the paragraph on Cr-serum, glucose tolerance and diabetes. No significant difference on urine Cr-output was observed versus the placebo's at the levels of significance $p < 0.005$ and $p < 0.01$. Their general conclusion was that their results cannot be interpreted as supporting the efficacy of treatment of human diabetes with oral Cr or yeast. Although this might be a right conclusion, caution about the analytical data is mandatory, as their normal urine Cr-values seem too high.

Saner et al.(25) recently studied a group of young patients with Turner's syndrome, a condition noted for its high incidence of diabetes. Their control group possess a Cr/creatinine ratio of 0.26 ± 0.14. Supposing 2.145 g creatinine and 1.5 l urine per day, the recalculated data become 0.37 ± 0.20 ng $Cr.ml^{-1}$ urine. The patients with Turner's syndrome exhibited very high urinary Cr-values of 6.14 ± 2.48 ng $Cr.mg^{-1}$ creatinine or 8.78 ± 3.55 ng Cr.

ml^{-1} urine. After 6 of these patients received a daily oral 50 µg Cr supplementation for 8 weeks, the paired Cr/creatinine ratios decreased, implying a decreased urinary output as a consequence of an increased Cr-intake. This finding is in flat contradiction with the results published by Anderson et al.(19), Kumpulainen et al.(4) and Uusitupa et al.(21). The statement by Saner et al.(25) that Cr-deficiency may have a role in the pathogenesis of the abnormal glucose tolerance tests encountered in patients with Turner's syndrome cannot be withheld on the basis of unconvincing analytical data.

Saner (26) also studied the urinary Cr-excretion during the different stages of pregnancy. The recalculated mean values ± S.D. were 6.82±2.74 ng Cr.ml^{-1} for 32 women in the 1^{rst} trimester of pregnancy, 9.62±3.24 ng.ml^{-1} (n=51) in the 2nd trimester and 11.85±4.15 ng Cr.ml^{-1} (n=62) in the 3rd trimester (recalculated data with a creatine per liter urine as given in publication). Nine pregnant women were subjected to a intravenous glucose tolerance test and in 8 out of the 9 cases the paired results appeared to decrease. The mean ± S.D. Cr-content prior to the test was 9.99±4.46 ng.ml^{-1} and afterwards 7.42±3.14 ng.ml^{-1}, which are, however, not statistically different. All the described values, as well as the relative Cr-increment and depletion are open to many doubts.

Urinary Cr and exercise

Anderson et al.(27) were also interested to learn the effect of exercise (running) on serum glucose, insulin, glucagon and Cr-excretion. Nine male runners were evaluated prior to and 2 h after a 6-mile run. The mean ± S.D. urinary Cr-concentration was increased nearly 5-fold 2 h after running : 0.72±0.63 ng.ml^{-1} 2 h postrun versus 0.15±0.09 ng.ml^{-1} prerun. These data demonstrated that accompanying the exercise-induced changes associated with increased glucose utilization, there was a significant increase in Cr-excretion. These findings add substance to the presumed role of Cr in glucose metabolism.

Speciation of Cr in urine

A few investigators ventured into the speciation of Cr in urine. Mianzhi and Barnes (28) found proof that Cr in human urine exists exclusively as Cr(III).

They measured a Cr(III)-concentration in urine below 1 $ng.ml^{-1}$. This finding is consistent with the view of Minoia et al.(29) who found no detectable traces of Cr(VI) in urine and hence decided that Cr is only present in the trivalent charge (n=26). A completely different opinion is expressed by Shimizu et al.(30), who claim that both Cr(III) and Cr(VI) are present in human urine, respectively for circa 58% and 40%. As the total urine Cr-content, described in this publication, amounts to 5.7 ± 0.65 ng $Cr.ml^{-1}$ (n=7) suspicion arises that the procedure may suffer from a major contamination problem and that the observations are erroneous.

The most reliable research on urinary Cr-content could be summarized as following. The normal mean content varies between 0.2 and 1.0 ng $Cr.ml^{-1}$. Increasing Cr-intake causes a similar increase in urinary excretion. The absorption of Cr from food is about 0.5%. Urinary Cr-data are no gauge for the Cr-status of the individual. After a glucose challenge there is no predictable influence on the Cr-excretion. The data for diabetics are not significantly different from those of normals. Vigorous exercise with healthy men causes a significant increase in Cr-excretion. Cr in urine seems to be solely present as Cr(III).

Chromium and the Glucose Tolerance Factor : Failure to Disclose the Link

The glucose tolerance factor (GTF) is to be described as a biologically active agent implicated in glucose metabolism of animals. Its existence was first postulated by Schwarz and Mertz as far back as 1957. A few years later they declared that GTF was a Cr-complex and since then this association of Cr and GTF-activity persisted (31).

Brewer's yeast has long been regarded as one of the best sources of GTF, but all trials to identify the Cr-GTF complex failed. The latest developments even give way to the unspoken presumption that the presence of Cr in GTF is fortuitous. Kumpulainen et al.(32) were first to express the opinion that the suggestion yeast cells possess a mechanism for the conversion of simple Cr(III) salts into a specific biologically active form was without foundation. By the aid of ^{51}Cr-labelling of yeast and of ^{125}I-insulin, they also demonstrated that ^{51}Cr-compounds were not bound to insulin.

Decisive proof of the annulment of the Cr-GTF-connection was recently given by Haylock et al.(33). They developed an assiduous separation scheme to isolate GTF of brewer's yeast grown on ^{51}Cr and showed that GTF can no longer be regarded as a Cr-complex. They made the reflection how the presence of Cr in the various GTF preparations, as described in the literature, is explicable on the basis of the complexing ability of contaminating Cr(III) ions in the commercial growth media and that with the benefit of hindsight this could have been expected. They did not succeed yet in identifying GTF, but seem determined to unlock the secret.

It was once thought that Cr-supplements needed to be in a biologically active form to obtain a physiological response. Brewer's yeast appeared to be the product of choice, but present findings unsettle this concept, as absorption of inorganic chromium happens to be the same (1). Anderson and Bryden (34) recently also demonstrated through measurement of the urinary Cr-content, that the absorption of Cr from beer is similar to that from Cr endogenously in foods. At the same time they concluded that beer ingredients exhibit no insulin potentiating activity.

Research Needs

This anecdotical citation of papers is meant to show how unsatisfactory the many attempts have been when it comes to define the physiological role of chromium as determined in serum and urine. Some researchers, whose data and conclusions are challenged in this paper, will disagree with the expressed criticism. Future work will hopefully help to unravel the truth.

To keep on the safe side, a rule of thumb to go by is to aim, at any cost, to obtain accurate analytical data. This implies reliable specimens with a Cr-content identical to that present in the substance to be analysed (35), as well as precise and accurate analytical procedures. The latter can be checked through standard biological reference materials, certified for Cr at levels similar to those of the unknowns. They do not yet exist but are looming on the horizon. The analyses of ng or subng $Cr.g^{-1}$ levels require a very special, expensive infrastructure, and are labour consuming. It is, however, a false economy to disregard the necessary precautions and produce uncertain and, more often than not, erroneous data. No doubt this endeavour can be successful, so that Cr can be revisited and will in the long run unveil more of its so intriguing characteristics. To this end, very promising complementary knowledge on the kinetics of Cr-transport in biological systems can be obtained with in vivo and in vitro experiments with ^{51}Cr-labelled compounds of extremely high specific activity.

References

1. Guthrie, B.,E.: The nutritional role of chromium, Chpt.6 in Biological and Environmental Aspects of Chromium, ed. Langård, S., pp. 117-148, Elsevier Biomedical Press, Amsterdam 1982.

2. Versieck, J., Cornelis, R.: Anal.Chim.Acta 116, 217-254 (1980).

3. Vanderlinde, R.E., Kayne, F.J., Komar, G., Simmons, M.J., Tsou, J.Y., Lavine, R.L.: in Chromium in Nutrition and Metabolism, eds. Shapcott, D., Hubert, J., pp. 49-57. Elsevier, North Holland Publ.Co., Amsterdam 1979.

4. Kumpulainen, J., Lehto, J., Koivistoinen, P.: Sci.Tot.Environm. 31, 71-80 (1983).

5. Kayne, F.J., Komar, G., Laboda, H., Vanderlinde, R.E.: Clin.Chem. 24, 2151-2154 (1978).

6. Versieck, J., Hoste, J., Barbier, F., Steyaert, H., De Rudder, J., Michels, H.: Clin.Chem. 24, 303-308 (1978).

7. Torgrimsen, T.: Analysis of chromium, Chpt.4 in Biological and Environmental Aspects of Chromium, ed. Langård, S., pp. 65-99, Elsevier Biomedical Press, Amsterdam, 1982.

8. Rabinowitz, M.B., Levin, S.R., Gonick, H.C.: Metabolism 29, 355-364 (1980).

9. Rabinowitz, M.B., Gonick, H.C., Levin, S.R., Davidson, M.B.: Biol. Trace Element Res. 5, 449-466 (1983).

10. Thompson, D.A.: Ann.Clin.Biochem. 17, 144-147 (1980).

11. Liu, V.J.K., Morris, J.S.: Am.J.Clin.Nutr. 31, 972-976 (1978).

12. Liu, V.J.K., Abernathy, R.P.: Am.J.Clin.Nutr. 35, 661-667 (1982).

13. Abraham, A.S., Sonnenblick, M., Eni, M., Shemesh, O., Batt, A.P. : Am.Heart J. 99, 604-606 (1980).

14. Wallaeys, B., Cornelis, R., Mees, L., Hoste, J., Lameire, N : Atomkernenergie - Kerntechnick (1984), in press.

15. Abraham, A.S., Sonnenblick, M., Eini, M.: Gerontol. 27, 326-328 (1981).

16. Manthey, J., Kübler, W.: Am.J.Cardiol. 45, 940-944 (1980).

17. Black, J., Maitin, E.C., Gelman, H., Morris, D.M.: Biomaterials 4, 160-164 (1983).

18. Lim, T.H., Sargent III,T., Kusubov, N.: Am.J.Physiol. 244, R 445-454 (1983).

19. Anderson, R.A., Polansky, M.M., Bryden, N.A., Roginski, E.E., Patterson, K.Y., Veillon, C., Glinsmann, W.: Am.J.Clin.Nutr. 36, 1184-1193 (1982).

20. Anderson, R.A., Polansky, M.M., Bryden, N.A., Roginski, E.E., Mertz, W., Glinsmann, W.: Metabolism 32, 894-899 (1983).

21. Anderson, R.A., Polansky, M.M., Bryden, N.A., Patterson, K.Y., Veillon, C., Glinsmann, W.H.: J.Nutr. 113, 276-283 (1983).

22. Uusitupa, M.I.J., Kumpulainen, J.T., Voutilainen, E., Hersio, K., Sarlund, H., Pyörälä, K.P., Koivistoinen, P.E., Lehto, J.T.: Am.J. Clin.Nutr. 38, 404-410 (1983).

23. Sherman, L., Glennon, J.A., Brech, W.J., Klomberg, G.H., Gordon, E.S.: Metab.Clin.Exp. 17, 439-442 (1968).

24. Ping, L., Matsumoto, K., Fuwa, K.: Anal.Chim.Acta 147, 205-212 (1983).

25. Saner, G., Yüzbaşiyan, V., Neyzi, O., Günöz, H., Saka, N., Çiğdem, S.: Am.J.Clin.Nutr. 38, 574-578 (1983).

26. Saner, G.: Am.J.Clin.Nutr. 34, 1676-1679 (1981).

27. Anderson, R.A., Polansky, M.M., Bryden, N.A., Roginski, E.E., Patterson, K.Y., Reamer, D.C.: Diabetes 31, 212-216 (1982).

28. Mianzhi, Z., Barnes, R.M.: Spectrochim.Acta 38B, 259-269 (1983).

29. Minoia, C., Colli, M., Pozzoli, L.: At.Spectrosc. 2, 163 (1981).

30. Shimizu, T., Kimoto, R., Shijo, Y., Sakai, K.: Bunseki Kagaku 30, 66-70 (1981).

31. Mertz, W.: in Chromium in Nutrition and Metabolism, eds. Shapcott, D., Hubert, J., pp. 1-14, Elsevier, North-Holland Publ.Co., Amsterdam 1979.

32. Kumpulainen, J., Koivistoinen, P., Lahtinen, S.: Bioinorg.Chem. 8, 419-429 (1978).

33. Haylock, S.J., Buckley, P.D., Blackwell, L.F.: J.Inorg.Biochem. 19, 105-117 (1983).

34. Anderson, R.A., Bryden, N.A.: J.Agric.Food Chem. 31, 308-311 (1983).

35. Versieck, J., Barbier, F., Cornelis, R., Hoste, J.: Talanta 29, 973-984 (1982).

D i s c u s s i o n Session III - 3

Subjects of the discussion

- normal Cr-levels in serum
- essentiality of Cr
- regional variations in Cr-levels
- tissue Cr-levels
- supplementation in intravenous fluids
- molecular basis of the biochemical reactivity

Contributors: Ganther, Oster, Ottaway, Simonoff, Weser

Session IV

MANGANESE IN NUTRITION AND MEDICINE

Plenary Lecture given by: L.S. Hurley (USA)

Chairman:

M. Kirchgeßner (FRG)

Poster Reports given by: M. Kirchgessner (FRG)

NUTRITIONAL ASPECTS OF MANGANESE

Lucille S. Hurley

Department of Nutrition, University of California, Davis, CA 95616, U.S.A.

Effects of Manganese Deficiency

Manganese has been known to be an essential element for over 50 years, but its specific biochemical and nutritional functions are not well understood. One reason for the relative lack of progress has been the difficulty of analyzing manganese in animal tissues where it is found in extremely small concentrations. Concentrations in rat tissues, for example, vary from under 2 mg/kg in liver to less than 0.1 mg/kg in muscle (Table 1) (1). The very low concentrations and the resulting problem of contamination during analysis have retarded studies of manganese metabolism and biochemistry. With the development of flameless atomic absorption spectrophotometry, the problem is

Trace Element - Analytical Chemistry in Medicine and Biology, Vol. 3

MANGANESE IN RAT TISSUE

TISSUE	Mn
	mg/kg
Liver	1.95 ± 0.15
Pancreas	1.35 ± 0.16
Testis	1.33 ± 0.10
Intestine	0.74 ± 0.07
Kidney	0.72 ± 0.05
Brain	0.47 ± 0.02
Heart	0.41 ± 0.04
Spleen	0.24 ± 0.09
Thymus	0.24 ± 0.04
Milk	0.21 ± 0.03
Lung	0.11 ± 0.02
Muscle (soleus)	0.07 ± 0.01
Whole blood	0.012 ± 0.002

Mean ± SEM; N=6

from ref. (1)

attenuated, but manganese analysis in animal tissues and fluids remains difficult and subject to many sources of error.

Skeleton

One of the major effects of manganese deficiency, during both prenatal life and postnatal growth, is on the skeleton (2). In chickens the skeletal disorder perosis or "slipped tendon disease" was a severe economic problem, until it was discovered in the early 1930's that this condition could be completely prevented by adequate amounts of manganese in the feed. Similar effects on the skeleton are seen in rats. In this species skeletal abnormalities include a chondrodystrophic type of disproportionate growth, curvature of the long bones and the spine, and dysplasia of the knee joint. These abnormalities are related to abnormal chondrogenetic bone growth demonstrated by depressed synthesis of mucopolysaccharides. In the cartilagenous models of tibia from fetal rats, the uptake of ^{35}S sulfate was much slower and lesser in quantity in manganese deficient fetuses than in controls; autoradiography demonstrated this depression as well. These skeletal defects can be ascribed to the function of manganese as a cofactor for the glycosyltransferase enzymes which are important in catalyzing the first step in the synthesis of glycosaminoglycans and glycoproteins.

Ataxia

When manganese deficiency is imposed during the prenatal period, the offspring show congenital irreversible ataxia characterized by lack of equilibrium, abnormal body righting reflexes and retraction of the head (2). The abnormal body righting reflexes are easily demonstrated by the inability of the animals to swim. Unlike normal animals, the ataxic offspring of manganese deficient animals swim below the surface of the water and will drown if not removed. This failure of body righting reflexes is caused by abnormal development of the otoliths, the calcified structures essential for balance found in the vestibular portion of the inner ears.

Genetic interactions

The same abnormal behavior shown by offspring of manganese deficient animals is also seen in a genetic mutant, the pallid mouse. In this strain, the inability of the animals to swim, as well as the abnormal development of the otoliths, can be completely prevented by giving pregnant females a diet supplemented with manganese. With manganese concentrations of 1500 to 2000 µg/g, otolith morphology was completely normal and all offspring were able to swim. The pallid mutant also showed lower than normal concentrations of manganese in some tissues (3). A gene analogous to pallid has been reported in mink, and in this case, also, the abnormal behavior and otolith morphology could be prevented by giving manganese supplements during pregnancy (4). These mutants demonstrate an interaction between specific genes and manganese so that a phenocopy of the mutant can be produced by manganese deficiency whereas the expression of the mutant gene can be prevented by manganese supplementation.

Another type of interaction of manganese with genetic factors was demonstrated in a study showing that various inbred strains of mice respond differently to manganese deficiency. Recently we have carried out a comparative study of the effect of manganese deficient diets on two strains of rats, the Sprague-Dawley and the Wistar (5). A great difference in the susceptibility of these strains to manganese deficiency was apparent. Seventy percent of offspring of Wistar rats fed a manganese deficient diet survived to 60 days of age, while only 27% of pups born to manganese deficient Sprague-Dawley females survived to this age. These differences in manganese requirements and/or metabolism leading to differential responses may explain some of the conflicting reports concerning effects of manganese deprivation.

Biochemical Functions of Manganese

Glycosyltransferase

As discussed above, manganese serves as a cofactor for the glycosyltransferase enzymes which are important for the synthesis of glycosaminoglycans and glycoproteins (11). Manganese deficient animals show a depression of the activity of this enzyme and this can account for the defects of the skeleton and otoliths observed in manganese deficient animals.

Superoxide dismutase

Manganese also functions as a part of the metalloenzyme manganese superoxide dismutase. This enzyme is thought to protect cells from damage by superoxide free radicals, as it catalyzes the dismutation of the superoxide anion to hydrogen peroxide. In its absence, hydroxyl radicals, which are very deleterious to cells, are formed (12). There are two types of superoxide dismutase enzymes. Manganese superoxide dismutase occurs primarily in mitochondria while copper zinc superoxide dismutase is found in the cytosol. In manganese deficient mice, activity of manganese-SOD is significantly lower than in controls (13). Paynter (14) has shown that in heart muscle the amount of manganese-SOD activity is correlated with the manganese concentration in the diet as well as with the manganese concentration in the tissue. In chickens, which are very sensitive to manganese deficiency, liver manganese-SOD was significantly lower than in controls after only one week of the manganese deficient diet given from hatching (13).

We have recently studied the enzymic development of this enzyme in rats and in mice. In rats the level of manganese-SOD activity in liver is very low after birth. Activity increased in control rats over 6-fold between day 1 and day 60. In manganese deficient rats, manganese-SOD activity was similar to that of controls until about day 15 of age, but

Convulsive disorders

Manganese deficiency also affects brain function. Manganese deficient rats are much more susceptible to convulsions than are normal rats, with a low threshold for electroshock, and epileptiform electroencephalograms (6). These observations led Tanaka (7) to measure the manganese in whole blood of children with convulsive disorders. He found that approximately 30% of these children had whole blood manganese concentrations significantly lower than normal, about 1/3 of the normal value. Papavasiliou et al. (8) measured whole blood manganese in adult epileptics and similarly found that the concentration was significantly lower than in controls. Furthermore, this effect could not be ascribed to the anticonvulsive drugs taken by these patients, as there was no correlation with blood levels of the drug. In addition, patients who had frequent seizures had lower blood manganese levels than those with infrequent seizures.

These observations on whole blood manganese levels are important. We have recently found in studies with manganese deficient and normal rats that whole blood manganese concentration correlates very well with the concentration of manganese in the liver (9). Thus, whole blood manganese appears to be a valid reflection of tissue manganese level and may be useful as an indication of manganese nutriture.

Ultrastructural defects

Another effect of manganese deficiency is on ultrastructural defects. Manganese is concentrated in mitochondria, and in mice that have been given a manganese deficient diet throughout their lives, abnormal mitochondria were observed with electron microscopy (10). The mitochondria showed evidence of membrane defects; the outer membrane was damaged and in some cases appeared to be missing. The reticuloendothelial system was also abnormal.

by day 60 it was only about half of the control value (15). A similar developmental curve was seen in mice.

Because of the membrane defects that had been observed in manganese deficient animals, we developed the hypothesis that membrane damage resulted from depressed activity of manganese-SOD leading to an increased concentration of superoxide anions or reaction products such as hydroxyl radicals. We therefore estimated lipid peroxidation in these animals by measuring malonaldehyde formation, using the thiobarbituric acid (TBA) technique. In isolated mitochondria with the addition of oxygen initiators to provide conditions of oxidative stress, lipid peroxidation was significantly higher at day 20 in deficient rats than in controls. Between day 20 and day 60, lipid peroxidation increased in deficient rats but not in control rats. The ratio of TBA reacting products to manganese-SOD activity was also higher in the deficient rats at all time points (15). Concomitant ultrastructural examination of liver from these animals showed that membrane damage became apparent in the manganese deficient animals by 60 days of age (16). These results are consistent with the hypothesis that damage to mitochondrial membranes produced by manganese deficiency may be due to depressed manganese SOD activity resulting in increased lipid peroxidation in biomembranes.

In order to test this hypothesis in a functional way, we have used ethyl alcohol as an agent to study the response of manganese deficient animals under conditions of increased oxidation (17). Manganese deficient rats and their controls were given 20% ethyl alcohol as drinking fluid for 14 days. Liver manganese was lower in manganese deficient animals both with and without alcohol than in the controls. However, manganese-SOD activity, although it increased in response to alcohol in both groups, was lower in deficient rats than in controls. Thus, although there was an increase in manganese-SOD activity in response to alcohol, the levels did not reach those of the manganese sufficient animals. By 14 days, the deficient animals given alcohol were extremely lethargic and in poor condition, unlike the alcohol fed controls (17). Thus, the manganese deficient animal may not be able to respond to the added stress of alcohol because of its inability to increase superoxide dismutase levels sufficiently. Another change that occurred as the result of alcohol feeding was a very high concentration of iron in the liver of manganese deficient rats. As iron is itself an oxidizing agent, this high liver iron may also provide additional oxidative stress and increase the rates of lipid peroxidation.

Carbohydrate metabolism

A relationship between manganese and abnormal carbohydrate metabolism was first suggested by Rubenstein and coworkers in 1962 (18). They reported the case of a diabetic patient resistant to insulin therapy who responded to oral doses of manganese chloride with a consistent drop in blood glucose levels. Everson and Shrader (19), studying manganese deficiency in guinea pigs, reported that those born to manganese deficient dams and fed manganese deficient diets from birth to 60 days, had abnormal glucose tolerance curves. These animals also showed hypertrophic pancreatic islet tissue with degranulated beta cells.

Pancreatic function. We have recently investigated the hypothesis that manganese deficiency might result in altered insulin secretion, which could explain, at least partially, the altered glucose metabolism observed in manganese deficient animals (20). The offspring of manganese deficient rats, also fed the deficient diet to maturity, showed a diabetic type of glucose tolerance curve when given oral glucose. Plasma insulin levels, however, were not commensurate with the high plasma glucose, although manganese concentration in the pancreas was only half the control value. Using an isolated perfused pancreas preparation, insulin output, was measured. The first phase insulin output, which reflects release of stored hormone, was only 76% of control values. The second phase of insulin output, representing release of newly synthesized as well as preformed hormone, was also significantly lower in the deficient rats than in the controls. The degree of inhibition of insulin release in deficient rats became progressively more pronounced with continued insulin stimulation. These findings suggest that dietary manganese deficiency can result in impaired insulin secretion, resulting in abnormal carbohydrate metabolism. In rats fed the deficient diet from weaning to maturity, but not exposed to manganese deficiency prior to weaning, carbohydrate metabolism was normal, even though manganese concentration of pancreas and other tissues was lower than control values. Thus, the timing of the deficiency may be a critical factor in the expression of abnormal carbohydrate metabolism (20).

Gluconeogenesis. Pyruvate carboxylase (PC), a manganese metalloenzyme, and phosphoenolpyruvate carboxykinase (PEPCK), a manganese activated enzyme, are involved in the first steps of gluconeogenesis. At birth, dramatic changes occur in both nutrition and in metabolic requirements, as the newborn animal must accomodate from a continuous nutrient supply to an intermittent one. The newborn rat rapidly utilizes the store of hepatic glycogen that it has accumulated before birth and then oxidizes lipid as an energy source. As rat milk provides a high fat diet, the neonatal rat depends on gluconeogenesis to maintain blood glucose levels. Increases in the activities of PC and PEPCK have been reported to occur during the first week of postnatal life (21). Thus, the high neonatal mortality of manganese deficient rats coincides with the period when gluconeogenesis begins to play a major role in glucose homeostasis (2).

In offspring of manganese deficient rats, plasma glucose levels were normal at birth, but were significantly lower than in controls on days 1 and 2 postpartum. After the first week of postnatal life, there were no significant differences with controls. From day 0 to day 3, PC activity increased 1-1/2 fold in controls and 2-1/2 fold in manganese deficient pups. However, at day 8, PC activity dropped in deficient pups to 50% of control levels, where it remained until day 10. After day 10 PC activity increased in both groups, but tended to be lower in the deficient animals (22).

PEPCK showed a different pattern. During the first 3 days after birth, activity increased 1-1/2 fold in manganese deficient rats. It remained higher in deficient rats than in controls until day 5 and then was generally lower than in controls (22).

These results suggest that glucose homeostasis is impaired in manganese deficient rats as evidenced by the lower blood glucose levels on day 1 and 2 postnatally. The developmental pattern of both PC and PEPCK correlated with the changes in blood glucose levels observed. PEPCK increased in both control and deficient pups on day one, but the increase was somewhat greater in deficient pups. PC, however, did not increase dramatically until day 2 and remained higher than control levels until day 6. Thus, the low plasma glucose resulting from either low glycogen stores or from insufficient gluconeogenesis, or both, is compensated by day 3 by the increased activity of PC and PEPCK. However, these activities rapidly fall, and blood glucose then declines below normal levels. One factor responsible for the high neonatal mortality of manganese deficient offspring may therefore be impaired carbohydrate homeostasis.

Even in adult rats, PC activity, although normal in fed animals, is significantly lower than normal in fasted animals (23). This observation differs from those of Scrutton et al. (24) in chickens. In this species, manganese deficiency does not cause a loss of PC activity, and in fact, the manganese in the metalloenzyme can be substituted by magnesium. Such does not appear to be the case for rats.

<u>Arginase</u>. The third manganese-containing metalloenzyme, arginase, is necessary for urea formation, and its activity is affected by dietary manganese intake. Kirchgessner and Heiseke (25) reported that in rats fed diets deficient in manganese (0.7 ppm), liver arginase activity was only 30% of control levels. In rats given a diet containing 6 ppm manganese, liver arginase was also considerably lower than in controls, 40% of control value. The functional significance of this low arginase

activity is not known, but at least in birds, it is difficult to see how arginase might be related to the lesions produced by manganese deficiency, since birds do not make urea.

Manganese-Iron Interaction

A number of indications in the literature suggest an interaction between manganese and iron (1). Recently, Lonnerdal and coworkers (26) have shown that in young mice given iron supplemented milk, the manganese concentration of liver was significantly lower than in mice that were not supplemented with iron. Thus, iron supplementation of infant formulas may influence manganese concentration in tissues.

This has some practical importance, as Lonnerdal et al. (27) in a recent investigation of a number of different infant formulas from various parts of the world, have shown that the ratio of iron to manganese varies greatly and is in most cases very much different from the ratio found in human milk. This observation takes on added significance when it is remembered that the metabolism of manganese appears to be very different in the young suckling from that in the more mature individual. Miller et al. (28) found that suckling mice did not lose any manganese from their bodies, that is, they retained all manganese that entered the body until day 17 of life. Recently, Zlotkin (29) of Toronto has made a similar observation in human infants.

The toxicity of oral manganese is said to be relatively low; that is, diets containing 25 to 50 times the normal level had no effect on growth in experimental animals (30). However, some of the infant formulas studied had manganese concentrations above this apparently nontoxic level.

References

1. Keen, C.L., Lonnerdal B., Hurley, L.S.: In: Frieden, E., ed., Biochemistry of the Elements, Plenum Publishing Co., New York, in press 1984.

2. Hurley, L.S.: Physiol. Rev. 61, 249-295.

3. Cotzias, G.C., Tang, L.C., Miller, S.T., Sladic-Simic, D., Hurley, L.S.: Science 176, 410-412 (1972).

4. Erway, L.D., Mitchell, S.E.: J. Heredity 64, 111-119 (1973).

5. Dungan, D.D., Zidenberg-Cherr, S., Keen, C.L., Lonnerdal, B., Hurley, L.S.: Fed. Proc. 43, 1054 (1984).

6. Hurley, L.S., Wooley, D.E., Rosenthal, E., Timiras, P.S.: Am. J. Physiol. 204, 302-304 (1963).

7. Tanaka, Y.: Amer. Chem. Soc. National Meeting (Chicago) (1977).

8. Papavasiliou, P.S., Kutt, H., Miller, S.T., Rosal, V., Wang, Y.Y., Aronson, R.B.: Neurology 29, 1466-1473 (1979).

9. Keen, C.L., Clegg, M.S., Lonnerdal, B., Hurley, L.S.: N. Engl. J. Med. 308, 1230 (1983).

10. Bell, L.T., Hurley, L.S.: Lab. Invest. 29, 723-736 (1973).

11. Leach, R.M.: Fed. Proc. Fed. Amer. Socs. Exp. Biol. 30, 991-994 (1971).

12. Fridovich, I.: Ann. Rev. Biochem. 44, 147-159 (1975).

13. de Rosa, G., Keen, C.L., Leach, R.M., Hurley, L.S.: J. Nutr. 110, 795-804 (1980).

14. Paynter, D.J.: J. Nutr. 110, 437-447 (1980).

15. Zidenberg-Cherr, S., Keen, C.L., Lonnerdal, B., Hurley, L.S.: J. Nutr. 113, 2498-2504 (1983).

16. Zidenberg-Cherr, S., Keen, C.L., Hurley, L.S.: In preparation.

17. Zidenberg-Cherr, S., Hurley, L.S., Lonnerdal, B., Keen, C.L.: Fed. Proc. Fed. Amer. Socs. Exp. Biol. 42, 1054 (1984).

18. Rubenstein, A.H., Levin, N.W., Elliott, G.A.: Lancet ii, 1348-1351 (1962).

19. Everson, G.J., Shrader, R.E.: J. Nutr. 94, 89-94 (1968).

20. Baly, D.L., Curry, D.L., Keen, C.L., Hurley, L.S.: J. Nutr. In press.

21. Ballard, F.J., Hanson, R.W.: Biochem. J. 104, 866-871 (1967).

22. Baly, D.L., Keen, C.L., Hurley, L.S.: Fed. Proc. Fed. Amer. Socs. Exp. Biol. 43, 1254 (1984).

23. Baly, D.L., Keen, C.L., Hurley, L.S.: In preparation.

24. Scrutton, M.C., Griminger, P., Wallace, J.C.: J. Biol. Chem. 247, 3305-3313 (1972).

25. Kirchgessner, M., Heiseke, D.: Internat. Z. Vit. Ern. Forschung 48, 75-78 (1978).

26. Lonnerdal, B., Keen, C.L., Ohtake, M., Tamura, T.: Am. J. Dis. Child 137, 433-437 (1983).

27. Keen, C.L., Fransson, G.-B., Lonnerdal, B.: J. Pediat. Gastroenterol. and Nutr. 3, 256-261 (1984).

28. Miller, S.T., Cotzias, G.C., Evert, H.S.: Am. J. Physiol. 229, 1080-1084 (1975).

29. Zlotkin, S.: In: Chandra, R., Trace Elements, Nestle Nutrition Workshop. In press.

30. Underwood, E.J.: Trace Elements in Human and Animal Nutrition, Fourth Edition, Academic Press, New York (1977).

D i s c u s s i o n Session IV - 1
Plenary Lecture

Subjects of the discussion

- Mn-deficiency in man (vitamin K deficiency)
- interaction of Mn with Fe and Al
- model for testing Mn-requirement
- connection to different enzymes
- bioavailability of Mn
- interactions with other trace elements

Contributors: Anke, Cochran, Kirchgessner, Koh, Kübler, Oster, Weser

INTERACTION BETWEEN MANGANESE, ZINC AND CALCIUM IN PIGS

Hans Schenkel, Bärbel Krehl, Margarete Müller, Christa Strobel
Institut für Tierernährung, Universität Hohenheim
D-7000 Stuttgart 70, FRG

Introduction

Interactions between inorganic dietary constituents may cause changes in absorption, excretion as well as distribution of the elements. Factors which influence the response to interactions include nutritional status, pattern of exposure, environmental conditions etc. Also differences in species and sex must be considered (1,2,3). Calcium and phosphorus supply of men and animals can vary over a considerable range under practical conditions. Interferences between calcium and phosphorus supply and metabolism of manganese and zinc are known even at higher intake of this minerals (4). For informations of an influence of marginal mineral intake on both trace elements are scare, in the following study manganese and zinc concentrations were determined in tissues of pigs, which received diets with different calcium and phosphorus supplementation.

Materials and Methods

Datas were presented from three feeding experiments with pigs (male castrated animals). In experiment I pigs were fattened with a diet based on barley and soybean meal. Diets for the three treatment group (13,15,14 animals each) differed with respect to calcium supplementation (Tab.1). Mean liveweight of the animals was at 30 kg on the beginning and 100 kg at the end of the experiment. At a mean duration of 97 days mean daily

Trace Element - Analytical Chemistry in Medicine and Biology, Vol. 3

weight gain was at 727 g. In experiment II pigs were fattened with a diet based on grains, soybeanmeal and fishmeal. Calcium: phosphorus ratio was constant (0,9:1) in both diets. Diet 1 contained both minerals at a lower, diet 2 at a higher level. Group 1 (n=5) in this experiment received diet 1 and group 3 (n=5) diet 2 throughout the whole experiment. Group 2 (n=5) were fed with diet 1 from liveweight 30 to 60 kg and diet 2 from 60 to 100 kg. Group 4 (n=5) were fed with both diets in an inverse manner (diet 2/diet 1). In accordance with experiment I mean daily weight gain during the 91 day experiment was at 773 g.
In experiment III young pigs were fed with diets based on barley, wheat and soybeanmeal. As could be seen from table 1 diets for the 5 treatment groups (6 animals each) differed with respect to calcium and phosphorus concentration. In the diets for group 1 and 2 calcium:phosphorus ratio was at 1.6:1 and in the diets for group 3 and 4 at 0.8:1, respectively. Animals of group 5 received during the first 15 days the same diet as group 1, thereafter a diet unsupplemented with calcium and phosphorus was fed. The experiment lasted 33 days. Mean liveweight was 15 kg at the beginning and 29 kg at the end of experiment. All animals were slaughtered at the end of the experiments. Minerals and trace elements in different tissues and in bile from gall-bladder were determined by flame atomic absorption spectrophotometry.

Table 1: Concentration of calcium, phosphorus, zinc and manganese in the diets (dry matter based)

Experiment	Diet	Calcium g/kg	Phosphorus g/kg	Ca:P ratio	Zinc mg/kg	Manganese mg/kg
I	1	4.8	6.8	0.7:1	176	94
	2	7.3	6.8	1.1:1	159	81
	3	10.6	6.8	1.6:1	176	98
II	1	6.2	7.0	0.9:1	88	58
	2	7.7	8.8	0.9:1	94	60
III	1	11.6	7.4	1.6:1	80	32
	2	9.2	5.8	1.6:1	83	33
	3	6.1	7.3	0.8:1	82	31
	4	4.0	5.5	0.7:1	78	31
	5*	1.2	5.3	0.2:1	77	31

* Fed to animals of group 5 during day 16-33

Results

Significant differences of manganese concentrations in organs between the treatment groups could only be stated in experiment III (Table 2). In this experiment lower manganese concentrations could be found in livers of group 3 and 4, which received diets with calcium:phosphorus ratios at 0.8:1. Therefore a close relationship between manganese concentration in liver and calcium:phosphorus ratio in the diet could be found ($r=0.513$, $p<0,01$, group 1-4, $n=26$). If ratios for manganese concentrations between kidney and liver and pancreas and liver, respectively, were calculated a close relationship could be stated between kidney/liver ratio and calcium concentration in the diet ($r= -0.670$, $p<0.001$, $n=26$, group 1-4) as well as between pancreas/liver ratio and dietary calcium:phosphorus ratio ($r= -0.502$, $n=26$, $p<0.05$).
Manganese concentration in bile from gall-bladder increased with decreasing dietary calcium in experiment I (Tab. 3). Also in experiment III for groups 3 and 4, which received diets with a calcium:phosphorus ratio of 0.8:1 a tendency to higher manganese concentrations could be stated. Within experimental groups with altered calcium and phosphorus supply in experiment II significant differences could be stated. Manganese concentration in bile of group 4 (higher-lower level) was lower than in group 2 (lower-higher level). Also manganese concentration in bile of group 5 in experiment III (without calcium and phosphorus supplementation during the second half of the experiment) was lower than in the other groups of this experiment.
Ratio of manganese concentration in bile and liver seemed to be related to calcium:phosphorus ratio in the diets. In experiment I and III significant linear correlations between these parameters could be calculated (expt.I: $r= -0.368$, $p<0.05$, $n=39$; expt.III: $r= -0.558$, $p<0.05$, groups 1-4, $n=25$). Therefore in experiment II where diets with the same calcium:phosphorus ratio were fed no difference between group 1 and 3 could be found.

Table 2: Concentration of calcium, manganese and zinc in different tissues (dry matter basis)

Organ	Group	Calcium g/kg	Manganese mg/kg	Zinc mg/kg
Experiment I				
Liver	1	195±17[a]	12.6±2.3[a]	240±41[a]
	2	201±25[a]	12.5±1.7[a]	241±56[a]
	3	208±35[a]	11.6±2.0[a]	241±70[a]
Kidney	1	388±29[a]	9.0±0.9[a]	164±19[a]
	2	368±24[a]	8.9±1.2[a]	144±19[b]
	3	387±65[a]	9.3±1.1[a]	144±20[b]
Pancreas	1	628±20[a]	7.3±1.1[a]	187±35[a]
	2	650±37[a]	7.1±1.1[a]	173±29[a]
	3	620±71[a]	6.7±1.3[a]	179±38[a]
Spleen	1	279±50[a]	1.5±0.2[a]	125± 5[a]
	2	274±54[a]	1.4±0.2[a]	127± 7[a]
	3	258±65[a]	1.5±0.2[a]	126± 6[a]
Experiment II				
Liver	1	168±11[a]	10.3±1.0[a]	276±78[a]
	2	168±12[a]	10.6±2.2[a]	281±62[a]
	3	176±39[a]	10.7±1.4[a]	246±34[a]
	4	162±35[a]	13.1±2.8[a]	219±20[a]
Kidney	1	387±33[a]	10.3±2.2[a]	180±40[a]
	2	406±22[a]	9.5±1.0[a]	156±23[a]
	3	377±25[a]	10.1±1.4[a]	158±17[a]
	4	380±22[a]	9.9±1.0[a]	152± 8[a]
Pancreas	1	560±41[ab]	8.0±0.7[a]	150±29[a]
	2	551± 8[a]	7.7±0.9[a]	146±15[a]
	3	560±138[ab]	7.9±1.3[a]	150±33[a]
	4	586±12[b]	8.1±2.0[a]	172±27[a]
Spleen	1	236± 9[a]	2.0±0.7[a]	118± 5[a]
	2	224±12[a]	1.5±0.4[a]	121±12[a]
	3	245±11[a]	1.4±0.2[a]	118± 4[a]
	4	238±11[a]	1.6±0.5[a]	119± 4[a]

Table 2 (continued): Concentration of calcium, manganese and zinc in different tissues (dry matter basis)

Organ	Group	Calcium g/kg	Manganese mg/kg	Zinc mg/kg
Experiment III				
Liver	1	179±31[ab]	12.8±4.9[ab]	155±32[a]
	2	178±13[ab]	12.6±3.6[a]	143±27[a]
	3	176±19[ab]	9.7±0.8[ab]	144±23[a]
	4	165±21[a]	9.1±1.0[b]	153±30[a]
	5	192±17[b]	10.1±1.1[ab]	197±57[a]
Kidney	1	411±27[a]	6.7±1.2[ab]	111± 8[a]
	2	406±14[a]	7.8±0.3[a]	110± 9[a]
	3	359±13[b]	7.3±0.3[b]	115± 8[a]
	4	378±22[b]	7.4±0.5[ab]	127±16[a]
	5	368±17[b]	6.6±1.3[ab]	122±18[a]
Pancreas	1	535±41[a]	7.8±2.3[ab]	140±20[a]
	2	580±45[a]	8.1±0.8[a]	133±16[a]
	3	546±28[a]	8.2±1.6[ab]	145±22[a]
	4	530±56[a]	7.3±1.1[ab]	130±35[a]
	5	584±73[a]	7.0±0.7[b]	134±12[a]

In experiment I zinc concentration in kidneys of group 1 was significantly higher than in the two other treatment groups. A similar increase of about 15 percent could also be observed in experiment II (group 1 vs 3) and III (group 4 vs 1 and 2). In experiment I and III changes in kidney zinc were followed by a corresponding alteration of copper concentration (expt.I: r=0.73, n=42; expt. II: r=0.87, n=30). In experiment I similar to manganese zinc concentration in bile tended to higher values with decreasing dietary calcium (r=0.414, $p<0.05$, n=38). A close relationship between manganese and iron concentration in bile could be stated in experiment I and III, respectively (expt.I: r=0.476, n=37; expt. III: r=0.502, n=26).

Table 3: Concentration of calcium, manganese and zinc in bile from gall-bladder (mg per kg dry matter)

Experiment	Group	Calcium	Manganese	Zinc
I	1	1470±381[ab]	9.7±3.0[a]	11.4±6.0[a]
	2	1561±216[a]	7.1±2.9[b]	9.5±3.4[a]
	3	1310±210[b]	6.1±0.9[b]	7.9±2.2[a]
II	1	1398±403[a]	9.7±5.3[ab]	4.3±2.9[a]
	2	1182± 99[a]	10.9±2.2[a]	6.9±1.9[a]
	3	1510±510[a]	11.1±5.0[ab]	9.5±4.3[ab]
	4	1122±265[a]	7.4±0.9[b]	11.7±3.8[b]
III	1	1547±370[a]	9.9±4.2[a]	6.5±2.9[a]
	2	1534±148[a]	7.9±2.6[a]	4.6±1.6[a]
	3	1548±327[a]	12.7±6.5[a]	5.3±3.6[a]
	4	1668±241[a]	10.1±2.2[a]	4.2±3.3[a]
	5	1828±595[a]	7.7±4.5[a]	4.6±2.5[a]

Discussion

Finishing pigs seemed to tolerate relatively wide variations in calcium supply without marked effects on tissue uptake of several trace elements (5, 6). Also different dietary calcium and phosphorus intake in experiments presented caused no marked changes of zinc and manganese concentrations in most of the tissues investigated, whereas in the same experiments distinct treatment effects with respect to cadmium concentrations in liver, kidney and other tissues could be found (7, 8). Decreasing dietary calcium:phosphorus ratios caused an increase of biliary manganese concentration. Unfortunately those datas gave no evidence on endogenous excretion rate. Nevertheless in experiment III a concomitant decrease of manganese concentration in liver was found. This was confirmed by experiments with weanling pigs (9). At dietary manganese levels of 40 mg/kg a decrease of calcium:phosphorus ratio from 2:1 to 0.5:1 caused a decrease of liver manganese concentration. High calcium and phosphorus levels in a 0.5:1 ratio caused an increase in manganese concentration as compared to lower dietary calcium and phosphorus levels at the same ratio. A similar tendency could be stated in groups 3 and 4 of experiment III with respect to different phosphorus levels.

Kirchgessner et al. (10) reported a balance experiment with pigs using dietary calcium:phosphorus ratios between 0.6:1 und 1.7:1 during initial fattening and between 0.3:1 and 0.8:1 during the final fattening period. Increase of calcium:phosphorus ratio by addition of $CaCO_3$ caused an increase of manganese retention followed to some extent by a decrease in faecal and urinary excretion. In experiments with rats a significantly lower excretion of ^{54}Mn in the faeces over a four day period after oral dosing and a higher ^{54}Mn deposition in tibia could be stated if a diet with 0.6% calcium were fed in comparison with a 0.1%Ca diet (11). Also significantly higher retention of orally administered ^{54}Mn was found, when diets contained 0.9% P compared with a 0.4% P diet (11). Decrease of Manganese retention after ingestion of high calcium supplemented diets was reported by Galik (12). Increase zinc concentration in kidneys stated at lower calcium intake were in accordance with experiments in pigs (12) and rats (14).

Acknowledgements

The authors wish to thank S.Schmidt and A.Marquardt for technical assistance. Part of the investigations were supported by the Ministerium für Ernährung, Landwirtschaft, Umwelt und Forsten, Baden-Württemberg.

References

1. Sandstead, H.H.: J.Lab.Clin.Med. 98, 457-462 (1981)
2. Davis, G.K.: Ann.New York Acad.Sci.355, 130-139 (1980)
3. Wise, A.: Nutr.Abstr.Rev., Ser.A, 50, 319-332 (1980)
4. Underwood, E.J.: Trace Elements in Human and Animal Nutrition. Third Edition, Academic Press, New York and London (1971)
5. Pond, W.G., Walker jr., E.F., Kirtland, D.: J.Anim.Sci. 41. 1053-1056 (1975)
6. Pond, W.G., Walker jr., E.F., Kirtland, D., Rounsaville, T.: J.Anim.Sci. 47, 1128-1134 (1978)
7. Schenkel, H., Müller, M., Strobel, C.: Landw.Forsch. SH39, 319-326 (1982)
8. Schenkel, H., Krehl, B. in: Trace Element Metabolism in Man and Animals - 4. Eds.: McC Howell, J., Gawthorne, J.M., White, C.L., Australian Academy of Science, Canberra, 588-589 (1981)
9. Kayongo-Male, H., Ullrey, D.E., Miller, E.R., Keahey, K.K.: Ghana J.Agric.Sci. 10, 39-46 (1977)
10. Kirchgessner, M., Munz, W., Oelschläger, W.: Arch.Tierernährung 10, 1-19 (1960)
11. Lassiter, J.W., Morton, J.D., Miller, W.J. in: Trace Element Metabolism in Animals, Ed.: Mills, C.F., Livingstone, Edinburgh, 130-132 (1970)
12. Galik, R.: Pol'nohospodarstvo 25, 593-599 (1979)
13. Morgan, D.P., Young, E.P., Earle, I.P., Davey, R.J., Stevenson, J.W.: J.Anim.Sci. 29, 900-905 (1969)
14. Lorentzon, R., Larsson, S.E.: Clin.Sci.Mol.Med. 53, 439-446 (1977)

D i s c u s s i o n Session IV - 2

<u>Subjects of the discussion</u>

- interactions between trace elements and mineral elements
- changes in trace element distribution
- levels of trace element administration

Contributors: Kirchgessner

Session V

SPECIATION OF TRACE ELEMENTS

Plenary Lecture given by: H.-J. Hartman, U. Weser (FRG)

Chairman:

H.E. Ganther (USA)

A STRUCTURE AND FUNCTION CORRELATION OF METAL-THIOLATE PROTEINS

Hans-Jürgen Hartmann and Ulrich Weser*

Anorganische Biochemie, Physiologisch-chemisches Institut der Universität Tübingen, Hoppe-Seyler-Str. 1, D-7400 Tübingen 1

*To whom all correspondence should be sent

Introduction

Proteins containing metal-sulphur or metal-thiolate chromophores are of widespread abundance in biological systems. Our knowledge on the low M_r iron-sulphur proteins with regard to their structural and functional side is well established. They are of central importance in electron transport reactions (1,2). By way of contrast, the biological role of the metallothioneins (M_r = ∿6500), another class of metal-thiolate proteins, is speculative. The polypeptide composition of these metalloproteins is rather unusual. They contain no aromatic amino acids and up to 33% of cysteine residues. Metallothioneins have been identified in many species and can be induced following administration of Zn, Cu, Cd, Hg and Ag (3). In general, hepatic metallothioneins contain the biologically essential metals Cu and Zn. The ratio Cu:Zn varies considerably (4-7). In all known metallothioneins the metals are tetrahedrally surrounded by 4 sulphur atoms (8,9). Surprisingly, such Zn-$(thiolate)_4$-tetrahedra which, in fact, could be considered a type of monomeric Zn-thionein, are also found in some zinc-enzymes. Two examples are known where isolated $Zn(SR)_4$-chromophores are present in the non-catalytic zinc binding centres of alcoholdehydrogenase (10) and in aspartate-transcarbamylase (11), respectively.

Trace Element - Analytical Chemistry in Medicine and Biology, Vol. 3

From baker's yeast a homogeneous Cu-thionein (M_r = 4800) containing 4 Cu atoms and 8 cysteine residues was isolated (12, 13). The primary structure (14) was different to that of the vertebrate metallothioneins (15). Nevertheless, this Cu-protein is continued to be named Cu-thionein attributable to the striking structural similarities of the copper binding centres with those found for vertebrate Cu-thionein.

Originally metallothionein was isolated as the Cd,Zn-species from mammalian kidney and liver (16,17). It was postulated to be a detoxifying agent for heavy metals (18-20). The Cd,Zn-thionein of chicken liver could be converted into Cu-thionein (21) and supported the suggestion of Porter that the insoluble cysteine-rich Cu-protein from neonatal hepatic tissue was related to the metallothioneins (22). In pre- and neonatal liver the Cu-content is significantly increased compared to the adult species (23). The isolation of soluble, cytosolic Cu,Zn-thionein from fetal liver (7,24) confirmed that most of the hepatic copper is bound to thionein in the Cu(I) state (7). After birth the liver Cu-concentration decreases within days to a "normal" level. This led to the assumption that metallothionein is assigned to the control of metal transport. It functions homeostatically as a reservoir, capable to sequester and/or donate its metal, i.e. Cu or Zn to metal requiring metalloproteins (3,7,10,12,25-29).

The molecular basis of Cu-transport deserves special attention. The high binding capacity to proteins, its ability to create reactive oxygen species (7,28) and its essential biochemical role in many oxidases demand for a subtle regulation of the cellular Cu-concentration. As free Cu(II)-ions can undergo many uncontrolled reactions Cu has to be kept in the stable Cu(I)-thiolate state. Different reaction mechanisms should be considered in the metabolism of Cu. Firstly, there is a direct transfer of Cu(I) which is facilitated by the hydrophobic nature of the Cu(I)-thiolate chromophore. Secondly, thionein-Cu can provide the organism with essential Cu(II) through a catabolic mechanism, i.e., an oxidative breakdown of the Cu(I)-thiolate chromophore.

Results

Structural aspects of yeast Cu-thionein and vertebrate Zn- or Cu,Zn-thionein

Although crystals of rat liver Cd,Zn-thionein II were recently obtained no complete X-ray diffraction data are available. Due to the limited success in crystallizing Cu-containing thioneins, attempts were made to reveal the molecular architecture of the Cu-chromphore employing indirect methods. Several spectrometric investigations gave rather encouraging results.

The simultaneous presence of Cu and Zn in variable concentrations resulted in the overlapping of many a characteristic spectroscopic feature of either metal-sulphur chromophore. To minimize obscuring of experimental data yeast Cu-thionein was used as this Cu-thiolate protein is known to be exclusively loaded with copper. In addition, its low relativ molecular mass proved to be highly advantageous to reach the vacant active site of many apo-copper-proteins.

From ^{1}H-NMR measurements it was established that the single histidine residue of the thionein chain is not involved in the Cu-coordination. X-ray photoelectron spectroscopy (XPS) revealed binding energy values of Cu $2p_{3/2}$ = 932.3 eV and S $2p_{1/2,3/2}$ = 162.0 eV which were almost identical to those of the reference compound Cu(I)(thiourea)$_3$Cl (10). XPS-measurements of Cu-containing vertebrate thionein resulted in similar binding energy values for Cu and S (7). This means that Cu is in the $3d^{10}$ state while S can be attributed to thiolate-S. The use of Cu-thiourea as a structural model, therefore, was encouraged especially as its tetrahedral Cu-S coordination is known (30). Extended X-ray absorption fine structure (EXAFS) measurements were carried out to obtain further insight on the type and number of first shell atoms surrounding the copper. The EXAFS oscillations of either Cu-compound revealed a striking similarity in both shape and amplitude (Fig. 1 and 2).

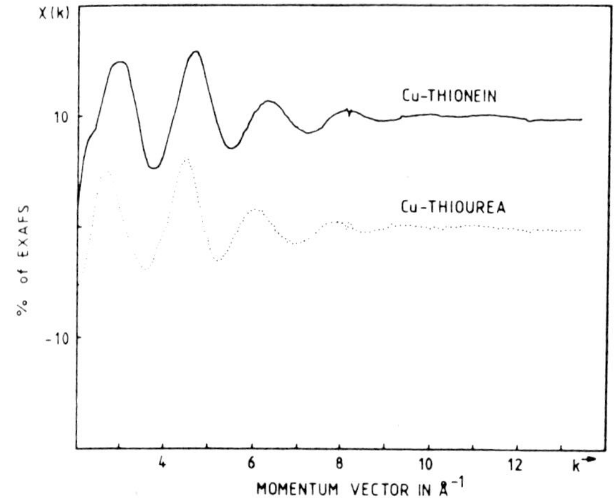

Fig. 1: EXAFS spectra of Cu-thionein and Cu-(thiourea)$_3$Cl.

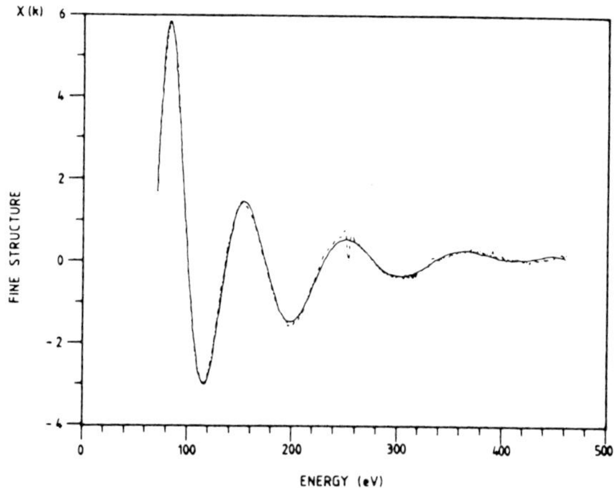

Fig. 2: First principle calculation fit of the EXAFS spectrum of Cu-thionein. (-•-•-•) experimental data, (——) best theoretical fit obtained with four sulphur atoms surrounding the copper, two atoms at 2.22 Å and another two at 2.36 Å.

The conclusion can be drawn that all atoms around the Cu are identical, i.e. sulphur. Furthermore, the same coordination number was assigned. A slight distortion of the Cu-thiolate tetrahedra in Cu-thionein is assumed due to the somewhat shorter distances between each of the two sulphur pairs, one at 2.22 Å, the other at 2.36 Å. The best proposal for the arrangement of four Cu-thiolate tetrahedra fitted the cubane type of many known Cu(I) compounds (8,10) (Fig. 3).

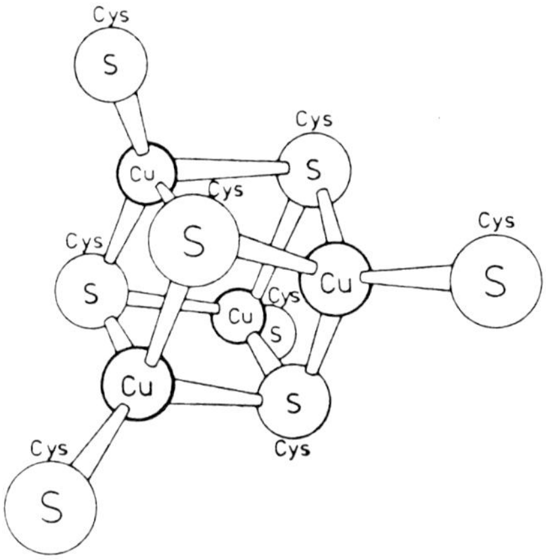

Fig. 3.: Proposed arrangement of the Cu-thiolate tetrahedra in yeast Cu-thionein (Fig. 1-3 with permission from the publishers (10)).

EXAFS of the Zn-sites in sheep liver metallothionein revealed also a tetrahedral surrounding of the metal by 4 sulphur atoms with a Zn-S distance of 2.29 Å (9). The tetrahedrally arranged metal-thiolate clusters including three-coordinate sulphurs throughout was already proposed in the authors' laboratory in 1978 (25,31).

Zn(II)-transfer

Zn-thionein was shown to transfer its metal ion into a series of apo-Zn-enzymes including aldolase, thermolysin, alkaline phosphatase and carbonic anhydrase (26,27). For example, the reaction of Zn-thionein with apo-carbonic anhydrase is depicted below (Fig. 4).

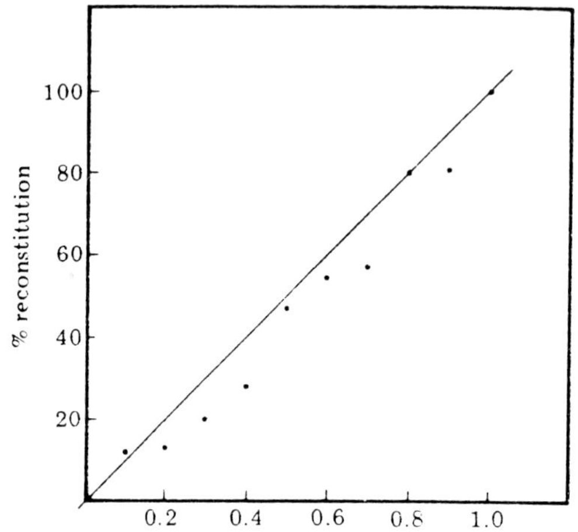

Fig. 4.: Stoichiometry of Zn-transfer from Zn-thionein into apo-carbonic anhydrase. With permission from (27).

Cu(I)-transfer

It was of interest whether or not Cu-thionein could be used to transfer thiolate-bound Cu directly into the empty Cu-binding site of the type I Cu or "blue" copper proteins. It was found that apostellacyanin was able to accept Cu(I) from Cu-thionein without any detectable unspecific Cu(II) intermediate both aerobically and anaerobically. A time-dependent incorporation of cupric Cu was deduced from electronic absorption at 604 nm, EPR spectrometry and circular dichroism measurement (32). The oxidation of Cu(I) to yield Cu(II) must have occurred after the Cu(I)-incorporation . No extraneous Cu(II) EPR-signal was seen when thionein was in excess of the amount required to reconstitute stellacyanin (Fig. 5). In order to exclude the possibility of an uncontrolled oxidation of Cu-thionein, the Cu(I)-transfer was also performed anaerobically (Fig. 6).

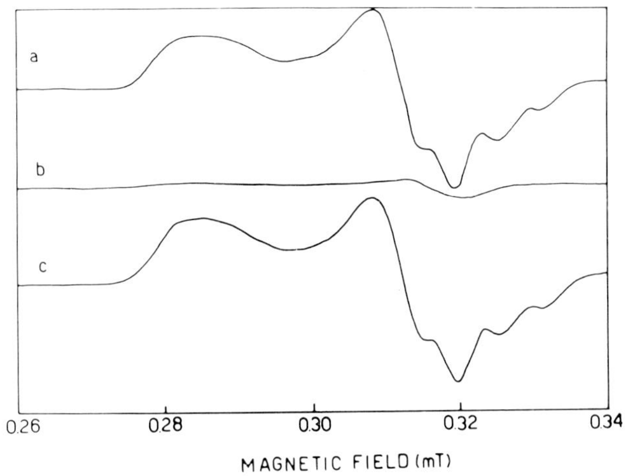

Fig. 5.: EPR spectra of stellacyanin. (a) holoprotein, (b) apoprotein, (c) apoprotein aerobically incubated with Cu(I)-thionein.

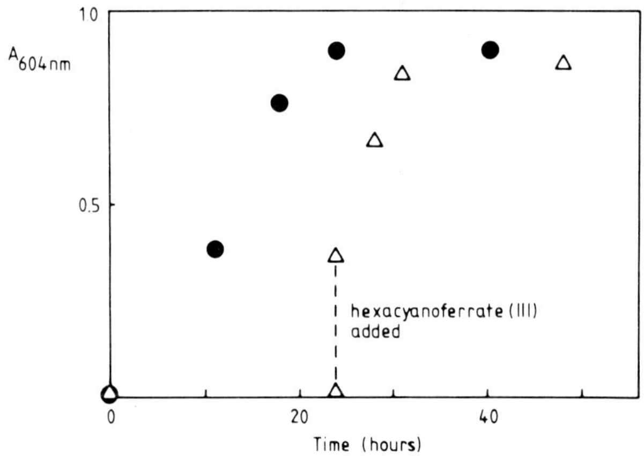

Fig. 6.: Reconstitution of apostellacyanin with Cu(I)-thionein. Incubation of apostellacyanin with Cu-thionein: (•) aerobic, (△) anaerobic, with subsequent addition of hexacyanoferrate(III). (Fig. 5 and 6 with permission from (32)).

After 24 h a molar equivalent of hexacyanoferrate(III) to apo-stellacyanin was added to the colourless solution which resulted in the immediate appearance of the major portion of A_{604nm}, the remaining portion being recovered in a slower phase. Apparently, an equilibrium was established between Cu(I)-thionein and reduced stellacyanin, which was shifted by hexacyanoferrate(III) due to the oxidation of stellacyanin. This amount of hexacyanonferrate(III) was not able to oxidize Cu-thionein alone. In contrast to apo-stellacyanin, spinach apo-plastocyanin, another representative for a type I Cu-protein, was only reconstituted to an extend of ∿15% with Cu-thionein (33). Attributable to the slow oxidative cleavage of the Cu-thiolate chromophore in the presence of a five fold excess of hexacyanoferrate(III) more than 24 h were required to yield 70% of Cu(II)-plastocyanin. However, unlike with $CuSO_4$ no spurious Cu(II) was seen in the EPR. A clear Cu(I)-transfer into apo-plastocyanin was noticed when the former structural analogue to Cu-thionein, Cu-thiourea was used. This Cu(I)-complex is stable for days in aqueous solution. Also Cu(I)-acetonitrile was successful as a Cu(I)-compound to reconstitute the apoprotein. In this case high concentrations of acetonitrile must be present for stabilization.

There is absolutely no detectable difference between the circular dichroism of native Cu(I)-plastocyanin and the respective Cu(I)-proteins reconstituted by Cu(I)-thiourea or Cu(I)-acetonitrile (Fig. 7).

The direct incorporation into the copper binding site is obvious. Oxidation with hexacyanoferrate(III) yielded the same Cotton bands as those usually obtained with native Cu(I)-plastocyanin.

The reason for the observed differences in the copper transfer of either type I copper protein might be attributed to the characteristic copper binding sites and the redox potentials at 180 and 370 mV, respectively.

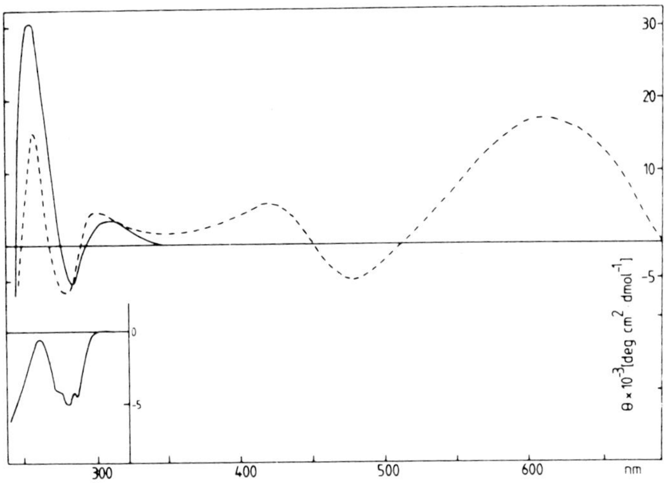

Fig. 7.: Circular dichroism of (---) native Cu(II)-plastocyanin, reconstituted and hexacyanoferrate(III)-oxidized Cu(I)-plastocyanin, (——) native and reconstituted Cu(I)-plastocyanin. Inset apoplastocyanin. With permission from (33).

Cu-transfer into type II Cu-proteins

Some Cu-proteins are successfully reconstituted in the presence of Cu(II) only. Thionein-Cu and -Zn are transferable into the apoprotein of Cu_2Zn_2superoxide dismutase (34). However, an oxidative breakdown of the metal thiolate chromophores must take place prior to the metal transfer. For example:

$$\mathrm{Cu(I)(^{-}SR)_2} \xrightarrow[-2e^-]{} \mathrm{Cu(I)} + \mathrm{RS\text{-}SR} \qquad (1)$$

$$\mathrm{Cu(I)} \xrightarrow[-e^-]{} \mathrm{Cu(II)} \qquad (2)$$

$$\mathrm{Zn(II)(^{-}SR)_2} \xrightarrow[-2e^-]{} \mathrm{Zn(II)} + \mathrm{RS\text{-}SR} \qquad (3)$$

This oxidative cleavage of metal thiolate binding centres could be termed the catabolic metal metabolism.

Cu-transfer into type III Cu-proteins

An attempt was made to use *Neurospora crassa* Cu-thionein as a copper transferring protein for the type III Cu in tyrosinase and haemocyanin (29). A direct transfer could be observed. Prior to incubation the Cu-thiolate centre of Cu-thionein had to be irreversibly destroyed by oxidation. The Cu(II)-thionein had to be reduced again before the reaction with either apo-protein. Following this pretreatment the less stable extra-neously bound Cu(I) could be transferred to the Cu-binding centres of apotyrosinase and apohaemocyanin. It should be emphasized that the mechanism cannot be listed under the direct Cu(I)-transfer reactions. The preceding catabolic breakdown of Cu(I)-thiolate chromophores supports the former suggested alternative mode for Cu-transport.

Discussion

The metallothioneins may play a vital role in general Zn- and Cu-metabolism. Reconstitution experiments using apo-Zn-proteins and Zn-thionein supported this conclusion. However, the data obtained from the Zn- and Cu-transfer studies provide no conclusive evidence that metallothionein is a specific Zn,Cu-transport protein in vivo. Nevertheless, this mode of Cu-transport in intact biological systems cannot be fully rejected. It has to be emphasized that, for example, non chelated Cu-ions can undergo many undesired reactions which are harmful to cellular life. The uncontrolled reactivity of so called "free copper ions" comprise the following mechanisms:

- There is virtually no free Cu(II)aq. ion at physiological

pH-values. Attributable to the precipitation of $Cu(OH)_2$ the concentration of soluble Cu(II) lies approximately at 10^{-20}M.

- Cu(II) ions belong to the strong chelating metals. Thus, unspecific bonding will create considerable confusion in the normal metabolism.

- Cu(II) ions can displace other metal ions including Mg(II), Ca(II), Mn(II), Zn(II) etc. The respective Cu-complexes are thermodynamically more stable.

- The Jahn-Teller distorted Cu(II)-complexes, though thermodynamically stable, exchange their ligands in axial position sometimes in less than 0.5 nsec. This kinetic lability may not be beneficial at all for many a biochemical reaction.

- Cu(II) reacts as an oxidizing agent on i.e. -OH, -SH and -CHO-groups.

$$RS^- + Cu(II) \rightarrow \frac{1}{n}\,(Cu(I)\cdot SR)_n$$

$$RS\text{-}SR_{1/2}Cu(I) \rightleftarrows \frac{1}{2}\,RS\text{-}SR + Cu(I) \qquad (4)$$

- Cu(I) ions disproportionate in aqueous solutions:

$$Cu(I) \rightarrow Cu^o + Cu(II) \qquad (5)$$

- Cu(I) can create reactive oxygen species according to the Fenton- or as usually termed the metal driven Haber-Weiss reaction (7,35).

Cu(II) and Cu(I) catalyze the conversion of $\cdot O_2^-$ and O_2^{2-} into $\cdot OH$. The $\cdot O^-$ is known to be highly reactive ($373\ kJ\cdot mol^{-1}$)(36) leading to the complete destruction of cellular life:

$$Cu(II) + \cdot O_2^- \rightarrow Cu(I) + O_2 \quad (6)$$

$$2\ \cdot O_2^- \rightarrow O_2^{2-} + O_2 \quad (7)$$

$$Cu(I) + H_2O_2 \rightleftharpoons Cu(II) + \cdot OH + OH^- \quad (8)$$

All the above uncontrolled reactions make the former drawn conclusion of a vital role of these thermodynamically stable metal-thiolate chromophores in general metal transport very attractive.

Evolutionary aspects of catalytically active Zn-binding centres

Apart from the possible control of Zn and Cu metabolism an attractive hypothesis considering the evolution of catalytically active Zn-binding centres found in many enzymes should be mentioned. For example, the two isolated Zn-binding centres of alcoholdehydrogenase may be seen as the transient state starting from a monomeric Zn-thionein cluster which is eventually converted into the catalytically active Zn-binding site. The non-catalytic $Zn(SR)_4$ complex resembles the rubredoxin iron centre and the monomeric unit of Zn- or Cu-thionein.

Zn-binding centres of alcoholdehydrogenase

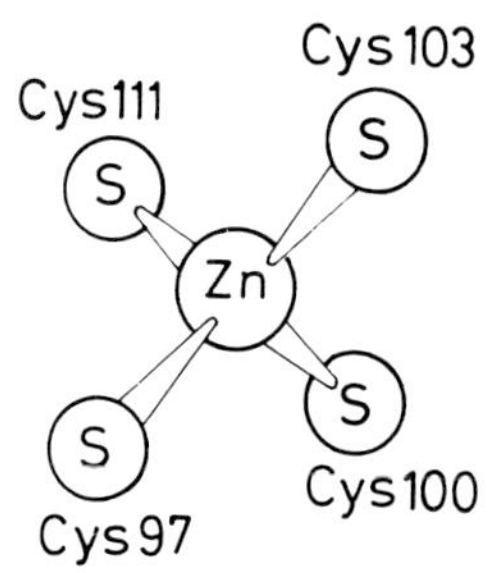

Non-catalytic $Zn(SR)_4$

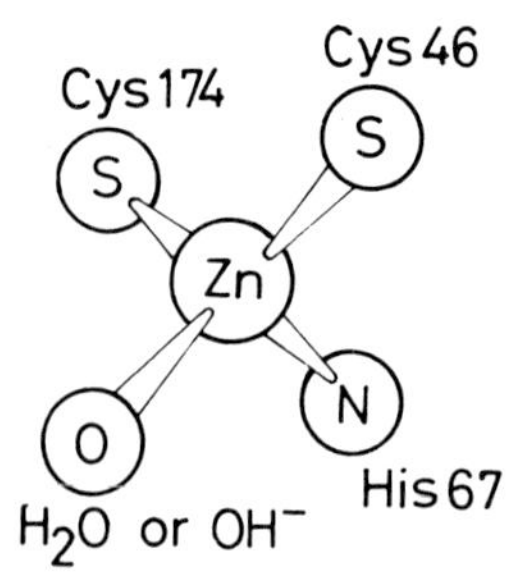

Catalytically active Zn

The catalytically active Zn is coordinated to two cysteine-sulphur atoms, one substrate binding site and one histidine-nitrogen. In other words, two thiolate-sulphur atoms of the metallothionein cluster have developed into somewhat more sophisticated ligands including the imidazolate moiety.

Acknowledgements

This study was aided by the Ministerium für Ernährung, Landwirtschaft, Umwelt und Forsten, Baden-Württemberg AZ 62-7990.2. Travel support was granted by the Breuninger Stiftung Stuttgart.

References

1. Lovenberg,W. (ed.): Iron-Sulphur Proteins, Academic Press, New York, Vols. 1 and 2 (1973), Vol. 3 (1977).

2. Orme-Johnson, W.H.: Iron-Sulphur Proteins: Structure and function, Annu. Rev. Biochem. 42, 159-204 (1973).

3. Kägi, J.H.R. and Nordberg, M. (eds.): Metallothionein, Experientia Supplementum 34, Birkhäuser Verlag, Basel (1979).

4. Bühler, R.H.O. and Kägi, J.H.R.: FEBS Lett. 39, 229-234 (1974).

5. Kägi, J.H.R., Himmelhoch, S.R., Whanger, P.D., Bethune, J.L. and Vallee, B.L.: J. Biol. Chem. 249, 3537-3542 (1974).

6. Bremner, I. and Young, B.W.: Biochem. J. 157, 517-520 (1976).

7. Hartmann, H.-J. and Weser, U.: Biochim. Biophys. Acta. 491, 211-222 (1977).

8. Bordas, J., Koch, M.H.J., Hartmann, H.-J. and Weser, U.: FEBS Lett. 140, 19-21 (1982).

9. Garner, C.D., Hasnain, S.S., Bremner, I. and Bordas, J.: J. Inorg. Biochem. 16, 253-256 (1982).

10. Bordas, J., Koch, M.H.J., Hartmann, H.-J. and Weser, U.: Inorg. Chim. Acta. 78, 113-120 (1983).

11. Phillips, J.C., Bordas, J., Foote, A.M., Koch, M.H.J. and Moody, M.F.: Biochemistry 21, 830-834 (1982).

12. Prinz, R. and Weser, U.: Z. Physiol. Chem. 356, 767-776 (1975).

13. Weser, U., Hartmann, H.-J., Fretzdorff, A. and Strobel, G.-J.: Biochim. Biophys. Acta. 493, 465-477 (1977).

14. Kimura, M., Otaki, N., Hartmann, H.-J. and Weser, U.: Regard sur la Biochimie 3, 101 (1981).

15. Kojima,Y. and Kägi, J.H.R.: Trends Biochem. Sci. 3, 90-93 (1978).

16. Kägi, J.H.R. and Vallee, B.L.: J.Biol. Chem. 235, 3460-3465 (1960).

17. Pulido, P., Kägi, J.H.R. and Vallee, B.L.: Biochemistry 5, 1768-1777 (1966).

18. Ito, T., Sawanchi, K. and Okijima, Y.: Fol. Anat. Jpn. 42, 107-117 (1966).

19. Nordberg, G.F.: Environ. Physiol. Biochem. 2, 7-36 (1982).

20. Weser, U., Rupp, H., Donay, F., Linnmann, F., Voelter, W., Voetsch, W. and Jung, G.: Eur. J. Biochem. 39, 127-140 (1974).

21. Rupp, H. and Weser, U.: FEBS Lett. 44, 293-297 (1974).

22. Porter, H.:Biochem. Biophys. Res. Commun. 56, 661-668 (1974).

23. Porter,H., Johnston, J. and Porter, E.M.: Biochem. Biophys. Acta. 65, 66-73 (1961).

24. Rydén, L. and Deutsch, H.F.: J. Biol. Chem. 253, 519-524 (1978).

25. Weser, U. and Rupp, H.: Metallothionein, Experientia Supplementum 34 (Kägi, J.H.R. and Nordberg, M. eds.) Birkhäuser Verlag, Basel, 221-230 (1979).

26. Udom, A.O. and Brady, F.O.: Biochem. J. 187, 329-335 (1980).

27. Li, T.-Y., Kraker, A.J., Shaw, C.F. and Petering, D.H.: Proc. Natl. Acad. Sci. USA 77, 6334-6338 (1980).

28. Weser, U. and Hartmann, H.-J.: Copper Proteins and Copper Enzymes Vol. 3, (Lontie, R. ed.), CRC Press, Inc., Boca Raton, in press (1984).

29. Beltramini, M. and Lerch, K.: FEBS Lett. 142, 219-222 (1982).

30. Okaya, Y. and Knobler, C.B.: Acta Crystallograph. 17, 928-930 (1964).

31. Rupp, H. and Weser, U.: Biochim. Biophys. Acta. 533, 209-226 (1978).

32. Hartmann, H.-J., Morpurgo, L., Desideri, A., Rotilio, G. and Weser, U.: FEBS Lett. 152, 94-96 (1983).

33. Brütsch, S., Hartmann, H.-J. Weser, U.: Inorg. Chim. Acta. 92, 147-149 (1984).

34. Hartmann, H.-J. and Weser, U.: Z. Physiol. Chem. 362, 207 (1981).

35. Haber, F. and Weiss, J.: Proc. Roy. Soc., Ser. A 147, 323-351 (1934).

36. Gaillard-Cusin, F. and James, H.: J. Chim. Phys. 70, 701-702 (1973).

D i s c u s s i o n Session V - 1
Plenary Lecture

Subjects of the discussion

- enzymatic function of metallothionein
- transport function of metallothionein
- stability of Cu-metallothionein
- synthesis of cytochromes

Contributors: Frank, Ganther, Hurley, Ottaway, White

A NEW CHEMICAL REACTION SYSTEM FOR FLOW THROUGH ANALYSIS AND POST COLUMN CHROMATOGRAPHIC DETECTION OF MANGANESE AND CHROMIUM SPECIES

Jochem Rüter and Bernd Neidhart
Institut für Arbeitsphysiologie an der Universität Dortmund (Geschäftsführender Direktor: Prof. Dr. Dr. J. Rutenfranz), Abteilung Toxikologie und Arbeitsmedizin (Direktor: Prof. Dr. Dr. H.M. Bolt), Ardeystr. 67, D-4600 Dortmund 1, W-Germany

Introduction

In nature, chemical elements exist in various forms which are called element species. These are, e.g., suspended or colloid oxides and sulfides or adsorbed inorganic and organic material bound by ion exchange or complex binding mechanisms. Element species in solution are free or hydrated metal ions, complex ions or organometal compounds (1).

For a critical examination of the biological (toxicological) and oecological behaviour it is necessary to determine not only the total content of the element in a sample but also the various species of the element. Therefore, in the last years the interest in speciation analysis has grown steadily and consequently the development of suiteable analytical determination methods is in progress.

In this paper, a new detector system for the continuous determination of manganese and chromium species on the ng-level is described. The chemical reaction detector contains, as its most essential part, a heterogeneous oxidation reactor, filled with

Trace Element - Analytical Chemistry in Medicine and Biology, Vol. 3

PbO_2-coated silica gel, in which manganese and chromium species with lower oxidation states than VII and VI, respectively, are oxidized in solution at 95°C to permanganate and chromate. The heterogeneous reactor is followed by a homogeneous reaction system in which the oxidized species react with reagents forming photometrically detectable compounds. The chemical reaction detector can be used either in the flow injection mode or, following a chromatographic separation, as a post column reaction detector.

Results

In commercially available autoanalyzer-manifolds for "Flow Injection Analysis" as well as for "Continuous Flow Analysis", the reaction of the sample takes place exclusively in the liquid phase by mixing it with defined volumes of reagent. Each sample reacts with fresh reagent solution. These equipments working in the homogeneous mode, however, are only one possible device for preparing derivatives. An alternative method is that with packed reaction tubes. The application of such a heterogeneous solid phase reactor on the basis of lead dioxide is demonstrated to be advantageous in element species analysis. The preparation of the lead dioxide coated silica gel which is used as packing material is described elsewhere (2). The homogeneous part of the chemical reaction detector in which permanganate oxidizes leucomalachite green to malachite green, possible interferences within the system, integration of an enrichment column for manganese as well as speciation measurements by the bypass technique, and systematical investigations of the system parameters have been discussed recently (2, 3). The experimental set-up used for chromium speciation measurements is the same as for manganese but the reaction conditions differ slightly. While the mobile phase in the manganese determination is an 0.1 M nitric acid, an 0.05 M sulfuric acid is used in the chromium measurements. Detection of the chromate which is eluted from the lead dioxide reactor is performed using 1,5-diphenylcarbazide as a reagent which reacts only with chromium (VI) but not with chromium (III). The detailed conditions of the measurement will be published elsewhere (4). Fig. 1 shows a calibration curve of chromium speciation measure-

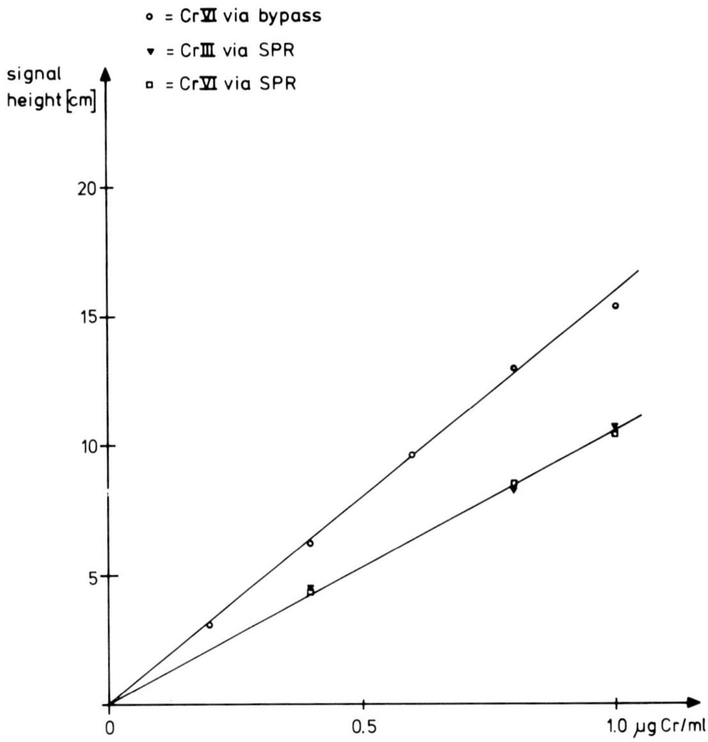

Fig. 1. Calibration of chromium speciation using the SPR in the flow injection mode.

ments. Two main conclusions can be drawn from this figure. It can be seen that the degree of conversion of chromium to chromate is complete under the reaction conditions because the data points of chromium III and VI are fitted by the same line. In addition, it becomes clear from the figure that the chromate, measured via the bypass, produces higher signals than the same amounts of chromate measured via the reactor. This is due to band broadening by diffusion inside the heterogeneous reactor. If signal evaluation is carried out by integration of the peak areas, the measurements via the bypass and via the solid phase reactor result in the same values. Fig. 2 shows speciation measurements of water solutions containing Cr(III) and Cr(VI).

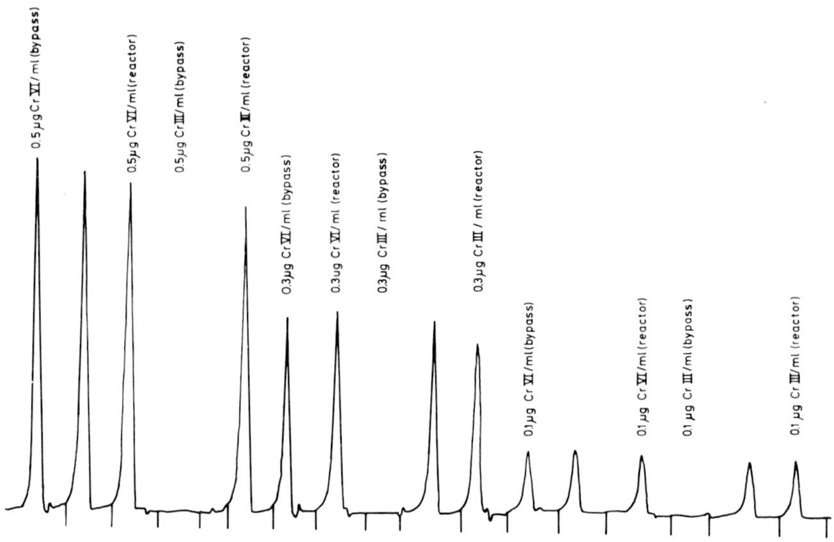

Fig. 2. Speciation measurements of chromium.

From this figure it becomes clear that, in agreement with theoretical considerations, Cr(III) gives only a signal when measured via the oxidation reactor but not when passing the bypass, while Cr(VI) samples are detected in each case.

In the two examples presented above the system was run in the flow injection mode. For the example of manganese Fig. 3 shows for comparison two post chromatograhic measurements, one with PAR and the other with the lead dioxide oxidation reactor followed by the malachite green reaction. This experiment demonstrates in a convincing way the selectivity of this kind

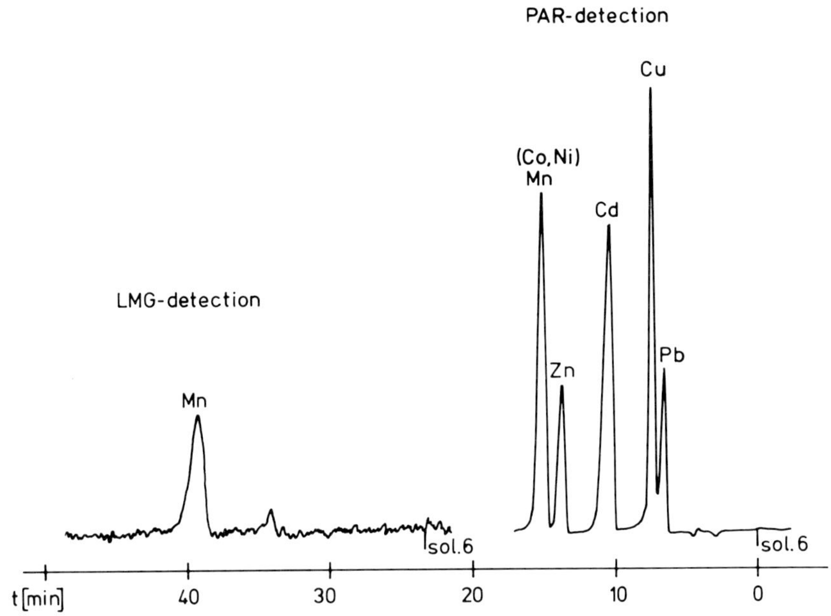

Fig. 3. Separation of metal ions by ion chromatography - comparison of two different detection systems.

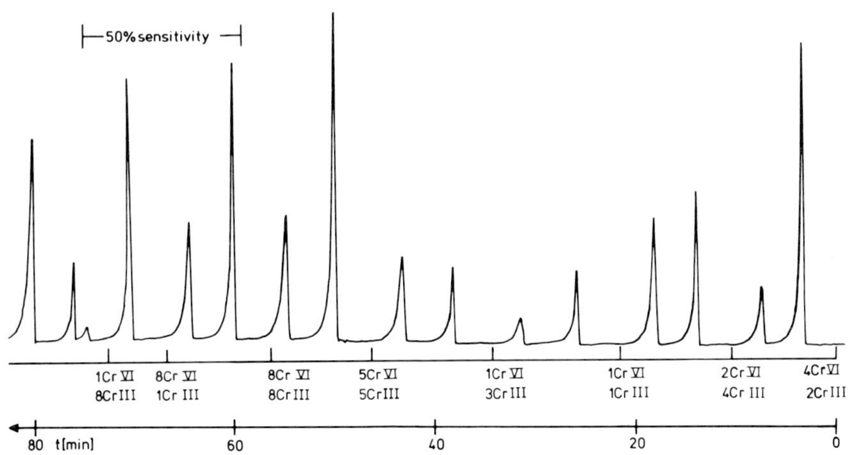

Fig. 4. Chromium speciation using the SPR as a post-chromatographic reaction detector.

of chemical reaction detector. Manganese is detected even in presence of large amounts of cobalt, nickel and zinc.

Fig. 4 shows the speciation of Cr(III) and Cr(VI) both dissolved in the same solution. The two chromium species are separated on the chromatographic column because trivalent chromium is retarded and chromate is not. This method allows the speciation of chromium in aqueous solutions down to 10 ng/ml of Cr(VI) and 25 ng/ml of Cr(III). For measurements in more complex organic matrices a suitable clean-up procedure of the sample has to be developed.

Although the investigation of the different applications of the solid phase reactor have just been started, it can be expected that, especially in the fields of medicine and biology, this type of reaction system will be of great practical use for metal detection and speciation.

References

1. G. Schwedt, in: Analytiker-Taschenbuch, Bd. 2, S. 255 (R. Bock, W. Fresenius, H. Günzler, W. Huber, G. Tölg eds.), Springer, Berlin-Heidelberg-New York 1981.

2. J. Rüter, B. Neidhart: Mikrochim. Acta 1984 I, 271-289 (1984).

3. J. Rüter, U.P. Fislage, B. Neidhart: Submitted to Mikrochim. Acta 1984.

4. J. Rüter, U.P. Fislage, B. Neidhart, in preparation.

IMMUNOLOCALISATION OF ZINC METALLOTHIONEIN IN RAT TISSUES

Johanna P. Clarkson, Margaret E. Elmes, B. Jasani
Department of Pathology,
Welsh National School of Medicine,
Heath Park, Cardiff, and

M. Webb,
M.R.C. Toxicology Unit,
Woodmansterne Rd.,
Carshalton, U.K.

Introduction

Early work on metallothionein indicated a detoxifying role for this low molecular weight metal binding protein. However, more recent studies have shown that it is important in the normal metabolism of zinc and copper (1). Reported studies have used biochemical extraction methods to isolate and identify metallothioneins in tissues, and the cellular distribution of zinc metallothionein has not been described so far.

Immunohistochemical studies have localised a cadmium metallothionein in rat tissues (2, 3, 4) and a metallothionein in human livers with copper retention (5). In the present paper we describe a sensitive immunohistochemical method which demonstrates zinc metallothionein in wax-embedded rat tissues, and the data presented support the theory that zinc metallothionein is a physiological protein which may have a transport function.

Trace Element - Analytical Chemistry in Medicine and Biology, Vol. 3

Methods

<u>Experimental animals</u> Male Wistar rats on stock laboratory diet, on a 7 am -7 pm light/dark cycle and weighing approximately 300 g were dosed at 0900 hours (time 0) and divided into 4 different groups as follows :-

i) intraperitoneal zinc injection 3-4 mg Zn^{++}/Kg body weight (1 ml volume)

ii) oral zinc, as i)

iii) oral water (1 ml, < 1.4 µg Zn^{++}/ml)

iv) undosed, fasted only

All rats were fasted with access to tap water from time 0 and housed in pairs in metabolic cages. Rats were killed in groups of 4 after 3, 6, 12, 18 and 24 hours. Controls consisted of 6 unfasted rats killed at time 0. Samples of liver, duodenum, jejunum and ileum were collected from the same region in each rat, fixed in neutral buffered formalin, and then wax embedded.

<u>Metallothionein and antiserum</u> Metallothionein was prepared from the livers of rats given 8-10 mg Zn^{++}/Kg body weight and killed 18 hours later. The livers were homogenised and metallothionein was isolated and purified by gel filtration followed by ion-exchange chromatography (6). Antibody to rat liver zinc metallothionein was prepared by the method described in (7) and was donated by Dr. Frank Brady.

<u>Immunohistochemistry</u> The anti-metallothionein antibody was used in conjunction with a hapten-antihapten sandwich technique (8, 9). This involved the application of dinitrophenyl (DNP)-labelled anti-metallothionein serum (1:800 dilution; 16 hrs. at 4°C), and its localisation by the sequential addition of an IgM monoclonal antibody used as a bridge reagent (1:500 dilution of 1 mg/ml ascites fluid), and a DNP labelled peroxidase conjugate as the enzyme marker (1:800 dilution of a standard preparation). The colour reaction was developed using diaminobenzidine and hydrogen peroxide (10). The slides were washed using 3 changes of phosphate buffered saline between each step in the procedure. Finally, sections were counterstained with haematoxylin and glass-mounted.

As controls, sections were incubated with non-immune rabbit serum in place of the anti-metallothionein serum, or with anti-metallothionein serum which had been pre-absorbed with purified metallothionein (30 µg MT/ml antiserum).

Results

In all cases the control sections showed no significant staining anywhere.

Liver In normal unfasted rats metallothionein immunoreactivity (MTI) as defined by distinct brown staining was widespread in the liver. It was present in both hepatocyte cytoplasm and nucleus, and showed a periportal distribution. The staining was also observed in the sinusoids, canaliculi, and the portal and hepatic blood vessels (Fig. 1).

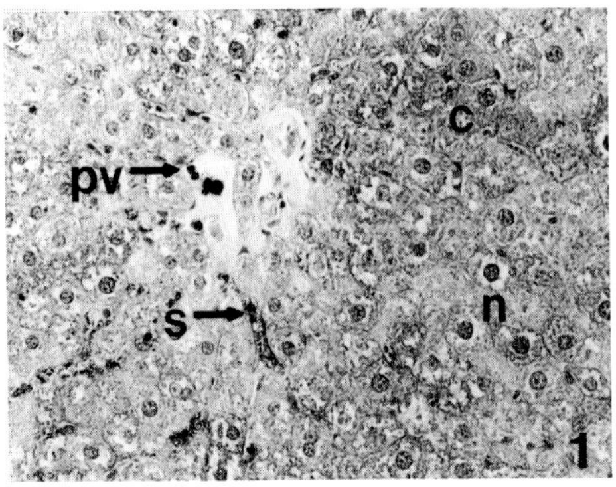

Figure 1. Liver from a normal control rat. Metallothionein immunoreactivity is present in a portal vessel (pv), sinusoids (s), hepatocyte cytoplasm (c) and nucleus (n). Counterstained with haematoxylin x 780.

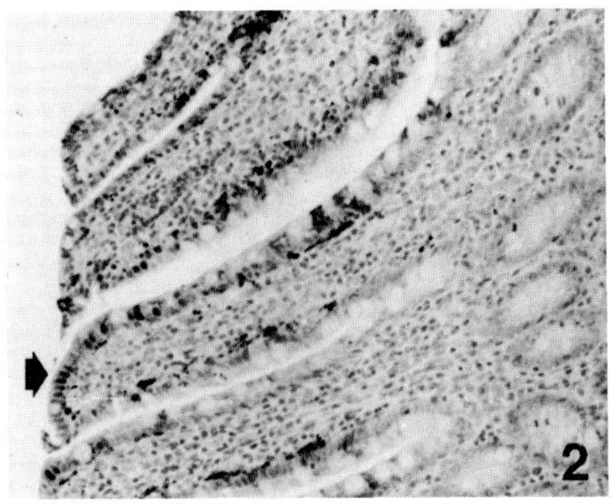

Figure 2. Ileum from a 12 hour fasted rat. Metallothionein immunoreactivity is present in the upper and mid villus epithelium. Arrow = nuclear staining. Counterstained with haematoxylin x 460

Small intestine MTI was observed in the upper and mid villus regions, but not in the crypts (Fig. 2). It was present in enterocyte cytoplasm and nucleus, and in some lamina propria elements, and was very variable in distribution. MTI was often very strong in the basement membrane region, and this was occasionally the only reactivity present, with no concomitant staining in the enterocytes.
Similar staining characteristics were observed in the duodenum, jejunum and ileum.

To assess the MTI on a semi-quantitative basis, the intensity and distribution of staining was scored subjectively on a 0 to 3 scale, where a score of 1 was given for light or locally confined MTI, 2 represented heavy or more widespread MTI and 3 referred to the maximum MTI observed. This was done separately for the different cellular features showing staining. Two sections from each organ from each animal were analysed, and the pooled data of the 4 animals in each group were assessed at each time interval. Some preliminary results from this study are summarised in Figs. 3 and 4; the complete results are to be published later.
In both non-dosed fasted and zinc injected animals, a consistent cyclical variation in MTI in the ileum was observed in both the enterocyte cytoplasm and the basement membrane region, and this extended between the 3 and 18 hour time periods (Fig. 3). A similar cyclical change was observed in the liver in fasted animals, in hepatocyte cytoplasm, sinusoids and canaliculi with a peak at 6 hours. Zinc injection appeared to abolish these cyclical changes in MTI in the liver producing a flat curve (Fig. 4). Nuclear staining in both the ileum and liver showed no obvious cyclical variation in either fasted or zinc injected animals.
Data from rats given oral doses of zinc or water is currently being assessed.

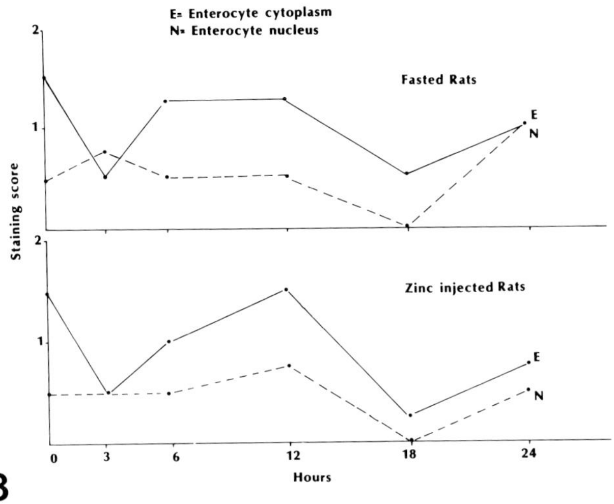

3

Figure 3. Metallothionein staining score in the ileum, fasted and zinc injected rats.

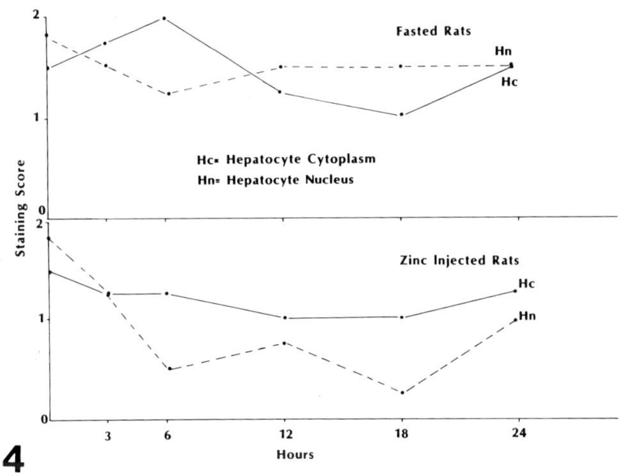

4

Figure 4. Metallothionein staining score in the liver, fasted and zinc injected rats.

Discussion

In this paper we have described a sensitive immunohistochemical technique for the localisation of metallothionein (MT) which distinguished the cytoplasmic, nuclear and extracellular compartments.

Metallothionein immunoreactivity (MTI) was observed in the tissues of normal, unfasted rats, and the distribution of MTI was altered both by fasting alone and in conjunction with the physiological levels of zinc administration. These results show that Zn metallothionein is a component of normal cells that responds to physiological stimuli.

A possible transport role for metallothionein has been suggested by our frequent observation of MTI in the liver sinusoids, canaliculi and blood vessels, and in the basement membrane region of the small intestinal epithelium. Although the precise localisation of some basal MTI is difficult to resolve at the light microscope level, the use of P.A.S. staining in conjunction with the immunohistochemical technique for MT has confirmed that in many sections this MTI was indeed extracellular. The presence of holes in the basement membrane, approximately 2 µm in diameter, with the passage of particulate material has been reported (11), supporting the possibility that MT may be transported across the membrane, either into or out of the enterocytes.

The MTI frequently observed in the hepatocyte and enterocyte nucleus may also reflect a transport function. Zinc is required for nucleic acid synthesis (12) and MT has a sufficiently low molecular weight to allow it to cross the hepatocyte nuclear envelope (2), thus MT may act as a translocation protein for zinc in the nucleus.

The cyclical variations in cytoplasmic MTI in the hepatocytes in fasted rats and in the enterocytes in both fasted and zinc injected animals, are suggestive of circadian rhythm-like changes. Zinc administration abolished this variation in the hepatocyte cytoplasmic MTI, suggesting that MT in this compartment may be sensitive to exogenous zinc administration. In both liver and ileum, the nuclear MTI showed no marked cyclical variations. Further work is in progress to examine the possible origins of such variations in MTI in rat tissues.

Johanna P. Clarkson was supported by the Welsh Scheme for the Development of Health and Social Research, and was in receipt of a Royal Society travel grant.

Subject Index :

Immunoperoxidase staining/metal binding protein/zinc/circadian variations

References

1. Webb, M., Cain, K. : Biochem. Pharmacol. 31 (2), 137-142 (1982).

2. Danielson, K.G., Ohi, S., Huang, P.C. : J. Histochem. Cytochem. 30 (10), 1033-1039 (1982).

3. Danielson, K.G., Ohi, S., Huang, P.C. : Proc. Natl. Acad. Sci. USA 79, 2301-2304 (1982).

4. Banerjee, D.,Onosaka, S., Cherian, M.G. : Toxicol. 24, 95-105 (1982).

5. Janssens, A.R., van den Hamer, C.J.A., van Noord, M.J., Bosman, F. : Liver (Copenhagen) 2, 318 (1982).

6. Webb, M. : Biochem. Pharmacol. 21, 2751-2765 (1972).

7. Brady, F.O., Kafka, R.L. : Anal. Biochem. 98, 89-94, (1979).

8. Jasani, B., Wynford-Thomas, D., Williams, E.D. : J. Clin. Pathol. 34, 1000-1002 (1981).

9. Jasani, B., Thomas, N.D., Newman, G.R., Williams, E.D. : Immunol. Commun. 12, 51 (1983).

10. Graham, R.C., Karnovsky, M.J. : J. Histochem. Cytochem. 14, 291-302, (1966).

11. Marsh, M.N. : In : Recent Advances in Gastroenterology second edn. (eds. J. Badenoch, B.N. Brooke), London and Edinburgh, Churchill Livingstone, (1972).

12. Cousins, R.J. : Nutr. Rev. 37 (4), 97-103 (1979).

THE SPECIATION OF ALUMINIUM IN HUMAN BLOOD SERUM

Philip Ekow Gardiner*), Markus Stoeppler, and
Hans Wolfgang Nürnberg

Institute of Applied Physical Chemistry, Nuclear Research Centre (KFA) Juelich, P.O. Box 1913, D-5170 Juelich, Federal Republic of Germany

1. Introduction

Aluminium intoxication maybe the probable cause of dialysis encephalopathy, vitamin D resistant osteomalacia and microcytic anaemia (1) found in some renal failure patients on maintenance haemodialysis. These patients had been exposed to aluminium present as a contaminant in the dialysis fluid and through the ingestion of aluminium-containing phosphate binders. The action of this metal at the vulnerable sites, the brain and bone, is not understood. Moreover, very little information is available about the form in which aluminium is transported in human blood. This information is essential to the understanding of its toxic action.

In 1977, Kaehny et al. (2) reported that aluminium was bound to a nondialysable plasma constituent. According to Lundin et al. (3) this protein-bound fraction accounted for about 60% of the total aluminium in serum. Subsequent work by Gardiner (4) showed that between 10-30% of the aluminium was present as low molecular mass species and thus ultrafiltrable. The reminder was distributed prinicipally among albumin and the globulins. King et al. (5) demonstrated that aluminium was associated

*) Author to whom correspondence should be addressed.

Trace Element - Analytical Chemistry in Medicine and Biology, Vol. 3

with albumin, high molecular mass protein, low molecular mass protein and/or bound to inorganic species. Graf et al. (6) confirmed that 20% of the aluminium in serum was ultrafiltrable. More recently, the affinity of aluminium for transferrin has been demonstrated by Cochran et al. (7).

The results of the above-mentioned authors, taken together, show that there are two major aluminium fractions in serum; protein-bound and that associated with low molecular mass (LMM) constituents. Although it is relatively easy to identify the protein-bound fraction, the identification of the LMM species presents some difficulties. First, a number of ligands, for example, phosphate, carbonate, citrate, oxalate, fluoride and some organic ligands are all possible aluminium binders. The situation is further complicated by the possible formation of mixed-ligand complexes. Second, aluminium complexes with the ligands mentioned-above are generally colourless, therefore identification by spectrophotometric methods is not possible. Third, separation of the individual complexes by conventional methods is tedious and time consuming. In addition, during the fractionation there is the possibility of altering the species. Therefore, in order to understand the nature of the metal-ligand association, and to be able to identify the individual species, it will be helpful to find a simple in-vitro model and to use this to study the effect of likely binding ligands on the aluminium distribution profile.

In this preliminary study, gel permeation chromatography is used to fractionate aluminium-bound species in sera obtained from dialysis patients, and the aluminium distribution is compared with that obtained from samples treated in-vitro. In addition, the effect of pH on the Al distribution is studied using ultrafiltration.

2. Materials and Methods

2.1 Apparatus

The equipment used for the gel permeation chromatography experiments has been previously described (8). For the fractionation by ultrafiltration a polycarbonate filtration apparatus, Type SM 165 10 was used together with a cellulose acetate filter, (MM cut off 10 000) Type SM 165 39-047 N (both supplied by Sartorius). Argon was used as the pressure gas.

Aluminium determination was performed by electrothermal atomic absorption spectrometry. The equipment included a heated graphite atomiser Model 500, atomic absorption spectrophotometer Model 4000, data station Model 3600, and autosampler Model AS-40 (all supplied by Perkin-Elmer).

2.2 Reagents

Reagents for the chromatography experiments are as described in an earlier publication (8), except the addition of physiological amounts of calcium, magnesium, potassium and sodium as the chloride salts to the elution buffer solution.

The standards for the atomic absorption measurements were prepared from a 100 mg/l stock solution of aluminium chloride (Fixanal, Riedel-de-Häen AG, Seelze-Hannover).

2.3 Procedures

2.3.1 The removal of metal contamination and packing the column.

The removal of aluminium contamination from the buffer, column and apparatus is as described for zinc, copper and iron in a previous publication (8).

The column of Sephacryl S-300 gel was packed as prescribed by the manufacturer.

The membrane filters used for ultrafiltration were soaked in HCl at pH 4 overnight, and stored in deionised water until used.

2.3.2 Sample collection

Blood samples, obtained by standard venepuncture, were transferred into 10 ml polyethylene sample tubes. After being centrifuged at 2500 rpm (22 000 g) for 15 min., the serum was transferred into 5 ml polystyrene sample tubes with tight-fitting polyethylene caps. Serum samples from normal persons were pooled and stored at $4^{o}C$.

2.3.3 Sample fractionation

A 2 ml portion of the serum was applied to the column for chromatographic separation. After eluting 100 ml of buffer, at a flow rate of 30 ml hr^{-1}, one hundred 5.0 ml fractions were collected.

For the ultrafiltration experiments 20 ml of pooled serum or buffer was measured into the chamber, 40 μg of a 100 μg Al ml^{-1} solution was added and thoroughly mixed. A pressure of 100 KPa (1 bar) was applied. Three consecutive 300 μl fractions of the ultrafiltrate were collected. The whole apparatus was rotated vigorously for 30 sec. after each fraction has been collected. Fractions were stored at $4^{o}C$ until analysis.

2.3.4 Aluminium determination

Standards containing 0, 5, 10, 15, 20, 30 and 40 μg Al l^{-1} were prepared in 0.1 M Tris-acetate buffer. For the determination of Al in serum and filtrate, the standards were prepared in pooled serum and ultrafiltrate, respectively. The samples and standards were diluted (1 + 4) with 0.1% $^{w}/v$ Triton x-100 in 0.001 M HNO_3. Details of the method will be published. The concentration of analyte in each fraction was read off a calibration graph (absorbance vs concentration).

Instrumental conditions were set as given in Table 1.

Table 1 Instrumental conditions for the determination of aluminium in serum, ultrafiltrate and gel permeation chromatography column fractions[a,b].

			Temp (°C)	Ramp time(s)	Hold time(s)
Wavelength (nm)	309.3	Dry	100	5	25[c]
Lamp Current (mA)	20	Ash	1400	5	25
Spectral bandwidth (nm)	0.7	Atomise	2400	0	4[d]
Sample size (μl)	10	Clean	2600	1	5
		Cool.	20	10	20

[a] Background correction used

[b] Samples injected on a L'vov platform

[c] A time of 40s was used for the analysis of Al in the column fractions

[d] No gas flow through the tube during atomisation, and an integragtion time of 5s was set.

3. Results and Discussion

The distribution of aluminium, and the UV profile at 280 nm obtained from the serum of a patient on dialysis is shown in Fig. 1; two areas of interest identified by A and B are shown.

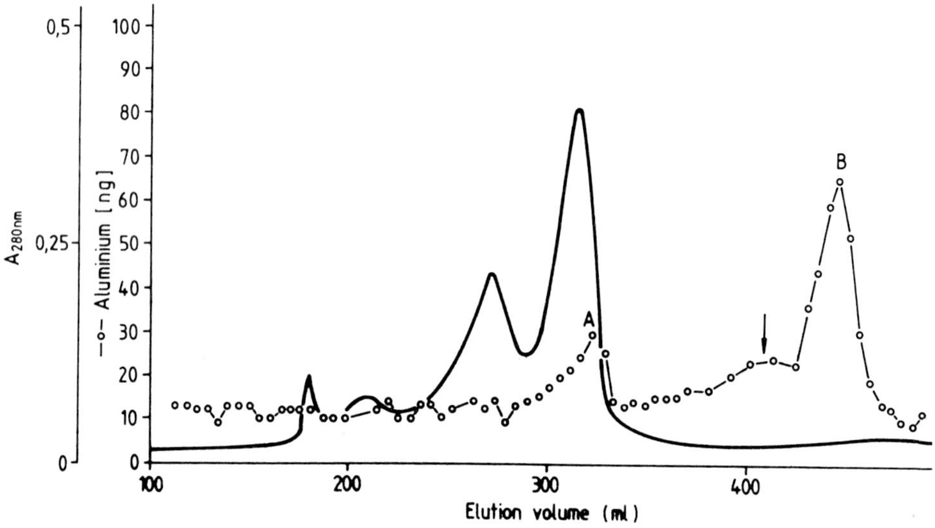

Figure 1 Distribution of aluminium (o) in human blood serum, obtained from a dialysis patient, after gel permeation chromatography on a Sephacryl S-300 column (2.6 x 94 cm). Mobile phase: 0.1 M Tris-Acetate, pH 7.4, Temp: 4°C. Flow rate: 30 ml h^{-1}. The continuous line is the UV profile at 280 nm.

Peak A is aluminium bound to albumin. However, because of the tailing of the rising edge of the peak, another protein may also be present. In other samples fractionated a shoulder was observed just before the aluminium-albumin peak. From the results of the analysis of protein in this region, and the work of Cochran et al. (7), transferrin maybe the likely protein. Peak B is aluminium bound to constituents with low molecular mass. Calibration of the column with proteins, the lowest with molecular mass of 13 700, and extrapolating the graph to cover this region, a molecular mass of less than 1000 was found. An accurate assessment of the molecular mass was not possible because of the lack of suitable calibration standards. Other techniques are being considered to help in the characterisation of this fraction. The presence of other minor aluminium species is shown by the arrow.

In the distribution shown here, and that from other samples so far fractionated, no evidence of a high molecular mass fraction (molecular mass greater than 100 000) was found. This is in disagreement with the findings of King et al. (4) who found two aluminium peaks. It is possible that the first of the two peaks is due to residual contamination on the column eluted in the void volume. A similar behaviour has been observed for copper, zinc and iron (9). Furthermore, a peak found in the low molecular mass region but attributed to aluminium bound to a protein (MM 60 000 - 80 000) that interacts with the gel material was also not observed in these samples.

In order to understand the nature of the interaction between aluminium and the serum constituents, in-vitro experiments were carried out. Aluminium was added to serum samples the pH of which had been adjusted to a predetermined value with 1.0 M HCl. In each sample the final concentration was 200 μg Al l^{-1} (7.4 μmol l^{-1}). The individual samples were subjected to gel permeation chromatography immediately after the addition of aluminium. A comparison between the aluminium profile found at pH 4.15 and that from a patient's serum at pH 7.4 is shown in Fig. 2. No significant protein binding was found in the in-vitro sample, and furthermore, the broadness of peak B was an indication that more than one aluminium-bound constituent was present in the sample. In addition, aluminium was most probably bound to inorganic ligands. It is interesting to note that as the pH was increased to about 6 only 10% of the aluminium added was recovered in this fraction, and no other association was observed. A similar experiment repeated at pH 7.4

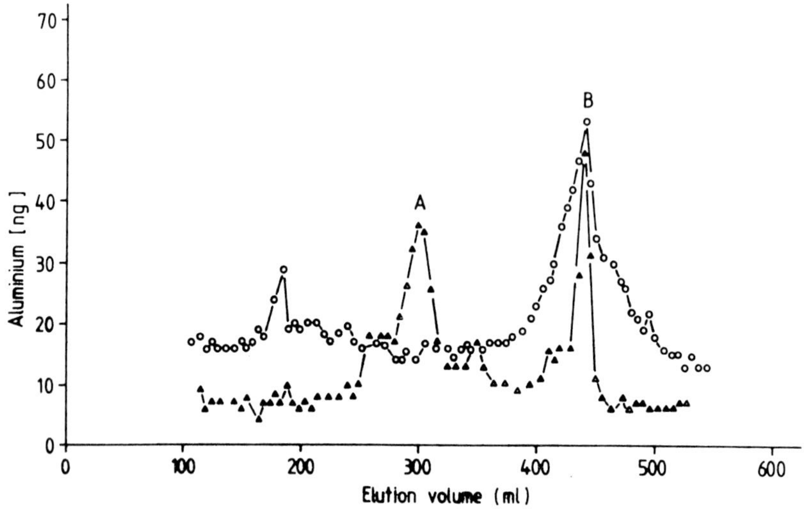

Figure 2 A comparison between the aluminium distribution in serum of a dialysis patient fractionated at pH 7.4, (▲) and pooled serum with aluminium added and fractionated at pH 4.15 (o).

gave the same results. No satisfactory explanation for this behaviour was found.

Further examination of the effect of pH on the aluminium distribution was carried out using ultrafiltration. The sera were prepared as described above. In addition the buffer solution used for the chromatography experiments was treated in a similar manner, except that the pH was adjusted with concentrated acetic acid. The results of these experiments are shown in Fig. 3. Some aluminium species that maybe present in the buffer solution are shown in this figure. At pH values between 4.0 and 5.0 all species present in the buffer solution are ultrafiltrable which is an indication that these are all low molecular mass species. In contrast, only about 50% of the

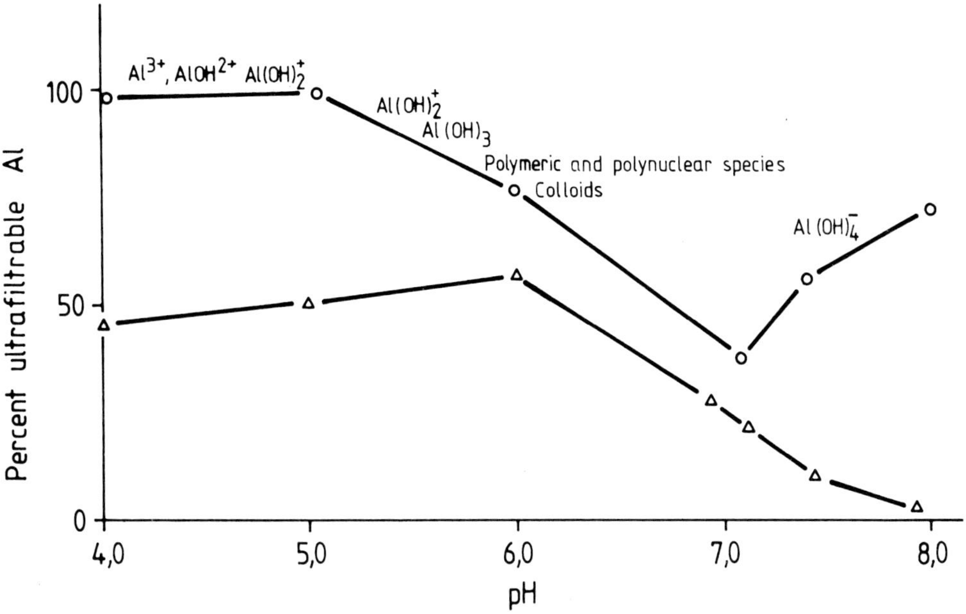

Figure 3 Changes in ultrafiltrable aluminium with pH in serum (Δ) and buffer solution (o). The most probable aluminium species present in the buffer solution are shown in the figure.

aluminium in serum is filtrable. This is in disagreement with the chromatography results which showed that at pH 4.15 all the aluminium was in the low molecular mass form. There are two possible explanations for this discrepancy. First, there maybe weak interactions between constituents in serum (molecular mass greater than 10 000) with various low molecular mass aluminium species. This interactions maybe too weak to survive the separation by chromatography. Second, under the conditions found in serum the aluminium species may have already started to polymerise.

The decrease in ultrafiltrable aluminium observed in the buffer solution between pH 5.0 and 7.0 is due to the formation of colloidial particles of $Al(OH)_3$, which are too large to permeate through the pores of the filter. Aluminium hydroxide starts to dissolve at pH values above 7.0 forming the aluminate, $Al(OH)_4^-$. This behaviour illustrates the well-known amphoteric nature of aluminium.

In contrast, between pH 6.0 and 8.0 the decrease in ultrafiltrable aluminium in serum is monotonic. Apparently the presence of some constituents in serum alter the behaviour of aluminium. It is interesting to note that at pH 7.4 the ultrafiltrable aluminium was 10% of the total aluminium in serum. This finding is in good agreement with previously published results (4,6).

After the ultrafiltration experiments it became clear that incubation of the serum after the addition of aluminium was essential. A comparison of the chromatogram obtained from a pooled serum sample with aluminium added and incubated at 37°C for 5 hrs with that from a dialysis patient is shown in Fig. 4. It can be seen that the major features are similar in both types of samples.

Four conclusions can be drawn from these series of experiments. Firstly, albumin and probably transferrin are the major proteins that bind aluminium. Secondly, added aluminium is taken up by serum constituents only after incubation. An indication that a slow exchange of aluminium between the species occurs in serum. Thirdly, the low molecular mass fraction is made up of mainly inorganic aluminium complexes. Finally, the pH of the serum determines the level of ultrafiltrable aluminium.

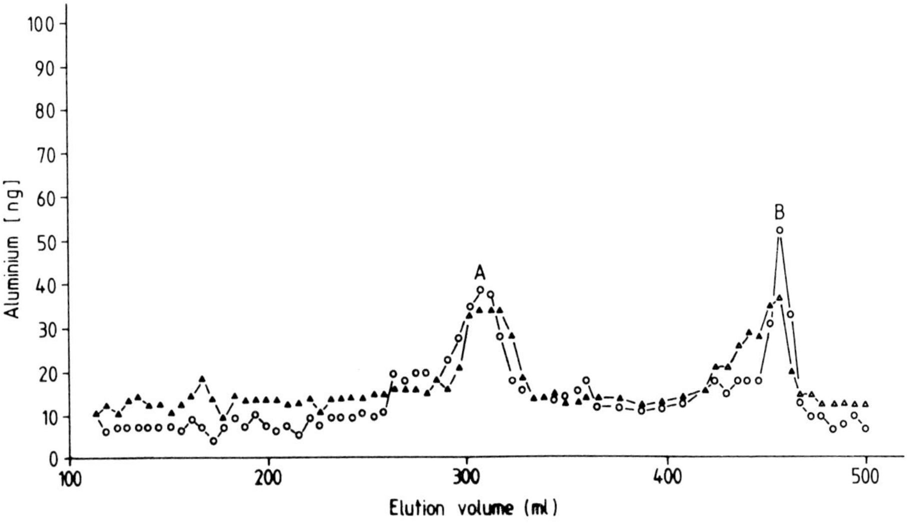

Figure 4 The elution profile of aluminium in the serum from a dialysis patient (o) and serum with aluminium added and incubated at 37 °C for 5 hrs (▲).

Acknowledgement

The authors wish to thank Drs. D.J. Halls, Glasgow Royal Infirmary, and O. Oster, Universitätsklinik, Mainz, for providing the samples from dialysis patients.

References

1. Elliott, H.L., MacDougall, A.I., Fell, G.S., Gardiner, P.H.E., Williams, E.D.: Dialysis and Transplantation 9 1027-1030 (1980)
2. Kaehny, W.D., Alfrey, A.C., Holman, R.E., and Shorr, W.J.: Kidney International, 12 361-365 (1977)
3. Lundin, A.P., Caruso, C., Sass, M., and Berlyne, G.M.: Clin. Res. 26 636A (1978)
4. Gardiner, P.H.E.: Ph.D. Thesis, The University of Strathclyde, (1979)
5. King, S.W., Savory, J., and Wills, M.R.: Ann. Clin. Lab. Sci. 12 143-149 (1982)
6. Graf, H., Stummvoll, H.K., Meisinger, U., Kovarik, J., Wolf, A., and Pinggera, W.F.: Kidney International 19 587-592 (1981)
7. Cochran, M., Patterson, D., and Coates, J.H.: Poster no. 73, 3rd International Workshop, Trace Element Analytical Chemistry in Medicine and Biology, April 16-18, (1984)
8. Gardiner, P.E., Rösick, E., Rösick, U., Brätter P., and Kynast, G.: Clin. Chim. Acta. 120 103-117 (1982)
9. Gardiner, P.E., Ottaway, J.M., Fell, G.S., and Burns, R.R.: Anal. Chim. Acta. 124 281-294 (1981)

PROTEIN BINDING OF ALUMINIUM IN PLASMA OF MAINTENANCE HAEMODIALYSIS PATIENTS.

M. Cochran,
Flinders Medical Centre, Adelaide.

D. Patterson,
Australian Mineral Development Laboratories, Adelaide.

J. H. Coates, P. T. H. Coates,
University of Adelaide, Adelaide

Introduction

Plasma aluminium usually increases during haemodialysis apparently against a concentration gradient (Fig 1) suggesting that there is binding of aluminium, most likely to a protein.

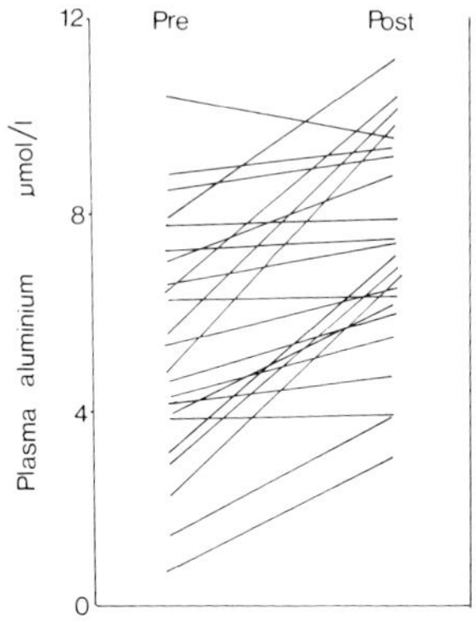

Figure 1. The change in plasma aluminium concentration during haemodialysis in 23 patients. Approximate dialysate aluminium concentration 2 umol/l.

Trace Element - Analytical Chemistry in Medicine and Biology, Vol. 3

To characterize the way in which the plasma carries aluminium, we have used standard gel filtration techniques with carefully minimized background aluminium, showing that transferrin is the major protein carrier of aluminium in plasma.

Methods

Post-dialysis plasma from 3 maintenance haemodialysis patients was pooled, concentrated to one-third volume on Amicon YM-5 membranes and then made up to two-thirds the original volume with buffer (25 mmol/l Tris HCl 100 mmol/l NaCl, pH 7.4) prepared from suprapure chemicals. Five successive 0.5 ml portions of the concentrated plasma were applied to a Sephacryl S-300 column (100 x 1.6 cm) and gel filtration carried out the protein content being monitored by the optical density at 280 nm. Between each procedure 0.5 ml EDTA (20 mmol/l) was filtered through the column to remove residual aluminium. Five sets of fractions (1.7 ml) were collected in an identical way and the aluminium estimated by electrothermal atomic absorption spectroscopy using a Varian-Tectron CRA-90 graphite furnace. Transferrin and albumin were measured by immuno-diffusion. Plasma (5 ml) was applied to a Sephadex G-50 column (50 x 2.5 cm) and after excluding of the void volume, the remaining fractions were pooled, concentrated as before, and again fractionated on a G-50 column (100 x 1.5 cm). The aluminium content of the fractions was determined.

Five samples of post-dialysis plasma were centrifuged for 90 minutes in a Beckman L5-65 centrifuge using a 50TI rotor to sediment all macroglobin. The aluminium concentration in the plasma and supernatant was compared.

Gel filtration was also carried out using pure human

transferrin with an equimolar quantity of $AlK(SO_4)_2$.

The binding capacity of transferrin for aluminium was examined by equilibrating portions of a solution of transferrin 35 umol/l in buffer (25 mmol/l Tris-HCl, 1.5 mmol/l $NaHCO_3$, 100 mmol/l NaCl, pH 7.4) against a range of concentrations of aluminium, such that the aluminium to transferrin molar ratio varied from 0-10. The mixtures were ultrafiltered by centrifugation using Amicon Centrifree membranes, and the aluminium in the filtrates estimated. The absorbance spectrum for transferrin and for the transferrin-aluminium complex were compared using a Pye-Unicam SP8-100 scanning spectrophotometer. A solution of pure transferrin (50 umol) in the buffer was then titrated with 10 umol aliquots of $AlK(SO_4)_2$ and the optical density at 240 nm monitored.

To estimate the affinity of transferrin for aluminium, by competitive inhibition of iron uptake, solutions of human transferrin in the buffer were equilibrated in a series of tubes with $AlK(SO_4)_2$ giving a range of concentrations from 0-200 nmol aluminium per nmol transferrin. Ferric ($^{59}Fe^{+++}$) citrate, in sodium citrate buffer (50 mmol/l, pH 7.4) was added to each tube to give 1.0 nmol Fe^{+++} per nmol transferrin. The mixture was shaken and left for exactly 30 minutes before being applied to Amicon Centrifree membranes and centrifuged. The radio-iron in the ultrafiltrate was measured, and the proportion of the total retained with transferrin calculated.

Transferrin was prepared by dissolving the crystalline material in buffered EDTA 20 mmol/l, pH 5.5, and then filtering through an Amicon YM5 membrane, followed by 3 washes with the Tris-NaCl buffer.

Results

In repeated fractionation of 5 portions of pooled plasma, analysis showed that elevated aluminium levels were present only in one zone where the concentration rose progressively in a clearly defined peak (Fig 2). A variable elevation of aluminium concentration occurred in the salts volume, but no clearly reproducible peak was observed. No peak of aluminium was identified following fractionation of plasma components on G-50.

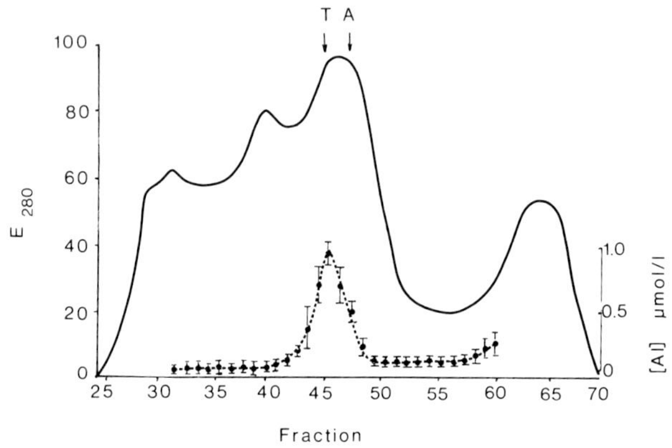

Figure 2. The aluminium concentration (mean $\pm$ SE) in fractions of post-dialysis plasma (broken line). The absorbance at 280 nm is shown (solid line). Transferrin (T) and albumin (A) peaks found by immunodiffusion are indicated.

The proteins present in the fractions containing aluminium were predominantly albumin and transferrin and could be readily distinguished by immuno-diffusion assay. Adjacent fractions showed a highly reproducible association between aluminium and transferrin with an almost constant mean value of 0.12 for the molar ratio (Table 1), but there was no consistent relationship between aluminium and albumin. The mixture of pure transferrin and aluminium yielded a peak of aluminium corresponding exactly to the eluted transferrin. A variable and inconstant peak of aluminium appeared with

Fraction Number	Transferrin G/l		Albumin G/l		Aluminium umol/l		Al/Transf. molar		Al/Albumin molar	
43	0.32	0	1.5	0.1	0.51	0.22	0.125	0.005	0.024	0.022
44	0.42	0.04	4.1	0.9	0.68	0.18	0.126	0.006	0.011	0.053
45	0.53	0.04	6.3	0.6	0.87	0.09	0.128	0.005	0.010	0.040
46	0.43	0.03	7.2	0.7	0.68	0.18	0.123	0.006	0.006	0.091
47	0.38	0.04	8.9	1.3	0.58	0.11	0.120	0.004	0.005	0.054
48	0.23	0.04	5.4	0.3	0.35	0.11	0.120	0.003	0.002	0.042

Table I. The concentration of transferrin, albumin and aluminium (mean and SE) in the fractions comprising the aluminium peak, with corresponding aluminium/protein molar ratios (mean and SE). From 5 analyses of pooled plasma.

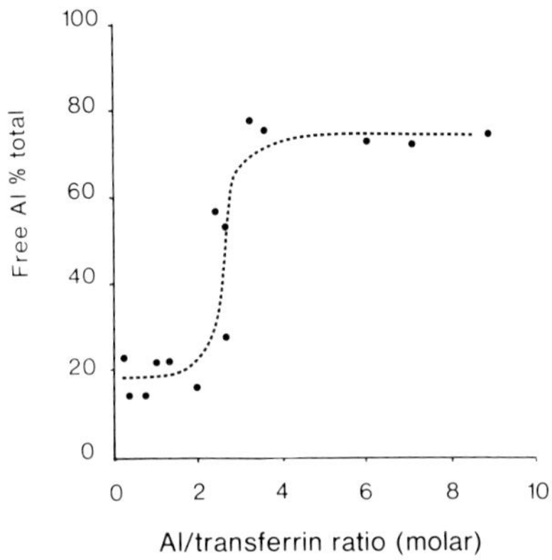

Figure 3. Binding of aluminium to transferrin following equilibration of increasing quantities of the metal to a constant amount of the protein. Free aluminium (% total) is shown in relation to the aluminium/transferrin molar ratio.

the void volume on both S-300 and G-50, though neither albumin nor transferrin was present. Following ultracentrifugation of plasma, the mean $\pm$ SE aluminium concentration decreased from 5.22 $\pm$ 2.09 umol/l to 4.50 $\pm$ 1.23 umol/l (not significant).

When solutions of a fixed concentration of transferrin were exposed to increasing concentrations of aluminium, at aluminium/transferrin molar ratios below 2, 17.8 $\pm$ 1.5% of the aluminium was free, compared to 71.0 $\pm$ 2.9% free when the aluminium/transferrin ratio exceeded 3 (Fig. 3), indicating that there are 2 binding sites on transferrin capable of being occupied by aluminium. The transferrin-aluminium complex showed a peak of optical density at 240 nm, whereas apotransferrin showed no such peak (Fig.4).

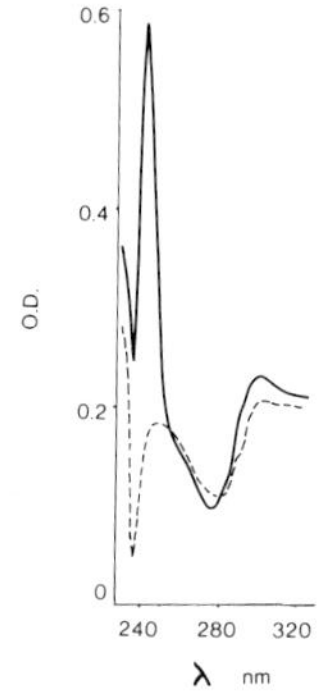

Figure 4. Absorbance spectrum of the transferrin- aluminium complex (solid line) and apotransferrin (broken line).

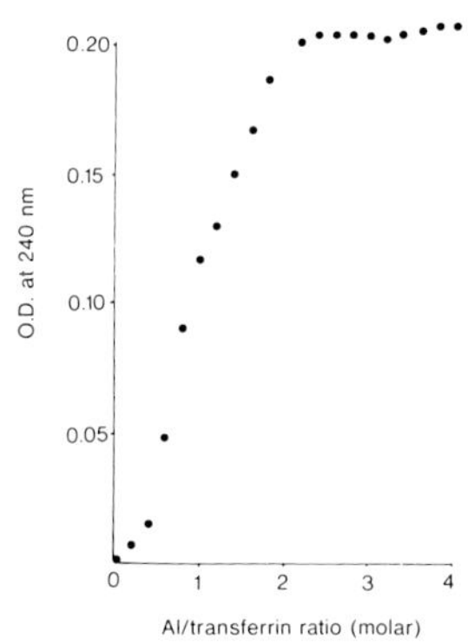

Figure 5. The change in optical density at 240 nm in relation to the aluminium-transferrin molar ratio.

When aluminium was titrated against transferrin in the presence of HCO_3^- a progressive rise in optical density at 240 nm was observed up to an aluminium- transferrin molar ratio of 2.0 when a plateau was reached (Fig. 5).There was also a small change in the gradient of absorbance on aluminium concentration at an aluminium-transferrin ratio of 1.0. When partly denatured protein was used, the same plateau occurred, but was followed by a further gradual increase in optical density at 240 nm. When the affinity of aluminium for transferrin was investigated by competitive interference with iron uptake, 50% inhibition was observed with an aluminium/iron molar ratio of 200-fold (Fig. 6).

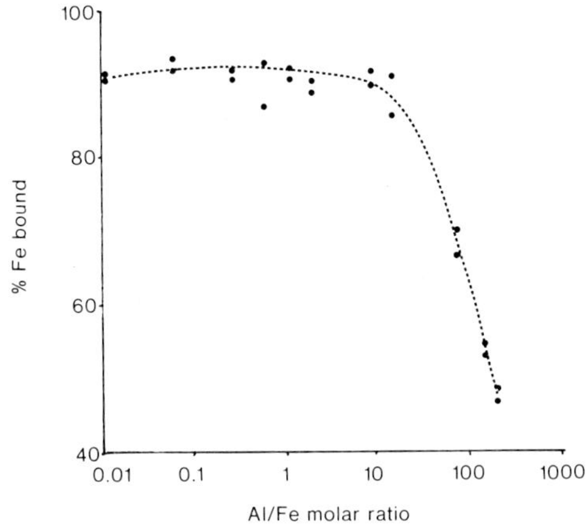

Figure 6. The effect on iron uptake of prior equilibration of transferrin with aluminium.

If the iron-transferrin binding constant is taken as 5 x 10^{23} (1), the affinity constant for transferrin and aluminium can be calculated from these data, with values between 2 x 10^{13} and 8 x 10^{13}, allowing for the hydrolysis of the citrate.

Discussion

These findings indicate that transferrin is the major, and probably sole, protein carrier of aluminium in plasma. Our data show no meaningful association between aluminium and albumin. However, non-specific binding may occur between aluminium and partly denatured transferrin. The data show also that both receptor sites are involved in aluminium uptake and permit an approximate value of 10^{13} to be derived for the effective association constant.

This binding is not sufficiently avid for aluminium to compete significantly with iron in the plasma but the affinity is high enough to make it unlikely that any loss of protein-bound aluminium would occur during haemodialysis and would also prevent urinary excretion of the metal (2). Thus, most of the plasma aluminium will exist as a transferrin-aluminium complex, which would account for all of the observations on plasma binding of the metal (3,4,5). We find no evidence of binding to any other plasma protein, but a small amount of aluminium would be associated in plasma with phosphate and citrate.

Acknowledgements

The authors thank Dr. D. Birkett for helpful advice, and Mr. A. Ciplys and Miss T. Mallick for technical assistance. The work was funded by Australian Kidney Foundation Grant G8/83.

References

1. Aasa, R., Malmstrom, B.G., Saltman, P., Vanngard, T.: Biochem. Biophys. Acta 75, 203-222 (1963).
2. Kovalchik, M.T., Kaehny, W.D., Hegg, A.P., Jackson, J.T., Alfrey, A.C.: J. Lab. Clin. Med. 92, 712-720 (1978).
3. Kaehny, W.D., Alfrey, A.C., Holman, R.E., Shorr, W.J.: Kidney Int. 12, 361-365 (1977).
4. Lundin, A.P., Caruso, C., Sass, M., Berlyne, G.M.: Clin. Res. 26, 363A (1978).
5. Hodge, K.C., Day, J.P., O'Hara, M., Ackrill, P., Ralston,A.J.: Lancet 2, 802-803 (1981).

D i s c u s s i o n Session V - 2

Subjects of the discussion

- clean up procedure for biological samples
- blanks caused by colour forming substances
- specificity of antibody
- integrity of Zn-metallothionein
- association of Al with transferrin
- nature of Al-containing low molecular mass fraction

Contributors: Cochran, Feinendegen, Ganther, Neidhart, Weser

Session VI

TRACE ELEMENTS IN CARDIOVASCULAR DISEASES

Chairman:

L.E. Feinendegen (FRG)

Poster Reports given by: L.E. Feinendegen (FRG)

TRACE ELEMENTS IN BREAST MILK IN ENDEMIC FLUOROSIS AREAS AND NORMAL AREAS.

K. Rajalakshmi Machari
National Institute of Nutrition, Tarnaka,
Hyderabad - 500 007.

Introduction

Breast milk constitutes the sole source of nourishment for infants in the developing countries, particularly in the rural population, in whom lactation is often prolonged, extending over 12 months. Thus the only source of trace element consumption during infancy in Indian infants is breast milk. Earlier studies have indicated that the major constituents of breast milk are not different in the breast milk of under nourished mothers. There have been no comprehensive studies on the trace elements in the breast milk of under nourished mothers. Also, it is important to know whether the trace element content of breast milk shows regional variation, particularly between areas where endemic genu valgum and endemic fluorosis are prevalent and non-endemic regions of rural India. It was felt important to assess whether breast milk trace element status could be considered as a reflection of the overall environmental and dietary status of trace elements. As a part of such programme, the following studies were carried out.

Materials and Methods

Subjects

1. Studies in non-endemic population groups
   a) Cross-sectional studies: The following groups of lactating mothers were included in the study.

Trace Element - Analytical Chemistry in Medicine and Biology, Vol. 3

Group I: High Income group mothers attending urban Nursing Homes - 100

Group II: Low Income group mothers from economically poorer segments of population attending Government Hospitals -412

Group III: Low Income rural mothers studied in their households in the villages - 208.

Group IV: Paired samples of breast milk and serum were studied in 151 women.

Group V: Paired samples of breast milk and serum of infants of the respective mothers - 41.

b) Semi-longitudinal studies: Twenty-four women belonging to urban low income group (Group II).

2. Studies in endemic fluorosis areas

a) Cross-sectional studies: The following group of lactating rural mothers were included in the study.

Group VI: Low Income group mothers from villages where endemic genu valgum was prevalent - 229.

Group VII: Low Income group mothers from villages where endemic fluorosis was prevalent but without genu valgum - 226.

Group VIII: Low Income group mothers from villages not endemic for fluorosis (Control rural) - 254.

b) Longitudinal studies: Twelve lactating mothers from group VI were investigated at weekly intervals for eleven weeks followed by monthly intervals for 10 months.

Methods

Information was obtained on maternal age, parity, duration of lactation and anthropometric indices.

Day to Day variation and diurnal variation were studied in 14 mothers in the group II and 12 mothers in group VI.

Sample Collection: Milk samples were collected by manual expression before the baby was due to be fed. Approximately 10 ml were delivered directly into pre-washed polythene bottles of 25 ml capacity. Blood samples of mothers were collected 3 hrs following usual breakfast. Serum samples from infants were collected by heel puncture.

Preparation of samples: Milk samples were centrifuged, fat layer removed and fat free portion was divided into 3 aliquots which were diluted by factors of 2, 5 and 100 for copper, zinc and magnesium estimations respectively. Serum samples of the mothers and infants were diluted with 8 per cent butonol made in deionized water and 1 in 5 dilution was used for both Cu and Zn determination.

Analysis: Care was taken at all stages to avoid contamination. Varian Techtron-100 AAS with air acetylene as fuel was used for Cu, Zn and Mg estimation. For the estimation of molybdenum, cobalt, chromium and manganese, samples were lyophilized and concentrated to 5 times. Flameless Varian Techtron Carbon-rod atomizer, Model AAS, CRA-90 was used for estimation. Instrument was checked with each estimation with standards prepared in serum equivalent solution as described earlier (1).

Observations

Studies in normal areas: In both high income and low income mothers Cu and Zn showed a similar trend. As the duration of lactation increased, the levels showed a downeard trend. Mean Cu content in colostrum was 0.46 μg/ml in the low income group. By 7-12 months, the value came down to 0.17 μg/ml. The corresponding values for zinc were 5.32 μg/ml and 1.12 μg/ml respectively. Magnesium, however, behaved differently. In colostrum the mean value was 35.1 μg/ml while it continued to be 30 μg/ml at the end of 13 months and above. This was observed both in urban and rural mothers.

The absolute values for Cu and Zn were significantly lower in the samples belonging to poorer income groups, especially between 1-3 months of lactation. Among the poorer mothers, no significant difference was observed between the urban and rural groups.

Multivariate analysis showed that age, parity and anthropometric measurements were not related to breast milk trace elements, Cu and Zn.

No correlation was observed between mature milk content of Cu and Zn and the corresponding mothers' serum values.

Molybdenum, chromium and cobalt showed a trend similar to copper and zinc. The mean values for the above three trace elements in colostrum were 35 ng/ml, 8 ng/ml and 96 ng/ml respectively. In mature milk the respective values were 14 ng/ml, 4.7 ng/ml and 9.4 ng/ml. In the case of manganese the values for colostrum and mature milk were 8.2 ng/ml and 4.7 ng/ml respectively. The values for the former three trace elements began to decline by 1-3 months while in the case of manganese, the levels were fairly high till 12 months of lactation.

Studies in Fluorosis areas: In all the study groups (Groups VI, VII and VIII) concentration of copper was high during the first few weeks of lactation. There was a slow decline in the values which in the case of copper remained at 50 per cent of colostrum values at the end of 18 months. Concentration of zinc remained 8 times greater than that of copper althrough the lactation period. Magnesium levels essentially remained unchanged till 18 months.

The mean values for Cu were significantly lower in milk samples of mothers of group VI, particularly among those whose parity was below 3. Also, mean Cu was lower among mothers below 30 years of age.

It was observed that 28.5 per cent of samples of group VI (endemic genu valgum) had Cu values below 100 μg/ml while in only 8.0 per cent and 12.0 per cent of samples in the other groups the values were below this level. It was also observed that only 15.4 per cent of samples of group VI had values for over 250 μg/ml in contrast to 23.6 per cent and 30 per cent observed in groups VII and VIII respectively. In the case of zinc and magnesium no significant differences were observed in the frequency distribution in the 3 study groups VIZ groups VI, VII and VIII.

Longitudinal study confirmed observations made in the cross-sectional study.

Discussion

Preliminary information was obtained regarding the frequency distribution of trace elements Cu, Zn and Mg in the breast milk of Indian women belonging to different socio-economic groups living in the endemic fluorosis areas and normal areas. In general, the highest contents of trace elements were observed in colostrum. There was a regular and slow decline in the breast milk contents of Cu and Zn with increasing periods of lactation. Mothers from lower socio-economic strata had quantitatively less Cu, Zn and Mg as compared to the counterparts belonging to higher strata of the society. This was a significant observation. Within a given socio-economic group, duration of lactation was the most important factor influencing the level of the trace elements in breast milk. All trace elements followed a similar pattern except magnesium and manganese.

Interestingly, in endemic genu valgum villages there was a tendency for copper values to be lower (2). More women had values towards the lower end of normal distribution and only a small proportion had values over 250 μg/ml. The reasons for this observation are not clear. In these villages children develop genu valgum. All children are exposed to high levels of fluoride from the time they start drinking water. It has been shown that subjects with endemic genu valgum have osteoporosis (3). They have been shown to have hyperparathyroidism (4). More interesting was the finding that children with genu valgum living in fluorosis areas have less copper content in their cortical bone compared to age matched children living in normal areas (5). Whether the observed low copper in the breast milk of mothers living in these villages in any way is a reflection of overall environmental copper deficiency is not known.

Mean values for Cu and Zn in mature milk reported here among normal mothers are similar to those reported for women from U.S.A. (6). Even among infants whose mothers show lower values for Cu, the serum levels are normal and they do not show any clinical manifestation of trace element deficiency clinically. It is important to assess quantitatively the actual consumption of copper by the infants living in endemic genu valgum areas. Earlier, an approximate estimate of copper intake by Indian infants born of low income mothers was was done (1). Similar studies are proposed in various population groups.

Acknowledgements

The author acknowledges with gratitude the financial support given by the Indian Council of Medical Research to attend the conference. Thanks are due to the Director, National Institute of Nutrition for his support. Secretarial assistance of Mrs. Lalitha Devi is thankfully acknowledged. Technical assistance was rendered by Mrs. V. Lakshmi K. Rao.

References

1. Rajalakshmi, K., Srikantia, S.G.: Amer. J. Clin. Nutr. 33, 664-669 (1980).

2. Krishnamachari, K.A.V.R., Rajalakshmi, K., Radhiah, C.: FLUORIDE 15, 81-87 (1982).

3. Krishnamachari, K.A.V.R., Kamala Krishnaswamy: Lancet 2, 877-879 (1973).

4. Sivakumar, B., Krishnamachari, K.A.V.R.: Hormone Metab. Res. 8, 317-319 (1976).

5. Krishnamachari, K.A.V.R.: FLUORIDE 15, 25-31 (1982).

6. Picciano, M.F., Guthrie, H.A.: Am. J. Clin. Nutr. 29, 242 (1976).

CORRELATION OF PLASMA CHROMIUM LEVELS WITH THE RESULTS OF CINEANGIOGRAPHIC STUDIES OF THE CORONARY ARTERIES

M. Simonoff, Y. Llabador, G.N. Simonoff
E.R.A. 144 - C.E.N.B.G. - Le Haut Vigneau - 33170 GRADIGNAN - France

Introduction

The possible implication of chromium in the etiology of cardiovascular disease is based on experimental and epidemiological evidence. Rats fed a chromium-deficient diet develop a diabetes-like syndrome and show increased incidence of aortic plaques (1). Chromium concentrations in aortas of subjects dying from atherosclerotic heart disease (2) are found to be significantly lower than in aortas of an unselected population (accident victims). Parenteral chromium on the other hand, has been shown to promote the regression of atherosclerotic plaques in cholesterol-fed rabbits (3).

The foregoing results do not establish a cause-effect relationship for the role of chromium in atherogenesis, but constitute presumptive evidence, reinforced by the role of chromium in the metabolism of glucides and lipids.

Trivalent chromium is an essential trace element when combined in an organic complex termed Glucose Tolerance Factor (GTF). Mertz (4) postulated that GTF acts as a cofactor for insulin. GTF is indispensable for normal glucide metabolism (5) and is also effective in the maintenance of serum lipids at normal levels (6). It raises the level of high-density lipoprotein (7), which is thought to have a protective action against the formation of atheroma (8, 9, 10).

The probable interrelations between the physiological role of chromium and atherogenesis induced us to measure plasma chromium levels for patients who were subjected to coronary artery cineangiography.

Trace Element - Analytical Chemistry in Medicine and Biology, Vol. 3

Proton-induced X-ray emission (PIXE) analysis was employed to measure the chromium concentrations in blood plasma from 150 patients.

Materials and Methods

1. Patients

The 150 subjects of this study were referred to the Division of Cardiology for selective coronary artery cineangiography. All of them suffered angor and the cineangiography was employed to distinguish those with true coronary artery disease from those with normal arteries.

The patients with known heart disease were subjected to both right and left catheterization. The others were subjected to left catheterization only.

Most of the heart disease patients had been treated and followed by the same department for several years and had recently developed angor in addition to their usual symptoms. This group of patients suffered from aortic insufficiency, mitral insufficiency, aortic stenosis with insufficiency, aortic insufficiency with mitral stenosis and myocardiopathy.

On the basis of the cineangiographic studies, the heart disease patients were divided into two groups : those with coronary artery disease (CAD) and those with heart disease but normal arteries (HD). Most of these patients had been under treatment with digitalin and diuretics over a prolonged period. The remaining subjects (no HD) fell into group CAD or were shown to have normal arteries (group N).

Detailed questionnaires and clinical data concerning the 150 patients showed that it was not possible to form groups characterized by levels of cholesterol, triglycerides, blood pressure, obesity or tabacco consumption, so that we could not attempt to correlate any of these factors with plasma chromium levels.

2. Blood Collection

The subjects were brought to the Cardiac Catheterization Laboratory in the fasting state and two or four blood collections were made in heparinized polyethylene test tubes.

2.1. Left catheterism (coronarography)

After local anaesthesis, a preformed teflon-covered catheter is introduced through a femoral artery. The tipe is positioned in the distal aorta and 19 to 20 mL of arterial blood are sampled for chromium analysis, the first 40 mL being rejected or stored for other purposes. Left ventricular and selective coronary artery cineangiography is then carried out with the same catheter. Again, 10 to 20 mL of blood are collected when the catheter is in the left ventricle.

2.2. Right catheterism

The catheter is introduced through a femoral vein and 10 to 20 mL of veinous blood are taken for chromium analysis. The catheter is then placed in the right ventricle and 10 to 20 mL of ventricular blood are collected. (The right catheterism is employed for the global cardiac evaluation of subjects with cardiac problems other than CAD).

3. Chromium Enrichement Procedure

We elaborated an extractive procedure (11) for the determination of trace chromium in blood plasma by PIXE, using less than 0.1 ng carrier-free ^{51}Cr for recovery determinations.

Wet ashing of 5 to 10 mL of plasma was carried out with perchloric and sulfuric acids. Complete oxidation from Cr(III) to Cr(VI) was made by $KMnO_4$(1%). Chromium VI was complexed with ammonium pyrrolidine dithiocarbamate-methyl isobutyl ketone at pH 2.4. After evaporation of the organic phase, the metallic residue was dissolved in nitric and hydrochloric acids. A known volume of a 50 ng/mL nickel solution was added as an internal standard for quantitative PIXE determinations and the solution reconcentrated. The few residual microliters were deposited on a polycarbonate foil which constituted the target. Recovery for this preconcentration step was measured by adding radiochromium (^{51}Cr,from the "Commissariat à l'Energie Atomique" C.E.A., Saclay).

Satisfactory values were found (11) for chromium in three NBS standard Reference Materials (tomato leaves, oyester tissue, water). The total recovery on the target varies considerably (usually 40 to 90%), owing to the difficulty of collecting and depositing the final few microliters on the target foil. The recovery determinations for chromium were accurate to within ± 4%. The detection limit was $\sim$ 0.3 ng/mL of plasma.

4. PIXE Method

The detailed PIXE analytical method used for this work has been published elsewhere (12).

A beam of 2.5 MeV protons was employed, using the 4 MeV Van de Graaff accelerator at the "Centre d'Etudes Nucléaires" in Bordeaux. The interaction between protons and target elements gives rise to X-rays with energies which depend on the atomic number of the element. Photon spectroscopy was carried out by means of an intrinsic germanium X-ray detector with a resolution of 180 eV (fwhm).

The sensitivity of the experimental arrangement is of the order of 10^{-9} to 10^{-10} g for all elements with $Z > 14$.

Results and discussion

The calculation of the concentration of chromium in a specimen is carried out by comparing the area of the characteristic chromium X-ray peak with that of an internal standard (e.g., nickel).

The spectra are analyzed on a graphic terminal connected to a CDC CYBER 750 computer.

Table Statistical analysis and distributions of plasma chromium levels in CAD, N and HD patients

Group	Weighted mean ± standard deviation (ng/mL)	Chromium level (ng/mL)	Fraction of group with chromium levels in given interval (%)
CAD (n = 67)	1.05 ± 0.11	⩽ 0.3	15
		0.3 - 1	30
		1 - 2	27
		2 - 4	15
		4 - 6	13
HD (n = 19)	1.72 ± 0.30	0.3 - 2	32
		2 - 4	11
		4 - 6	37
		6 - 8	16
		> 8	4
N (n = 23)	8.51 ± 0.47	4 - 6	20
		6 - 8	25
		8 - 10	35
		>10	20

At the Cardiac Catheterization Laboratory, the coronary arteries are made visible in multiple roentgenogram projections by injecting a solution of Telebrix. X-ray films are exposed with a cine camera. The films are reviewed by specialized interpreters. When coronary artery disease is present, the degree of obstruction is assessed in terms of the percentage of coronary artery lumen that is occluded. The results showed that the patients fell into three groups : coronary artery disease (CAD) ; heart disease, but relatively normal arteries (HD) and normal arteries with no heart disease (N).

The present results concern the analysis of 350 samples from 150 subjects, two-thirds of whom belonged to group CAD. The remaining third was divided approximately equally between groups HD and N. The results obtained are shown in the Table. Measurements for about twenty subjects were not carried to completion for a variety of reasons ; the first few samples analyzed showed a manganese peak (from $KMnO_4$) which partly masked that of chromium ; samples with yields < 40 % were eliminated (these low yields resulted from sputtering losses during the final concentration to micro-liter volumes). For ten CAD subjects, the chromium peak was not distinguishable from the "chemical" background due to wet ashing and the oxidation of Cr III to Cr VI. The plasma chromium level for these cases was probably less than 0.3 ng/mL, our limit of detection, and these values have not been included in the program for the calculation of the mean for group CAD ; inclusion of these values would have slightly lowered the weighted means shown.

The observed chromium levels showed no significant dependence on the location from which the blood sample had been taken. The chromium level assigned to each patient is the mean of the two or four measurements (depending on whether a single catheterism, or both left and right catheterism had been carried out). The spread between the 2 or 4 values for a single patient has been taken as representing the error of the measurement for that patient.

The table shows the distribution of chromium for each group of patients. These distributions (histograms) have been treated by the XFIT program (13) (giving a least-square fit for a single variable) in order to obtain a weighted mean and a standard deviation for each group.

Over the last twenty years, the measured levels of serum chromium have fallen from 100 to 0.1 ng/mL. The present results lie in a range starting below our limit of detection (0.3 ng/mL) and rising to an upper limit of about 10 ng/mL.

It seems improbable that our measurements suffer from systematic chromium pollution from the materials employed for blood collection, since no systematic dependence of chromium concentration on the location (method) of sampling was observed. As a further check, we measured serum chromium levels in samples from nine volunteers (doctors) with no heart disease. The samples were taken by venipuncture with a polypropylene-over-the-needle catheter (Intranule-Vygon). The serum chromium values obtained (8 ± 2ng/mL) agreed with those of group N, who had been subjected to catheterization and cineangiography.

Whether the present results represent absolute or only relative plasma chromium levels, they clearly distinguish between two populations : group CAD + group HD on the one hand, and group N on the other. No subject in group N had a chromium level < 4 ng/mL, whereas 87 % of the CAD group showed Cr levels less than that value. These results suggest that an upper limit may be established (6 ng/mL in this study) beyond which CAD may be considered to be extremely unlikely. Such an upper limit would make it possible to eliminate a proportion of cineangiographic examinations which presently serve only to demonstrate the absence of CAD.

Acknowledgements

The authors wish to thank Prof. P. Besse (Catheterization Laboratory, Hopital Cardiologique de Bordeaux) for his kind cooperation in providing the blood samples analyzed in this work. Many thanks are also to Prof. C. Conri for his encouragement and much useful discussion. They wish to thank Dr. A. Mackenzie Peers for help with the english text and M. C. Hamon for his careful technical assistance in chemistry.

References

1. Schroeder, H.A. : Am. J. Clin. Nutr. 21, 230 (1968)
2. Schroeder, H.A. : J. Chronic. Dis. 23, 123 (1970)
3. Abraham, A.S., Sonnenblick, M., Eini, M., Shemesh, O., Batt, A.P. : Am. J. Clin. Nutr. 33, 2294 (1980)
4. Mertz, W. : Nutr. Rev. 5, 129 (1975)
5. Committee on Dietary Allowances, Food and Nutrition Board. National Research Council Recommended Dietary Allowances 9th rev. ed. National Academy of Sciences, DC, Washington (1980)
6. Offenbacher, E.G., Pi-Sunyer, F.X. : Diabetes 29, 919 (1980)
7. Riales, R.P.D., Albrink, J.M. : Am. J. Clin. Nutr. 34, 2670 (1981)
8. Miller, G.J., Miller, N.D. : Lancet i 16 (1975)

9. Gordon, T., Castelli, W.P., Hjortland, M.C., Kannel, W.C., Dawber, T.R. : Am. J. Med. 62, 707 (1977)

10. Miller, N.E., Forde, O.H., Thelle, D.S., Mjos, O.D. : Lancet i, 965 (1977)

11. Simonoff, M., Llabador, Y., Hamon, C., Simonoff, G.N. : Anal. Chem. 56, 454 (1984)

12. Simonoff, M., Llabador, Y., Simonoff, G.N., Besse, P., Conri, C. : Nucl. Instr. Meth. (1984) (in press)

13. Bevington, P.R. : "Data Reduction and Error Analysis for the Physical Sciences" McGraw-Hill Book Co., New York (1969)

PLATELETS AND THROMBOSIS: SELENIUM AND OTHER MINERAL ELEMENTS

Josef Kiem and Ludwig Emil Feinendegen
Institute of Medicine, Nuclear Research Center Jülich GmbH, D-5170 Jülich, F.R.Germany

Introduction

Arteriosclerosis and thrombosis are responsible for about 50% of the deaths in the highly industrialized countries of today. Since nine years our laboratory is, among many other groups all over the world, trying to find out whether there are risk factors derived from trace element imbalances regarding these diseases. In 1978 we reported the concentrations of 13 trace elements in normal human platelets (1). Later on we detected several systematic errors which had affected those results (2, 3). Fig. 1 recalls the correct mean concentrations on a wet weight basis. The selenium concentration of normal human platelets is in the order of 500 ng/g. Therefore our observation made in 1978 (1) is still valid, namely that there is no other cell in the human body with a selenium concentration as high as in the platelets.

Fig. 1

Mean Concentrations on a Wet Weight Basis of Normal Human Platelets

Ag	20	ng/g	Fe	7.7	µg/g
Au	0.15	ng/g	K	2.7	mg/g
Cd	3.6	ng/g	Mo	2.1	ng/g
Co	4.7	ng/g	Rb	6.7	µg/g
Cr	4.0	ng/g	Sb	11.0	ng/g
Cs	35	ng/g	Se	0.50	µg/g
Cu	0.80	µg/g	Zn	30	µg/g

Trace Element - Analytical Chemistry in Medicine and Biology, Vol. 3

In 1980 we reported that platelets may tend to retain their selenium firmer during selenium depletion than do other cells (4). Fig. 2 shows that the lowest selenium concentration in platelets is still higher than the highest selenium concentration in erythrocytes and plasma. Stimulated by our observation that platelets are exceedingly rich in selenium and our suggestion that they might be the link between selenium deficiency and cardiovascular disease also other investigators started to study these cells under this aspect.

Fig. 2

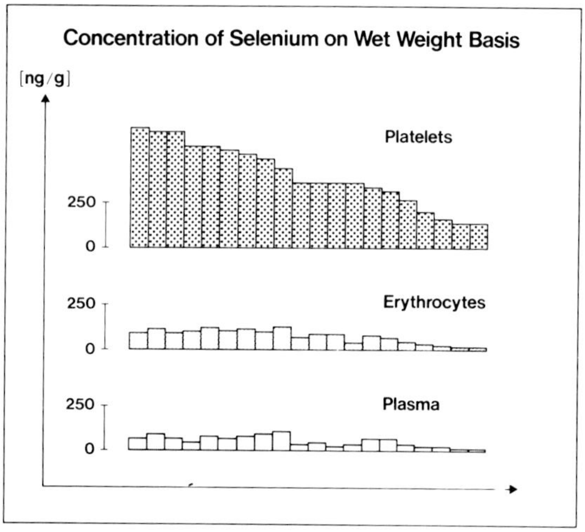

In 1980 Bryant et al. reported an altered platelet glutathione peroxidase activity and arachidonic acid metabolism after initiation of a selenium deficient diet and subsequent selenium repletion in a controlled human study (5). Fig. 3 recalls the data of these investigators from six subjects. The abscissa shows the duration in weeks of selenium depletion and repletion, adjusted to the minima of measured enzyme activity, given by the vertical line. The ordinate shows the activity of platelet glutathione peroxidase. The question now arises: could it be that the platelets which showed the low enzyme activities were sampled within a few days or even at the same day for all the six subjects ? We pose this question because in our laboratory we observed that the activity of platelet glutathione peroxidase and other biochemical parameters for platelets such as β-thromboglobulin in plasma, change on certain days without any clear reason after exclusion of methodological errors. Thus alterations in enzyme activity on single days may not only be caused by selenium imbalance but also by some other influences such as were reported from meteorological-geophysical characteristics of a particular day.

Fig. 3

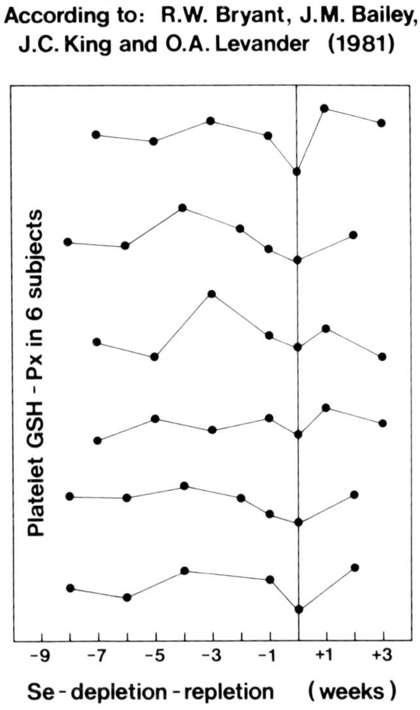

In 1981 we reported the activities of glutathione peroxidase in platelets from three groups of people, using three substrates in each (Fig. 4). Group A represents younger normal subjects, group B patients with acute myocardial infarction, group C older normal subjects. About 50% of the infarction patients had decreased activities of the enzyme. But it remained obscure whether the lowered enzyme activity in platelets was cause or effect of the disease (6).

Around 1980 Chinese investigators reported the successful treatment with selenium of the Keshan disease, an endemic cardiomyopathy in selenium-deficient areas of China (7). At present it is believed that heart muscle cells are directly affected by selenium deficiency in Keshan disease. Yet our own work permits us to put forward the hypothesis that platelets also play a role in the pathogenesis of Keshan disease, and that it is not the heart muscle cells which suffer from selenium deficiency but the platelets, which in turn would affect the myocardium by causing microthrombosis. The known histological features and specific cardiac symptoms do not contradict the triggering involvement of altered platelets in this disease.

Fig. 4

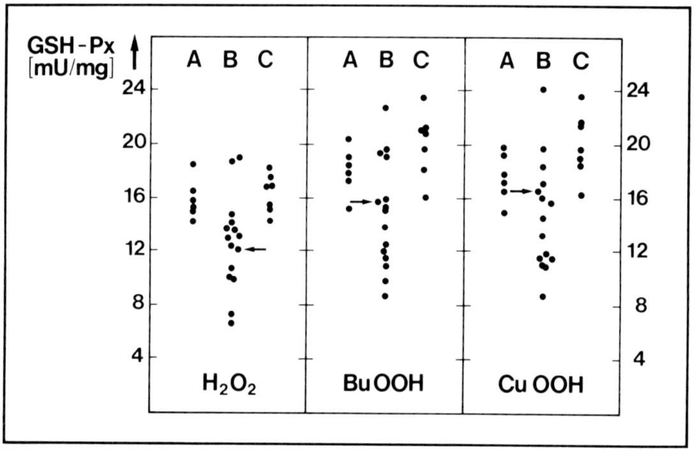

Results and Discussion

We have chosen to measure selenium in platelets on a wet weight basis using the technics described previously (1, 3, 6). Fig. 5 shows on the left side the values from 13 normal subjects, on the right side those from 8 patients with acute myocardial infarction which appear spread. If we assume that there is, in this group, no double-peak distribution, which

Fig. 5

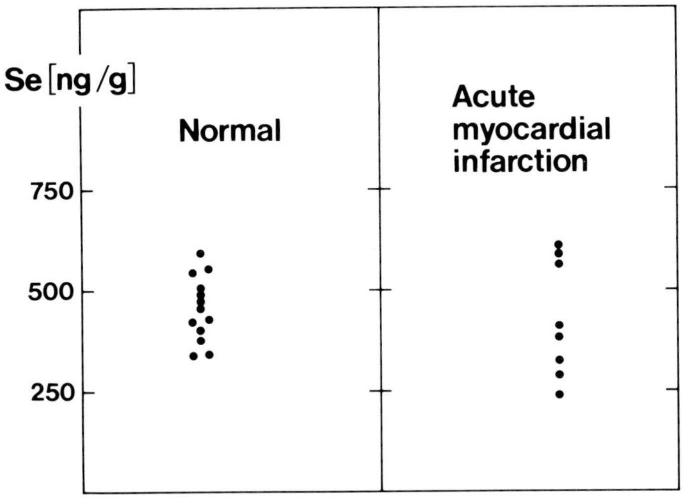

can not be excluded, there is no significant difference between the two groups (P(t) = 0.73).

Fig. 6

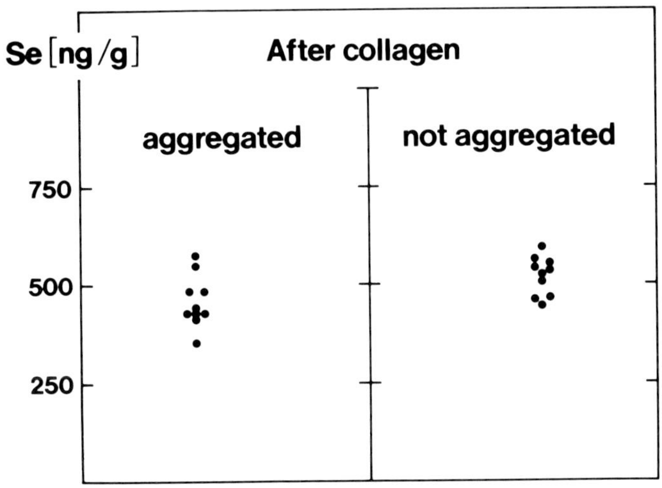

Fig. 6 shows the effect of adding collagen in a concentration of 1 µg per 1 ml of platelet rich plasma from 10 normal subjects. It is known that collagen is an important constituent of the blood vessel wall and often exposed to the blood cells in arteriosclerosis. Therefore it is the most physiological stimulator for platelet aggregation. After stimulation of the platelets by collagen they were separated into two populations by differential centrifugation. Platelets which aggregated after adding collagen have a significantly (P(t) = 0.98) lower selenium concentration, as shown on the left, compared to the right side listing the data from platelets which did not aggregate under identical conditions. These results could be due either to collagen-induced loss of selenium from platelets or due to an aggregating effect of collagen on selenium-deficient platelets. In order to differentiate between these two mechanisms another measurement was made (Fig. 7).

Fig. 7

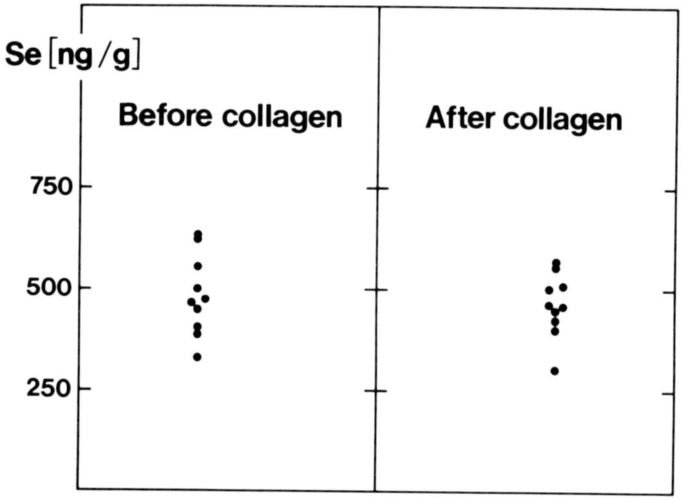

On the left there are data for platelets from the same 10 subjects as before, not treated with collagen, and on the right, for the same population of platelets treated with collagen. By phase contrast microscopy we confirmed that collagen-treatment aggregated most of the platelets whereas there was no significant aggregation without collagen. There is no significant difference between these two groups (P(t) = 0.65) and thus no indication of selenium loss from platelets due to aggregation. Therefore it appears that collagen preferentially acts on selenium-deficient platelets, or in other words, platelets with a higher selenium concentration are less prone to aggregate. Possible systematic errors or pitfalls as they may derive from changes and heterogeneity of platelet size, shape, turgor and protein fraction need further consideration.

Conclusions

The unusually high concentration of selenium in platelets suggests a particular function. It appears that selenium rich platelets are relatively resistent to the aggregating effect of collagen, provided that systematic errors are excluded.

References

1. Kiem, J., Iyengar, G.V., Borberg, H., Kasperek, K., Siegers, M., Feinendegen, L.E., Gross, R.: Proc. Nuclear Activation Techniques in the Life Sciences 1978, IAEA-SM-227/53, 143-164, IAEA, Vienna, 1979.
2. Kiem, J., Feinendegen, L.E.: Proc. Schadstoffe in Nahrungsketten 1982, AGF, Bonn, 1982.
3. Kiem, J., Weese, H., Feinendegen, L.E.: Fresenius Z. Anal. Chem. 317, 438-440 (1984).
4. Kiem, J., Iyengar, G.V., Kasperek, K., Lombeck, I., Wang, Y.X., Feinendegen, L.E.: Proc. Mineral Elements '80, Part I, 253-257, Academy of Finland, Helsinki, 1981.
5. Bryant, R.W., Bailey, J.M., King, J.C., Levander, O.A.: Proc. Selenium in Biology and Medicine 1980, 395-399, AVI Publishing Company, Westport, Connecticut, 1981.
6. Wang, Y.X., Böcker, K., Reuter, H., Kiem, J., Kasperek, K., Iyengar, G.V., Loogen, F., Gross, R., Feinendegen, L.E.: Klin. Wschr. 59, 817-818 (1981).
7. Chen, X., Yang, G., Chen, J., Chen, X., Wen, Z., Ge, K.: Biol. Trace Elem. Res. 2, 91-107 (1980).

ZINC AND COPPER CONCENTRATIONS IN PLASMA AND ERYTHROCYTES DURING OPERATIONS WITH THE CARDIOPULMONARY BYPASS

Dietrich Gulba

Medizinische Hochschule Hannover, Abtl. f. Kardiologie, Konstanty-Gutschow-Straße 8, D-3000 Hannover 61, FRG

Wolfgang Heller, Hans-Eberhard Hoffmeister

Eberhard-Karls-Universität Tübingen, Abtl. f. Thorax-, Herz- und Gefäßchirurgie, Calwerstraße 7, D-7400 Tübingen 1, FRG

Introduction

In recent papers we reported from the influence of the cardiopulmonary bypass (CPB) on the 2+ trace element serum levels. From our findings we concluded a pre and post bypassal, stress dependent influx of zinc into tissue cells (5, 8). From increasing serum zinc concentrations during the total CPB however an efflux of zinc ions out of ischemic injured cells was concluded (5, 7). On the other hand copper levels during CPB dropped only as a matter of dilution while the p. o. increase in serum copper was due to supplementation (7). However, these studies were only undertaken on serum levels.
Blood consits of two 'compartments' - plasma and cells, mainly erythrocytes. Thus, the observed changes of the serum zinc levels might not have been caused by tissue effects, as was interpreted. It also seems resonable, that these changes might be due to fluxes between the two blood compartments. On the other hand fluxes between blood cells and serum might as well have vailed tissue accumulation or loss of tissue copper. The present studies were undertaken to elucidate, if fluxes between the two blood compartments really do not need to be considered. Thus, we measured zinc and copper levels in serum

Trace Element - Analytical Chemistry in Medicine and Biology, Vol. 3

and in whole blood simultaneously. From these data and from the hematocrit it is possible to calculate the concentrations of these trace elements in blood cells.

Materials and methods

Investigations were done in 11 patients (1 ♀, 10 ♂) suffering from coronary artery disease (CAD) and undergoing aorto coronary bypass grafting (ACBG) with a mean age of 54.3 ± 8.7 years and a mean weight of 79.5 ± 8.7 kg. The mean time of extracorporal circulation was 62.3 ± 26.5 min. and the total time of ischemia was 33.2 ± 15.2 min. in the mean. The lowest blood pressure during CPB was measured 46.2 ± 12.2 mm Hg. Investigations were done using a heart lung maschine type 3000 supplied by SARNS Inc., Ann Arbor, Michigan, USA. The cooling unit was supplied by SARNS Inc., Ann Arbor, Michigan, USA too. Filters Pall Ultipor from TRAVENOL GmbH, Munich, FRG and Bently Q 220 from BENTLY Inc., Irvine, California, USA were used in combination. Oxygenators Spiroflo BOS 10 were supplied by BENTLY Inc., Irvine, California, USA. A period of at least 3 min. with partial bypass was kept before the total bypass was set up, because the bypass thus was better tolerated (16). Anticoagulation was done with heparin. The dose was calculated as the mean of 400 U/kg body weight and 9000 U/m^2 body surface (1.93 ± 0.11 m^2); after one hour on bypass an additional 1/3 to 1/2 of the initial dose was given, depending on the expected bypass time. The heparin used was Thrombophob^R supplied by NORDMARK WERKE GmbH, Hamburg, FRG.

In the first 24h after the CPB approx. 1000 ml/m^2 body surface were infused. Always included in this volume were 500 ml of Inzolen Infusio containing K^+-, Mg^{2+}-, Cu^{2+}-, Zn^{2+}-, Mn^{2+}- aspartate and sorbit. For cardioplegia solution LK 352 was used. LK 352 as well as Inzolen Infusio^R were supplied by DR. FRANZ KÖHLER CHEMIE GmbH, Alsbach, FRG. The prime volume consisted of one half of whole blood and one half of Ringer's lactate solution.

Blood sampling was performed at six defined times. The first sample was collected pre narcosis, the second 10' into operation, the third at the beginning of CPB, the forth 10' after CPB and two samples in the p. o. period (24h p.o. and 48h p.o.). For serum electrolyte determinations fresh whole blood was centrifuged at 3000 g and the serum such gained was stored at -20°C until the measurements were performed. For determinations in whole blood, fresh whole blood was stored at -20°C until the determinations were performed. Complete decomposition of the blood cells was achieved by dilution with aqua dest., for zinc determinations blood was diluted 1 : 20 and for copper determinations 1 : 1 with aqua dest.. This simple decomposition procedure was chosen, as more sophisticated procedures did not result in higher accuracy, and values obtained by this procedure and by other decomposition procedures were comparable (6, 10). Atomic absorption measurements were performed on a PERKIN ELMER 4000 atomic absorption spectrophotometer. Cellular concentrations were calculated using the following formula:

$$c_c = c_b - (c_s \cdot Hk)/100$$

c_c: conc. in blood cells; c_b: conc. in whole blood; c_s: conc. in serum; Hk: hematokrit.

For statistics Fischer's t-Test was used. Calculations were done on a HP 85 HEWLETT PACKARD calculator.

Results

Pre operative zinc serum levels are measured 82.82 ± 15.43 µg/dl. They dropped and 68.55 ± 13.18 µg/dl were measured at the beginning of CPB ($p < 0,02$). During and in the first ten minutes after CPB they rise slightly to 73.64 ± 13.85 µg/dl (n.s.). In the p.o. period up to 48h p.o. they drop

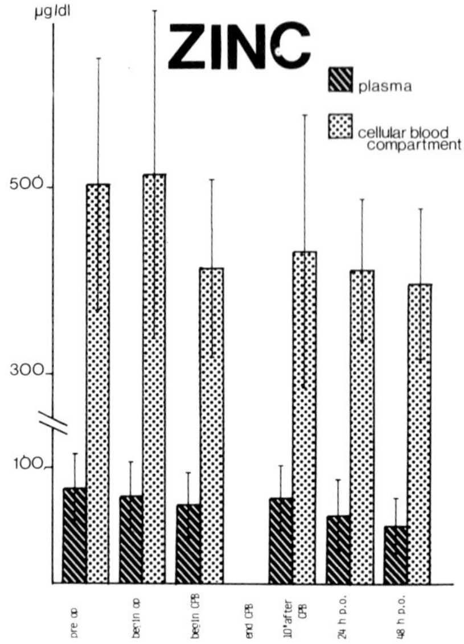

Fig. 1: Zn^{2+} conc. in serum and erythrocytes during operations with CPB.

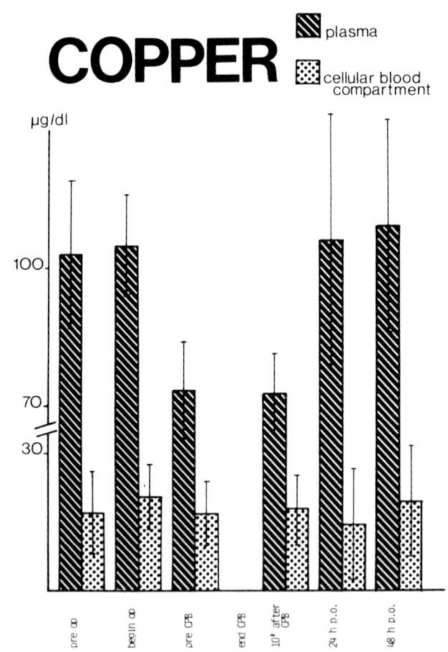

Fig. 2: Cu^{2+} conc. in serum and erythrocytes during operations with CPB.

to 50,45 ± 11,33 μg/dl ($p < 0.001$). In blood cells zinc values are measured 503,45 ± 134,84 μg/dl pre narcosis. This levels drop to 411,27 + 94,43 μg/dl at the beginning of CPB ($p < 0,05$). During CPB and in the first 10' after CPB these levels rise slightly to 429,27 ± 145,35 μg/dl (n.s.). In the p.o. period up to 48h p.o. zinc conc. in blood cells drop slightly to 397,82 ± 79,03 μg/dl (n.s.). (Tab. 1, Fig. 1).

sample		pre op	beg op	beg CPB	10' p CPB	24h po	48h po
Zn^{2+}							
serum	x	82,62	75,45	68,55	73,64	58,91	50,45
	s	15,43	16,2	13,18	13,85	16,28	11,33
ery	x	503,45	512,82	411,27	429,27	410,45	397,82
	s	134,84	172,16	94,43	145,35	75,56	79,03
Cu^{2+}							
serum	x	103,63	94,82	73,45	72,72	106,27	109,73
	s	16,1	11,06	10,64	8,32	26,95	23,12
ery	x	17,27	20,64	16,91	18,	14,55	19,36
	s	8,89	6,82	6,85	7,59	11,61	16,25

Tab. 1: Zn^{2+} and Cu^{2+} conc. in serum and erythrocytes (ery) during operations with CPB.

Copper levels at the beginning of CPB are measured 103,64 ± 16,1 μg/dl. They drop to 73,45 ± 10,64 μg/dl at the begining of the CPB ($p < 0,001$). During CPB they are almost unchanged and 10' after CPB they are measured 72,73 ± 8,32 μg/dl. In the first 24h p.o. they increase up to 106,27 ± 26,95 μg/dl ($p < 0,001$) and are almost unchanged after the following 24h. Copper conc. in the erythrocytes before narcosis are measured 17,27 ± 8,89 μg/ml and all differences in copper levels throughout the whole observation time are random. (Fig. 2., Tab. 1).

Discussion

In polytraumatized patients (15), after severe burns (9), after myocardial infarctions (14, 15) and after heart operations (9, 15), a rapid decrease of zinc serum levels has been reported. At the same time tissue zinc conc. increase (9, 15). This effect is caused by a stress related release of corticotrophins, which stimmulate the uptake of zinc into tissue cells (4, 17, 18).

In our previous studies we found increasing zinc levels during CPB instead of a stress dependent decrease (5, 8). In fatal cases, in the p.o. period too, we could demonstrate increasing zinc levels instead of the expected decrease of serum zinc conc. (5, 8). Flynn et al. reported from a positive relation between serum ACTH conc. and serum zinc levels in patients suffering from severe CAD. From our previous studies we concluded that the increasing zinc levels are caused by ischemic cell injury, and that the real zinc 'wash out' out of the ischemic injured cells - which is biased by the continuing, stress dependent tissue zinc uptake - might be much higher than can be read from our data (5, 8). These findings are in accordance with those of Spiekermann et al. who, when setting artificial infarctions in hearts of bastard dogs, found zinc conc. in the coronary lymph, which exceeded the zinc conc. measured in the arterial blood up to 100 fold (11).

In the present studies serum zinc levels decrease until the begining of the CPB ($p < 0,02$), increase slightly during CPB (n. s.), and decrease again in the p.o. period ($p < 0,001$). Cellular zinc levels at the beginning of the operation slightly increase (n. s.) and decrease some 20 % at the beginning of the CPB ($p < 0,05$). They increase slightly but insignificantly during the CPB and decrease again slightly in the p.o. period (n. s.).

Taking into account the hematocrit, the real effects are elucidated. The decrease of serum zinc levels in the pre op. period is due to fluxes between the two blood compartments to a great extend. As at the same time a comparable increase of the zinc conc. in the blood cells occurs, the pre op. stress effect on zinc is over estimated, when only serum levels are regarded. On the other hand in the period at the beginning of the CPB fluxes of zinc out of the erythrocytes are more than one magnitude higher than that measured in serum (100,5 µg/dl compared to 6,9 µg/dl). Thus when only serum levels are regarded the real effect on blood zinc conc. is underestimated for magnitudes. During CPB a slight but insignificant increase in serum zinc as well as in the cellular zinc is detected.

In our previous studies, when only the total CPB was studied, the increase in serum zinc levels was more than double the effect shown in the present studies (5, 8). It seems plausible, that the massive efflux of zinc out of the erythrocytes and out of serum would continue throughout the whole operation time, if they were not compensated by other effects. During CPB circulation conditions are comparable to a cardiogenic shock syndrom (12, 13). Thus the oxygen supply of tissues, during this period, might not be sufficient. Cells, thus undersupplied with oxygen, can not maintaine their membrane functions. As intra cellular zinc conc. are 10 fold the conc. in serum, an efflux of zinc out of these cells is the result of the ischemic injury. Especially the heart muscle seems to be jopardized, which we could already demonstrate in our comparative studies between coronary and peripheral serum zinc, magnesium, and calcium levels (5, 8). When recirculation of blood through the coronary arteries is partly restalled, the oxygen supply of the heart muscle is improved. It seems plausible, that, when the effects at the end of the CPB are compared to the effects at the end of the total CPB, the later effects must be higher to some extend.

The p.o. stress dependent zinc decrease is seen in both, serum and blood cells. The tissue zinc uptake in the p.o. period is much higher than that in the pre op. period. However, this effect is overestimated as compared to the operational stress effects, when only serum levels are measured.

Serum copper levels decrease some 20 % at the beginning of the CPB which is due to dilution from the prime volume. They rapidly normalize in the p.o. period. This is due to supplementation by Inzolen Infusio[R], which is part of our p.o. standard treatment. The observed effects on the intra cellular copper conc. are random. Thus, as already stated in previous studies (7), a spezial influence of the CPB on copper conc. in blood and tissues can be excluded.

Summary

During heart operations and in the p.o. period zinc, stress dependently, shifts from blood to tissues. These effects can only be estimated in the correct proportions, when serum levels and the conc. in the cellular blood compartment are regarded simultaneously. Thus, when only serum levels are measured, the operational stress effects on zinc is under- and the pre and post operational stress effects are overestimated for magnitudes. During CPB the stress related zinc decrease is overcompensated. This effect presumably is caused by an efflux of zinc out of tissues due to ischemic cell injury. The heart muscle seems to be mostly jopardized by the ischemic conditions during CPB. Serum copper levels and cellular copper conc. seem to be not influenced by the CPB.

References

1) Dieter, R.A.jr.; Neville, W.E.; Piffarré, R.: J. Thorac. Cardiovasc. Surg. 59, 168 (1970)

2) Flynn, A.: Lancet I, 598 (1972)

3) Flynn, A.; Pories, W.J.; Strain, W.H.; Hill, O.A.: Lancet II, 235 (1972)

4) Fodor, L.; Dölp, R.; Eschner, J.; Ahnefeld, F.W.: Anaesthesist 22, 393 (1973)

5) Heller, W.; Gulba, D.; Hoffmeister, H.-E.: Med. Welt 34, 1184 (1983)

6) Gulba, D.; Heller, W.; Hoffmeister, H.-E.: unpublished

7) Gulba, D.; Heller, W.; Hoffmeister, H.-E.; Huth, C.: In: Proceedings of the International Congress on Health Effects and Interactions of Essential and Toxic Elements, in press

8) Gulba, D.; Heller, W.; Hoffmeister, H.-E.? Walker, H.: In: Trace Element - Analytical Chemistry in Medicine and Biology, Vol. 2; Brätter, P.; Schramel, P. (eds.); Walter de Gruyter, Berlin, pp. 206 (1983)

9) Heller, W.; Domres, B.: In: Trace Element - Analytical Chemistry in Medicine and Biology Vol. 1; Brätter, P.; Schramel, P. (eds.), Walter de Gruyter, Berlin, pp. 117 (1980)

10) Köhler, F.: personal communication

11) Kahles, H.; Fuchs, C.; Nordbeck, H.; Preusse, C.J.; Spieckermann, G.: In: Kalium-, Magnesium-, Zink- Aspartat; Kronschwitz, H. (Hrg.); pp. 207 (1975)

12) Lillehei, R.C.: Surgery 56, 247 (1964)

13) Lillehei, R.C.: In: Schock, Pathogenese und Therapie; Euler, U.S.v.; Bock, K.D. (Hrg); Springer, Berlin, p. 348 (1962)

14) Lindemann, R.D.; Bottomley, R.G.; Cornellison, R.L.; Jacobs, L.A.: J. Lab. Clin. Med. 79, 452 (1972)

15) Seeling, W.; Ahnefeld, F.W.; Dick, W.; Fodor, L.: Anaesthesist 24, 329 (1975)

16) Stunkat, H.: Habilitationsschrift, Tübingen (1975)

17) Tawodzera, P.B.P.; Bell, R.M.S.; Jones, J.J.: Lancet I, 1072 (1972)

18) Tawodzer, P.B.P.; Bell, R.M.J.; Jones, J.J.: Lancet II, 767 (1972)

ZINC SUBSTITUTION UNDER EXTREME EXCRETION DURING AND AFTER EXTRACORPOREAL CIRCULATION

A.Schroll, A.Brehm, E.Struck, N.Mendler, J.Richter*, F.Sebening

Klinik für Herz- und Gefäßchirurgie, Institut für Anästhesiologie*, Deutsches Herzzentrum München, D-8000 München 2, Lothstr. 11

Introduction

A wide spectrum of symptoms related to electrolyte deficiency is known to medicine today and has been extensively described in the literature. Clinical experience is best documented for potassium and its critical relation to cardiac function (1,4). Although adequate supply of potassium is therefore generally considered, in recent studies a serious potassium deficiency was demonstrated as early as one week after the onset of diuretic therapy for arterial hypertension (7). Following the availability of routine magnesium determinations by the introduction of commercial AAS equipment in the last decade much knowledge about the physiological importance of this cation has been accumulated (3). Its relation to myocardial function comes close to potassium. Deficiency of magnesium may effect cardiac arrhythmias, muscular and vascular spasms including coronary angiospasm (5). Commonly potassium and magnesium deficiency will present concomitantly. Latent magnesium deficiency is likely to exist in an appreciable subpopulation of the industrial societies (6). It is probably due to its slow and chronic development that symptoms are generally rare and become manifest only in response to stress. Clinical symptoms under hypomagnesiemia affect most organs: the heart and circulation (tachycardia, arrhythmia, hypertonia, ventricular fibrilla-

Trace Element - Analytical Chemistry in Medicine and Biology, Vol. 3

tion, stenocardia, cardiospasm), the central nervous system (vertigo, depression, fear, halluzination, paresthesia), the intestinal system (vomitus, pylorospasm) and the muscle (convulsions, tremor, hyperreflexia, tetanic spasm, athetotic movements, laryngospasm, premature uterine contractions).

Not only for these physiological electrolytes in the mmolar concentration range many symptoms related to deficiency have been demonstrated, but also for trace elements in the µmolar range like Iron, Copper and Zinc. Medical terminology defines trace elements as elements, whose plasma concentrations are in the range of iron or below. The list of classical trace elements (Fe, Cu, Zn) has been recently extended by others (Co, Mn, Se, V, Mo, W, Al, Cr) for which physiological knowledge rapidly accumulates. Environmental and toxicological aspects have initiated extensive studies on Al, Cd and Pb. Little is known as yet about the role of elements like Si, Ti, Ni, Zr, As, Rb, Li, probably due to their difficult determination in the physiological concentration range. It is well appreciated, that iron deficiency effects anemia, it is less well known that the same is true for copper. In addition, lack of copper is related to growth and development of bones and connective tissue, infertility, alopecia, dermal pigmentation, adequate function of respiratory and thermal regulation and disorders of lipid metabolism. Cu is a structural part of Ceruloplasmin, the transport protein for iron, and of various enzymes like cytochromoxidase, superoxiddismutase, monoaminooxidase, uricase, dopaminhydroxilase. The biochemical reactivities of Cu and Zn are very similar and therefore competitive effects may develop, e.g. in a sense that symptoms of Cu-deficiency may be aggravated by hig uptake of Zn. Both elements are essential for biosynthesis (2).

Situations leading to Zn-deficiency are (8,9): preprocessed food, unbalanced alimentation, communal board and pregnancy may affect the otherwise healthy person, while inhibited resorption during intestinal disorders or diseases, any stress in connection with the surgical trauma, high renal exretion under induced diuresis, longterm parenteral alimentation, insufficient or no substitution under prolonged hospitalisation are observed in the ill patient. In addition, bioavailability of Zn may be reduced by binding to antibiotics and under a diet high in phytin.

Storage of Zinc

The total body Zn is preferentially stored in muscle (65%) and bone (25%). The highest storage capacity is observed with slow and continuous uptake. Conversely, other organs with a high Zn content but fast uptake (kidney, liver, pancreas) posses a limited storage capacity and are early affected by deficiency. An intermediate position is taken by the lungs, intestines, lymph nodes, heart and brain. These organs show a moderately fast uptake and an average storage capacity.

The Zn-metalloenzymes exhibit different Zn-affinities. For glutamate-DH, malonic-DH and lactate-DH activity remains unchanged in Zn deficiency, whereas it is found reduced for ethanol-DH, glyceraldehyd-DH, RNA- and DNA-polymerease, alcaline phosphatase, carboxipeptidase and carboanhydase (9). Decreased concentrations in Zn deficiency is found for the retional-binding-protein essential for the synthesis of Vit.A.

Some of the known clinical signs and symptoms of Zn deficiency develop rapidly. Among these are: altered platelet aggregation, visual disorder, loss of taste, smell and appetite, impaired wound healing, eczematoid dermatitis. Others manifest only after prolonged Zn deficit: decubitus, weight loss, diarrhoea, catabolism, cachexia, lethargia, depression, immune deficiency, acrodermatitis, hyperceratosis, alopecia, paronchia, sexual and fertility disfunction.

Zinc metabolism

In presence of an adequate Zn supply of 10 to 15 mg/d (153-230 umol/d), the urinary concentration of Zn is very constant and independent of the uptake. Only about 11% of the alimentary Zn are excreted renally, about the same amount by sweating. The main portion of approximately 80% is excreted with the faeces. In Zn-deficiency retention of alimentary Zn will occur depending on the magnitude of the deficit to be replaced in the depots. Orally supplied ^{65}Zn will occur in the blood after 2-4 hours, but only a portion is absorbed: After 3 days 66% of the dose will have appeared in the faeces. Only 2% are excreted in the urine and approximately 30% will be retained. A better retention is observed with parenteral dose of ^{65}Zn: Only 4% appear in the faeces within 5 days, after 45 days this amounts to 20%, again only minimal renal excretion is observed (11).

Electrolytes and trace elements in cardiac surgery

As a consequence of cardiac surgery servere unbalances of electrolyte metabolism are regularly observed. The underlying pathophysiological mechanisms are manifold:

Transcellular shift by:	- hypoxidosis and acidosis - aldosterone & insulin secretion - energetic deficit - hypothermia - increased membrane permeability - surgical stress and trauma - duration of surgery
Plasma concentration decrease by:	- hemodilution - inadequate substitution
Electrolyte loss by:	- increased diuresis - hyperaldosteronism - variable renal retention

The primary reasons for a decrease of plasma electrolyte concentrations are enhanced diuresis and transcellular shifts. Preferentially intracellular cations: K, Mg, Zn are lost as well. An extreme challenge for fast and efficient electrolyte substitution is presented by the patient undergoing open heart surgery in extracorporeal circulation (ECC). Duration of surgery, hypothermia, hemodilution with resultant forced diuresis, circulatory instability and endocrine responses contribute to a severe homoiostatic disorder.

This study was aimed to improve cation substitution in these patients by the develoipment of a polyelectrolyte solution. For practical reasons the dosage of this solution has to be based on frequent determinations of the K demand. The composition of the solution must therefore be adjusted in a way to provide adequate substitution for Mg, Zn, and Cu as well.

Patients and methods

In 129 patients with coronary heart desease undergoing open heart surgery with ECC for an aortocoronary vein bypass the plasma cation levels including Zn were studied. In 5 patients the electrolyte substitution was performed with conventional K,Mg,Zn,Cu - aspartate solution (Inzolen^R), in 14 patients with a newly designed solution (Inzolen HM^R), where Zn was reduced to 60%. The dosage given for simultaneous intravenous substitution was according to plasma potassium values. In a balance the substitution and the renally excreted amounts of zinc were compared (10). Zinc in plasma and urine was determined with a Perkin-Elmer AAS 300 at 214 nm.

Results

Following conventional zinc substitution a transient elevation of plasma Zn levels was observed lasting up to three hours after the operation (fig.1). On the day of operation mean urinary excretion of Zn amounted to 64 umol. This is equivalent to 23% of the total Zn input of 280 umol. On the first postoperative (po) day 27% of the cumulative supply were excreted (fig.2). Renal elimination was fastest shortly after ECC with a peak value of 10 umol/h (fig.3).
In contrast, substitution using the improved cation aspartate solution (K:Mg:Zn:Cu = 1:0,25:0,0018:0,0013 mmol) resulted in an intraoperative time course of plasma Zn levels within the physiological range (fig.1).

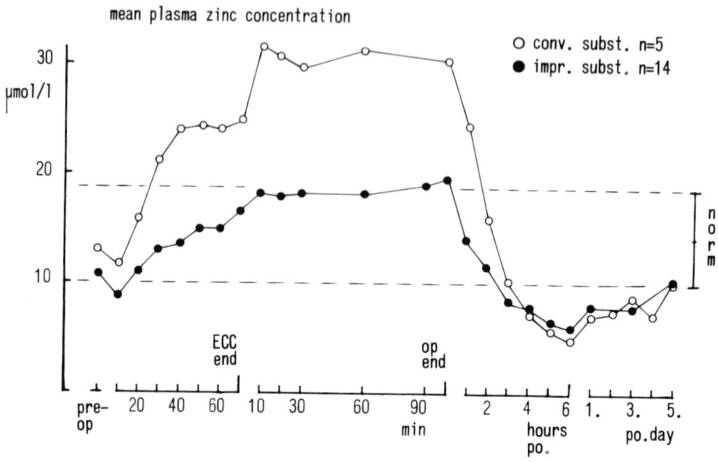

fig.1

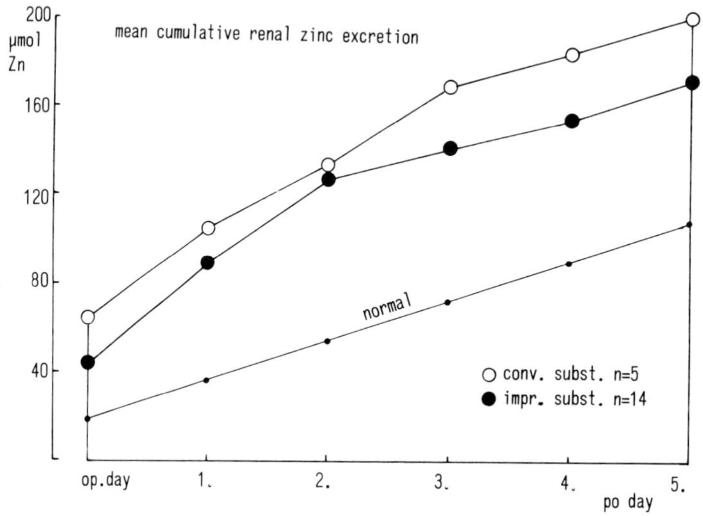

fig.2

As compared to conventional substitution there was no difference in slightly decreased po plasma concentrations. Reduction of Zn input to 234 umol on the day of operation resulted in a decrease in output to 19% and to 25% on the first po day. The intraoperative peak value of Zn excretion was reduced to 7,5 umol/h (fig.3). Total renal elimination was similar in both groups suggesting that even under elevated parenteral substitution zinc excretion can be only slightly enhanced (fig.3,4). Under induced diuresis and high intravenous zinc supply the renal zinc excretion can be raised only for a short time (fig.3).

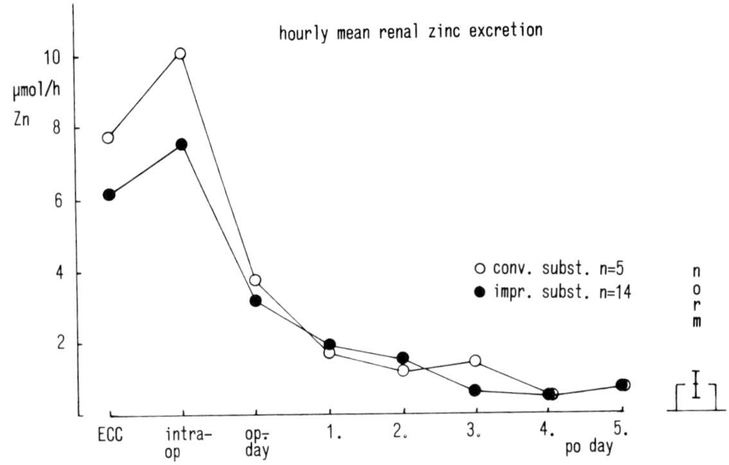

fig.3

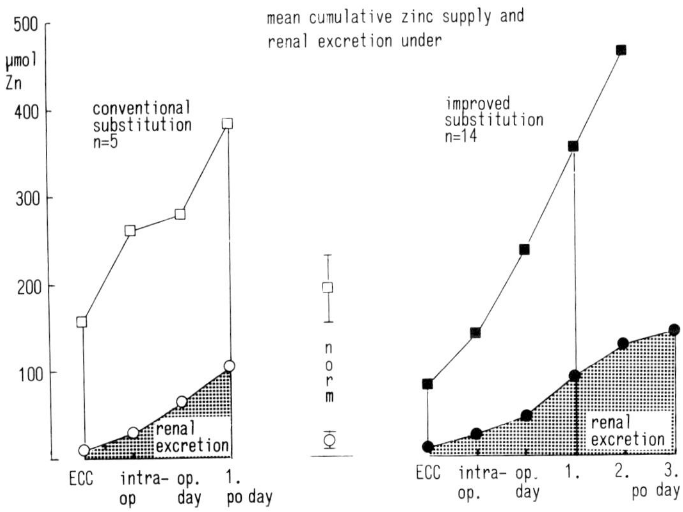

fig.4

Conclusion

Open heart surgery effects pronounced renal losses of zinc which amount to the fourfold of normal under induced diuresis on the day of operation and stay elevated until the third po day. A well adjusted substitution from the onset of surgery is mandatory to avoid consequences of acute zinc depletion and to allow recovery from the early po catabolism. This is provided by a polyelectrolyte aspartate solution with a cation composition balanced to keep zinc (magnesium and copper) within the normal plasma concentration range when given according to potassium demands. In daily clinical practice zinc definicy with the indicated symptoms as well as supply are: inhibition of zinc enzymes, inhibition of collagen synthesis, retention in liver and muscle, acute renal failure, acute liver failure, intoxication.

References

1. Bolte H.D., Lüderitz B., Riecker G.: Klin. Wschr. 49, 306 (1971)
2. Cordano A.: in "Zinc and Copper in Clinical Medicine", ed. Hambidge K.M., Nichols B.L., SP Medical and Scientific Books, New YOrk - London 1978
3. Ebel H., Günther T.: Med. Klinik 75 257 (1980)
4. Gunning J.F., Harrison C.E., Coleman H.N.: Journal Molecular and Cellular Cardiology 4, 139 (1972)
5. Iseri L., Freed J., Bures A.: Am. J. Med. 58, 837 (1975)
6. Marier C.R.: Magnesium 1,3 (1982)
7. Münchner Blutdruckstudie 1982: Jahresbericht 1982 p. 23, Gesellschaft für Strahlen- und Umweltforschung, München-Neuherberg
8. Prasad A.S.: in "Zinc and Copper in Clinical Medicine" as 2
9. Roth H.P., Kirchgeßner M.: in "Zinkstoffwechsel, Bedeutung für Klinik und Praxis", ed: Kruse-Jarres J.D., TM-Verlag, Bad Oeynhausen 1979
10. Schroll A., Mendler N., Brehm A., Struck E., Richter J., Sebening F.: in "Spurenelemente: Bedeutung für Chirurgie, Anästhesiologie und Intensivmedizin" ed: Staib I., Schattauer Verlag Stuttgart 1982
11. Spencer H.: Radiation Research 24, 432 (1965)

THE USE OF ^{56}MN FOR MEASUREMENTS OF MANGANESE UPTAKE IN ISOLATED LIVING MYOCARDIAL TISSUE

A.Stampfl

Gesellschaft für Strahlen- und Umweltforschung
Abteilung für Pharmakologie
D-8042 Neuherberg, West Germany

W.Vierling

Institut für Pharmakologie und Toxikologie
der Technischen Universität München
Biedersteiner Str.29
D-8000 München 40, West Germany

Introduction

In the guinea-pig papillary muscle, manganese ions produce a negative inotropic effect followed by a positive one as shown by Vierling and Reiter (1). The positive inotropic action could be explained by the assumption that the sequestration of released calcium is inhibited by intracellularly accumulated manganese. The increase in force of contraction was found to be accelerated by an increase in contraction frequency. Later, it was observed that also addition of noradrenaline enhanced the positive inotropic effect (2). Noradrenaline is known from electrophysiological experiments to increase the slow inward current during the action potential which is carried mainly by calcium ions (3). Since Ochi (4) found that manganese ions can also pass through the channel of the slow inward current, it was postulated that noradrenaline increases an action-dependent uptake of manganese. Therefore, we tried to prove this hypothesis by studying the uptake of manganese into isolated papillary muscles with ^{56}Mn (half-time: 2.58h) as a tracer. We preferred this method, because it was possible to measure the emitted γ-rays of this isotope without further treatment of the muscles.

Trace Element - Analytical Chemistry in Medicine and Biology, Vol. 3

Methods

Isolated papillary muscles with a mean weight of 1.5 mg from the right ventricles of guinea-pig hearts were mounted in a two-chambered vessel (volume 50 ml) with internal circulation of the bath medium (5). The muscles were incubated in a phosphate-free solution of the following composition (in mmol/l):NaCl 115; $NaHCO_3$ 25; KCl 5.9; $CaCl_2$ 2.4; $MgSO_4$ 1.2; glucose 10.0. A phosphate-free solution was used in order to prevent precipitation of manganese salts after addition of $MnCl_2$. The bath medium was vigorously bubbled and circulated by a mixture of 95% O_2 and 5% CO_2. The temperature was kept constant at 35° C. The muscles were stimulated at their mural end by square wave pulses of 1 ms duration and an intensity slightly above threshold. The resting force was automatically kept constant at 3.9 mN (6). After an equilibration period of 1 h at a contraction frequency of 1 Hz, Mn^{++} (0.6mmol/l) was added with ^{56}Mn as a tracer (specific activity: 4MBq/mg Mn). The isotope was produced in the FRM (Forschungsreaktor München) nearby our laboratory. In some experiments noradrenaline (10 µmol/l) was added 5 min before addition of manganese. To prevent inactivation of noradrenaline, ascorbic acid was added to the bath to reach a final concentration of 0.28 mmol/l (7). After the incubation time, the muscles were removed and then the mural end and the tendon was cut away. Thereafter, the muscles were submerged in an inactive solution for a few seconds, dabbed on plastic foil and weighed. The radioactivity of each muscle and a sample of the corresponding bath solution (50 µl) was measured in a 3 inch gamma detector (Gamma 4000, Beckman Instruments, Irvine, California) without dissolving the tissue. For counting, we used the maximum emission peak of ^{56}Mn (99% at 847 KeV). The counts were corrected for background and radioactive decay ($t_{1/2}$: 2.58 h). The T/M ratio of the activity between tissue(T) and bath medium(M) was calculated as an indicator of the manganese content of the muscle. This is possible, because the muscle under normal conditions contains very small concentrations of manganese (< 10 pmol/mg wet weight).

Results and Discussion

The manganese uptake from a solution containing 0.6 mmol/l Mn^{++} was studied at various contraction frequencies with and without noradrenaline (Fig.1). In the absence of noradrenaline, the T/M ratio (2.4 at rest) was not significantly different between rest and 0.5 Hz. However, an increase in contraction frequency to 1 Hz increased the T/M ratio by 58% to 3.8. This demonstrates that a manganese influx at rest is superimposed by an action-dependent uptake, detectable only at a relative high contraction frequency. An extra uptake of manganese due to stimulation has also been observed by Delahayes (8), who used high manganese concentrations (10 mmol/l) which produced no positive inotropic effect but contracture.

Under the influence of 10 µmol/l noradrenaline, the T/M ratio of the unstimulated muscle was significantly reduced by 30% to 1.7. A possible explanation for this unexpected effect would be that manganese can be eliminated from the cell by a mechanism which is enhanced by noradrenaline via an increase of the intracellular cAMP concentration. Such a mechanism was shown to exist for calcium involving a magnesium-dependent Ca ATPase which can be stimulated by cAMP (9). In the presence of noradrenaline, stimulation with the low frequency of 0.1 Hz increased the Mn content reaching the rested-state value in the absence of noradrenaline. With the rise of the contraction frequency to 0.5 Hz and 1 Hz the Mn uptake increased further. From this, we conclude that the action-dependent in-

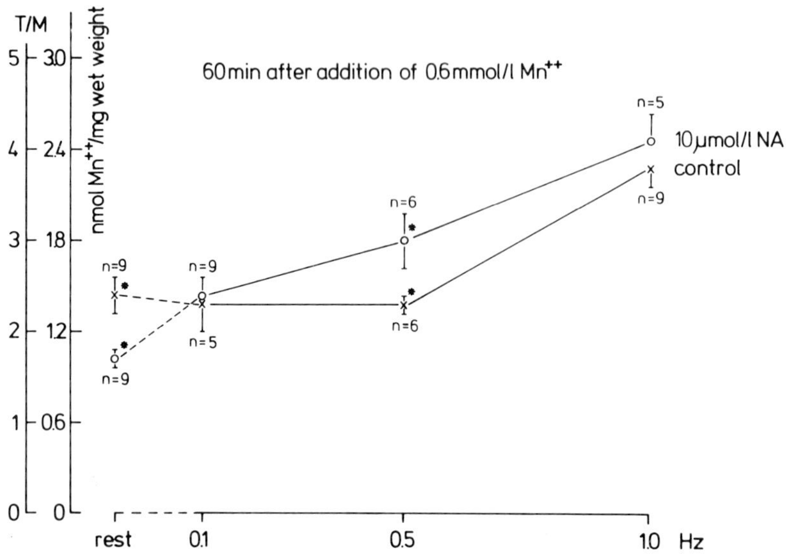

Figure 1: Mn content of guinea-pig papillary muscles as influenced by an increase in contraction frequency in presence or absence of noradrenaline.

After preequilibration at the desired frequency with or without noradrenaline, the Mn content of papillary muscles was determined after 1 h incubation time in a solution containing 0.6 mmol/l Mn (^{56}Mn: specific activity: 4 MBq/mg). Each point represents the mean value $\pm$ standard error from a number of papillary muscles as indicated in the figures.
x significant differences ($p < 0.05$).
Ordinates: Tissue/medium ratio, and Mn content per mg wet weight
Abscissa: Frequency of contraction in Hz

flux of manganese is augmented by noradrenaline, probably by an enhancement of the slow inward current. An increase of Mn uptake in isolated perfused rat hearts under the influence of isoprenaline was also demonstrated by Hunter et al. (10),

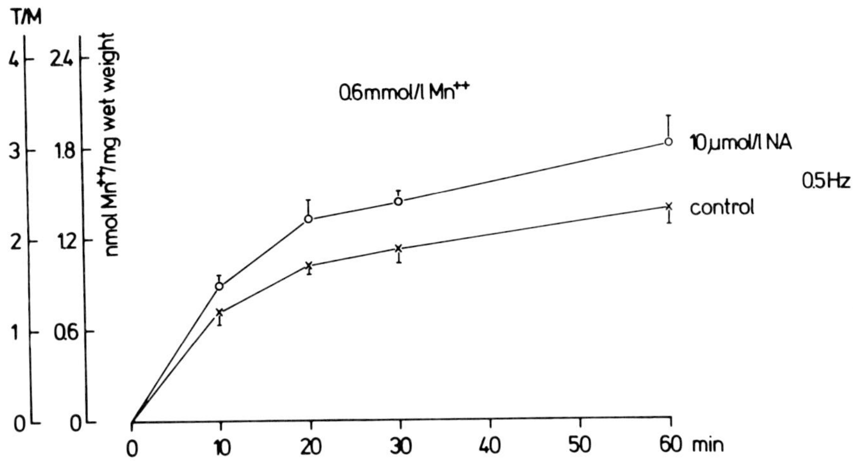

Figure 2: Time course of the Mn content at 0.5 Hz contraction frequency under control conditions and in presence of noradrenaline.
Muscles were incubated for different times in a solution containing 0.6 mmol/l Mn^{2+} with ^{56}Mn as a tracer (specific activity: 4 MBq/mg). Each mean value ± standard error was obtained from one group of 6 muscles.
Ordinates: Tissue/medium ratio,and Mn content per mg wet weight
Abscissa: Time after addition of Mn

who, however, did not keep the contraction frequency constant.
The time course of the Mn uptake at 0.5 Hz contraction frequency is shown in figure 2. It is evident, that the difference in the uptake velocity appears already within the first 20 min after addition of manganese. Later on, the curves with and without noradrenaline run nearly parallel.

Summary

Our experiments show, that ^{56}Mn can be successfully used to measure the cellular uptake of manganese by isolated contracting papillary muscles of guinea-pig hearts under various experimental conditions. It was possible to verify assumed mechanisms of the influence of noradrenaline on Mn uptake either during rest or in dependence on contraction frequency which had been previously deduced from experiments using force of contraction or electrophysiological parameters as indicators for the uptake of manganese in the heart cell.

References

1. Vierling,W., Reiter,M.: Naunyn-Schmiedeberg´s Arch. Pharmacol. 306, 249-253 (1979)

2. Vierling,W., Stampfl,A.,: Naunyn-Schmiedeberg´s Arch. Pharmacol. 321, R 38 (1982)

3. Reuter,H.: J.Physiol. 242, 429- 451 (1974)

4. Ochi,R.: Pflügers Arch. 316, 81-94 (1970)

5. Reiter,M.: In: Greeff,K.(ed.): Handbook of experimental pharmacology, Vol.56/I Springer, Berlin Heidelberg New York, 153-159 (1981)

6. Stampfl,A.: Elektronik 25, 82-84 (1979)

7. Löffelholz,K., Scholz,H.: Experientia 26/6, 637-638 (1970)

8. Delahayes,J.F.: Circ.Res. 36, 713-718 (1975)

9. Carafoli,E.: J.Mol.Cell.Cardiol.13 (Suppl.1):14 (Abstract) (1981)

10. Hunter,D.R., Haworth,R.A., Berkoff,H.A.: J.Mol.Cell.Cardiol. 13, 823-832 (1981)

Discussion Session VI

Subjects of the discussion

- different levels of Cu in breast milk
- relationship between Cr and coronary artery disease
- functional analysis of the platelets
- effect of polyunsaturated fatty acids on platelet aggregation
- stress dependent decrease of Zn
- high renal excretion during cardiopulmonary bypass
- trace elements - especially Se - and heart diseases

Contributors: Cochran, Feinendegen, Gulba, Koh, Oster, Westermark

Session VII

APPLICATION OF DIFFERENT ANALYTICAL TECHNIQUES

Chairman:

J.M. Ottaway (UK)

Poster Reports given by: J.M. Ottaway (UK)

LASER MICROPROBE MASS ANALYSIS (LAMMA) IN NEPHROLOGICAL INVESTIGATIONS

A.H. Verbueken[1], F.L. Van de Vyver[2], G.J. Paulus[3], W.J. Visser[4], G.A.Verpooten[5], M.E. De Broe[5] and R.E. Van Grieken[1]

Departments of Chemistry[1], Nuclear Medicine[2], Pathology[3], and Nephrology-Hypertension[5], University of Antwerp, B-2610 Wilrijk, Belgium

Laboratory of Bone Metabolism[4], University of Utrecht, 3500 CG Utrecht, The Netherlands

Introduction

In renal research and in general nephrotoxicity studies, increasing efforts are being made towards a better qualitative and possibly even quantitative knowledge of the distribution of deposited metals and diffusible ions in normal and pathological tissue. Elements to be identified may be naturally occurring, accumulated by some pathological process, or added to the tissue as tracers or histochemical staining reagents. Information generated by the ultrastructural localization of insoluble elemental deposits in kidney tissue and other organ systems is becoming available due to the advent of sophisticated technical equipment for microanalysis. Electron microscopy in either conventional transmission mode (TEM) or scanning configuration (SEM) adapted with suitable X-ray detecting systems has already successfully been applied in renal investigations (1). More recently, laser microprobe mass analysis (LAMMA) was introduced as a sensitive mass spectrometric microanalytical method providing information about the elemental constituents of microvolumes (e.g. 1 cubic micron) of thin histological sections (2, 3). A relative detection sensitivity down to the ppm range is obtained, particularly for easily ionizable elements such as most physiological cations. In the present review, the analytical potential of LAMMA as a microprobe to study biomedical specimens for application to important problems in nephrology is discussed.

Experimental

Tissue specimens.

Small needle biopsies (Tru-Cut[R] needle) were taken from human liver and dog kidney. From the dog, renal cortical tissue was also obtained at

Trace Element - Analytical Chemistry in Medicine and Biology, Vol. 3

autopsy. Transiliac bone biopsies were taken from patients with a 7 mm diameter trephine of Bordier-Meunier.

Specimen preparation.

Common procedures for tissue fixation using glutaraldehyde and post-osmication as used routinely in electron microscopy, were applied for the liver and kidney samples. Dehydration was achieved in stages with acetone followed by embedding in epoxy resin (Epon). After block polymerization, sections of the desired thickness (0.2-0.35 μm) were cut using glass knives on a Reichert ultramicrotome and mounted on copper grids without any supporting film to avoid additional organic mass interferences in LAMMA. In the dog kidney experiments (4), the use of heavy metals for staining and also of osmiumtetroxide for postfixation was preferably avoided. On the contrary, for the liver samples under study (5, 6) the sections were also stained with alcoholic uranyl acetate and lead citrate to improve the ultrastructural information provided by electron micrographs recorded before and after the LAMMA analysis.
The bone biopsies (6) were fixed for 24 hours in Burkhardt's solution and then transferred to absolute methanol. The undecalcified specimens were embedded in methyl methacrylate and sections were cut on a Jung K sledge microtome. For LAMMA, 2 μm thick unstained sections were prepared. After the LAMMA analysis, the sections were stained with toluidine-blue to enable the identification of the analysed regions with light microscopy. In recent experiments, LAMMA was also performed on 2 μm thick bone sections stained with aluminon (ammonium aurintricarboxylate), a dye which forms a bright-red lake with aluminium ions and is used as a specific stain for the demonstration of aluminium in undecalcified bone sections (7).

Electron microscopy.

The samples were examined by TEM with a Zeiss EM 109 electron microscope. The TEM electron micrographs taken before and after LAMMA analysis provided records of the nature of the analysed subcellular fragments and of the evaporated area. Energy-dispersive electron probe X-ray microanalysis was carried out on uncoated and carbon-coated liver sections using a JEOL Superprobe 733 apparatus.

Laser microprobe mass analysis (2, 3).
The laser-induced mass spectra were obtained by using the LAMMA-500[R] instrument of Leybold-Heraeus (Köln, F.R. Germany). The histological section is viewed with a binocular light microscope. The system is equipped with two lasers : the first is a low-powered (2 mW) red helium-neon laser which can be focused to a micrometer-sized spot to indicate the subcellular region to be analysed. Collinear with this search laser is a frequency-quadrupled neodymium-YAG laser which provides a high power pulse (Q-switched) and is able to vaporize and ionize a selected area of a few square microns of a thin section. After perforation, the created positive or negative ions are extracted and separated into a time-of-flight mass spectrometer to produce a complete mass spectrum for each laser pulse.

Results and Discussion

Detection of platinum in the kidney.
Platinum complexes used alone and in combination with other drugs have emerged as effective agents in cancer chemotherapy. Cisplatin (cis-dichlorodiammineplatinum (II)) is one of the best known and clinically most widely used of these antineoplastic compounds. However, the drug exhibits various toxic side effects in man and animal, a cumulative nephrotoxicity being the most important and dose-limiting factor. The information on the distribution of platinum in the different tissues is still scanty. In this context, it is indispensable to study possible sites of accumulation of the heavy metal at a cellular or even ultrastructural level. For this purpose, renal biopsies were taken from a dog that had been administered intravenously the antitumour drug cisplatin; a total dose of 5 mg per kg body-weight was given (4). By means of LAMMA, we were able to detect platinum in proximal tubular cells of the dog kidney (4).
Although platinum was proven to be present within these cells, and not in the extra-cellular renal material, further subcellular tracing appeared unfeasible up to now, because the low concentration and the high first ionization potential of platinum necessitated the use of high laser power, resulting in larger section perforations ($\sim$ 6 μm) and consequently also a limited spatial resolution of analysis.

Making use of the calibration system, recently introduced by us (8), and using appropriate morphometric data for the renal cortex, we found the LAMMA-results, obtained by scanning some proximal tubular cells, to be consistent with bulk analysis by energy-dispersive X-ray fluorescence, giving a platinum concentration of 20 to 30 μg/g of fresh kidney tissue (4).

Detection of aluminium in liver and bone.

Already in 1970 the potential hazards and possible pathology of administering aluminium compounds to uremic patients were recognized. Still, there was actually little knowledge regarding aluminium metabolism and its systemic toxicity until the last years. Aluminium toxicity has been implicated in the pathogenesis of a number of clinical disorders occurring in patients with chronic renal failure (9). Patients on long-term dialysis treatment (hemodialysis, peritoneal dialysis, hemofiltration) are above all exposed to aluminium, due to its presence in the dialysate, in parenterally administered fluids, and in orally given phosphate binders. Since renal function is necessary to eliminate aluminium from the body, considerable accumulation of this element may occur in the tissues of uremic patients, resulting in bone disease such as vitamin-D-resistant osteomalacia and in damage of the brain leading to encephalopathy.

Liver.

Making use of the LAMMA microprobe we established the ultrastructural localization of aluminium in fresh liver biopsies of non-demented dialysis patients suffering from osteomalacia; the results have been compared to those for control patients (5, 6, 10). In these aluminium localization experiments the laser power could be reduced to provide a satisfactory spatial resolution of analysis, optimally about 1 μm (see Fig. 1), while sufficient detection efficiency was still maintained. Optimal spatial resolution is thus obtained at the cost of reduced sensitivity. In these osteomalacia patients, frequently also suffering from iron overload (10), aluminium was persistently detected together with iron in electron-dense structures, compatible with lysosomes, within the liver cells, parenchymal as well as Kupffer cells (5). A LAMMA mass spectrum illustrating this lysosomal co-trapping of iron and aluminium is presented in Figure 2. Significant iron and aluminium signals are seen at the level of the hepatic lysosomes.

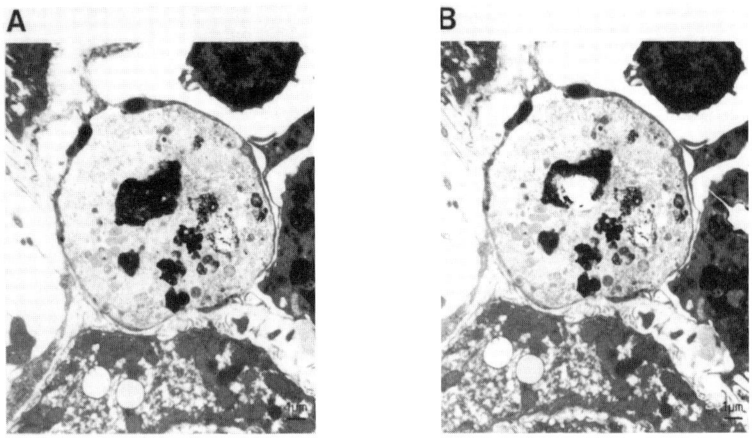

Fig. 1. Electron micrographs (TEM) of a liver cell from a dialysis patient before (A) and after (B) LAMMA microanalysis.

Using electron probe X-ray microanalysis, we could only confirm the lysosomal iron accumulation in the liver tissues under examination, illustrating the abundant presence of iron in the liver samples of these osteomalacic patients under study.

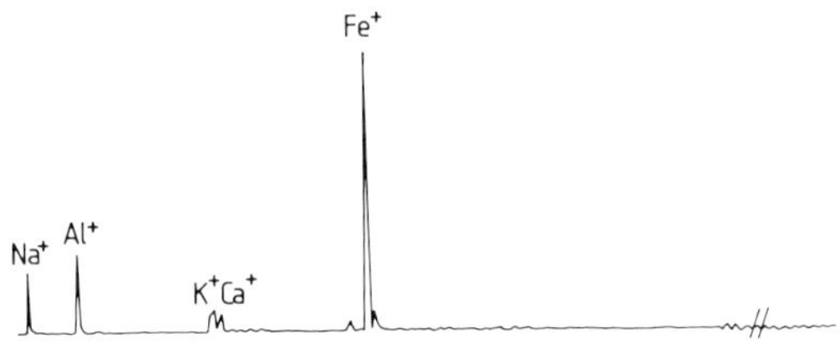

Fig. 2. LAMMA mass spectrum originating from the hepatocyte lysosome illustrated in Figure 1.

In some mass spectra, also magnesium and barium were present next to calcium, aluminium and iron at the level of the hepatocyte lysosomes. Negative-ion LAMMA spectra of the electron-dense bodies demonstrated abundant phosphorus-oxy ions (PO_2^-, PO_3^-) (6). No evidence was found for aluminium accumulation in the nucleus, mitochondria, cytoplasm, intercellular space, and also erythrocytes.

Bone.

In the unstained bone sections of the dialysis osteomalacia patients, aluminium was found to be localized at the osteoid/calcified-bone boundary (6). The LAMMA mass spectra originating from this osteoid/calcified-bone interface also showed the presence of magnesium, sodium and potassium. In patients with iron overload, a concomitant iron signal was observed at this same location (10).

In recent experiments we introduced LAMMA as an ideal method to uncover or confirm the localization of several elements (Al, Fe, Pb) in bone sections obtained by histochemistry. Taking advantage of the light-microscopic visualization aid, we performed LAMMA directly on aluminon-stained undecalcified bone sections, in order to compare histochemical results with LAMMA microanalysis, and to evaluate the specificity of the histochemistry used. In this context, it was e.g. found that in dialysis osteomalacia patients with severe iron overload, not only iron, but also aluminium was concomitantly detected by LAMMA in a large number of histochemically aluminon-negative bone marrow cells (probably of the macrophage system) (11). With LAMMA, aluminium has been demonstrated in osteocytes and bone marrow fibroblasts as well (12).

Some of the hemodialysis patients with osteomalacia presented a purplish line at the osteoid/calcified-bone interfaces instead of the classic bright-red line indicative of aluminium. LAMMA performed in one of these cases demonstrated large amounts of iron present concurrently with aluminium at the purple-stained location (Fig. 3). Multi-element microanalysis, which is provided by LAMMA, thus appears very useful in evaluating histochemical methods and in assessing possible element interferences.

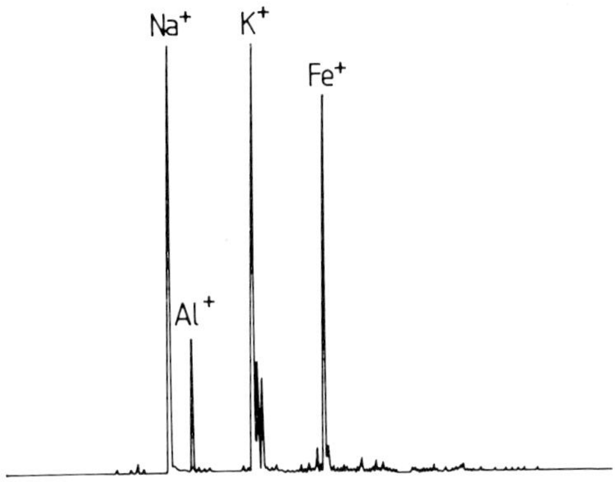

Fig. 3. LAMMA mass spectrum of an aluminon-stained bone section from a patient with iron and aluminium overload and dialysis osteomalacia. The region analysed corresponds to the osteoid/ calcified-bone interface.

Conclusion

We have shown the usefulness of LAMMA for sensitive probing of pathologically important elements in man and animal. Due to its more sensitive elemental detection capability as compared to other micro-analytical techniques such as the electron probe X-ray method, LAMMA is a promising aid for investigations in medicine and biology.

Acknowledgements

This work was supported in part by the Interministerial Commission for Science Policy, Belgium, through research grants 80-85/10 and 82-87/47.

References

1. Bulger, R.E. : Scanning Electron Microsc. II, 511-516 (1980).
2. Denoyer, E., Van Grieken, R., Adams, F., Natusch, D.F.S. : Anal. Chem. 54, 26A-41A (1982).
3. Proceedings of the 1st LAMMA Symposium held in Düsseldorf, F.R.G., October 8-10, 1980 (Organizers : F. Hillenkamp and R. Kaufmann), Fres. Z. Anal. Chem. 308, 193-320 (1981).
4. Verbueken, A.H., Van Grieken, R.E., Paulus, G.J., Verpooten, G.A., De Broe, M.E. : Biomed. Mass Spectrom., in press.
5. Verbueken, A.H., Boelaert, J., Paulus, G.J., Van de Vyver, F.L.J., Roels, F., Van Grieken, R., De Broe, M.E. : Néphrologie 4, 95 (1983).
6. Verbueken, A.H., Van de Vyver, F.L., Van Grieken, R.E., Paulus, G.J., Visser, W.J., D'Haese, P., De Broe, M.E. : Clin. Chem., in press.
7. Buchanan, M.R.C., Ihle, B.U., Dunn, C.M. : J. Clin. Pathol. 34, 1352-1354 (1981).
8. Verbueken, A.H., Van Grieken, R.E., Paulus, G.J., de Bruijn, W.C. : Anal. Chem., in press.
9. Parkinson, I.S., Ward, M.K., Kerr, D.N.S. : J. Clin. Pathol. 34, 1285-1294 (1981).
10. Van de Vyver, F.L., Vanheule, A.O., Verbueken, A.H., et al. : Contr. Nephrol. 38, 153-166 (1984).
11. Van de Vyver, F.L., Verbueken, A.H., Visser, W.J., Van Grieken, R.E., De Broe, M.E. : J. Clin. Pathol. (Letter), in press.
12. Zumkley, H., Schmidt, P.F., Bertram, H.P., et al. : Lancet, submitted.

SERUM TRACE ELEMENTS IN HEMOPROLIFERATIVE DISEASES.

Yves Beguin (1), Jeanne-Marie Delbrouck, Iwan Roelandts, Georges Weber (2)

Department of Medicine and Department of Experimental Nuclear Physics, University of Liège, Sart-Tilman B 15, B-4000 Liège, Belgium.
Research Assistant (1) and Research Associate (2) of the National Fund for Scientific Research (Belgium).

Abstract

The PIXE method has been used to compare a reference group (N=100) and eight groups of patients suffering from hematological malignancies (Hodgkin's disease, lymphosarcoma, chronic lymphocytic leukemia, acute lymphocytic leukemia, acute granulocytic leukemia, chronic granulocytic leukemia, myeloproliferative disorders and myeloma) (N=140).
The serum Cl, K, Ca, Fe, Cu, Zn, Se and Br contents have been measured. Significative differences were noted in several groups. A statistical analysis has been performed to correlate these trace element values with several clinical and biological parameters such as blood cells counting, clinical stage, histology, inflammatory tests,...

Introduction

Trace elements metabolism has already been extensively studied in various medical fields (1, 2). In hematology, many authors have reported on serum copper increase in Hodgkin's disease and lymphosarcoma and return to normal values when complete remission is achieved (3, 4, 5, 6). Much less has been described about other trace elements, especially about bromine whose role in human metabolism remains obscure.
We report here on our experience with serum trace elements dosage and confirm our previous communication (7) on abnormal values in several hemoproliferative diseases.

Trace Element - Analytical Chemistry in Medicine and Biology, Vol. 3

Patients and methods

The PIXE method has been used to measure the serum content of chlorine (Cl), potassium (K), calcium (Ca), iron (Fe), copper (Cu), zinc (Zn), selenium (Se) and bromine (Br), in normal subjects and in patients suffering from hemoproliferative diseases.

1) Normal subjects. This group was composed of 100 healthy subjects taking no drug. Although this group was rather younger than our patient group, no significant difference with age was noted in trace element values. We found a statistically significant difference between sexes in calcium (103 in male versus 100 ug/ml in female, $p = 0.03$), copper (1.05 in male versus 1.17 ug/ml in female, $p = 0.02$), zinc (1.14 in male versus 1.01 ug/ml in female, $p < 0.001$).
2) Patients. 319 samples were taken from 150 patients, many of them being studied on several occasions, both in active disease and in remission. They are divided up as follows : myeloma (MY : 20), Hodgkin's disease (HD : 12), lymphosarcoma (LS : 28), chronic lymphocytic leukemia (CLL : 26), acute lymphocytic leukemia (ALL : 6), acute myelocytic leukemia (AML : 26), chronic myelocytic leukemia (CML : 11) and myeloproliferative disorders (MPS : 21).

   We also studied serum trace elements values according to clinical status and some biological parameters usually regarded as representative of disease activity. So blood cell counting, erythrocyte, sedimentation rate, fibrinogen, serum lactate deshydrogenase (LDH) and serum proteins were recorded on the same day as the trace elements.

   Thirteen patients presenting with newly diagnosed or relapsing AML and undergoing intensive induction chemotherapy, have been serially followed for up to three months. The chemotherapy regimen generally consisted in a seven-day course associating high doses of daunomycine, vincristine and cytosine-arabinoside. Samples were taken before treatment and then about twice a week.

Table 1 - Serum copper, zinc and bromine (ug/ml) in controls and different patient groups. P-value for Student t-test is given when significant.

TABLE 1.

	COPPER	ZINC	BROMINE
N (100)	1.10	1.10	5.36
AML (26)	1.34 ($P < 0.001$)	0.94 ($P < 0.001$)	4.00 ($P < 0.001$)
LS (28)	1.40 ($P < 0.001$)	0.82 ($P < 0.001$)	4.70 (NS)
CLL (26)	1.65 ($P < 0.001$)	0.96 ($P < 0.001$)	4.10 ($P < 0.005$)
CML (11)	1.26 ($P < 0.05$)	0.99 ($P < 0.05$)	4.80 (NS)
MPS (21)	1.34 ($P < 0.001$)	1.00 ($P < 0.01$)	7.10 (NS)
HD (12)	1.34 ($P < 0.002$)	0.83 ($P < 0.001$)	6.17 (NS)
MY (20)	1.29 ($P < 0.003$)	0.87 ($P < 0.001$)	4.13 ($P < 0.002$)
ALL (6)	1.18 (NS)	0.88 ($P < 0.002$)	4.81 (NS)

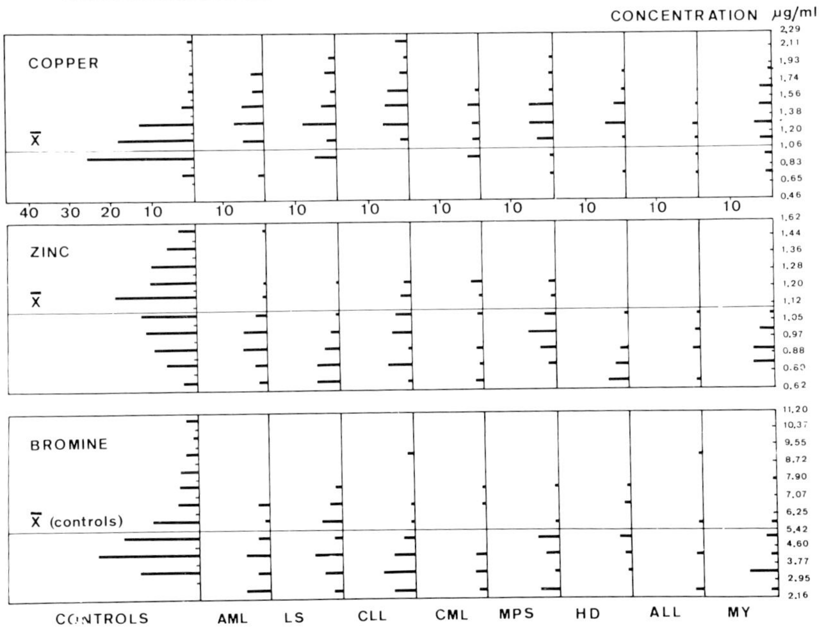

Fig. 1 - Serum copper, zinc and bromine (ug/ml) in controls and different patient groups, with range demonstrated.

Results

The mean values for copper, zinc and bromine in normal subjects and in patients are listed in Table 1. P-values for student t-test between controls and patients are mentioned when significant. Ranges are shown in Figure 1, but some very high values for bromine are outside this picture.

Copper was always increased when compared with controls, and this reached statistical significance in all groups but ALL which was too small. Zinc was significantly reduced in all patient groups, though the absolute difference wasn't important in every groups.

Bromine displayed some less systematic pattern. It was significantly decreased in AML, CLL and myeloma, but in AML this feature is dependent upon treatment (see below); it was slightly reduced in lymphosarcoma, CML and ALL, and slightly enhanced in HD and MPS, but in these groups most values were in fact very low and the mean artificially normalised by very high values.

Selenium showed no difference neither in absolute value nor in range.

Chlorine and calcium remained almost constant from group to group. Potassium was often increased, reflecting extensive cellular turn-over. Iron was sometimes reduced, mostly in CLL and LS groups where copper reached the highest levels.

Because of the small size of some groups, one must be careful in interpreting apparent correlations between trace elements and biological parameters. So we shall report only on the most important ones. Within the patient group as a whole, copper is correlated with white blood cell counts, inflammatory tests, serum LDH and potassium; zinc disclosed a weak correlation with calcium, iron and selenium, and bromine with potassium; potassium was related to whole blood cell counts, serum proteins and chlorine; calcium increased with serum proteins, chlorine, potassium and selenium.

In CLL, copper was correlated with WBC counts ($R=0.71$, $p < 0.001$), serum LDH ($R=0.796$, $p < 0.001$), ESR ($R=0.504$, $p < 0.05$) but not with fibrinogen; on the opposite, selenium was inversely correlated with WBC counts ($p < 0,01$). In lymphosarcoma, copper showed significant correlation with

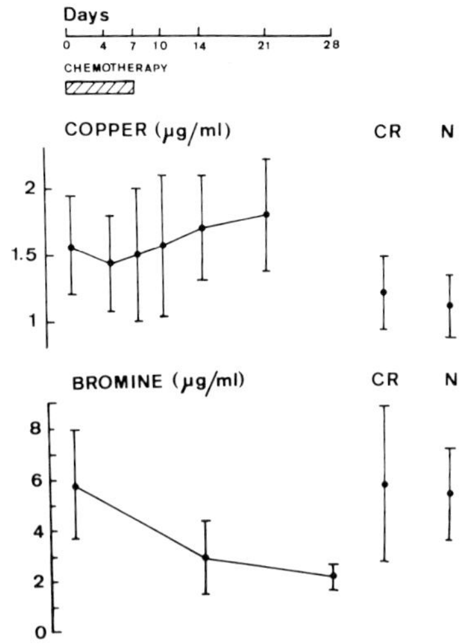

Fig. 2 - Copper and bromine evolution in patients undergoing chemotherapy for AML? Mean values $\pm$ σ are given with comparison with AML in complete remission (CR) and normal controls (N).

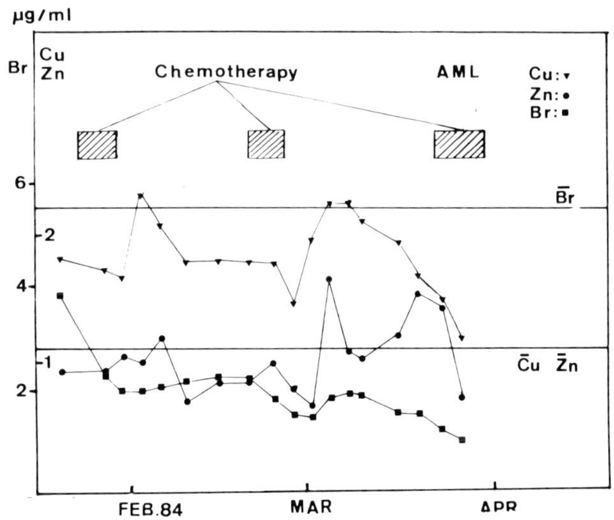

Fig. 3 - A typical pattern of bromine, copper and zinc evolution during chemotherapy in a 20 year old woman with AML.

ESR and fibrinogen, and inverse relationship with hemoglobin.
The AML patients undergoing intensive chemotherapy bore a characteristic evolution (Figure 2). The mean values before treatment were 5.79 ± 2.07 ug/ml for bromine and 1.57 ± 0.37 ug/ml for copper. Bromine dropped dramatically to 2.74 ± 1.45 on day 14 and 2.10 ± 0.48 ug/ml on day 28; the level remained low for several weeks and then progressively reached normal pre-treatment values. Copper displayed a less constant pattern but usually cut down during chemotherapy and thereafter reaugmented to significantly a higher level (1.79 ± 0.43 ug/ml on day 21). A few weeks later, while in complete remission, dosage reduced to 1.21 ± 0.28 ug/ml, thus only slightly more than in controls. A typical pattern is presented in Figure 3 with a 20 year old woman undergoing three chemotherapy courses for refractory AML.

Conclusions

1) Zinc is systematically reduced in hemoproliferative diseases. This has not always been confirmed (8) and doesn't seem to be related to nutrition status, treatment or other biological abnormalies. It may be due to tissular catabolism in cancer patients (1, 2).
2) Copper is usually enhanced in our patients but considerable overlap with normal values prevents us from its excessive clinical use. Nonetheless, we found a significant correlation between copper, and inflammatory tests in LS (3-6), and WBC chiefly in CLL.
3) Bromine remains misunderstood in human metabolism. Although we observed a general tendency towards decreased values in patients, very high values were noted, that were not related to any peculiar disease status nor to drug intake. The fall noticed in AML chemotherapy was independent of digestive tolerance to disease and treatment. This was supported by the absence of such a phenomenon in patients undergoing a slimming fast (unpublished data). This drop seemed to be due to chemotherapy itself.
4) Selenium is much the same in our patient groups as in controls. But measured concentration is very low and errors may be relevant.

Bibliography

1. Prasad, A.S., Ed. : Trace Elements in Human Health and Disease, Academic Press, New York, 1976.

2. Browner, F., Coburn, J.W., Ed. : Disorders of Mineral Metabolism, Vol. 1., Trace Elements, Academic Press, New York, 1981.

3. Hrgovcic, M.J., Schullenberger, C.C., Ed. : Copper and lymphomas, CRC Press, Bora Raton, Florida, 1984.

4. Shah, I., Lewkow, L.M., Khilawani, U. : Cancer 51, 851-854 (1983).

5. Wilimas, J., Thompson, E., Smith, K.L. : Cancer 42, 1929-1935 (1978).

6. Asbjörnsen, G. : Scan. J. Haematol. 22, 193-196 (1978).

7. Weber, G., Robaye, G., Bartsch, P., Collignon, A., Beguin, Y., Roelandts, I., Delbrouck, J.M. : Nucl. Instr. Meth. (In the press) (1984).

8. Gobbi, P.G., Scarpellini, M., Minoia, C., Pozzoli, L., Perugini, S. : Haematologica 63, 143-155 (1978).

DISTRIBUTION OF ELEMENTS IN HUMAN KIDNEY BY PIXE ANALYSIS

Masae Yukawa, Kensuke Kitao

National Institute of Radiological Sciences
9-1, Anagawa-4-chome, Chiba-shi 260 Japan

Minoru Terai

Tokyo Metropolitan University
2-1-1, Hukasawa, Setagaya, Tokyo, 158 Japan

Introduction

Studies on distributions of trace elements present in the human bodies have been recently carried out in our laboratory to get information about the physiological roles of these elements and to establish a monitoring method of environmental pollutions[1),2)]. Particle Induced X-ray Emission (PIXE) analysis is available for the purpose of these kinds of studies, since PIXE has high sensitivity enough to detect even trace elements in a small sample and determines many elements simultaneously and non-destructively. One more advantage of PIXE is a microbeam scanning characteristic which makes it possible to research a very small area continuously without separation of a sample into many small pieces[3)-6)]. In this study, the scanning PIXE analysis by proton microprobe was applied to study the microscopic details of the element distribution in a human kidney. The kidney is roughly composed of two parts, the cortex and the medulla, which have different physiological functions and are different histologically to each other. The distribution of trace elements in a kidney were observed across the cortex and the medulla along the scanning line. Trace elements in the cortex or in the medulla of the same kidney used for PIXE analysis were determined

Trace Element - Analytical Chemistry in Medicine and Biology, Vol. 3

using the technique of neutron activation analysis to compare with the results obtained from PIXE analysis.

Experimental procedure

1) Sample. A freeze-dried human kidney was sliced using a clean razor into about a 10 mm x 10 mm segment with a thickness of 1 mm. A piece of the sliced kidney sample was mounted on a thin plastic film of 2 μm thick Makrofol, and the mounted sample was attached to the scanning holder for irradiation (Fig. 1).

2) Proton irradiation and X-ray spectrometry. The microprobe system used for proton irradiation of samples is shown in Fig. 2. This system has been installed at 3 MV Van de Graaff accelerator. The proton beam was mechanically collimated to 500 μm to the scanning direction in the range of 500 μm to 2000 μm in width. The sample was irradiated with 2.0 - 2.3 MeV proton beam at a rate of a few nA under 10^{-6} torr, moving the sample by a stepping motor in 500 to 2000 μm increments within about 10 mm for micro-beam scanning. Each position was certified by reading a dial guage attached to the shaft of the target holder and bombarded for 15 min or less.

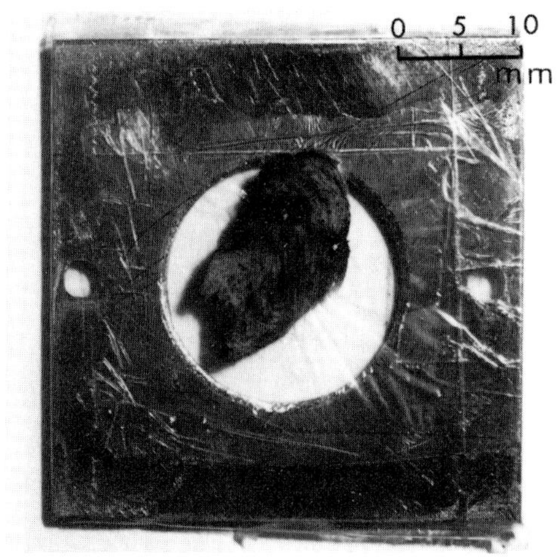

Fig. 1. A segment of a kidney mounted on Makrofol film attached to an Alminum plate

Induced X-ray was detected by a Si(Li) detector with 165 eV FWHM at 5.9 KeV. Relative X-ray intensity in each scanning was normalyized with beam current.

3) Neutron activation analysis. The cortex and the medulla separated from the freeze-dried kidney were powdered using an agate mortar. A few hundred mg of these powders were sealed in a clean polyethylene bag and irradiated in neutron flux of $5x10^{11}$ n x cm^{-2} x sec^{-1} for 18 hours. Activated Fe, Zn and Cd were detected by Ge(Li) detector.

Fig. 2. Schematic layout of the proton microprobe system for the scanning PIXE analysis

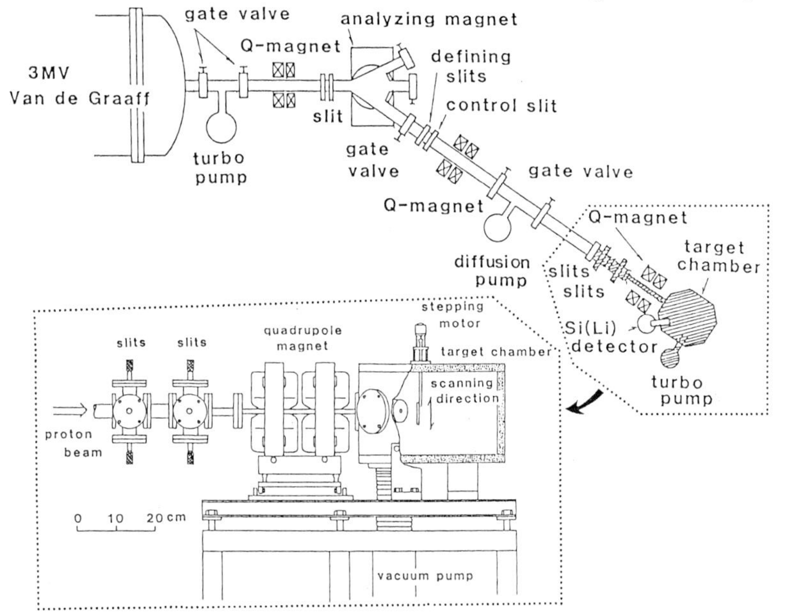

Results and discussion

A typical X-ray spectrum of human kidney obtained by proton beam bombardment is shown in Fig. 3. The elements detected from the spectrum were Cl, K, Ca, Mn, Fe, Cu, Zn, Cd, Br, Rb

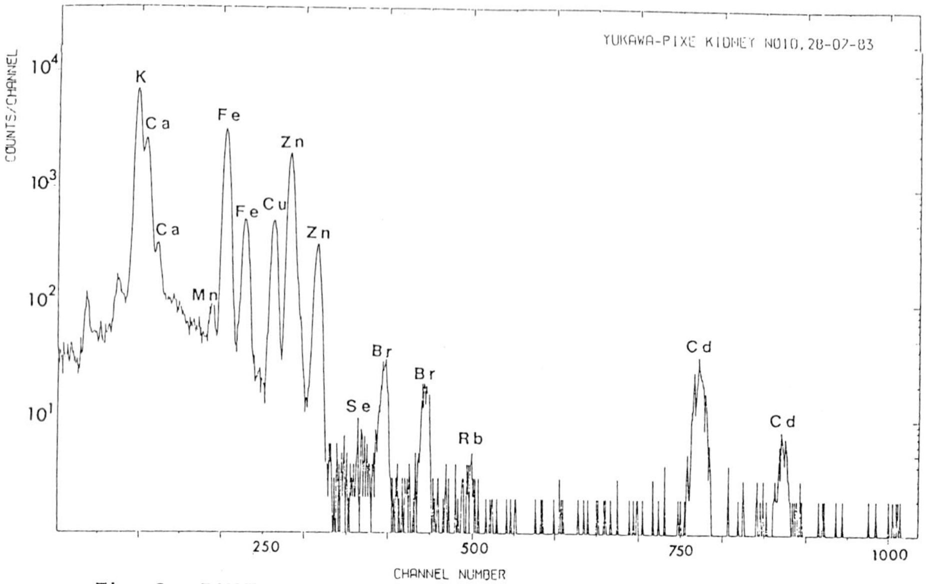

Fig. 3. PIXE spectrum of a renal cortex obtained with 2.5 MeV protons for an accumulated charge of 6μC

and Se. Fig. 4 shows the results of microbeam scanning. Fig. 4a illustrates a section of the sliced kidney and the irradiation positions along the scanning line, inwhich the boundary area of the cortex and the medulla is visualized between irradiation points Nos. 8 and 9. Characteristic X-ray intensities of Cu, Mn, Ca, Zn, Fe and Cd are varied along the scanning line from the cortex area to the medulla area (Fig. 4b). The results of neutron activation analysis on the renal cortex and the renal medulla shown in Table 1 supported

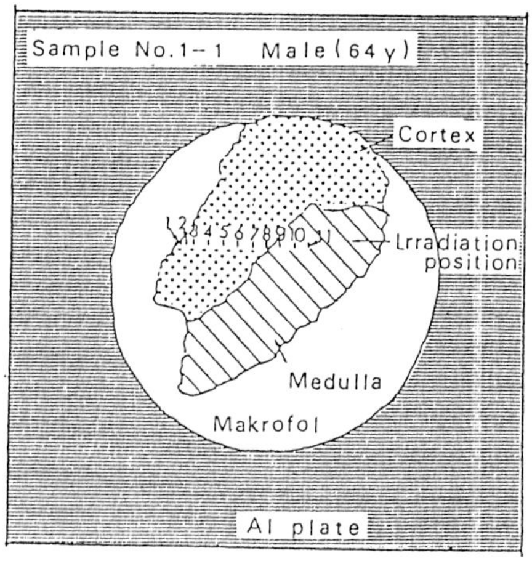

Fig.4a. Irradiation positions of the renal segment along the scanning line of the proton beam

this difference in Zn and Cd. The X-ray intensity ratios of Fe to Zn along the scanning line were plotted to clarify the distribution profiles of Fe and Zn (Fig.5). In this figure, three analyses of one human kidney (male, 64 years old) were illustrated. Nos. 1-1-1 and 1-1-2 indicate the results along the different scanning lines on the same sample segment. No.1-2 indicates the result of the segment taken from a different part in the same kidney. These results showed that the concentration ratios of Fe to Zn change obviously at the boundary of the cortex and the medulla. In addition, the ratios in the cortex were similar among these three determinations. However, the ratios were considerably different from each other in the medulla. The X-ray intensity ratios of Cd to Zn and Cd to Fe were also plotted (Fig.6). The ratio of Cd to Fe showed a clear change at the boundary of the cortex and the medulla, though the ratio of Cd to Zn was only slightly higher in the cortex than

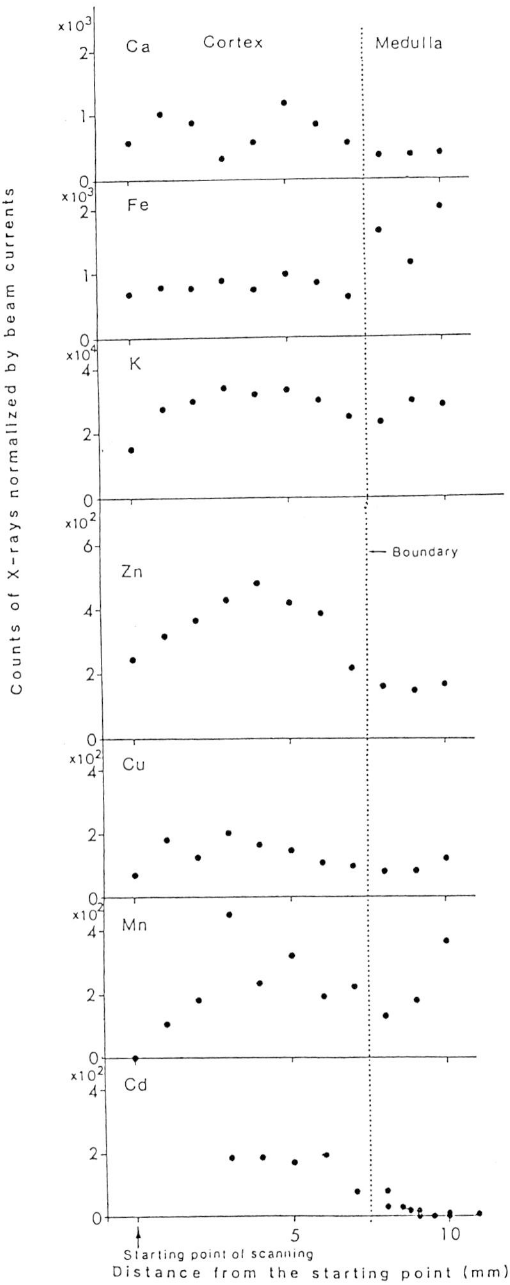

Fig.4b. Concentration profiles of elements in the kidney across the cortex and the medulla

Table 1. Concentrations of Fe, Zn and Cd in Human Kidney by Neutron Activation Analysis (ppm in Wet Tissue)

Element	Cortex	Medulla
	Conc./S.D.(N)	Conc./S.D.(N)
Fe	40.7/4.4 (3)	36.2/5.0 (3)
Zn	33.9/1.2 (3)	12.9/2.4 (3)
Cd	42.1/3.7 (3)	11.3/2.8 (3)

S.D.;Standard deviation N;Number of specimen

that in the medulla. These results indicated that distributions of Fe, Zn and Cd are different between the renal cortex and medulla. To investigate the physiological roles of trace elements in the human body, it is important to demonstrate the interaction among these elements, since some toxic elements to the human body were influenced in their toxicity by the presence of other elements. It has been reported, for example, that the toxicity of methyl-mercury is reduced by Se, and Zn containing metalothionein present in the renal cortex has the

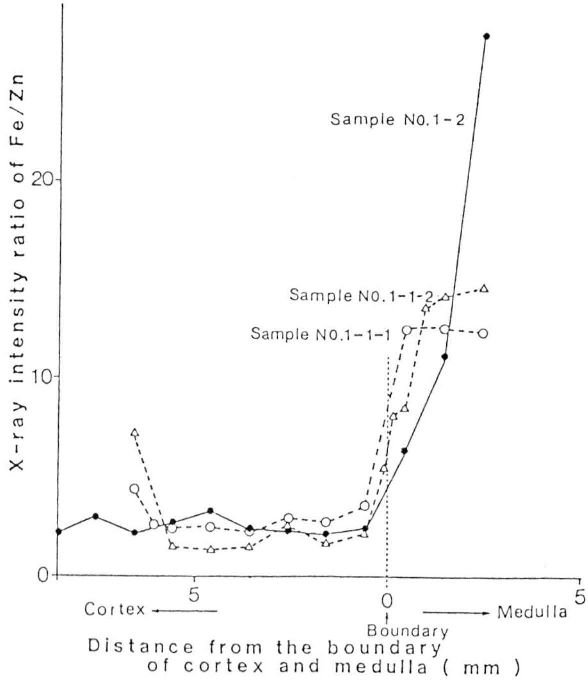

Fig. 5. Changes in the ratio of Fe/Zn in the kidney along the scanning line

the effect of inhibiting the Cd toxicity. PIXE analysis allows to determine distributions of elements in small and complicated tissues or organs using microbeam scanning technique in greater detail than in neutron activation analysis, in addition to simultaneous determinations of multielements in the same sample. Therefore, PIXE technique is very useful for this kind of research.

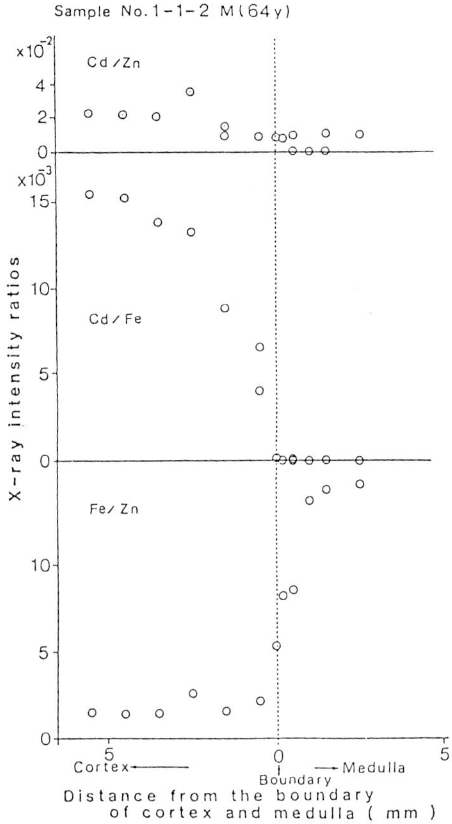

Fig.6. Changes in the ratios of Cd/Zn, Cd/Fe and Fe/Zn in the kidney along the scanning line

References

1) Yukawa, M., Amano, K., Suzuki-Yasumoto, M., Terai, M. : Archives of Environmental Health 35, 36-44 (1980).

2) Yukawa, M., Tanaka, S., Suzuki-Yasumoto, M. : To be published.

3) Imaseki, H., Ishikawa, M., Kitao, K.: Radiochem. Radioanal Letters 55, 49-56 (1982).

4) Ishikawa, M., Kitao, K., Imaseki, H.,Ishii, T., Uchida, S. : J. Radioanal. Nucl. Chem. 82, 189-200 (1984).

5) Bos, A.J.J., Stap, C.C.A.H. v.d., Lenglet, W.J.M., Vis, R.D., Valkovic, V.: IEEE Transactions on Nuclear Science NS-30, 1294-1251 (1983).

6) Sen, P., Mehta, G.K.:IEEE Transaction on Nuclear Science NS-30, 1302-1304 (1983).

BRONCHIAL CARCINOMA AND METAL CONTENT OF THE LUNG STUDIED BY PIXE

P. BARTSCH, G. WEBER*, G. ROBAYE, J.M. DELBROUCK-HABARU,
A. COLLIGNON, I. ROELANDTS.

DPT Medicine, DPT Nuclear Physics, DPT Geochemistry,
University of Liège, Sart-Tilman B-15
Liège, Belgium.

In previous studies (1) (2) we have shown that the elements currently detected in the human lung by the Particle Induced X-Ray Emission (PIXE) method (3) could be divided into two sets.
The first set : Potassium, calcium, iron, copper, zinc and rubidium are indeed homogeneously distributed within the different regions of the lung and of course in the five lobes of the two lungs. On the contrary, titanium, chromium, nickel and strontium which constitute the second set exhibit a positive gradient from base to apex, and are also strongly concentrated in the bronchial lymph nodes.

We raised the hypothesis that this fact could be related to the environmental pollution, peculiar to the industrial specificity of Liege area. This area has been devoted for centuries to coal mining, and then to steel industry. It is only in the last decade, that coal extraction was progressively abandoned, but the fly ashes, until a very recent time, contained high levels of titanium in this area (4), due to coal burning in industry and at home.

Moreover, when looking at element concentrations in the lungs of autopsy control patients, we were struck by the very high levels of the "second set" elements in two lung cancer patients and in one silicosis,

* Research Associate of the National Fund for the Scientific Research.

Trace Element - Analytical Chemistry in Medicine and Biology, Vol. 3

which were, all three, excluded from the mean value calculation of controls. We thus investigated systematically, the lung cancer of different histological types obtained during surgery at the University Hospital (5). We were able to confirm the results of the first two cases on 21 lung cancer patients of all kinds. We report here on 44 patients with the two major types of lung cancer to investigate the possibility of difference in lung mineral content according to histology.

Material and methods

The PIXE set-up has been described in detail in previous papers (6). Diagnosis of cancer was obtained histologically in all 49 patients during pre-surgery investigation for operability. Only the patients with epidermoid carcinoma (N = 30) and adenocarcinoma (N = 14) are reported here. The other subclasses are in limited number which allows no significant statement as a group.

Results and discussion

Fig. 1 gives the individual results on a semi-log graph of all patients (N = 49) and controls (N = 35) for Ti, Cr (2nd set) and Cu, Zn (1st set) for the lung tissue sampled at several centimeters from the tumor.
The values of copper in controls and patients are in the same range. This is also true for Zn, with a tendency for slightly higher values in the patients. On the contrary, an obvious deviation towards high content in chromium, and to very high levels of titanium is observed in the cancer cases.
Table I displays the results for all the elements detected in each of the 44 cases comprising 30 squamous cell carcinoma (epidermoïd) and 14 adenocarcinoma compared to the controls.

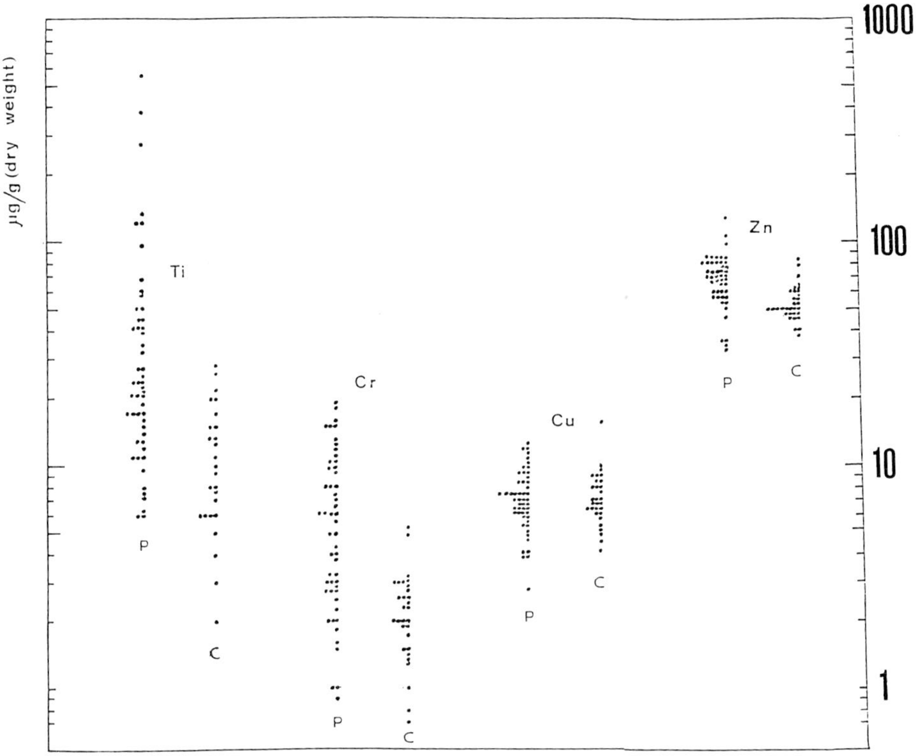

Fig. I - Distribution of Ti, Cr (Set II) and Cu, Zn (Set I) in peritumoral tissue : (C) = Controls (N = 35) (P) = Patients with lung cancer (N = 49).

Set I.

Potassium is always more abundant in the tumor than in the surrounding parenchyma. The increase is more striking in the squamous cell type (+ 30 % , $P < 0.001$) than in the adenocarcinoma (+ 10 % , $P = 0.5$).

A similar pattern is observed for copper and calcium. The values of zinc are very similar in all types of tumor and normal surrounding tissue but are slightly higher than controls ($P < 0.001$).
Copper is significantly higher in the squamous cell tumor ($P < 0.015$)
Rubidium follows a similar tendency.
In the two cell types iron is much less abundant in the tumoral tissue than in the lung parenchyma ; this could be due to the peculiar richness in capillaries and blood, caracteristic of normal lung tissue.

Set II.

Titanium content is strikingly high in peritumoral tissue of cancerous patients compared to controls and this is particularly true for squamous cell. Chromium is more than double of the reference values and also in a noticeable way in squamous cell type.
Nickel follows also a 2 fold mean increase but some cases seem to contribute highly to this effect.

We also made an attempt to correlate some elements of the normal peritumoral tissue, of the tumor itself and of the tumor with the surrounding tissue (Table II).
One must be very careful, when interpreting such data.
So we shall emphasize only some of these.

1) In the tumor itself four pairs are disclosed :
   K-Cu, K-Zn, K-Rb, Cu-Zn.
   and the same can be seen in the normal surrounding tissue.
   The couple K-Rb is well known by geologists and biochemists : here in the tumor we can notice a R value of 0,733.

		Non Cancerous (N=35)	Squamous cell (N=30)		Adeno (N=14)	
			Peritumoral tissue	Tumoral tissue	Peritumoral tissue	Tumoral tissue
	K	7450±1500	6830±1920	9218±2980	7118±2440	7844±2840
	Ca	1050± 380	742± 817 574± 180(N=28)	970±1390 723± 330(N=29)	580± 218	893± 790 710± 400(N=13)
	Fe	1050± 630	976±1115 704± 420(N=28)	420± 250 387± 174(N=29)	743± 322	390± 230
Set I	Cu	7,5± 2,1	7,26± 2,0	9,94±4,55 9,28±2,82(N=29)	7,3± 2,9	8,34±2,64
	Zn	54,5±10,5	66,8± 14,4	70,4±14,2	70,9± 22	69,8±12,6
	Rb	13,5± 5,5	15,8± 6,0 15,1± 4,7(N=29)	19,3± 8,7 18,4± 7,3(N=29)	17,3± 5,8	17,1± 5,9
	Ti	12,5± 7,5	73,8± 129 34 ± 33,5(N=27)	41,5± 55	27,5±20,5	13,6±11,6
Set II	Cr	2,85± 1,8	6,8± 5,2	5,23± 4,4 4,74± 3,6(N=29)	5,75± 4,5	5,28± 8,9 3,0 ± 2,5(N=13)
	Ni	0,40±0,25	1± 1,85 0,50± 0,45(N=27)	1,0 ±1,65 0,52±0,66(N=27)	0,78±0,90	1,32± 2,7 0,63±0,90(N=13)

Table I - A comparison of elemental concentrations in μg/g (dry weight) between cancerous and non cancerous lungs. Lower N values are given, when some individual values have been rejected ($/ x - \bar{x} / < 3$).

Peritumoral tissue (rows) vs. peritumoral tissue and tumoral tissue (columns)

PERITUMORAL TISSUE	K	Ca	Ti	Cr	Fe	Ni	Cu	Zn	Rb	Sr	K	Ca	Ti	Cr	Fe	Ni	Cu	Zn	Rb	Sr
	PERITUMORAL TISSUE										TUMORAL TISSUE									
K							{0.465}	0,515	0,641											
Ca																				
Ti				0,593	0,550				{0,383}				0,810							
Cr					0,547				{0,409}				{0,463}							
Fe									0,611						0,592				0,513	
Ni																				
Cu								0,692												
Zn																				
Rb																			0,661	
Sr																				

Tumoral tissue (rows) vs. tumoral tissue (columns)

TUMORAL TISSUE	K	Ca	Ti	Cr	Fe	Ni	Cu	Zn	Rb	Sr
K							0,556	0,634	0,733	
Ca										0,822
Ti										
Cr						0,867				
Fe										
Ni										
Cu								{0,410}		
Zn										
Rb										
Sr										

Table II - Correlation factors for elemental concentrations in the lung tissue. Correlations are given between elements in the tumor, around a tumor, and by elements inside the tumor compared to outside.

2) Ti, Cr, Fe and Rb bear a multiple correlation with a R value between ± 0,4 to 0,6 in the peritumoral tissue.

3) Ti, Fe, and Rb of the peritumoral tissue are correlated with the same elements within the tumor.
A high R value is observed for titanium (R = 0,81).
A weak correlation is seen for Cr (0,3).

4) A very significant correlation(R = 0,82) is obvious in the tumor for Ca-Sr and an even greater one for Cr-Ni (0,867).

Conclusions

1°/ We confirm our previous studies showing a significant raise of Ti, Cr and Ni in normal lung tissue of lung cancer patients, compared with our normal values in control patients which are in good accordance with old and recent data from others (7) (8).

2°/ The accumulation of titanium on the tumor is particularly evident in the squamous cell type, with some patients exhibiting several tens increase.

3°/ Some correlations are observed (K-Rb) related to chemical similitude between elements, or linked to a common origin as pollutants (Cr-Ni).

4°/ We cannot answer until now the question on the pathogenesis of lung cancer : are the metals with higher concentrations, the cause of lung cancer together with organic compounds known as potents carcinogens ? If so, are they accumulated in the lung because of atmospheric pollution or by tobacco smoke or both ?

Bibliography

1. Bartsch, P., Collignon, A., Weber, G., Robaye, G., Delbrouck, J.M., Roelandts, I., Yujie, J. : Arch. Environ. Health 37, 111-117 (1982).

2. Weber, G., Robaye, G., Bartsch, P., Collignon, A., Roelandts, I., Delbrouck, J.M., : IEEE Trans. Nucl. Sci. 2, 1313-1315 (1983).

3. Johansson, T.B., Akselsson, R., Johansson, S.A.E. : Nucl. Instr. Meth. 84, 141-143 (1970).

4. Priest, P., Desaedeleer, G., : Anal. Chem. 51, 2112-2116 (1979).

5. Weber, G., Robaye, G., Bartsch, P., Collignon, A., Beguin, Y., Roelandts, I., Delbrouck, J.M. : Nucl. Instr. and Meth. (In the Press) (1984).

6. Robaye, G., Weber, G., Delbrouck, J.M., Roelandts, I., Bartsch, P., Collignon, A., : Nucl. Instr. and Meth. 181, 59-62 (1981).

7. Tipton, I.H., Schoeder, H.A., Perry, J.R., H.M. : Health Physics 11, 403-451 (1964).

8. Vanoeteren, J., Cornelis, R., Versieck, J., Hoste, J., Deroose, J. : Proceedings of Trace Element Analytical Chemistry In Medecine And Biology, Munchen (1983).

SCANNING ELECTRON MICROSCOPY AND X-RAY FLUORESCENCE ANALYSIS OF PARTICULATE MATTER DEPOSITED IN HUMAN LUNG

Carine Vanoeteren, Rita Cornelis, Richard Dams, Rudi Ryckaert*

Rijksuniversiteit Gent, Laboratory for Analytical Chemistry
Proeftuinstraat 86, B-9000 Gent, Belgium

* Academic Hospital, Department of Pneumonology, De Pintelaan 185, B-9000 Gent, Belgium

Introduction

The technological character of modern society increases the risks of exposure to a variety of materials, especially among industrial workers. Because of their anatomical structure and physiological function, the lungs are constantly exposed to substances that are potentially hazardous. Inorganic materials such as silica, asbestos and certain heavy metals are known to produce interstitial lungdiseases. In spite of recent advances in diagnostic techniques, the specific etiology in individual cases of interstitial lungdiseases remains in many instances undetermined. Identification of foreign materials deposited in pulmonary tissue may reveal a link with the observed clinical picture.

Scanning electron microscopy (SEM) in combination with energy dispersive X-ray fluorescence spectrometry (EDXRF) enables the detection and analysis of exogenous particles deposited in the lungtissue (1-5). Shape, size and elemental composition of the particles can be studied simultaneously.

Materials

Three types of lungmaterial were qualitatively analysed by the SEM-EDXRF technique.

1) Particles were studied in the pulmonary tissue. An attempt was made to determine the deposition place of the particles in the lung structure. Tissue samples were collected from 6 patients at the occasion of a

Trace Element - Analytical Chemistry in Medicine and Biology, Vol. 3

pneumectomy or an open lungbiopsy. Only one of the patients (H.E.H., table 1) is known to have been previously exposed to industrial dust. The tissue samples of this patient were analysed both by SEM-EDXRF and neutron activation analysis (NAA).

2) Particles were studied in the residue remaining after enzymatic digestion of the lung tissue. The main disadvantage of this technique is that the relationship to the anatomical structure is lost. Another set-back may be a partial dissolution of the particles by the chemicals used to digest the tissue (6). The collected insoluble residue was filtered and studied with the SEM-EDXRF technique. The findings were compared to the analytical data obtained by NAA on the same lungsamples and their accumulated dust (7, 8). The lungs were obtained at autopsy from individuals with no occupational history.

3) Particles were studied in the broncho-alveolar lavage (BAL)-fluid. This fluid provides samples of the endo-alveolar lungsection. The element-analytical and/or mineralogical information on the particulates found in the fluid can be determinant in the study of the relationship between inhaled mineral particles and lungdiseases (9). Broncho-alveolar lavage fluid was collected from 4 patients and microscopically analysed with the SEM-EDXRF system. These patients suffered from several types of lungdiseases probably linked to their occupational history.

Sample Preparation

1. Lungtissue. On the occasion of a pneumectomy or an open lungbiopsy an additional piece of ca. 1 cm^3 tissue was removed for subsequent SEM-EDXRF analyses. The sample was rinsed in 0.13M phosphate-buffer (Millonig) (pH 7.3) followed by a preliminary fixation in a 1% glutaraldehyde solution (10). After dissection in smaller fragments, it was fixed in a 2.5% glutaraldehyde solution for at least 48 h. The samples were rinsed with destilled H_2O, dehydrated in a graded alcohol series, plunged in a mixture of solid and liquid N_2 and freeze-broken. The specimens were critical point dried from CO_2. The surfaces were examined under an optical microscope to trace the areas with anthracotic pigmentation. For the SEM examination, the samples were attached to the Al-holder with carbon cement, and coated with 35-70 nm carbon.

2. Enzymatic digestion of the lungtissue. Autopsy samples were freeze-dried and 50-100 mg dry tissue was ground and digested with the proteolytic enzyme ficin as can be read in an earlier publication (7). The solution was filtered on a 0.8 µm Millipore AA filter and thoroughly washed with 1.10^{-5}N NH_4OH to eliminate the cystine crystals that remained from the digestion. After drying, part of the filter was mounted to the Al-stub and carbon coated.
3. Broncho-alveolar lavage. Broncho-alveolar lavage was performed by bronchofibroscopy, and washing of a segmental or subsegmental bronchus with 5 x 50 ml physiological saline solution (9 ‰). The fluids were stored at -30°C. At the moment of examination, the solution was thawn and 5, 10, or 20 ml of the fluid was centrifuged and filtered on a 0.2 µm nuclepore filter. In one case, part of the fluid was previously treated with an equal volume of a filtered 5% NaOCl solution to digest the excess organic residue.

Instrumentation

The analyses were carried out with a Cambridge Stereoscan 150 scanning electron microscope fitted with a United Scientific Si(Li) detector and a TN 2000 multichannel analyser. The samples were analysed for minimum 60 s (usually 300 s) at a 20 kV operating potential, and with a specimen current of 10^{-10}A. The specimens were kipped over 45°. In some cases, a spotanalysis was performed. Mapping of X-rays over one or more regions of interest was frequently done. Only a restricted amount of particles was analysed in each sample. All elements above Z=10 could be identified, if present at concentrations above ± 0.5%.

Results and Discussion

1. Analysis of lungtissue.

Macrophages were analysed in all lungsamples. The elements most frequently detected were P, S, Ca and Fe. P and S are major elements of lungtissue. P may however also originate from the phosphate buffer. Fe and Ca can be of both exogenous or endogenous origin, the latter as a result of reactions of the organism on exogenous material. It is well known that inhaled particles, such as asbestos and fibrous glass, are coated by the living tissue, giving rise to the formation of Fe and Ca bodies in the lung. Some macrophages contained the element Si which, as SiO_2 or in silicates is one of the most abundantly inhaled substances.

One patient working as an engraver of sporttrophies (H.E.H., table 1), had been exposed to several types of metal dust. The macrophages in his lungs contained also the elements Cu and Zn.

Since it is very difficult to discern exogenous particles from lung tissue on the basis of their morphology, and since no free particles were found in the alveoles, this study was oriented towards the pulmonary black pigmentation whenever present. These deposits are usually found in peribronchial tissue or beneath the pleura. There, several particulates and cell remainders were spotted. Small particles, containing one or a few metalic elements could be detected with the mapping technique. Fig. 1, 2 and 3 show a particle in the lungs of a 26 year old man (V.D.B., table 1), the corresponding X-ray spotanalysis and the mapping of the Cu Kα-line. Very often, size and shape of the particles were difficult to judge because they were coagulated or partially covered with the tissue. Table 1 summarizes all elements detected in the particles deposited in the six lungsamples. Some of these were found in all lungs : Ca, Fe, P, S, Si. A second group of elements was frequently detected, although not in all tissues : Al, Cr, Cu, Mg, Ni, Pb, Ti, Zn. Other elements were only discovered in one lung : V and Zr in the lungs of patient V.D.B., K in the lungs of patient V.H.S. and particles containing Cd in the lungs of patient H.E.H. Some elements could be associated with one another. The Fe, Cr and Ni Kα-peaks were often present in a constant ratio : 1 : 0.3 : 0.12. These peak-ratios are not to be confound with mass-ratios but yet strongly suggest the presence of stainless steel particulates. Al and Mg were usually detected in combination with Si. Some Al-Si particles were small (≤ 1 µm) and spherical, others were fibrous such as most of the Mg-silicates. The elements Cu and Zn were also often associated. Most particles contained only a few elements in fairly high concentrations.

The samples of the previously mentioned engraver (H.E.H., table 1) were studied extensively. The elements Cu and Zn found in the macrophages, were

TABLE 1 : Elements detected in the particles deposited in the lungs

Patient	Sex	Age j	Occupation	Elements detected in all lungs					more or less frequently								rarely				
V.H.S.	♀	26	housewife	P	S	Fe	Ca	Si	Al	Cu	Zn	-	-	-	-	Mg	-	-	-	-	K
D.W.J.	♂	52	warehouse-keeper	P	S	Fe	Ca	Si	Al	-	-	-	-	-	-	Mg	-	-	-	-	-
V.D.B.	♂	26	nurse	P	S	Fe	Ca	Si	Al	Cu	Zn	Cr	Ni	Pb	-	Mg	-	V	Zr	-	-
F.J.	♂	65	mechanician	P	S	Fe	Ca	Si	-	Cu	Zn	-	-	Pb	Ti	-	-	-	-	-	-
D.H.	♂	45	military man	P	S	Fe	Ca	Si	Al	-	-	-	-	-	Ti	-	-	-	-	-	-
H.E.H.	♂	45	engraver sport-trophies	P	S	Fe	Ca	Si	Al	Cu	Zn	Cr	Ni	-	-	Mg	Cd	-	-	Na	-

also detected in free particles in the anthracotic pigmentation zones. Other typical particulates contained Fe, Cr and Ni. Lungsamples of this patient were also analysed by NAA. The concentrations of the elements Co, Cr, Cs, Fe, Mn, Rb, Sb, Se and Zn were compared with those of a 33 year old man without history of occupational exposure and who died in an accident (Table 2). The concentrations of Co, Cr, Cs, Fe and Rb were somewhat higher whereas the Sb-concentrations were lower. These findings correspond roughly with those of the SEM analyses.

TABLE 2 : Comparison of element-concentrations analysed in lung material of a 45 y old metal engraver (H.E.H.) with those of a 33 y old man (C.R.) without any occupational exposure.

Element	µg/g wet tissue H.E.H.	C.R. mean ± s
Co	0.022	0.016 ± 0.004
Cr	0.285	0.128 ± 0.043
Cs	0.020	0.008 ± 0.0009
Fe	168	98 ± 20
Mn	0.42	-
Rb	3.45	1.91 ± 0.18
Sb	0.012	0.023 ± 0.006
Se	0.24	0.25 ± 0.10
Zn	12.5	11.5 ± 1.87

2. Analysis of particles after tissue removal

The analyses of several dust filters proved that many of the exogenous elements, found in the tissue by total element analysis can be traced back to the residues. The most prominent examples were : Al, Cr, Cu, K, Mg, Si, Ti, Zn besides Ca, Cl, Fe, P, S. The presence of P, S and Cl can eventually be due to the reagents used for the enzymatic digestion. Most of the particles were < 5 µm in size although some may have coagulated into larger fragments. The fibres contained Si, Al, Mg and Ca, but also Ti (probably as TiO_2). Diatom shells were found in the lung residues of a 73 y old man (a textile worker) who died of a braintumor. They were identified by morphology (Fig.4) and the pure Si Kα X-ray. The presence of diatom shells in lungtissue is also mentioned by other authors (5).

3. Analysis of broncho-alveolar lavage fluid

In search for a possible blank, the complete sampling procedure for the BAL-technique was simulated in vitro and the fresh saline solution was filtered and checked. The same was done with the double destilled water, used to wash the filters. Silicates (Al, Si and traces of Fe, K) appeared to be regularly present. These findings emphasize the fact that all solutions should be filtered prior to use, at the risk of introducing a blank. The elements detected in the BAL-fluid of the four patients are compiled in table 3.

TABLE 3 : Elements present in the broncho-alveolar lavage fluid of 4 patients

Patient	Sex	Age y	Disease	Occupation	Elements														
H.E.H.	♂	45	lungfibrosis	engraver of sporttrophies	Si	Al	Ca	Fe	Cr	Ni	Ti	K	Cu	Zn	-	Pb	-	-	-
D.A.	♂	66	pneumoconiosis	worked in - coal mines (5 y) - brick-works and as a stoker > 20y	Si	Al	Ca	Fe	Cr	Ni	Ti	K	Cu	-	S	-	-	-	-
J.M.	♂	53	pulmonary hemosiderosis	worked in rubber vulcanisation industry	Si	Al	Ca	Fe	-	-	Ti	K	Cu	-	S	-	Ba	-	-
D.L.G.	♂	62	hemoptoe unknown etiology	village policeman	Si	Al	Ca	Fe	-	-	Ti	K	-	-	S	-	-	Na	Mg

The first patient in this table (H.E.H.) is the same metal engraver as in table 1. The most abundant element in this sample was Si, often in combination with Al, Fe and Ca. The particles containing Al-Si were usually small ($\leq$ 1 µm) and spherical, possibly fly ash particles. Other elements identified in the solution were Ti, K, Pb, Cu and Zn. Again particles containing Fe, Cr and Ni in a very constant ratio (1 : 0.3 : 0.12) were frequently found, in agreement with the previously mentioned findings concerning this patient. In the sample of the second patient (D.A.), three main groups of particles were discovered. Particles containing Fe, Cr and Ni were irregularly structured and ca. 5 µm large. A second group of 5 µm particles contained Cu in combination with S. The majority of the particles were silicates. Most of them were small (1-2 µm) and contained both Si and Al in varying ratios, and sometimes less important amounts of Fe and K (clay?), Ca, Ti and S. Some particulates contained almost exclusively Si as SiO_2. The presence of silicates was in accordance with the patient's disease and occupational history. It should however be kept in mind that any lung contains considerable amounts of silicates and that solutions are easily contaminated with them. In the lavage fluid of the third patient (J.M.) no particles containing the Fe, Cr, Ni combination were detected. The other elements Al, Ca, Cu, Fe, K, S, Si, Ti, as well as Ba, were also identified in this material. In the solution of the fourth patient (D.L.G.) the elements Al, Ca, Fe, K, Na, Si and Ti were detected. Apart from these, long, straight fibres, coated with ferroproteins, as could be inferred from the Fe-peaks in the X-ray spectrum, were discovered on the filter (Fig.5). As it has been proved that regularly shaped and segmented ferruginous bodies with a fine central core almost certainly represent asbestos bodies, they were considered as such (11). Elemental analysis of the central fibre was often difficult because the segmentation was not interrupted. A few naked fibres were found. They consisted indeed of Si, Mg and to a lesser extent of Ca and Fe. The length of the fibres varied from 10 to more than 50 µm, the diameter from 1 to 2 µm. Another part of this BAL-fluid was treated with NaOCl during 12 h. This period seemed too long since the ferroprotein coatings disappeared, leaving but naked fibres.

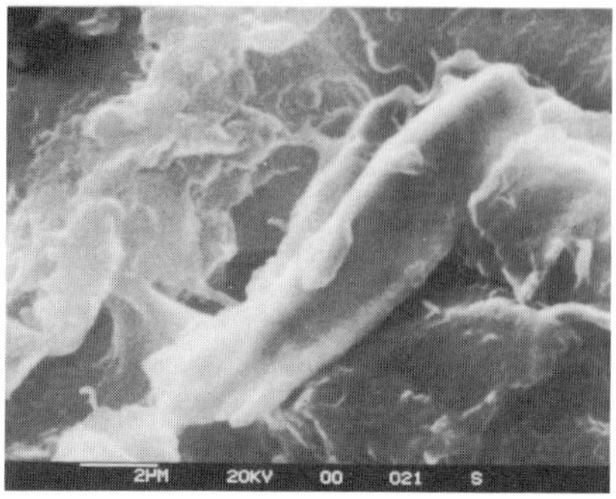

Fig. 1

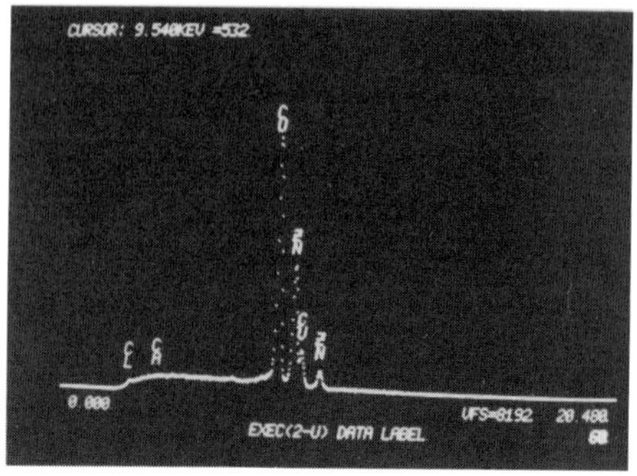

Fig. 2

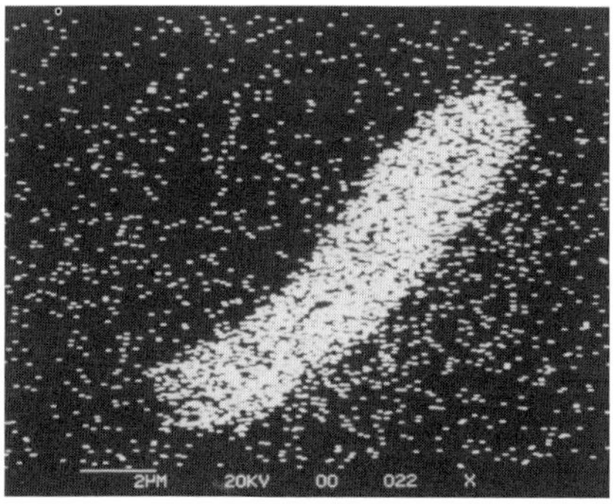

Fig. 3

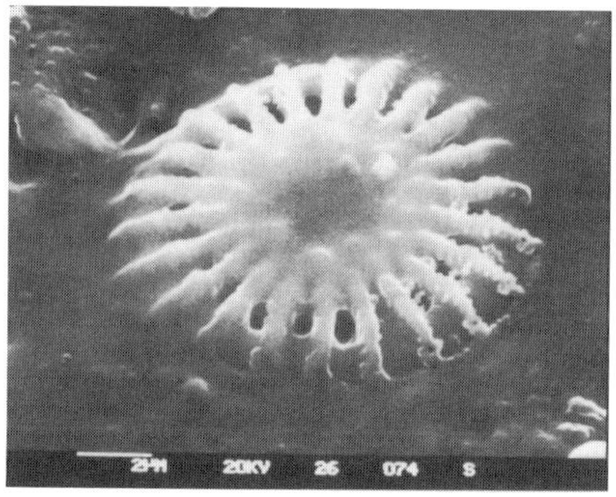

Fig. 4

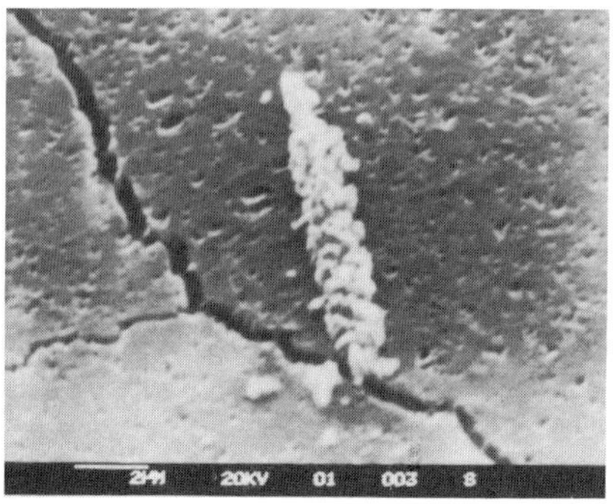

Fig. 5

Fig. 1, 2, 3 : particle deposited in the lungs of a 26 y old man (V.D.B.,table 1), X-ray spot-analysis and X-ray mapping of the Cu Kα-line.

Fig. 4 : diatom shell in the lung residue remaining after enzymatic tissue digestion.

Fig. 5 : Ferruginous body in broncho-alveolar lavage (BAL) fluid (D.L.G., table 3).

Conclusion

Scanning electron microscopy coupled to energy dispersive X-ray fluorescence spectrometry was used for qualitative analysis of particles deposited in lung tissue both in situ or after removal of the tissue, and in broncho-alveolar lavage fluid. A great number of elements : Al, Ba, Ca, Cd, Cr, Cu, Fe, K, Mg, Na, Ni, P, Pb, S, Si, Ti, V, Zn and Zr could be detected, even when the patient had not been exposed to any kind of occupational dust. Most particles contained only a few of these elements, estimated to be present in fairly high concentrations.

Particles were traced in the anthracotic pigmentation, present in most lung specimens and in the macrophages. The same types of particulates were found after enzymatic tissue digestion. The elements detected in the tissue and BAL-fluid of patients with an occupational history were in accordance with the inhaled materials. Their presence does not however constitute unequivocal prove for a cause-effect relationship between the particulates in the working environment and the developed disease.

The amount of particles present in the fluid or in the tissue and the elemental concentrations, can yield additional evidence for a causal relation with the pathological process.

Acknowledgment

The authors acknowledge with gratitude the invaluable help of Dr.J.Versieck and the financial support of the Inter-University Institute for Nuclear Sciences.

References

1. Funahashi, A., Pintar, K., Siegesmund, K.A.: Arch.Environ.Health, 30, 285-289 (1975).

2. Pariente, R., Berry, J.P., Galle, P., Cayrol, E., Brouet, G.: Thorax 27, 80-82 (1972).

3. Berry, J.P., Pariente, R., Watchi, J.M.: Rev.Franç. Etudes Clin. et Biol. 14, 915-918 (1969).

4. Abraham, J.L., Barton, R.T., Høgetveit, A.C., Andersen, I., Smith,G., Nickel Toxicology, Academic Press, London (1980). Ed. S.S.Brown, F.W.Sunderman Jr, 155-158.

5. Ophus, E.M., Mowe, G., Osen, K.K., Gylseth, B.: Br.J.Ind.Med. 37, 375-381 (1980).

6. Denee, P.B., Proc. of the Workshop on Techniques for Particulate Matter Studies in SEM, SEM, IITRI, Chicago, Illinois (1976).

7. Vanoeteren, C., Cornelis, R., Versieck, J., Hoste, J., De Roose, J.: J.Radioanal.Chem. 70, 219-238 (1982).

8. Vanoeteren, C., Cornelis, R., Hoste, J., Vanderhaeghen, L., Versieck, J.: Trace Element Analytical Chemistry in Medicine and Biology, Vol.2 (1983). Ed. W.De Gruyter & Co., Berlin, 315-332.

9. Brochard, P., Bignon, J., Sebastien, P.: Acta Endos. 12, 421-427 (1982).

10. Millonig, G.: J.Appl.Phys. 32, 1637-1639 (1961).

11. Churg, A., Warnock, M.L., Green, N: Lab.Invest. 40, 31-38 (1979).

D i s c u s s i o n Session VII

Subjects of the discussion

- significance of Br-analysis
- use of statistical methods in comparing results
- localisation of Al in bone
- possibilities of LAMMA-technique
- PIXE-analysis
- trace elements and lung diseases

Contributors: Cochran, Cornelis, Koh, Ottaway

Session VIII

NEWER ESSENTIAL TRACE ELEMENTS

Plenary Lecture given by: M. Anke, B. Grappel,
H. Kronemann (GDR)

Chairman:
P. Brätter (FRG)

Poster Reports given by: M. Anke (GDR)

SIGNIFICANCE OF NEWER ESSENTIAL TRACE ELEMENTS (LIKE SI, NI, AS, LI, V ...) FOR THE NUTRITION OF MAN AND ANIMAL

Manfred Anke, Bernd Groppel, Hella Kronemann
Wissenschaftsbereich Tierernährungschemie der Karl-Marx-Universität Leipzig
GDR-6900 Jena, GDR

Introduction

The classic trace elements Fe, J, F, Cu, Mn, Zn, Co, Mo, Se and Cr are essential components of nutrition. Humans and animals need between 0.1 and 100 mg/kg ration dry matter of them. It is more important for their characterization that deficiency symptoms were detected in animals in practice.

Since 1970 a number of experimental works have been published which point to the essentiality of new trace elements (Si, Sn, V, Ni, As, Cd, Li, Pb).

The following criteria were elaborated to prove essentiality (124, 130, 206, 78, 29):

- growth deceleration
- reduced reproduction performance (rut symptoms, conception rate, abortions, birth weight, rate of survival)
- decreased life expectancy
- the trace element must be a component or activator of an essential body substance
- pathological changes
- reproducibility of results in at least 3 different experiments in different species
- disappearance of deficiency symptoms after supplementation of the element and renewed occurrence when animals stop taking the element

These criteria are also valid for the essentiality of new trace elements. The requirements of humans and their domestic and laboratory animals amounts to about 50 ng/g food-stuff

Trace Element - Analytical Chemistry in Medicine and Biology, Vol. 3

(136) in case of their essentiality.

1. Silicon

In the first experiments to prove the essentiality of Si in rats (52,188,183) and chicks (53) synthetic rations with crystalline amino acids were used as protein source, which did not allow an optimal growth of rats and chicks. Under these conditions the animals with Si-poor rations put on less weight than control rats and chicks. These results could not be confirmed with semi-synthetic rations and natural protein as protein source, which allowed optimal growth (57, 58).
In all experiments on Si-essentiality in chicks and rats the importance of Si for the normal development of connective tissue and skeleton was correspondingly to be seen (51,52,53, 57,58,59,188,181,182,56). Si-deficiency led to changes at the skull and the bones of limbs characterized by poorly formed joints and defective endochondral bone growth.
Si-deficient chick tibiae exhibited depressed contents of articular cartilage, water, hexosamine, and collagen. The importance of Si for the formation of connective tissue and skeleton in chicks and rats is unmistakable. Si is essential for it. But a clearer characterization of Si-deficiency is necessary.
The biochemistry of Si is also known only partially. Si is a component of several glycosaminoglycans and polyuronides. It is apparently bound to the polysaccharide matrix (181,182). Si is assumed to occur as sinolate in the mucopolysaccharides, and it could link portions of the same polysaccharide chain, different polysaccharides to each other, or acid mucopolysaccharides to protein. The assumed structures are still to be identified. The collagen and elastin of connective tissue contain Si or are influenced by it (54,58,59,60,186,117). Si is also said to have a preventive effect against atherosclerosis (184,116,209,38,41,40,180). In rabbits with an atherogenous diet the fibres of aortic tissue were depleted thin and fragmented. After Si-supplementation the elastic fibres of aortic tissue

were dense, regular, and often thickened. Si plays a role in the cross-linking mechanism in collagen and elastin.
Si also plays a role in bone calcification, but the mechanism is not clear yet. In any case the formation of the organic bone matrix is more strongly influenced by Si-deficiency than the mineralization process (59). Si seems to be effective above all at the beginning of mineralization, since the Ca- and Si-content of bone tissue firstly increases congruently. At the advanced stage of bone formation the Si-content decreases significantly (55). The Si-influence on bone mineralization might be performed via mitochondria, which contain Si (119,120) and are involved into the calcification process (112).
Metabolism and Si-requirement are insufficiently examined as well. The estimated requirement of chicks for Si, as sodium silicate, is in the range of 100-200 ppm (131). Other Si-compounds are said to be essentially effective in the prophylaxis of Si-deficiency (182). In case of their application the Si-requirement would be significantly lower than 100 ppm.
The form of dietary Si is a control factor in its absorption (49).
In case of Si-deficiency absorbable Si is produced in the gastro-intestinal tract. Furthermore, the raw fibre content of food-stuffs might influence Si-absorption (90). The Si-intake of humans varies between 21 and 46 mg/day (90).
It can be summarized that further investigations on the effect and the essentiality of Si are necessary. There are, nevertheless, a number of indications that Si is essential in humans and atherosclerosis, osteoathritis and chronic hypertension. The experiments carried out in rats and chicks, in which different Si-preparations reduced the lipid level in vivo (39) and increased the collagen biosynthesis (190), are very impressive, but they cannot be transferred to humans without any problems.
Si is essential for rats and chicks. It might be essential for humans as well; further investigations on the influence

of Si-deficiency on growth, reproduction, diseases and life expectancy are necessary.

2. Nickel

At the end of the sixties and the beginning of the seventies the biological interest in the heavy metal Ni increased since it was possible to reduce the Ni-offer considerably by means of semi-synthetic rations and to produce Ni-deficiency symptoms in chicks, rats, goats and mini-pigs (129,130,1,14,197, 142). The effects of Ni-deficiency were summarized under different aspects (102,17,8).

In rats, mini-pigs and goats Ni-deficiency led to significantly reduced growth. In Ni-deficiency experiments with growing and gravid goats, which were repeated ten times, Ni-deficiency goats gained 11 % less weight than control goats (table 1). Mini-pigs (1, 14, 25), rats (139,167,174) and lambs (193,194) reacted on Ni-deficiency similarly to kids (1,25,28).

Table 1: Daily weight gain of female control and Ni-deficiency goats on the average of 164 experimental days (g per goat and day) (11)

n	control goats s	$\bar{x}$	Ni-deficiency goats $\bar{x}$	s	p	%
49; 49	33	99	88	42	$< 0,05$	89

Intrauterine Ni-deficiency reduced fetal live weight development as well (table 2). At birth Ni-deficiency kids weighed 10 % less than control kids on the average. During the first

Table 2: Live weight development of control and Ni-deficiency kids from birth to 91st day of life in kg (11)

day of life	control kids s	$\bar{x}$	Ni-deficiency kids $\bar{x}$	s	p	%
1st	0,77	3,1	2,8	0.80	< 0.05	90
28th	1.9	7.6	7.1	1.6	> 0.05	93
56th	3.0	12.4	11.2	2.3	< 0.05	90
91st	4.2	19.1	16.3	3.8	< 0.001	85

28 days of life with an almost exclusive milk diet the difference between the live weights decreased and became insignificant. The difference between the two groups increased again and reached 15 % in the following period with a higher intake of semi-synthetic feed-stuffs instead of milk. Though the Ni-content of the milk of Ni-deficiency goats is significantly Ni-poorer than that of control goats, it contains more Ni than the semi-synthetic ration with < 100 µg Ni/kg (table 3). For that reason the growth of Ni-deficiency kids improved first in order to get worse with the increasing consumption of semi-synthetic rations.

Table 3: Nickel content of the milk of control and Ni-deficiency goats (µg/kg dry matter) (11)

n	control goats		Ni-deficiency goats		p	%
	s	$\bar{x}$	$\bar{x}$	s		
colostral milk (82; 47)	166	233	140	105	< 0.01	60
mature milk (115; 78)	132	233	189	101	< 0.001	81

Both in mini-pigs (1,14,31) and in goats (1,31, 28, 11) and rats (167,174) Ni-deficiency influenced the reproduction performance compared to other essential trace elements (e.g. As) only little (32) (table 4). Success of first insemination, conception and abortion rate were significantly worse in Ni-deficiency goats than in control animals.

As a rule Ni-deficiency goats and Ni-deficiency mini-pigs are in heat later than control animals.

In intrauterinely Ni-poor he-goats frequent cases of small testicles together with lacking sperm production and libido sexualis were registered.

The influence of Ni-deficiency on life expectancy is differently described by different research teams. In intrauterinely Ni-poor rats both increased (139) and uninfluenced mortality (174) were registered. In long-term Ni-deficiency ex-

Table 4: The influence of Ni-deficiency on the reproduction performance of female goats (11) (n 89; 96)

	control goats	Ni-deficiency goats	p
success of first insemination, %	70	55	< 0.05
conception rate, %	83	71	< 0.05
non-pregnant goats, %	17	29	< 0.05
pairings per gravidity	1.4	1.9	< 0.05
abortions, %	1	9	< 0.001

periments significantly increased mortality was found in she-goats as well as in their off-springs (table 5).

Table 5: The influence of Ni-deficiency on the mortality of she-goats and their off-springs (11)

parameter	control goats	Ni-deficiency goats	p
died mothers,%	24	43	< 0.01
died kids,%	9	38	< 0.001

In Ni-deficient chicks (143,138), mini-pigs (1,14, 25), goats (21,28, 16) and rats (139,167,168) (rough coat) changes of epithelia and hair were observed, which became above all visible at the udder of goats during lactation (28). The hair of Ni-deficiency goats was rough, there were isolated cases of dwarfism. In mini-pigs Ni-deficiency resulted in an increased renal Ca-excretion and a reduced Ca-content of the skeleton (1,14). Instead of Ca Mg was increasingly incorporated into bones in rats (91). Under the influence of the disturbed Ca-metabolism there was also a Zn-deprivation in goats (28), mini-pigs (1, 14) and rats (169), which was explained by the reduced size of organs in the latter case (101). The decelerated Zn-incorporation into the organs and the milk of Ni-deficiency goats could be demonstrated by means of ^{65}Zn. Rumen and intestine content of Ni-deficiency animals was ^{65}Zn richer than that of control goats, which may point to a simultaneously disturbed Zn-resorption (18). There are interactions as well between Ni and Ca and Zn as between Ni and Fe. Rats (168, Apart from the interactions between Ni and Fe. Rats (168,170, 171,101,137,145,133, 134, 135) and goats (28, 26) with Ni-poor rations fell ill with anemia, which resulted in reduced hemoglobin and hematocrit values. Ni is essential for the enzymatic formation, or structural integrity, of a molecule involved in ferric ion absorption (136).

Via several Ni-dependent or Ni-containing phytogenic and bacterial enzymes the Ni-offer can influence the microbiological feed disintegration above all in ruminants and herbivores with marked caecum digestion. The possibilities of this influence are summarized in detail (8). Ni is essential for the biosynthesis of the hydrogenase of "Knallgas" bacteria (36, 71), of the carbon monoxide dehydrogenase of bateria producing acetate (63, 204) of the factor F 430 of methanogenic bacteria (64,212) and of the plant urease (65,150, 67). These findings are interesting for the rumen and caecum activities of the flora and also explain several differences between monogastric species of animals and ruminants, which occurred in Ni-deficiency experiments (8).

Ni-deficiency goats, which received a part of the necessary N as urea, had only 2 % of the urease activity of control goats in rumen flüid (79). In lambs with Ni-poor rations and in fattening cows the urease activity could be increased by Ni-supplementation (192, 193, 194).

In rats Ni-deficiency also led to reduced activity of different dehydrogenases and transaminases and above all of α-amylase (96,99,100,102,103,104,172, 175). In ruminants the findings could be confirmed only principally. Goats died from Ni-deficiency only reduced their sorbite dehydrogenase activity in liver and blood serum significantly. The other dehydrogenases and transaminases were not affected by Ni-deficiency (201). Unlike rats α -amylase activity of the blood serum of goats remained uninfluenced by Ni-deficiency (11).

As it is to be expected after the changes of the enzyme status, Ni-deficiency above all influences the carbohydrate metabolism

(103). A marked decrease was observed for the energy sources, fat, glucose, and glycogen. On the other hand, there seems to be the trend of an increasing cholesterol level of the blood plasma in case of Ni-deficiency (11).

An Ni-containing macroglobulin, nickeloplasmin, has been found in human and rabbit serum (196). Unfortunately there is no clear indication as to the physiological significance or function of Ni-plasmin.

Since no Ni-dependent enzyme system similar to plant urease is known in mammals, the capacity of different organs to indicate Ni-status had to be studied. These investigations were carried out in mini-pigs and goats (19, 20) (table 6). In both species skeleton best reflected the Ni-status. The same result was found in rats (102). The hair of goats also reflected the Ni-status.

Table 6: Reflection of the Ni-status by different organs of control and Ni-deficiency adult female goats in µg/kg dry matter (11)

organ, n	control animals		Ni-deficiency animals		p	%[1)]
	s	$\bar{x}$	$\bar{x}$	s		
carpal 29; 29	778	876	158	128	<0.001	18
rib 24; 28	438	611	174	112	<0.001	28
hair 10; 14	361	756	255	128	<0.001	34
uterine 10; 11	672	985	357	215	<0.01	36
kidney 26; 32	645	684	254	158	<0.001	37
liver 29; 33	560	604	281	163	<0.01	47
cerebrum 26; 28	540	654	312	165	<0.01	48
heart 25; 36	347	399	197	130	<0.01	49
lungs 23; 31	436	464	283	191	<0.01	61
skeletal muscle 27; 28	205	344	218	128	<0.05	64
spleen 27; 30	294	427	277	121	<0.05	65
pancreas 16; 17	816	845	575	440	>0.05	68

[1)] control animals ≙ 100 %, Ni-deficiency ≙ x %

Apart from embryos and babies age did not influence the Ni-status in humans (176). Ni passes through the placenta quickly and apparently in large amounts as it was described for rats (198). The teratogenic effect of Ni in case of an oversupply is based on that. The high Ni-content of the embryo in the 3rd and 4th month is surprising (table 7).
Apart from ribs, sex does not influence the Ni-content of human organs (table 8).

Table 7: Nickel concentration of the human embryo dependent on the stage of development in µg/kg dry matter (17)

	Month of pregnancy			Smallest limit difference	%
	3 - 4	5	6 - 7		
$\bar{x}$	3.573	1.620	1.646	1.226	45
s	931	690	316		

Table 8: Normal nickel concentration of different organs of men and women in µg/kg dry matter (17)

organ, n	women		men		p	%
	s	$\bar{x}$	$\bar{x}$	s		
rib 109; 128	1.343	1.796	1.262	792	<0.001	70
cerebellum 12; 28	325	643	618	352	>0.05	96
kidney 109; 127	372	639	609	423	>0.05	95
liver 104; 125	412	626	571	440	>0.05	91
thyroid gland 8;11	456	595	636	336	>0.05	107
lungs 14; 27	231	485	555	288	>0.05	114
testicle 11	-	-	549	491	-	-
cerebrum 16; 31	192	402	371	169	>0.05	92
heart 26; 37	188	300	372	220	>0.05	124
prostate gland 18	-	-	320	164	-	-
pancreas 10; 20	134	284	300	223	>0.05	106
F_p		<0.001	<0.001			

Most ingested Ni remains unabsorbed by the gastrointestinal tract. Some fecal Ni may come from the bile (147). Limited studies indicate that, generally, less than 10 % ingested Ni are absorbed (113). The Ni-ions use the Fe transport system located in the proximal part of the small intestine (37). The Ni-transfer from the mucosal to the serosal side of Fe defi-

cient rat intestinal segments was elevated (68).
The Ni-requirement of rat, chicken, goat, sheep and mini-pigs might be > 50 μg/kg ration dry matter in the first two species (104, 16, 17). In goats and mini-pigs semi-synthetic rations with 100 μg Ni/kg produced Ni-deficiency after long-term consumption (14,25,31, 28, 17). Goats reacted on Ni-deficiency rations with 100 μg/kg more sensitively than mini-pigs. Experiments with fattening bulls showed that rations with 500 μg Ni/kg ration dry substance covered the Ni-requirement of growing ruminants (157). By means of the α -amylase activity of the rat a Ni-offer of 60 to 500 μg Ni/kg ration dry substance was calculated. In gravid sows it amounted to 700 μg/kg ration dry matter (98). After the experiences hitherto available it can be proceeded from the fact that in case of a Ni-requirement of < 500 μg/kg ration dry matter, which might be considerably lower in humans, the Ni-content of food- and feed-stuffs covers the primary Ni-requirement of animals and men (33, 72).
The Ni-content of feed- and food-stuffs is species- and site-specific (17). All leafy species (vegetables) are Ni-rich (> 1000 μg/kg dry matter), potatoes, carrots e.g. supply more than 500 μg/kg and cereals > 200 μg Ni. Meat contains > 100 μg Ni/kg, milk and milk products > 200 μg Ni/kg (12,108,199, 17). The geological origin of the area influences the Ni-content of the flora significantly (table 9).
Hence follows that the Ni-offer for humans and animals is considerably higher in Hungary than in the GDR (189). In the USA the rations contained 270 μg Ni/kg dry matter (127). Further details of the Ni-content of feed- and food-stuffs were summarized (17).
A Ni-deficiency in humans could be reckoned with only in case of parenteral nutrition. 39 tested preparations for parenteral nutrition contained between 6 and 39 μg Ni/l (2). Ni-deficiency in humans has not been described yet. Ni-nutrition may be disturbed in individuals with disease that results in excessive levels of thiols in the body. Hypogeusia

Table 9: The site-specific Ni-content of indicator plants (acre red clover, meadow red clover, wheat, rye) in per cent of the Ni-richest site of the GDR (x = relative number) (Anke et al., unpublished)

geological origin of the site	x
weathered soils of the lower strata of new red sandstone	98
syenite weathered soils	87
slate weathered soils	82
phyllite weathered soils	81
gneiss weathered soils	69
granite weathered soils	68
alluvial riverside soils	67
diluvial sand	66
boulder clay weathered soils	66
shell lime weathered soils	64
loess	62
new red sandstone weathered soils	61
keuper weathered soils	60
moor, peat	55

found in such an individual was corrected by Ni (77). Cardiac infarction is not caused by Ni-deficiency (34). Diseases reducing Fe-absorption might reduce Ni-absorption as well and thus induce Ni-deficiency. In spite of the missing primary Ni-deficiency symptoms in humans Ni is essential according to the results obtained in animals.

3. Arsenic

The essentiality of As for animals and human beings has not only been experimentally tested in the last years. Previous investigations were not successful due to the high As-content of feed-stuffs (87, 177, 191). It has, however, been known for a long time that As might stimulate the growth of an organism or poison it, if it occurs in a certain quantity (125, 70, 3). Therefore it must be distinguished between As-deficiency doses (< 50 µg/kg ration dry matter), normal As-supply (350-500 µg/kg), therapeutic doses 3.5-5.0 mg/kg and toxic As-amounts > 5.0 mg/kg. The different effects of As are important for the experiments since the supplementation of the control rations with therapeutic As-offers can lead

to wrong data.
The essentiality of As in gravid, pregnant and lactating goats, in mini-pigs and their offsprings has been systematically investigated with <10 and 350 µg As/kg ration dry matter in our department since 1973 (22, 23, 4). The influence of As-deficiency on growth was identified with certainty only after intrauterine As-depletion in mini-pigs, goats, and rats (22,23,4,32, 140). On the average of 10 experiments As-deficiency goats gained 20 % less weight than control goats (table 10). The difference between the groups corresponds with

n	control goats s	control goats $\bar{x}$	As-deficiency goats $\bar{x}$	As-deficiency goats s	p	%
54; 55	29	95	76	43	< 0.01	80

Table 10: Growth of control and As-deficiency goats on the average of the first 167 experimental days (g per goat and day) (32)

the difference found in Li-deficiency goats (7) and is greater than that between control and Ni-deficiency goats (32). Hence follows that As-deficiency can considerably influence the growth of ruminants. Chicks (207) and rats (185) also reacted on As-deficient rations with significantly reduced growth. In accordance with previous investigations in kids and piglets maternal As-deficiency led to a lower mean birth weight (table 11). At birth As-deficiency lambs weighed 10 %, As-deficiency piglets 13 % less than the offsprings of control goats and control mini-pigs. The difference in live weight gain between control and As-deficiency kids exists till the 70th day of life and increases with the beginning exclusive intake of synthetic rations. The milk of As-deficiency goats contains more As than synthetic rations with < 10 µg As/kg dry matter.
The > 100-day-old As-depleted female goats and mini-pigs and the surviving two and several years old goats and mini-pigs had normal heat symptoms. In spite of this fact the success of the first insemination was highly significantly worse in

Table 11: Live weight of the kids of control and As-deficiency goats in kg (32)

day of life	control kids s	$\bar{x}$	As-deficiency kids $\bar{x}$	s	p	%
1st	0,77	3.1	2.8	0.77	< 0.05	90
49th	2.6	11.1	10.2	2.5	> 0.05	92
91st	4.3	19.0	16.5	4.3	< 0.01	87

As-deficiency goats (table 12). The influence of As-deficiency on the abortion rate is most impressive. These findings correspond well with those obtained in As-deficiency rats, in which an increased perinatal mortality was registered (132).

Table 12: Influence of As-deficiency on the reproduction of female goats (32)

parameter	control goats	As-deficiency goats	p
success of first insemination, %	69	47	< 0.001
conception rate, %	84	67	< 0.05
non-gravid goats, %	16	33	< 0.05
services per gravidity	1.4	2.1	< 0.001
abortion rate, %	1	33	< 0.001

As-deficiency had no significant influence on the milk production of goats. Only the milk fat production of As-deficiency goats was significantly reduced by 25 % in the first 56 days of lactation (32).
As already described (23, 4) there were cases of sudden death in lactating goats. As a rule the goats died suddenly and without previous symptoms. In the last years these cases of death were concentrated on the period between the 17th and 35th day of their second lactation. No goat survived the second gravidity. Non-gravid goats, however, can get very old (> 5 years). About one third of the kids died between the 8th and 91st day of life (table 13).
Histologically detectable changes at the cardiac muscle of

Table 13: Influence of As-deficiency on the mortality of goats and their kids (32)

parameter	control goats	As-deficiency goats	p
died kids, %	3	32	< 0.001
died mothers, %	25	40	< 0.001

As-deficiency kids were registered very early (23). The systematic investigation of the skeleton and cardiac muscles and of the liver of As-deficiency goats shortly before death resulted in a defective differentiation of the types of fibres by means of NADP-diaphoresis. It was found out ultrastructurally that there was a deposition of electron dense material in the mitochondrion membrane of the skeleton muscles, the myocardium and the liver. At an advanced stage this material is removed from the mitochondrion membrane and is freely detectable in the cytoplasma of the corresponding cells afterwards (Fig. 1) (166). Further investigations are necessary in order to clarify the effect of As-deficiency on mitochondria.

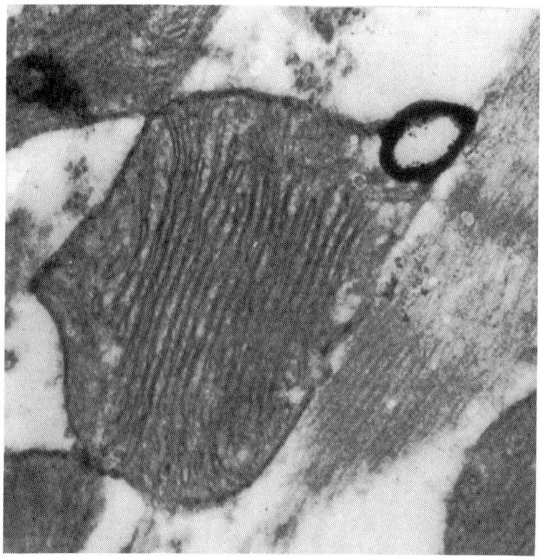

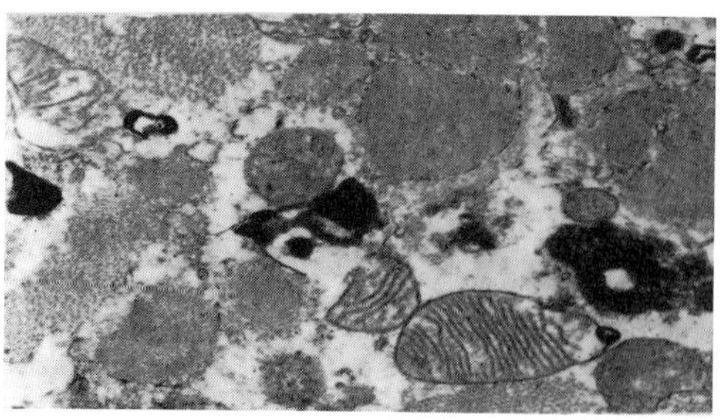

Fig. 1a Fig. 1b

Further possibly pathologically effective changes were found in As-deficiency chicks in form of reduced plasma uric acid concentrations, an As-Zn interaction and a decreased arginase content of

the kidneys. As-deprivation depressed kidney arginase and plasma urea only in Zn-deficient chicks (144,207,208). There is no biochemical explanation for these findings.
As-deficiency led to a significantly reduced ash content of the skeleton of goats as well (22). Leg damages were registered in As-deficiency chicks (207).
Special attention was paid to the influence of As-deficiency on the blood picture (hemoglobin, hematocrit, etc.) and the Fe-content of the body. Deficiency effects were neither registered in As-deficiency goats nor in As-deficiency pigs, if control animals were given 350 µg As/kg ration dry matter (22, 23).
In As-deficiency goats a significantly reduced triglyceride content was registered as well. As might have a role related to lipid phosphorus in biological systems. A novel phospholipid, O-phosphatidyltrimethylarsonium lactic acid, has been identified in algae (61, 44). The precursor of this phospholipid is apparently trimethylarsonium lactic acid. This arsonium phopholipid is essential to marine algae.
No biological role for, or specific action of, As has been found. The As-requirement for goats, mini-pigs, chicks and rats is < 50 µg/kg dry matter (136). Probably it amounts to < 25 µg/kg as the experiments with mini-pigs and goats prove.
As for other elements, any requirement for As could be altered by several factors. Absorption, retention and excretion are influenced by the chemical form and level of the As that is ingested. All forms of As are well absorbed. Hens incorporated 34 % of the given amount 45 minutes after the ^{76}As-intake in the crop. The percentage decreased to 1 % within 96 hours (80, 26).
The As-content of the feed- and food-stuffs is regularly > 50 µg/kg dry matter in Middle Europe. New investigations on the daily As-intake in Belgium, the USA and Japan show that it ranges between 20 to 130 µg As (47, 88, 86). Diets high in dairy products, certain vegetables, and fruits would contain much less As than diets high in fish and seafood (211). As is

essential for goats, pigs, rats and chicks. Under the aspect of sudden cases of death in case of animals with As-poor rations it might be of importance for humans as well.

4. Lithium

Li as the lightest alkali metal has been of importance for medicine and many other scientific disciplines not only since the discovery of the positive influence of this element in the treatment of manic depressions (50, 46, 45). But experiments to investigate the essentiality of this element began in the middle of the seventies only. In experiments with growing, gravid and lactating goats, which have been repeated seven times since 1976, a significant influence of Li-poor rations on the growth of growing and gravid female goats could be demonstrated within the first 176 experimental days (6, 15, 10 7). Li-deficiency goats gained less weight than corresponding control goats (table 14).

Table 14: Growth of control and Li-deficiency goats during the first 176 experimental days on the average of 7 experimental years (g per goat and day) (   )

experimental period	n	control goats s	$\bar{x}$	Li-deficiency goats $\bar{x}$	s	p	%
1976 - 1983	38; 40	18	103	76	47	< 0.05	74

In figure 2, the dynamics of live weight gain of growing, gravid and lactating goats of the first year of life is presented. Li can apparently not be stored on a large scale. Li-deficiency highly influenced the intrauterine live weight gain of kids. On the average, Li-deficiency kids were born with a 13 % lower birth weight (6, 15, 10, 7). In a further experiment 5 Li-deficiency kids, which got Li-poor rations intrauterinely and during the lactation period, were given Li-rich rations during 84 days. Afterwards they were again fed on Li-poor rations (fig. 3).

While they were given the Li-rich ration, they gained more weight than control goats and thus reached their normal live

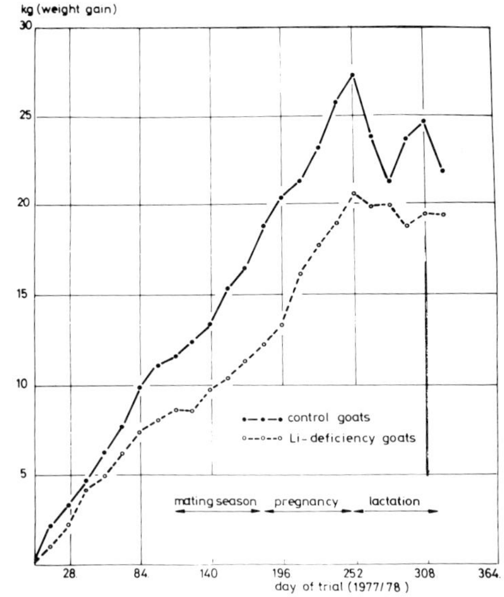

Fig. 2: Influence of Li-deficiency on the live weight gain of goats within the test period 1977/78

weight. After the change-over to the Li-deficiency ration they gained almost no weight during 65 days. On the 210th experimental day the difference in live weight gain again amounted to 20 %.

In Li-deficiency experiments with growing rats, which began in 1976 as well, no influence of Li-deficiency on growth could be found (149, 148). There are no investigations on the effects of Li-deficiency on the growth of other species of animals up to now.

The influence of Li-deficiency on the reproduction of goats is shown in table 15. Li-deficiency did not influence heat symptoms of goats. In spite of that the first mating resulted in a significantly worse insemination. Repeated matings improved the conception of Li-deficiency goats. For that reason more services per gravidity were

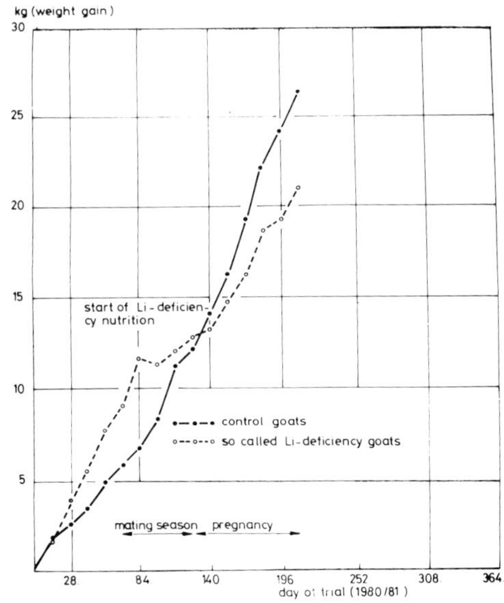

Fig. 3: Influence of Li-deficiency on the live weight gain of goats within the test period 1980/81

necessary. A high percentage of goats with Li-poor rations aborted their fetuses in the last third of gravidity. A longer conception time of two weeks was registered in Li-deficiency rats as well. The litter size was 20 to 30 % smaller in all generations (149).

Table 15: Influence of Li-deficiency on the reproduction of female goats (10)

parameter	control goats	Li-deficiency goats	p
success of first insemination,%	72	45	< 0.001
conception rate, %	85	73	>0.05
non-gravid goats, %	15	27	>0.05
services per gravidity	1.4	2.0	< 0.01
abortions, %	2	14	< 0.01

The milk production of Li-deficiency goats was reduced only insignificantly during the first 42 days compared to control goats. Between the 43rd and 56th day of lactation they produced only 61 % and thus significantly less milk than control animals (10).

In spite of their low birth weight the mortality of Li-deficiency kids during the lactation period did not differ from that of control kids (table 16). In that connection only the

Table 16: Mortality of control and lithium deficiency goats during the first experimental year and of their offsprings (10)

parameter	control goats	Li-deficiency goats	p
died kids, 8th-91st day,p.p., %	5	10	> 0.05
died mothers, 1st experimental year, %	12	29	< 0.05

mortality between the 8th and 91st day of life was taken into consideration since the cause of death cannot always be found out unobjectionably in the first week of life. In rats, compared to controls, only 53 to 60 % of the offsprings with

Li-poor rations survived up to one week (149).
Apart from the described Li-deficiency symptoms, no other symptoms could be found. One Li-deficiency goat showed massive changes in epithelium. It died with these epithelial damages. Some other goats showed the same symptoms, but not such massive ones (7).
Li-deficiency did not influence the blood picture of either young or old goats (7). The same is true for the content of free glycerine, glucose, volatile fatty acids, lactat, cholesterol, neutral fats and total lipids in the blood serum. The found differences were either statistically not relevant or not reproducable (10, 7).
But Li-deficiency significantly influenced the activity of the blood serum enzymes SDH and GLDH resp. connected with the carbohydrate and N-metabolism, of the ALD involved into glycolysis and of the MDH and IDHC resp. important for the Krebs cycle (202,203). Furthermore, the MAO-activity and the serotonin metabolism in the liver of Li-deficiency goats were reduced significantly (202). The activity of CK was increased, which could be explained by an increased stress sensitivity.

Table 17: Activity of serum (U/l) and liver enzymes (nmol/mg/h) in control and Li-deficiency goats (202)

enzyme	control goats		Li-deficiency goats		p	%
	s	$\bar{x}$	$\bar{x}$	s		
serum, SDH; EC1.1.1.14	4.2	12	7.4	3.3	<0.05	62
" GLDH;EC1.4.1.3	5.3	14	5.7	3.4	<0.001	41
" ALD;EC4.1.2.13	6.3	17	11	3.3	<0.05	65
" MDH;EC1.1.1.37	65	258	183	53	<0.05	71
" ICDH;EC1.1.1.42	2.5	4.0	1.4	1.1	<0.05	35
" CK; EC2.6.1.2	8.2	20	35	16	<0.05	175
liver, MAO	3.8	36	26	1.3	<0.01	72

The changes of the serum enzyme and MAO-activity point to the fact that there are disorders of metabolism in Li-deficiency animals.
The Li-requirement of animals and humans is unknown. The rea-

son for that is that Li-deficiency experiments were carried out in rats with 5-15 ng/g and in goats with $<$ 1 mg (149,10). These differences must be due to analytical difficulties, which both research teams try to overcome at the moment. The Li-content of the flora is strongly influenced by the geological origin of the site (165,200,205) and the species of plant (6,109,118,151, 110). The Li-content of feed- and foodstuffs from the GDR, Belgium, Hungary, Poland, Czechoslovakia and the Baltic Republics of the USSR varied between 1 and 40 mg/kg dry matter dependent on species and site. The vegetation of saliferous soils reached Li-concentrations of $>$ 200 mg/kg dry matter (205). Therefore, Li-deficiency in animals and humans is not to be reckoned with.
Li-supplementation of barley and soya rations in pigs and of concentrate rations in fattening bulls did not result in an improvement of weight gain in both species (13). In other experiments with fattening pigs improvements of live weight gain could be achieved (158). These results have to be confirmed.
Under Li-deficiency conditions the Li-status is best reflected by blood serum and hair. Lungs, spleen, several parts of the skeleton, liver and kidney also indicate the Li-status significantly (106). Thus, the identification of Li-deficiency by means of blood serum and hair is easy.
It can be summarized that Li-poor semi-synthetic rations had a significantly negative influence on the reproduction performance of goats and rats and their life expectancy. Li-poor nutrition led to growth deceleration in goats, but not in rats. The enzyme level of the blood serum was influenced by Li-deficiency. Further investigations must clarify the Li-requirement of animals and humans and confirm the effects of Li-poor nutrition in other species. Evidence is given suggesting that Li may be essential in goats and rats.

5. Vanadium

After the first investigations on the effect of V-supplementation on the life expectancy of rats and mice had not shown any results because of the high V-content of rations (178, 179), the first data about a possible essentiality of V come from chicks, in which a ration with < 10 µg V/kg resulted in a reduced growth of wings- and tail-feathers (83). In further investigations a V-offer of 30-35 µg V/kg ration dry matter led to a decreased growth of chickens (140). V-supplementation of 250-500 µg V/kg feed-stuffs led to significant growth improvements in rats. The V-content of these rations amounted to 50-100 µg V/kg (187,195,182). In a 168-day test period, which was repeated three times, growing and gravid goats did not react on V-offers of < 10 µg V/kg ration dry matter with decreased growth. Intrauterine V-deficiency did also not lead to a reduced birth weight. There were also no differences in growth between control and V-deficiency kids during the milking period up to the 91st day of life (9).

In chicks and rats there was also no V-induced growth reduction in every V-deficiency experiment (140). These changing experimental results may among others, be caused by feed co-sumption (table 18).

Table 18: Mean feed-consumption of young control and V-deficiency goats during growth and lactation on the average of three experiments (g per animal and day) (9)

period	control goats s	control goats $\bar{x}$	V-deficiency goats $\bar{x}$	V-deficiency goats s	p	%
growth	161	742	737	102	> 0.05	99
lactation	193	832	682	149	< 0.001	82

During growth there were no differences between control and V-deficiency animals, but they were obvious during lactation.

In rats V-poor nutrition (< 10 µg V/kg) led to a reduced success of first insemination, a smaller number of offsprings and a significantly increased mortality of pups in the 3rd and 4th generations (82, 85). Goats with < 10 µg V/kg in the ration had a worse success of first insemination ($p > 0.05$).

Their conception rate, the percentage of non-gravid goats and the necessary services per gravidity were the same in both groups. But the abortion rate showed significant differences. About one third of goats already aborted their kids in the 4th or 5th month of gravidity ($p \leq 0.001$) (9).
In accordance with the reduced feed-intake lactating V-deficiency goats produced significantly less milk during the first 56 days (table 19). The low V-offer did not at all influence

Table 19: Influence of V-deficiency on milk and milk fat performance of goats during the first 56 days of lactation (9)

parameter	control goats s	control goats $\bar{x}$	V-deficiency goats $\bar{x}$	V-deficiency goats s	p	%
milk, l/day	0.81	1.6	1.1	0.42	<0.001	69
fat, %	1.2	4.4	4.3	0.80	>0.05	98
fat, g/day	45	71	48	22	<0.001	68

the fat concentration of milk. V-deficiency led to decreased milk production at all stages of lactation.
The influence of V on life expectancy was observed during 4 years (table 20). As it was described in V-deficiency rats there was a high mortality of V-deficiency kids. More than 50 % of them died during the first 91 days of life, mostly after the 70th day. They died suddenly and in a good nutritional condition.

Table 20: Influence of V-deficiency on the mortality of goats and their kids (9)

parameter	control goats	V-deficiency goats	p
died kids, %	3	52	<0.001
died mothers, %	25	54	≤ 0.01

Their death was accompanied by strong cramps. About 50 % of lactating mothers died as well. The majority of them died at the end of the lactation period. At this time they were in a

state of malnutrition, but did not show any particular deficiency symptoms. Of all criteria life expectancy was most intensively influenced by the low V-offer in goats (9).
In rats (195) and chicks (141) V-deficiency led to a significantly increased packed cell volume of blood and a high Fe-content of blood and bones. In other experiments with chicks V-deficiency did not lead to growth and hematocrit changes (214). The reason might be the different Fe- and protein offer of experimental rations, which can also influence feed-intake. The connections between V- and Fe-metabolism were already mentioned. The blood picture of anemic rats could be normalized faster by means of Fe- and V-supplementation than with Fe alone (136). V is important for non-enzymatic Fe-containing proteins (transferrin, liver ferretin, lactoferrin), which bind V in vivo (159,160,161,162,163,164). In unicellular green algae (Scenedesmus obliquus) V was able to overcome a limited Fe-deficiency completely (123). Nielsen et al. (1983) finally examined the relations between V, Fe and cystine in rats and found out that there were interactions between them. In moderately Fe-deficient rats, V-deprivation apparently depressed hematocrit when supplemental cystine was 3.0 mg/g, and slightly elevated when supplemental cystine was 8.5 mg/g. V-deprivation did not affect hematocrit in severely Fe-deficient rats, nor in rats fed adequate Fe.
The influence of V-deficiency on the serum cholesterol plasma content showed differing results (84,141). V-deficiency led both to lower and higher cholesterol values of the serum, whereas plasma triglyceride levels were greatly increased in V-deficient chicks (85). V-deprivation elevated the total lipids in the plasma of rats. The elevation was influenced by the Fe and cystine status of the rats. In severely Fe-deficient rats, V-deprivation more markedly elevated total lipids in the plasma of rats fed the supplemental 8.5 mg cystine than of rats fed the supplemental 3.0 mg cystine. Also V-deprivation more markedly elevated total lipids in the plasma of Fe-adequate (100 µg/g diet) than moderately Fe-deficient

investigated in rats only.

With a synthetic ration (200 μg Pb/kg) it was possible to show in 1974 that Pb-supplementation of 1.0 and 2.5 ppm resp. improved the growth of rats significantly (182). After intrauterine deficiency by means of a ration with 18 μg Pb/kg these rats gained 11 % less weight than control rats with 1 mg Pb/kg. Further depletion after weaning increased the lower weight gain to 22 %. Intrauterine Pb-deficiency did not influence the birth weight of offsprings (156). Rations with 45 μg Pb/kg ration dry matter did not result in reduced growth of offsprings (155). But these Pb-deficient rats had an 11 % lower hematocrit, a 15 % lower hemoglobin and MCV values reduced by 9 % and they suffered from microcytar hypochromic anemia (155). The Pb-induced changes of the blood-picture could be normalised by means of repletion. Intrauterine Pb-deficiency is responsible for them (154) which leads to Fe-deficiency in mothers. Pb-depleted rats also absorb less Fe than corresponding control animals whereas the intermediary availability of Fe is higher in Pb-depleted rats (93).

Table 21: Absorbability and intermediary availability of Fe in growing rats (93)

criteria	control	depletion	p
Fe-absorbability, %	53.0	23.9	<0.001
intermediary availability of Fe,%	77.7	93.0	<0.001

Pb-depleted rats also suffered from a reduced Fe-content in the serum whereas the iron binding capacity was increased (94). The same was true for the Fe-content of liver and spleen, which was reduced under the conditions of Pb-deficiency (153). Catalase-activity of blood was only reduced in the blood of the most Pb-poor rats. It can be explained by the induced Fe-deficiency (152). Enzyme changes in serum and several organs caused by Pb-deficiency are not very characteristic (152). The same is true for the protein, glucose and glycogen content of the liver of young Pb-deficiency rats. But consi-

(15 µg/g diet) rats. When supplemental Fe was 500 µg/g diet, V-deprivation did not elevate total lipids in the plasma of rats fed supplemental 8.5 mg of cystine, but did elevate total lipids in the plasma of rats fed supplemental 3.0 mg cystine (146). These findings also indicate that differences in level of dietary cystine (perhaps amino acid composition) and Fe may explain some of the inconsistencies in V-deprivation signs found in earlier studies.

The available data found in three species of animals by four different research teams point to the essentiality of V in animals and probably also in humans.

V is essential for green algae (122,121) and for several marine macroalgae (69), but may be not or in extremely, low amounts only for higher developed plants (210).

Further investigations must clarify the specific tasks of V in metabolism. V may play a role in RNase (215). There are lots of publications on the significance of V as efficient inhibitor of phosphotransferases and phosphohydrolases, e.g. alkaline phosphatase, glucose-6-phosphatase and Na, K-ATPase (76). V-surplus is said to lead to manic-depressive psychoses (128, 63).

The V-requirement might amount to 10 to 25 µg V/kg ration or diet dry substance according to the available data and under the consideration of the different composition of the ration. Further extensive investigations are necessary. V-absorption by the body might not be high. The absorption rate varies between 1 and 5 to 12 % (48, 66). It is influenced by many nutritional factors (136,213). The V-content of feed- and foodstuffs is low. As a rule 12 to 30 µg (126,127) or about 10 µg are daily taken in by humans (49). V might be of importance for animals and humans because of the normally low offer.

6. Lead

In the consciousness of people Pb belongs to the mainly toxic elements. There are lots of publications about its toxicity. On the other hands, problems of the essentiality of Pb were derable decreases were found in the triglyceride content of the serum and the liver of Pb-deficiency animals, whereas their cholesterole content increased (95).

The available findings point to the essentiality of Pb in rats (92). Further investigations in long-lived species should confirm these results. Pb-resorption is high, above all in young animals and babies. Children at the age of 2 years absorbed 42 % of the Pb taken in in case of a low Pb-offer. The net retention reached 32 % of intake (214). The Pb-offer of animals and humans is certainly higher than 100 µg P/kg ration dry matter, which might cover the needs according to the experiments with rats (79). The Pb-status is easy to determine by means of the Pb.content of blood, rib, kidneys, liver, pigmented hair, faeces and δ-aminolavulin acid dehydratase (74), i.e. control can be carried out without difficulties.

7. Cadmium

Though almost all publications on Cd deal with its toxicity, it cannot be excluded that this element has essential tasks as well.

Growth could be improved by the supplementation of 50 to 100 µg Cd/kg to a synthetic ration with 5 µg/kg in rats (189). Similar findings were obtained in growing and gravid goats (fig. 4) (30).

After depletion goats with 15 µg Cd in the semi-synthetic ration gained less weight than control goats with 304 µg Cd/kg dry matter (24). In contrast to rats 68 µg Cd/kg ration dry matter met the Cd-requirement and made a normal growth of 118 g per animal and day possible. Control goats gained 122 g/day weight on the average (30).

Intrauterine Cd-deficiency also influenced the birth weight of kids (table 22). Offsprings of goats with 304 and 68 µg Cd/kg ration dry matter weighed significantly more than those of goats with 15 µg Cd/kg ration dry matter (30). The experiments with goats were repeated five times.

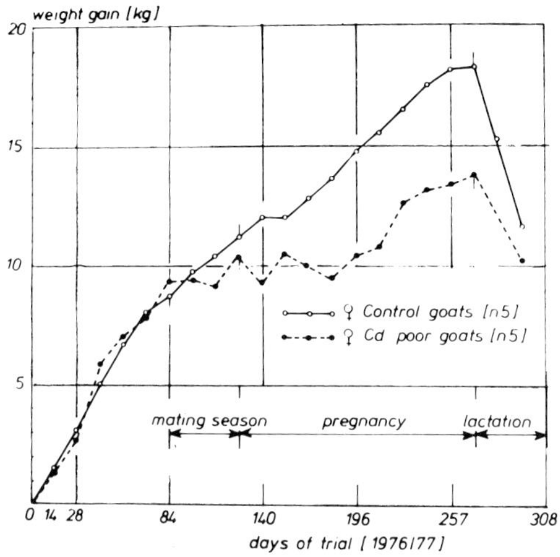

Fig. 4: The influence of < 20 µg Cd/kg ration dry matter on the weight gain of female goats

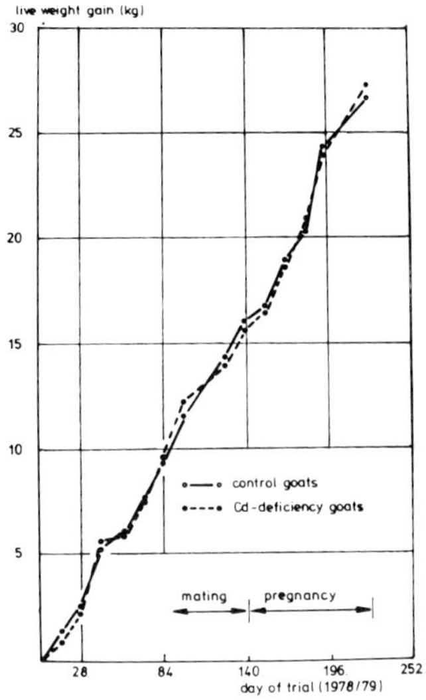

Fig. 5: The influence of 304 and 68 µg Cd/kg feed dry matter on the weight gain of female goats

Table 22: Birth weight of control and Cd-deficiency kids in kg (30)

Cd-offer,	n	control goats s	control goats $\bar{x}$	Cd-deficiency goats $\bar{x}$	Cd-deficiency goats s	p	%
304 µg/kg	34	0.9	3.1	2.3	0.3	< 0.05	74
68 µg/kg	24	0.9	3.0			< 0.05	77

With regard to the parameters of reproduction performance Cd-deficiency influenced the success of first insemination, the conception and abortion rate and the percentage of non-gravid goats significantly. Again 68 µg/kg met the Cd-requirement for reproduction (table 23).

The mortality of kids with different Cd-offers differed significantly as well (30). In case of this criterion of essentiality 68 µg Cd/kg ration dry substance also met the Cd-requirement (table 24).

Table 23: Influence of Cd-deficiency on the reproduction performance of female goats (30)

parameter	control goats 304 µg/kg	68 µg/kg	Cd-deficiency goats	p
success of first insemination, %	74	54	36	<0.01
conception rate, %	83	79	43	<0.01
services per gravidity	1.4	1.9	3.7	>0.05
abortions, %	3	16	33	<0.05
non-gravid goats, %	17	16	57	<0.01

Table 24: The mortality of control and Cd-deficiency kids from the 8th to the 91st day of life (30) in per cent

Cd-offer	control kids	Cd-deficiency kids	p
304 µg/kg	4	50	<0.001
68 µg/kg	0		<0.001

Apart from some speculations, the functions of Cd in the body are unknown. In vitro Cd can replace Zn in the activation or as component of several enzymes (e.g. carboxypeptidase B). It is not proved if these possibilities are of biological importance.

The analysis of human embryos at different stages of development showed, however, that in case of rising ash content only their Cd-content increased whereas the Fe-, Zn-, Cu-, Mn-, Mo-, Ni-, and Li-concentration decreased from trimester to trimester (table 25).

The Cd-status is best reflected by kidneys. They are followed by liver and hair with a by far worse indicating capacity. Cerebrum and spinal cord, ribs and carpals, reflect the Cd-status worst or not at all (105). Four days after the oral intake of Cd-amounts in the ration Cd-absorption varied between 1.9 % in mini-pigs, 2.3 % in laying hens and 4.0 % in monkeys (27, 81).

The Cd-requirement, if it exists, may be considerably

Table 25: The Ni-, Li-, and Cd-content of human embryos in µg/kg dry matter (n 15) (Anke et al., unpublished)

element		month of pregnancy 3rd-4th	5th	6th-7th	p	minimum limit difference
nickel	$\bar{x}$	3573	1620	1646	< 0.001	1226
	s	931	690	316		
lithium	$\bar{x}$	6900	2100	1900	< 0.05	4469
	s	4200	1000	900		
cadmium	$\bar{x}$	269	404	443	< 0.05	-
	s	75	206	97		

$<$ 50 µg/kg ration or diet dry matter. It might be within a range of 25 µg/kg dry matter. This Cd-amount is contained in Cd-unexposed feed- and foodstuffs in Middle Europe (107). The essentiality of Cd is suggested by the available experiments in goats and rats. It is probable, but it must be secured in further experiments. The strong reaction of the body on Cd-exposure and the continuous concentration beginning in the embryo point to this trend. It might be of special importance for humans.

8. Tungsten

W stands in the 6th sub-group of the periodical system of elements together with the essential Mo. Because of the similarity of both elements it was tried to replace the Mo of Mo-dependent enzymes by W. This was possible in case of nitrogenase (42), nitrate reductase (75, 111) and sulfite oxydase (89). The "W enzymes", however, do not have the specific activity of Mo-enzymes (5). But in the midseventies the biological significance of W as a component of the formiate dehydrogenase of Clostridium thermoaceticum and C. formicoaceticum, which catalyses the reaction $CO_2 + NADPH \rightleftharpoons HCOO^- + NADP^+$, was experimentally demonstrated (115, 114).
Since the biological significance of W for the fauna was completely unknown, experiments with goats with W-poor rations began in 1974, which were repeated 6 times up to 1980.

W-poor rations with 60 μg W/kg dry matter in the semi-synthetic ration did not influence growth significantly in the six experiments (5, 39). On the average goats with W-poor rations gained 7 % less weight than control goats ($p > 0.05$). The growth curves of both groups were approximately identical (fig. 6). Intrauterine W-deficiency did also not lead to a significant growth deceleration, though there were differences up to 10 % (5). On the other hand, the supplementation of 13 μg W per kg body weight led to a growth improvement of 12 % in guinea-pigs. There is no information about the W-content of the ration (35).

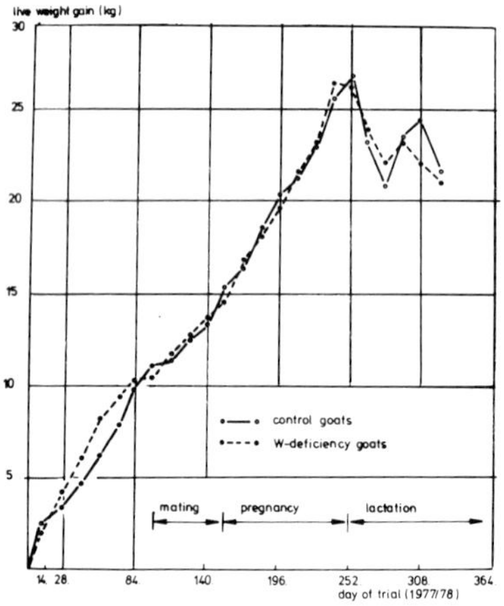

Fig. 6: Live weight gain of control and W-deficiency goats (1977/78) (5)

Success of first insemination, conception and abortion rate of W-deficiency goats were also only insignificantly worse than those of control goats (39). The same is true for the mortality of offsprings. Only the mortality of their mothers and the reticulocytes content of the blood of young and old goats with W-poor rations were significantly higher than those of control goats (39). Compared to other new trace elements these findings are not sufficient to postulate the essentiality of W. At the present level of knowledge 60 μg Mo/kg ration dry substance are sufficient for normal growth and normal reproduction. W is either not essential or the W-requirement of goats is considerably < 60 μg/kg ration dry matter. Independent of the results of further investigations, W-deficiency in ruminants can be excluded, since their feed-stuffs contain > 100 μg/kg dry matter in Middle Europe (39).

Conclusion

It can be summarized that Si, Ni, As, V and may be other elements such as Li, Pb and Cd are essential for animals and humans. The offer of these elements is influenced by the golological origin of the site, the age of plants, the species of plants, emission, processing and treatment. According to available data there are no indications of a primary deficiency of these elements in Europe and the USA.

Under the influence of antagonists their retention can get worse as well. Then secondary deficiency cannot be excluded. Furthermore, it must be underlined that deficiency of new trace elements might be caused by metabolic and genetic disorders such as those shown for Zn, Cu, and Fe.

It is to be assumed that in the course of the development of animals and humans many elements got tasks in the body and therefore are essential. The clarification of these functions still requires a lot of scientific work.

1. Anke, M.: Tag.-Ber., Akad. Landw.-Wiss. DDR, Berlin, 132, 197 (1973).
2. Anke, M.: In: Abdulla, M. (ed) Health Effects and Interactions of Essential and Toxic Elements. (Abstracts) Studentliteratur Lund p. 25 (1983).
3. Anke,M., Anke, I.: Jb. Tierernähr. Fütterung 8, 365 (1972).
4. Anke, M., Groppel, B., Grün, M., Hennig, A., Meissner, D.: In: Anke, M., Schneider, H.-J., Brückner, Chr. (eds) Arsenic, Spurenelement-Symposium 3 - Wissenschaftl. Publikationen F.-Schiller-Univ. Jena p. 25 (1980).
5. Anke, M., Groppel, B., Grün, M., Kronemann, H.: In: Anke, M., Brückner, Chr., Gürtler, H., Grün, M. (eds) Mengen- und Spurenelemente 3 - Karl-Marx-Univ. Leipzig p. 37 (1983).
6. Anke, M., Groppel, B., Grün, M., Kronemann, H., Riedel, E.: In: Szentmihályi, S. (ed) International Conference on Food Additives. 1 - Hung. Soc. Agricult. Sci. Budapest p. 245 (1981).
7. Anke, M., Groppel, B., Grün, M., Kronemann, H., Riedel, E.: Wiss. Z. Karl-Marx-Univ. Leipzig, Math.-Naturwiss. R. 32, 260 (1983).
8. Anke, M., Groppel, B., Kronemann, H.: in press. WHO Nickel-Symposium 1983 in Lyon
9. Anke, M., Groppel, B., Kronemann, H., Führer, E.: In: Anke, M., Baumann, W., Bräunlich, H., Brückner, Chr. (eds) Spurenelement-Symposium 4 - Wissenschaftl. Publikationen F.-Schiller-Univ. Jena p. 135 (1983).
10. Anke, M., Groppel, B., Kronemann, H., Grün, M.: In: Anke, M., Baumann, W., Bräunlich, H., Brückner, Chr. (eds) Lithium, Spurenelement-Symposium 4 - Wissenschaftl. Publikationen F.-Schiller-Univ. Jena p. 58 (1983).
11. Anke, M., Groppel, B., Nordmann, S., Kronemann, H.: In: Anke, M., Baumann, W., Bräunlich, H., Brückner, Chr. (eds) Spurenelement-Symposium 4 - Wissenschaftl. Publikationen F.-Schiller-Univ. Jena p. 19 (1983).
12. Anke, M., Groppel, B., Riedel, E., Schneider, H.-J.: In: Brown, S., Sunderman, F.W.Jr. (eds) Nickel Toxicology, Academic Press, London, New York, Toronto, Sydney, San Francisco p. 65 (1980).
13. Anke, M., Groppel, B., Schwarz, S.: Der Einfluß zusätzlicher Lithiumgaben auf das Wachstum, den Futterverzehr und das Verhalten von Mastschweinen und Mastbullen. Unveröffentl. Untersuchungen (1984).

14. Anke, M., Grün, M., Dittrich, G., Groppel, B., Hennig, A.: In: Hoekstra, W.G., Suttie, J.W., Ganther, H.E., Mertz, W. (eds) Trace Element Metabolism in Animals 2- University Park Press, Baltimore, London, Tokyo p. 715 (1974).

15. Anke, M., Grün, M., Groppel, B., Kronemann, H.: In: Anke, M., Schneider, H.-J. (eds) Mengen- und Spurenelemente 1- Karl-Marx-Univ. Leipzig p. 217 (1981).

16. Anke, M., Grün, M., Groppel, B., Kronemann, H.: Zbl. Pharmazie, Pharmakotherapie und Laboratoriumsdiagnostik, 121, 474 (1982).

17. Anke, M., Grün, M., Groppel, B., Kronemann, H.: In: Sarkar, B. (ed) Biological Aspects of Metals and Metal-Related Diseases. Raven Press, New York p. 89 (1983).

18. Anke, M., Grün, M., Hoffmann, G., Groppel, B., Gruhn,K., Faust, H.: In: Anke, M., Schneider, H.-J. (eds) Mengen- und Spurenelemente 1 - Karl-Marx-Univ. Leipzig p. 189 (1981).

19. Anke, M., Grün, M., Kronemann, H.: In: Anke, M., Schneider, H.-J., Brückner, Chr. (eds) Nickel, Spurenelement-Symposium 3 - Wissenschaftl. Publikationen F.-Schiller-Univ. Jena p. 237 (1980).

20. Anke, M., Grün, M., Kronemann, H.: In: Brown, S., Sunderman, F.W.Jr. (eds) Nickel Toxikology, Academic Press, London, New York, Toronto, Sydney, San Francisco p. 70 (1980).

21. Anke, M., Grün, M., Partschefeld, M.: Arch. Tierernährung, 26, 740 (1976).

22. Anke, M., Grün, M., Partschefeld, M.: In: Hemphill, D.D. (ed) Trace Substances in Environmental Health 10 - Univ. Missouri, Columbia, USA p. 403 (1976).

23. Anke,M., Grün, M., Partschefeld, M., Groppel, B., Hennig, A.: In: Kirchgessner, M. (ed) Trace Element Metabolism in Man and Animals 3 - Arbeitskreis Tierernährungsforschung Weihenstephan p. 248 (1977).

24. Anke, M., Hennig, A., Groppel, B., Partschefeld, M., Grün, M.: In: Kirchgessner, M. (ed) Trace Element Metabolism in Man and Animals 3 - Arbeitskreis Tierernährungsforschung Weihenstephan p. 540 (1977).

25. Anke, M., Hennig, A., Grün, M., Partschefeld, M., Groppel, B., Lüdke, H.: Arch. Tierernährung, 27, 25 (1977).

26. Anke, M., Hoffmann, G., Grün, M., Groppel, B., Riedel, E.: JAEA, TECDOC 267, 135 (1982).

27. Anke, M., Hoffmann, G., Hennig, A., Groppel, B., Kronemann, H., Grün, M.: In: Anke, M., Brückner, Chr., Gürtler, H., Grün, M. (eds) Mengen - und Spurenelemente 2 - Karl- Marx-Univ. Leipzig p. 85 (1982).

28. Anke, M., Kronemann, H., Groppel, B., Hennig, A., Meissner, D.,Schneider, H.-J.: In: Anke,M., Schneider, H.-J.; Brückner, Chr. (eds) Nickel, Spurenelement-Symposium 3 - Wissenschaftl. Publikationen F.-Schiller-Univ. Jena p. 3 (1980).

29. Anke, M., Kronemann, H., Groppel, B., Partschefeld, M.: In: Anke, M., Brückner, Chr., Gürtler, H., Grün, M. (eds) Mengen- und Spurenelemente 3 - Karl-Marx-Univ. Leipzig p. 45 (1983).

30. Anke,M., Kronemann, H., Groppel, B., Riedel, E.: Wiss. Z. Karl-Marx-Univ. Leipzig, Math.-Naturwiss. R. 33, 157 (1984).

31. Anke, M., Partschefeld, M., Grün, M., Groppel, B.: Arch. Tierernährung 28, 83 (1978).

32. Anke, M., Schmidt, A., Groppel, B., Kronemann, H.: In: Anke, M., Baumann, W., Bräunlich, H., Brückner, Chr. (eds) Spurenelement-Symposium 4 - Wissenschaftl. Publikationen F.- Schiller-Univ. Jena p. 97 (1983).

33. Anke, M., Schneider, H.-J., Brückner, Chr.: In: Anke, M., Schneider, H.-J., Brückner, Chr. (eds) Nickel, Spurenelement-Symposium 3 - Wissenschaftl. Publikationen F.-Schiller-Univ. Jena p. 375 (1980).

34. Anke, M., Schneider, H.-J., Grün, M., Lösch, E., Fuchs, M.: Nickel und Herzinfarkt. In Press (1984).

35. Balaeva, M.S.:In: Kirchgessner, M. (ed) Trace Element Metabolism in Man and Animals 3 - Arbeitskreis Tierernährungsforschung Weihenstephan p. 617 (1977).

36. Bartha, R., Ordal, E.J.: J. Bacteriol., 89, 1015 (1965).

37. Becker, G., Dörstelmann, U., Fromberger, U., Forth, W.: In: Anke,M., Schneider, H.-J., Brückner, Chr. (eds) Nickel, Spurenelement-Symposium 3 - Wissenschaftl. Publikationen F.-Schiller-Univ. Jena p. 79 (1980).

38. Becker, C.H., Engler, E., Hoebbel, D., Rühlmann, K.: In: Anke, M., Brückner, Chr., Gürtler, H., Grün, M. (eds) Mengen- und Spurenelemente 3 - Karl-Marx-Univ. Leipzig p. 194 (1983).

39. Becker, C.H., Jánossy, A.G.S.: Micron 10, 267 (1979).

40. Becker, C.H. , Matthias, D., Wooßmann,H., Schwartz, A., Engler, E.: In: Anke,M., Baumann, W., Bräunlich, H., Brückner. Chr. (eds) Spurenelement -Symposium 4 - Wissenschaftl. Publikationen F.-Schiller-Univ. Jena p. 142 (1983).

41. Becker, C.H., Schwartz, A., Matthias, D., Wooßmann, D., Engler, E.: In: Anke, M., Brückner, Chr., Gürtler, H., Grün, M. (eds) Mengen- und Spurenelemente 2 - Karl-Marx-Univ. Leipzig p. 211 (1982).

42. Benemann, J.R., Smith, G.M., Kostel, P. J., Mc Kenna, C.E.: FEBS Lett. 29, 219 (1973).

43. Benke, G.M., Osborne, T.W.: Food Chem. Toxicol. 17, 123 (1979).

44. Benson, A.A., Cooney, R.V., Summons, R.E.: In: Anke, M., Schneider, H.-J., Brückner, Chr. (eds) Arsenic, Spurenelement-Symposium 3 - Wissenschaftl. Publikationen F.-Schiller-Univ. Jena p. 139 (1980).

45. Börtitz, S., Däßler, H.-G.: In: Anke, M., Baumann, W., Bräunlich, H., Brückner, Chr. (eds) Lithium, Spurenelement-Symposium 4 - Wissenschaftl. Publikationen F.-Schiller-Univ. Jena p. 10 (1983).

46. Brückner, Chr., Holzapfel, G., Hanf, G.: In: Anke, M., Baumann, W., Bräunlich, H., Brückner, Chr. (eds) Lithium, Spurenelement-Symposium 4 - Wissenschaftl. Publikationen F.-Schiller-Univ. Jena p. 6 (1983).

47. Buchet, J.P., Lauwerys, R., Vandevoorde, A., Pycke, J.M.: Food Chem. Toxicol. 21, 19 (1983).

48. Byrne, A.R., Kosta, L.: Sci. Total Environm. 10, 17 (1978).

49. Byrne, A. R., Kosta, L.: Sci. Total Environm. 13, 87 (1979).

50. Cade, J. F. I.: Med. J. Aust. 36, 349 (1949).

51. Carlisle, E.M.: Science 167, 279 (1970).

52. Carlisle, E. M.: Fed. Proc. 29, 565 (1970).

53. Carlisle, E.M.: Science 178, 619 (1972).

54. Carlisle, E.M.: Fed. Proc. 33, 704 (1974).

55. Carlisle, E.M.: Fed. Proc. 33, 1758 (1974).

56. Carlisle, E.M.: In: Hoekstra, W.G., Suttie, J.W., Ganther, H.E., Mertz, W. (eds) Trace Element Metabolism in Animals 2 - University Park Press, Baltimore, London, Tokyo p. 407 (1974).

57. Carlisle, E.M.: In: Kirchgessner, M. (ed) Trace Element Metabolism in Man and Animals 3 - Arbeitskreis Tierernährungsforschung Weihenstephan p. 253 (1978).

58. Carlisle, E.M.: J. Nutr. 110, 352 (1980).

59. Carlisle, E.M.: J. Nutr. 110, 1046 (1980).

60. Carlisle, E.M.: Fed. Proc. 39, 787 (1980).

61. Cooney, R.V., Mumma, R.O., Benson, A.A.: Proc. Natl. Acad. Sci. USA 75, 4262 (1978).

62. Dick, D.A.T., Naylor, G,J., Dick, E.G.: Psychological Med. 12, 533 (1982).

63. Diekert, G.B., Graf, E.G., Thauer, R.K.: Arch. Microbiol. 122, 117 (1979).

64. Diekert, G.B., Klee, B., Thauer, R.K.: Arch. Microbiol. 124, 103 (1980).

65. Dixon, N.E., Gazzola, C., Blakely, R.L., Zerner, B.: J. Amer. Chem. Soc. 97, 4131 (1975).

66. Faulkner-Hudson, T.G.: In: Browning, E. (ed) Vanadium: Toxicology and Biological Significance. Elsevier Monographs on Toxic Agents. Amsterdam, Elsevier (1964).

67. Fishbein, W.N., Smith, M.J., Nagarajan, K., Scurzi, W.: Fed. Proc. Fed. Amer. Soc. Exp. Biol. 35, 1643 (1976).

68. Forth, W., Rummel, W.: In: Skoryna, S.C., Waldron-Edward, D. (eds) Intestinal Absorption of Metal Ions, Trace Elements and Radionuclides. New York, Pergamon Press p. 173 (1971).

69. Fries, L.: Planta 154, 393 (1982).

70. Frost, D.V.: Poultry Sci. 32, 217 (1953).

71. Graf, E.-G., Thauer, R.K.: FEBS Lett. 136, 165 (1981).

72. Groppel, B., Anke, M., Grün, M.: In: Anke, M., Schneider, H.-J., Brückner, Chr. (eds) Nickel, Spurenelement-Symposium 3 - Wissenschaftl. Publikationen F.-Schiller-Univ. p. 269 (1980).

73. Grün, M., Anke, M., Hennig, A., Kronemann, H.: In: Anke, M., Brückner, Chr., Gürtler, H., Grün, M. (eds) Mengen- und Spurenelemente 2 - Karl-Marx-Univ. Leipzig p. 159 (1982).

74. Grün, M., Hennig, A., Kronemann, H., Anke, M., Zintl, F.: Wiss. Z. Karl-Marx-Univ. Leipzig, Math.-Naturwiss. R. 32, 645 (1983).

75. Guerrero, M.G., Vega, J.M.: Arch. Microbiol. 102, 91 (1975).

76. Hansen, O.: Acta Pharmacol. Toxicol. 52, Suppl. I 1 (1983).

77. Henkin, R.I., Bradley, D.F.: Life Sci. 9, 701 (1970).

78. Hennig, A.: In: Anke, M., Baumann, W., Bräunlich, H., Brückner, Chr. (eds) Spurenelement-Symposium 4 - Wissenschaftl. Publikationen F.-Schiller-Univ. Jena p. 6 (1983).

79. Hennig, A., Jahreis, G., Anke, M., Partschefeld, M., Grün, M.: Arch. Tierernährung 28, 267 (1978).

80. Hoffmann, G., Anke, M., Grün, M., Groppel, B., Riedel, E.: In: Anke, M., Schneider, H.-J., Brückner, Chr. (eds) Arsenic, Spurenelement-Symposium 3 - Wissenschaftl. Publikationen F.-Schiller-Univ. Jena p. 41 (1980).

81. Hoffmann, G., Anke, M., Partschefeld, M., Groppel, B., Kronemann, H.: In: Anke, M., Schneider, H.-J. (eds.) Cadmium, Spurenelement-Symposium 2 - Wissenschaftl. Publikationen F.-Schiller-Univ. Jena p. 58 (1979).

82. Hopkins, L.L.Jr.: In: Hoekstra, W.G., Suttie, J.W., Ganther, H.E., Mertz, W. (eds) Trace Element Metabolism in Animals 2 - University Park Press Baltimore, London, Tokyo p. 397 (1974).

83. Hopkins, L.L.Jr., Mohr, H.E.: In: Mertz, W., Cornatzer, W.E. (eds) Newer Trace Elements in Nutrition, Marcel Dekker, New York p. 195 (1971).

84. Hopkins, L.L.Jr., Mohr, H.E.: Fed. Proc. 30, 462 (1971).

85. Hopkins, L.L.Jr., Mohr, H.E.: Fed. Proc. 33, 1773 (1974).

86. Horiguchi, S., Teramoto, K., Kurono, T., Ninomiya, K.: Osaka City Med. J. 24, 131 (1978).

87. Hove, E., Elvehjem, C.A., Hart., E.B.: Amer. J. Physiol. 124, 205 (1938).

88. Jelinek, C.F., Corneliussen, P.E.: Environm. Health Perspect. 19, 83 (1977).

89. Johnson, J.L., Cohen, H.J., Rajagopalan, K.V.: J. Biol. Chem. 249, 5046 (1974).

90. Kelsay, J.L., Behall, K.M., Prather, E.S.: Amer. J. Clin. Nutr. 32, 1876 (1979).

91. Kirchgessner, M., Perth, J., Schnegg, A.: Arch. Tierernährung 30, 805 (1980).

92. Kirchgessner, M., Reichlmayr-Lais, A.M.: In: McC Howell, J., Gawthorne, J.M., White, C.L. (eds) Trace Element Metabolism in Man and Animals 4 - Austral. Acad. Sci. Canberra p. 391 (1981).

93. Kirchgessner, M., Reichlmayr-Lais, A.M.: Int. J. Vit. Nutr. Res. 51, 421 (1981).

94. Kirchgessner, M., Reichlmayr-Lais, A.M.: Biol. Trace Element Res. 3, 279 (1981).

95. Kirchgessner, M., Reichlmayr-Lais, A.M.: Ann. Nutr. Metab. 26, 50 (1982).

96. Kirchgessner, M., Roth-Maier, D.A., Schnegg, A.: In: McC Howell, J., Gawthorne, J.M., White, C.L. (eds) Trace Element Metabolism in Man and Animals 4 - Austral. Acad. Sci. Canberra p. 621 (1981).

97. Kirchgessner, M., Roth-Maier, D.A., Spörl, R.: Z. Tierphysiol., Tierernährung u. Futtermittelkde. 44, 233 (1980).

98. Kirchgessner, M., Roth-Maier, D.A., Spörl, R.: Arch. Tierernährung, 31, 21 (1981).

99. Kirchgessner, M., Schnegg, A.: Bioinorg. Chem., 6, 155 (1976).

100. Kirchgessner, M., Schnegg, A.: Nutr. Metab., 23, 62 (1979).

101. Kirchgessner, M., Schnegg, A.: In: Anke, M., Schneider, H.-J., Brückner, Chr. (eds) Nickel, Spurenelement-Symposium 3 - Wissenschaftl. Publikationen F.-Schiller-Univ. Jena p. 27 (1980).

102. Kirchgessner, M., Schnegg, A.: In: Nriagu, J.O. (ed) Nickel in the Environment. J. Wiley and Sons, New York, Chichester, Brisbane, Toronto p. 635 (1980).

103. Kirchgessner, M., Schnegg, A.: In: Anke, M., Schneider, H.-J., Brückner, Chr. (eds) Nickel, Spurenelement-Symposium 3 - Wissenschaftl. Publikationen F.-Schiller-Univ. Jena p. 23 (1980).

104. Kirchgessner, M., Schnegg, A.: Ann. Nutr. Metab., 25, 307 (1981).

105. Kronemann, H., Anke, M.: In: Anke, M., Brückner, Chr., Gürtler, H., Grün, M. (eds) Mengen- und Spurenelemente 3 - Karl-Marx-Univ. Leipzig p. 141 (1983).

106. Kronemann, H., Anke, M., Groppel, B., Riedel, E.: In: Anke, M., Baumann, W., Bräunlich, H., Brückner, Chr. (eds) Lithium, Spurenelement-Symposium 4- Wissenschaftl. Publikationen F,-Schiller-Univ. Jena p. 85 (1983).

107. Kronemann, H., Anke, M., Grün, M., Partschefeld, M.: In: Anke, M., Schneider, H.-J. (eds) Cadmium, Spurenelement-Symposium 2 - Wissenschaftl. Publikationen F.-Schiller-Univ. Jena p. 230 (1977).

108. Kronemann, H., Anke, M., Thomas, S., Riedel, E.: In: Anke, M., Schneider, H.-J., Brückner, Chr. (eds) Nickel, Spurenelement-Symposium 3 - Wissenschaftl. Publikationen F.-Schiller-Univ. Jena p. 221 (1980).

109. Lambert, J., Sapek, A., Sapek, B.: In: Anke, M., Baumann, W., Bräunlich, H., Brückner, Chr. (eds) Lithium, Spurenelement-Symposium 4 - Wissenschaftl. Publikationen F.-Schiller-Univ. Jena p. 32 (1983).

110. Latvietis, J.J., Ruwalds, I.W., Osols,J.A., Miesite, I. A.: In: Anke, M., Brückner, Chr., Gürtler, H., Grün, M. (eds) Mengen- und Spurenelemente 3 - Karl-Marx-Univ. Leipzig p. 186 (1983).

111. Lee, K.K., Erickson, R., Palm, S., Jones, G., Mai, F., Nason, A.: J. Biol. Chem. 249, 3953 (1974).

112. Lehninger, A.L.: Biochem. J. 119, 129 (1970).

113. Linder, M.C.: In: Stave, U. (ed) Perinatal Physiology, 2nd Ed, New York, Plenum p. 425 (1978).

114. Ljungdahl, L.G.: TIBS p. 63 (1976).

115. Ljungdahl, L.G., Andreesen, J.R.: FEBS Letters 54, 279 (1975).

116. Loeper, J., Goy-Loeper, J., Rozensztajn, L., Fragny, M.: Atherosclerosis 33, 397 (1979).

117. Loeper, J., Loeper, J., Fragny, M.: Nobel Symp. 40, 281 (1978).

118. Malaiškaite, B.S.: In: Anke, M., Baumann, W., Bräunlich, H., Brückner, Chr. (eds) Lithium, Spurenelement-Symposium 4 - Wissenschaftl. Publikationen F.-Schiller-Univ. Jena p. 52 (1983).

119. Mehard, C.W., Volcani, B.E.: J. Histochem. Cytochem. 23, 348 (1975).

120. Mehard, C.W., Volcani, B.E.: Cell Tiss. Res. 174, 315 (1976).

121. Meisch, H.-U., Benzschawel, H.: Arch. Microbiol. 116, 91 (1978).

122. Meisch, H.-U., Benzschawel, H., Bielig, H.-J.: Arch. Microbiol. 114, 67 (1977).

123. Meisch, H.-U., Bielig, H.-J.: Arch. Microbiol. 105, 77 (1975).

124. Mertz, W.: Fed. Proc. Fed. Amer. Soc. Exp. Biol. 29, 1482 (1970).

125. Most, K.-H.: Diss. Graz, Philosoph. Fakultät (1939).

126. Myron, D.R., Givand, S.H., Nielsen, F.H.: Agric. Food Chem. 25, 297 (1977).

127. Myron, D.R., Zimmerman, T.J., Shuler, T.R., Klevay, L.M., Lee, D.E., Nielsen, F.H.: Amer. J. Clin. Nutr. 31, 527 (1978).

128. Naylor, G.J., Smith, A.H.W.: Psychol. Med. 11, 249 (1981).

129. Nielsen, F.H.: In: Mertz, W., Cornatzer, W.E. (eds) Newer Trace Elements in Nutrition Dekker, New York p. 215 (1971).

130. Nielsen, F.H.: In: Hoekstra, W.G., Suttie, J.W., Ganther, H.E., Mertz, W. (eds) Trace Element Metabolism in Animals 2 - Baltimore, University Park Press p. 381 (1974).

131. Nielsen, F.H.: Food Technol. 28, 38 (1974).

132. Nielsen, F.H.: In: Draper, H.H. (ed) Advances in Nutritional Research 3 - New York, Plenum p. 157 (1980).

133. Nielsen, F.H.: Biol. Trace Element Res. 2, 199 (1980).

134. Nielsen, F.H.: J. Nutr. 110, 965 (1980).

135. Nielsen, F.H.: In: Anke, M., Schneider, H.-J., Brückner, Chr. (eds) Nickel, Spurenelement-Symposium 3 - Wissenschaftl. Publikationen F.-Schiller-Univ. Jena p. 33 (1980).

136. Nielsen, F.H., Myron, D.R.: Proc. ND Acad. Sci. 34, 31 (1980).

137. Nielsen, F.H.: In: Liss, A.R. (ed) Clinical, Biochemical, and Nutritional Aspects of Trace Elements, New York p. 379 (1982).

138. Nielsen, F.H., Myron, D.R., Givand, S.H., Ollerich, D.A.: J. Nutr., 105, 1607 (1975).

139. Nielsen, F.H., Myron, D.R., Givand, S.H., Zimmerman, T.J., Ollerich, D.A.: J. Nutr. 105, 1620 (1975).

140. Nielsen, F.H., Myron, D.R., Uthus, E.O.: In: Kirchgessner, M. (ed) Trace Element Metabolism in Man and Animals 3 - Arbeitskreis Tierernährungsforschung Weihenstephan, p. 244 (1977).

141. Nielsen, F.H., Ollerich, D.A.: Fed. Proc. 32, 929 (1973).

142. Nielsen, F.H., Ollerich, D.A.: Fed. Proc. 33, 1767 (1974).

143. Nielsen, F.H., Sauberlich, H.E.: Proc. Soc. Exper. Biol. Med. 134, 845 (1970).

144. Nielsen, F.H., Shuler, T.R.: Fed. Proc. Fed. Amer. Soc. Exper. Biol. 34, 923 (1978).

145. Nielsen, F.H., Shuler, T.R., Zimmerman, T.J., Collings, M.E., Uthus, E.O.: Biol. Trace Element Res. 1, 325 (1979).

146. Nielsen, F.H., Uhrich, K.E., Shuler, T.R., Uthus, E.O.: In: Anke, M., Baumann, W., Bräunlich, H., Brückner, Chr. (eds) Spurenelement-Symposium 4 - Wissenschaftl. Publikationen F.-Schiller-Univ. Jena p. 127 (1983).

147. Onkelinx, C., Becker, J., Sunderman, F.W.Jr.: Res. Commun. Chem. Pathol. Pharmacol. 6, 663 (1973).

148. Patt, E.L., Pickett, E.E., O'Dell, B.L.: Bioinorganic Chem. 9, 299 (1978).

149. Pickett, E.E.: In: Anke, M., Baumann, W., Bräunlich, H., Brückner, Chr. (eds) Lithium, Spurenelement-Symposium 4 - Wissenschaftl. Publikationen F.-Schiller-Univ. Jena p. 66 (1983).

150. Polacco, J.C.: Plant Sci. Lett. 10, 249 (1977).

151. Regius, A., Pavel, J., Anke, M., Szentmihályi, S.: In: Anke, M., Baumann, W., Bräunlich, H., Brückner, Chr. (eds) Lithium, Spurenelement-Symposium 4 - Wissenschaftl. Publikationen F.-Schiller-Univ. Jena p. 45 (1983).

152. Reichlmayr-Lais, A.M., Kirchgessner, M.: Zbl. Vet. Med. A 28, 410 (1981).

153. Reichlmayr-Lais, A.M., Kirchgessner, M.: Z. Tierphysiol., Tierernährung u. Futtermittelkde. 46, 8 (1981).

154. Reichlmayr-Lais, A.M., Kirchgessner, M.: Ann. Nutr. Metab. 25, 281 (1981).

155. Reichlmayr-Lais, A.M., Kirchgessner, M.: Arch. Tierernährung 31, 731 (1981).

156. Reichlmayr-Lais, A.M., Kirchgessner, M.: Z. Tierphysiol. Tierernährung u. Futtermittelkde. 46, 1 (1981).

157. Riedel, E., Anke, M., Schwarz, S., Regius, A., Szilagyi, M., Löhnert, H.-J., Flachowsky, G., Zenker, G., Glös, S.: In: Anke, M., Schneider, H.-J., Brückner, Chr. (eds) Nickel, Spurenelement-Symposium 3 - Wissenschaftl. Publikationen F.-Schiller-Univ. Jena p. 55 (1980).

158. Ruvalds, I., Latvietis, J.: In: Anke, M., Baumann, W., Bräunlich, H., Brückner, Chr. (eds) Lithium, Spurenelement-Symposium 4 - Wissenschaftl. Publikationen F.-Schiller-Univ. Jena p. 78 (1983).

159. Sabbioni, E., Marafante, E.: Bioinorg. Chem. 9, 389 (1978).

160. Sabbioni, E., Marafante, E.: J. Toxicol. Environm. Health 8, 419 (1981).

161. Sabbioni, E., Marafante, E., Amantini, L., Ubertalli, L.: Bioinorg. Chem. 8, 503 (1978).

162. Sabbioni, E., Marafante, E., Pietra, R., Goetz, L., Girardi, F.: IAEA-SM-227/115 p. 179 (1979).

163. Sabbioni, E., Rade, J.: Toxicol. Lett. 5, 381 (1980).

164. Sabbioni, E., Rade, J., Bertolero, F.: J. Inorg. Biochem. 12, 307 (1980).

165. Sapek, B., Sapek, A.: In: Anke, M., Baumann, W., Bräunlich, H., Brückner, Chr. (eds) Lithium, Spurenelement-Symposium 4 - Wissenschaftl. Publikationen F.-Schiller-Univ. Jena p. 25 (1983).

166. Schmidt, A., Anke, M., Groppel, B., Kronemann, H.: Exper. Pathol. in press (1984).

167. Schnegg, A., Kirchgessner, M.: Z. Tierphysiol., Tierernährung u. Futtermittelkde. 36, 61 (1975).

168. Schnegg, A., Kirchgessner, M.: Nutr. Metab. 19, 268 (1975).

169. Schnegg, A., Kirchgessner, M.: Arch. Tierernährung 26, 543 (1976).

170. Schnegg, A., Kirchgessner, M.: Int. Z. Vit. Ernährungsforschung 46, 96 (1976).

171. Schnegg, A., Kirchgessner, M.: Zbl. Vet. Med. 24, 242 (1977).

172. Schnegg, A., Kirchgessner, M.: Int. Z. Vit. Ernährungsforschung 47, 274 (1977).

173. Schnegg, A., Kirchgessner, M.: Z. Tierphysiol., Tierernährung u. Futtermittelkde. 38, 200 (1977).

174. Schnegg, A., Kirchgessner, M.: In: Anke, M., Schneider, H.-J., Brückner, Chr. (eds) Nickel, Spurenelement-Symposium 3 - Wissenschaftl. Publikationen F.-Schiller-Univ. Jena p. 11 (1980).

175. Schnegg, A., Kirchgessner, M.: In: Anke, M., Schneider, H.-J., Brückner, Chr. (eds) Nickel, Spurenelement-Symposium 3 - Wissenschaftl. Publikationen F.-Schiller-Univ. Jena p. 17 (1980).

176. Schneider, H.-J., Anke, M., Klinger, G.: In: Anke, M., Schneider, H.-J., Brückner, Chr. (eds) Nickel, Spurenelement-Symposium 3 - Wissenschaftl. Publikationen F.-Schiller-Univ. Jena p. 277 (1980).

177. Schroeder, H.A., Balassa, J.J.: J. Chron. Dis. 19, 85 (1966).

178. Schroeder, H.A., Balassa, J.J.: J. Nutr. 92, 245 (1967).

179. Schroeder, H.A., Mitchener, M., Nason, A.P.: J. Nutr. 100, 59 (1970).

180. Schwartz, A., Becker, C.H., Matthias, D., Wooßmann, H., Engler, E.: In: Anke, M., Brückner, Chr., Gürtler, H., Grün, M. (eds) Mengen- und Spurenelemente 2 - Karl-Marx-Univ. Leipzig p. 201 (1982).

181. Schwarz, K.: Proc. Nat. Acad. Sci., USA 70, 1608 (1973).

182. Schwarz, K.: In: Hoekstra, W.G., Suttie, J.W., Ganther, H.E., Mertz, W. (eds) Trace Element Metabolism in Animals 2 - University Park Press, Baltimore, London, Tokyo, p. 355 (1974).

183. Schwarz, K.: Fed. Proc. 33, 1748 (1974).

184. Schwarz, K.: The Lancet, Febr. 26, 454 (1977).

185. Schwarz, K.: In: Brown, S.S. (ed) Clinical Chemistry and Chemical Toxicology of Metals, Amsterdam, Elsevier/ North Holland p. 3 (1977).

186. Schwarz, K., Chen, S.C.: Fed. Proc. 33, 704 (1974).

187. Schwarz, K., Milne, D.B.: Science 174, 426 (1971).

188. Schwarz, K., Milne, D.B.: Nature (London), 239, 333 (1972).

189. Schwarz, K., Spallholz, J.E.: In: Anke, M., Schneider, H.-J. (eds) Cadmium, Spurenelement-Symposium 2 - Wissenschaftl. Publikationen F.-Schiller-Univ. Jena p. 188 (1977).

190. Simkhovich, B.Z., Slutsky, L.I., Dombrovskaya, L.E., Gilev, A.P., Zelchan, G.I., Lukevich, E.Ya.: Bjull. Eksper. Biol. Mediz. 2, 153 (1979).

191. Skinner, J.T., Mc Hargue, J.S.: Amer. J. Phys. 145, 500 (1946).

192. Spears, J.W., Hatfield, E.E.: In: Anke, M., Schneider, H.-J., Brückner, Chr. (eds) Nickel, Spurenelement-Symposium 3 - Wissenschaftl. Publikationen F.-Schiller-Univ. Jena p. 47 (1980).

193. Spears, J.W., Hatfield, E.E., Forbes, R.M.: J. Anim. Sci. 48, 649 (1979).

194. Spears, J.W., Hatfield, E.E., Forbes, R.M., Koenig, S.E.: J. Nutr. 108, 313 (1978).

195. Strasia, G.A.: Purdue Univ. Ann. Arbor. Mich. USA (1971).

196. Sunderman, F.W.Jr., Decsy, M.I., Mc Neely, M.D.: Ann. NY Acad. Sci. 199, 300 (1972).

197. Sunderman, F.W.Jr., Nomoto, S., Morang, R., Nechay, M. W., Burke, C.N., Nielsen, S.W.: J. Nutr. 102, 259 (1972).

198. Sunderman, F.W.Jr., Shen, S.K., Reid, M.C., Allpass, P.R.:In: Anke, M., Schneider, H.-J., Brückner, Chr. (eds) Nickel, Spurenelement-Symposium 3 - Wissenschaftl. Publikationen F.-Schiller-Univ. Jena p. 301 (1980).

199. Szentmihályi, S., Regius, A., Anke,M., Grün,M., Groppel, B., Lokay, D., Pavel,I.: In: Anke, M., Schneider, H.-J., Brückner, Chr. (eds) Nickel, Spurenelement-Symposium 3- Wissenschaftl. Publikationen F.-Schiller-Univ. Jena p. 229 (1980).

200. Szentmihályi, S., Regius, A., Lokay, D., Anke, M.: In: Anke, M., Baumann, W., Bräunlich, H., Brückner, Chr. (eds) Lithium, Spurenelement-Symposium 4 - Wissenschaftl.Publikationen F.-Schiller-Univ.Jena p.18 (1983).

201. Szilágyi, M., Anke, M., Szentmihályi, S.: In: Szentmihályi, S. (ed) International Conference on Feed Additives 1 - Hung. Soc. Agricult. Sci. Budapest p. 257 (1981).

202. Szilágyi, M., Anke, M., Szentmihályi, S.: In: Anke, M., Baumann, W., Bräunlich, H., Brückner, Chr. (eds) Lithium, Spurenelement-Symposium 4 - Wissenschaftl. Publikationen F.-Schiller-Univ. p. 71 (1983).

203. Szilágyi, M., Anke, M., Szentmihályi, S. : 16th Meeting of the Federation of European Biochem. Soc., June 25 - 30, 1984 Moscow, USSR

204. Thauer, R.K., Diekert, G., Schönheit, P.: Trends in Biochemical Sci. 5, 304 (1980).

205. Tölgyesi, G.: In: Anke, M., Baumann, W., Bräunlich,H., Brückner, Chr. (eds) Lithium, Spurenelement-Symposium 4 - Wissenschaftl. Publikationen F.-Schiller-Univ. Jena p. 39 (1983).

206. Underwood, E.J.: Trace Elements in Human and Animal Nutrition. Academic Press, Inc. New York, San Francisco, London (1977).

207. Uthus, E.O., Nielsen, F.H.: In: Anke, M., Schneider, H.-J., Brückner, Chr. (eds) Arsenic, Spurenelement-Symposium 3 - Wissenschaftl. Publikationen F.-Schiller-Univ. Jena p. 33 (1980).

208. Uthus, E.O., Nielsen, F.H.: In: Anke, M., Baumann, W., Bräunlich, H., Brückner, Chr. (eds) Spurenelement-Symposium 4 - Wissenschaftl. Publikationen F.-Schiller-Univ. p. 105 (1983).

209. Voronkov, M.G., Zelchan, G.I., Lukevitz, E.: Silizium und Leben. Akademie-Verlag Berlin (1975).

210. Welch, R.M., Huffman, E.W.D.: Plant Physiol. 52, 183 (1973).

211. Westöö, G., Rydälv, M.: Var Föda 24, 21 (1972).

212. Whitman, W.B., Wolfe, R.S.: Biochem. Biophys. Res. Commun. 92, 1196 (1980).

213. Wiegmann, T.B., Day, H.D., Patak, R.V.: J. Toxicol. Environm. Health 10, 233 (1982).

214. Williams, D.L.: Ph. D. Thesis Purdue Univ., Ann. Arbor. Mich , USA

215. Wlodawer, A., Miller, M.,Sjölim, L.: Proc. Natl. Acad. Sci. USA 80, 3628 (1983).

216. Ziegler, E., Edwards, B.B., Jensen, R.L., Mahaffey, K.R., Fomon, S.J.: Pediat. Res. 12, 29 (1978).

D i s c u s s i o n Session VIII - 1
Plenary Lecture

Subjects of the discussion

- preparation of a trace element deficient diet
- sufficient supplementation
- chemical forms of the different trace elements supplemented to the diet (speciation)
- trace element levels during fetal development
- analytical techniques for testing the diet

Contributors: Aito, Brätter, Gökman, Koh, Oster

URINARY LITHIUM EXCRETION IN ENDEMIC (EG) AND NON ENDEMIC GOITER AREAS (NEG) IN VENEZUELA, PRELIMINARY REPORT.

Benzo Z, Fraile R.
Chemistry Center, Instituto Venezolano de Investigaciones Científicas. Apartado 1827, Caracas 1010-A, Venezuela.

Cevallos J.L., Wallis V, Drayer R, López M., De los Ríos V, Salas M.C.
Laboratorio de Investigaciones Clínicas, Luis Razetti School of Medicine. Universidad Central de Venezuela. Apartado 3007, Caracas 1010-A. Venezuela.

Introduction

Several facts have been desmonstrated, concerning the biological action of lithium (Li^+). First, due to its physical-chemical properties it behaves within the biological environment, very similar to Na^+ and K^+, being the reason why it may partially and imperfectly susbtitute them (1), entering the cell by Na^+ exchange, flowing along with sugar and aminoacids instead of Na^+ and directly by protein chanells (2,3). It interfers with a variety of cellular functions such as ionic membrane selectivity and hormonal responses (1). Second, Li^+ actions upon the Central Nervous Systems interfering with the neurons ion flux and cathecolamines metabolism, which explains its usefulness in the treatment of manic disorders (1,2,4). Third, Li^+ also interfers with the adenyl-cyclase production induced by polypeptides hormones like antidiuretic (ADH) and thyroid stimulating (TSH) ones (1,2,5). This is one of the proposed explanations for the well stablished antithyroid and goitrogenic effects of Li^+ at therapeutical doses, observed in some patients treated with Li^+ salts for manic disorders (6,7,8),

Trace Element - Analytical Chemistry in Medicine and Biology, Vol. 3

this associated with the influence in the Iodine thyroid kinetic, facilitates the Iodine entrance to the gland independently from the TSH (9,10).

Fourth, it has been demonstrated that the lithium goitrogenic effect not only occurs when administered in human at therapeutical doses (average doses of 90 mg/d of Li^+, equivalent to approximately 0.77 mg per kg of body weight) (6). We have experimentally demonstrated in rats that Li^+ in doses as low as 1.0 to 1.5 μg/d (equivalent to approximately 4 to 6 μg of Li^+ per kg of body weight), administered for prolonged period of times, was able to determine goiter, under low iodine, low protein and normocaloric diets(11).

Concerning our main topic of interest namely: trace elements in medicine and biology, we found Li^+ concentrations which varied from 0.005 to 0.08 ppm in the drinking waters of certain endemic goiter communities of the Venezuelan Andes region, located in the Trujillo state (Fig. 1).

Fig. 1,Situation of Trujillo State, in the Andean region of Venezuela.

In addition, these montaneous communities has a well established iodine deficiency condition. However, the prevalence of endemic goiter varies among them, which have been attributed to additional factors: Li^+ presence as trace element (11) and hypoproteic diet (12). This hypothesis was also experimentally proved by feeding male rats with drinking water containing Li^+ (0.05ppm) from these goitrogenic regions and under hypoiodic, hypoproteic and isocaloric diet, during a three months period. It was noticed a significant thyroid weight difference of 7.5 ± 1.6mg/100g of rat weight vs 4.9 ± 1.3mg/100g from the control group ($p<0.005$). Another possible goitrogenic contaminant of these waters was ruled out by repeating the experiment, this time placing another group, under the same diet conditions, and feeding it with low Li^+ content waters (0.005ppm) from the same geological area, founding no statistical significant differences in thyroid weight between this group and the control group: 5.3 ± 1.1 vs $4,5 \pm 1.4$mg/100g of rat weight (13). The previous mentioned evidences led us to measure the drinking water Li^+ content and the urinary excretion of Li^+ in human urine samples, taken from people living in small towns of the endemic goiter area (EG) of the Trujillo State, and compare it with Li^+ levels excreted in inhabitants of Caracas, (NEG) where Endemic goiter does not exist, and there is a very low Li^+ contents of its waters.

Methods:

Sampling:

Water samples were obtained directly from rivers and wells of five communities of the Trujillo State. They were collected in polyethylene flasks thoroughly washed and previously tested for Li^+ contamination.

Seventeen tap water samples coming from different aqueduct systems of the metropolitan area in Caracas were collected directly as control.

Thirty six urine samples were collected from single voided sam-

ples taken during home survey of three different small communities of the Trujillo state.
Twenty four urine control samples collections were carried out in Caracas from subjects living in this area.

Apparatus:
A Varian Techtron Model AA-5 Atomic Absorption Spectrophotometer was used in the flame emission mode. The instrumental parameters were optimized as recommended by the manufacturer.

Reagents:
A 1000 µg/mℓ Li stock solution was prepared from lithium nitrated (BDH). A secondary standard solution of 5 µg/mℓ was made from this stock solution. Working standards in the range of 5-60ng/mℓ were prepared containing 2000µg/mℓ potassium.

Procedure:
Water samples were read directly after a 1:1 dilution with deionized water, containing 2000 ppm potassium. Urine samples were diluted 1:4 in de-ionized water also containing 2000 ppm potassium. This dilution factor is used in order to correct matrix viscosity effects, allowing in this way, the use of aqueous standards (14).
Urinary creatinine was measured by the method of Owen and Tausky based in the reaction of Jaffe (15).
The 24 hours Li^+ excretion was calculated on the basis of the Li^+/ creatinine ratio, in single urine samples, which was multiplied by a constant of creatinine excretion per 24 hours. This was corrected according to the age, sex and weight, as it is customary used for urinary Iodine excretion from casual samples (16).

Results:
The water lithium concentration in five communities of the Trujillo state and in the control area (Caracas) is shown in Ta-

ble I., and the urinary lithium excretion expressed as μg per mℓ of Li (casual samples) and μg of Li per 24 hours from the same places above mentioned is shown in table II. As we can see, lithium urinary excretion in 24 hours is much lower in the control group (NEG) than in the endemic (EG) area as we expected.

TABLE I

WATER Li^+ CONCENTRATION (μg/mℓ)

Trujillo State:	
Valera (n=2)	0.01
Fundo J. de Dios (n=2)	0.02
San Antonio (n=2)	0.025
La Quebrada (n=2)	0.005
Motatan (n=2)	0.08
Caracas (n=17)	0.0025±0.0005

TABLE II

Li^+ URINARY EXCRETION

	TRUJILLO	CARACAS	p
Li^+ (μg/ml) (casual samples)	0.14±0.08 (n=34)	0.028±0.009 (n=24)	<0.0001
Li^+ (μg/24h)	143.6±137.1 (n=36)	33.2±12.4 (n=24)	<0.005

n= number of samples

Discussion:

Several aspects are interesting, concerning Li^+ as contaminant in trace amount. First: It has been shown the significative difference in Li^+ concentration, between drinking waters from Trujillo and Caracas. Second: The same happens with urinary Li^+ excretion of the studied inhabitants of the mentioned areas. Third: Let us assume that a subject from Trujillo usually drinks one liter of water per day, containing an average of 0.05 ug/ml of Li^+, which means an intake of 50 ug/d, which will generate a similar amount of daily Li^+ urinary excretion. Nevertheless, our findings showed a greater daily Li^+ urinary excretion of $143.6 \pm 137,1$ ug ; this make us to think in the existence of additional sources of Li^+ intake, such as vegetable, able to concentrate biologically this element, or Li^+ contamination from processed foods. Same observation applies to the Caracas group , whose Li^+ daily intake is approximately 2ug/d, excreting 33.2 ± 12.4 ug/d of Li^+ in the urine . Consequently, the systematic search for other sources of Li^+ intake, should be stressed.

Fourth: Taking into consideration, that Li^+ is a very uncommon element on the earth's crust (0.005%) (17), founding it in a silicate form in granitic rocks (Pegmatites), the question arises: How it is able to contaminate the water sources?. One answer could be-at least in the Venezuelan Andean region-, the report of Li^+ presence in hot water springs, in concentration up to 0.86ug/ml (18). Additionaly this region has also mica mines, mineral particularly rich in Li^+. This led us to point out the need of measuring Li^+ in hot springs water particularly when this waters contribute to supply the drinking water of a determinated community.

This fact could be another explanation to the prevalence of endemic goiter in areas in which iodine defficiency does not exist at all, or persistence of endemic goiter far after salt iodination programs has been established in well known EG, as has been demonstrated by Gaitan in Colombia (20) namelly the existence of goitrogenic sulfurated hydrocarbons compounds con taminating the drinking waters. On the other hand, up to now Li^+ biological effect as a contaminant, has been mainly related with its action upon the thyroid gland, nevertheless, it has also been associated with an apparent benefitial action: lowering death-rate from arteriosclerotic heart disease (19). All this data makes very appealing the systematic search for Li^+ presence in waters and in natural and processed food, as a matter of public health interest.

References:

1. Singer I, Rotenbery D.: New Eng. J. Med. 289,254-260(1973)
2. Tosteson DC.:Sci. Am. 244, 130-137 (1981).
3. Pandey GN, Dorus E, Davis JM, Tosteson DC.: Arch Gen.Psychiatry 36, 902-906 (1979).
4. Bunney WE, Pert A, Rosenblatt J, Pert CB, Callafor D.: Arch Gen. Phsychiatry 36, 898-901 (1979)
5. Cox M., Singer I.: Am. J. Med.,59, 153-157 (1975)
6. Schou M, Amdisen A, Eskjaer Jensen S., Olsen T.: Br. Med.J. 3, 710-713 (1968).
7. Luby ED, Schwartz D, Rosenbaum H.: JAMA 218, 1298-1299(1971).
8. Candy J,: Br. Med. J. 3, 277-(1972).
9. Lazarus J.H., Fell GS, Robertson JM, Millar WT, Bennie EH.: Archs Oral Biol. 18, 329-335 (1973).
10. Schou M.: Acta Pharmacol et Toxicol 15,115-124 (1958)
11. Fossi M, Cevallos J.L. in Stockigt JR, Nagataki S (eds) Thyroid Research VIII Australian Academy Science, Camberra Vol 1. 206-209 (1980).
12. Ingenbleek Y, De Wisscher H.: Metabolism 28, 9-19 (1979)
13. Rojas JA, Cevallos J.L., López M., De Los Ríos V., Fraile R., Wallis V., Ordoñez M., Salas MC., Rojas JR, in Niepomniszcze H. Pisarev M (eds) Tiroides 82, Ed. Med. Panamericana Buenos Aires 270-271 (1982)
14. Delves H.T.: Prog. Analyt. Atom. Spectrosc. 4, 1-48 (1981)
15. Henry RJ, Cannon DC., Wickelman JW.:Clinical Chemistry Harper & Row Publishers Inc. Hagonstown USA (Spanish translation) Ed. Jims. Barcelona 546-548 (1980)
16. Jolin T, Escobar del Rey F.: J. Clin Endocrin 25,540-541 (1965).

17. Stecher PG. Ed. The Merck Index. Eds. Merck Co. Inc. New Jersey p: 621 (1969)

18. Burguera JL, Burguera M, Andressen R.: Acta Científica Venz. 34, (sup 1) 528 (1983).

19. Blachly PH.: N. Eng. J. Med. 281, 682 (1969)

20. Gaitan E.: World Re Nutr. Diet. 17, 53-90 (1973).

LOCALIZATION OF ALUMINUM IN BONE OF PATIENTS WITH DIALYSIS OSTEOMALACIA

Peter Fritz Schmidt[1], Heinz Zumkley[2], Hans Bertram[3], Arno Lison[2], Bernd Winterberg[2], Rudolf Barckhaus[1]

[1] Institut für Medizinische Physik
[2] Medizinische Universitäts-Poliklinik
[3] Institut für Pharmakologie und Toxikologie

Universität Münster
D-4400 Münster

Abstract

In patients suffering from renal insufficiency and undergoing intermittent hemodialysis treatment an increased aluminum content in tissue particularly in bone was found combined with an increased osteoid volume and a reduced mineralization front (dialysis osteomalacia).

To get more information on the pathogenetic mechanisms between aluminum intoxication and development of dialysis osteomalacia the aluminum content has been determined directly in bone cells of patients without kidney disease and of patients with dialysis osteomalacia. These studies have been performed by means of a laser microprobe mass analyser (LAMMA 500), which permits a morphologically controlled analysis of small volumes with an outstanding low limit of detection. By LAMMA an increased aluminum content in bone cells of patients with dialysis osteomalacia could be detected. This result supports the hypothesis of a direct toxic effect of aluminum on bone cells with a consecutive development of dialysis osteomalacia.

Trace Element - Analytical Chemistry in Medicine and Biology, Vol. 3

Introduction

In patients suffering from renal insufficiency and undergoing intermittent hemodialysis treatment, an increased concentration of aluminum in blood and tissue is caused by the ingestion of aluminum containing drugs or from the use of aluminum rich dialysis fluids (1,2). One of the consequences of this aluminum intoxication is an accumulation of aluminum in bone, combined with an increased osteoid volume and a reduced mineralization front (dialysis osteomalacia) (3). The mechanisms by which aluminum causes abnormal bone mineralization are still unknown. It is supposed, that aluminum may have a direct toxic effect on bone mineralization and may induce osteomalacia (4).
The diagnosis of alterations of aluminum concentrations in dialysis osteomalacia is based on the interpretation of clinical symptoms and on the determination of the aluminum contents in blood and bone tissue using an integrating analysing technique like atomic absorption spectrometry (AAS)(2,3). Although this method permits the analysis of relatively small volumes, only integrated results can be obtained for cellular or subcellular compartments, a distinct localization of elements within the tissue is impossible. But to get more information on the pathogenetic mechanisms between aluminum intoxication and development of dialysis osteomalacia, the determination of the aluminum content in bone cells is of pronounced importance for the estimation of effects of an aluminum intoxication on the mineralization process. Therefore an analytical technique is required, which can localize aluminum in trace concentration at a cellular and subcellular level. This possibility is given by the Laser Microprobe Mass Analyser (LAMMA)(6, 7).

Methods

The principle of the laser microprobe mass analysis is based on laser beam induced ion production by evaporating of microvolumes of a specimen and the mass spectrometric analysis of the ions.

The instrument LAMMA 500 (Leybold-Heraeus, Köln) is a combination of an optical microscope, a laser source, a specimen chamber and a time-of-flight mass spectrometer (Fig. 1). Specimens are mounted on grids used in electron microscopy and placed in vacuum. The specimen holder can be manually shifted in x-y-direction. The optical microscope allows the observation of the specimens (1200 x, phase contrast) and focusing of the laser light. The area to be analysed is indicated by a red spot of a collinear Helium-Neon-laser and is then evaporated and ionized by a short pulse of a high power laser (a frequency quadrupled Neodymium doped Yttrium Aluminum oxide Garnet (Nd-YAG) laser, λ = 265 nm, pulse time τ = 15 ns). The ionized part of the evaporated material is analysed in the time-of-flight mass spectrometer in which ions of different m/e ratios are discriminated according to their different flight times: $t \approx \sqrt{M}$. The spectrometer can be used either for positive or negative ions depending on the polarity of the potentials of the electrodes. The detector provides output signals corresponding to the

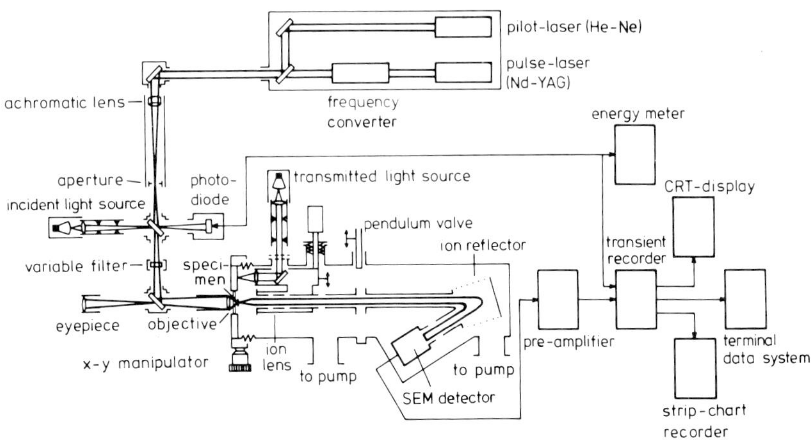

Fig. 1 Schematic diagram of a Laser Microprobe Mass Analyser LAMMA 500

amount of each ionic species. These signals are fed to a fast digital transientrecorder and stored. The analysis time is in the range of seconds, limited only by the time needed for localization of the area of interest.

This system permits the analysis of mass from m/e = 1 to m/e = 2000. Thus elements, isotopes and also molecules or molecular fragments can be detected simultaneously in one spectrum. The amplitude of the ion signal corresponds to the number of ions present in the evaporated specimen volume. The ion yield depends on the concentration of the analysed element, but also on the ionization potential, the binding state of the ions to the surrounding organic matrix and the bond strength to inorganic molecules.

The most important feature of the laser microprobe is the very low detection limit 10^{-18} to 10^{-20} g, combined with a sensitivity in the ppm range. The spatial resolution is in the order of 0.8 µm to 1 µm. With a section thickness of 0.5 µm the evaporated volume is about 10^{-12} to 10^{-13} cm^3. This corresponds to 10^{-12} to 10^{-13} g specimen matter. At the present state of investigations only a relative quantitation is possible (8), an absolute quantitation is being developed.

Preparation

Bone samples were taken by needle puncture from the pelvic bone from patients without kidney disease and patients with dialysis osteomalacia. One part of the tissue provided for the analysis by LAMMA had been prepared by cryotechniques in order to prevent any redistribution of aluminum. The tissue had been shock frozen by immersion into melting N_2, freeze dried in a freeze drying device at below -80° C, plastic embedded in an epoxy resin (Spurr's low viscosity medium), and sectioned dry on a conventional ultramicrotome (section thickness 1 µm). The sections were mounted on 3 mm grids used in electron microscopy.

The other part of the tissue was provided for determination of aluminum concentration by atomic absorption spectrometry (2).

Results and conclusions

The bone aluminum content of a patient with dialysis osteomalacia was measured by atomic absorption spectrometry and resulted in 43.8 μg aluminum/g fresh weight, while the aluminum concentration of an individuum without renal failure resulted in 6.9 μg/g fresh weight.

Using LAMMA the distribution of aluminum had been determined (Fig. 2): LAMMA could reveal an increased aluminum content in

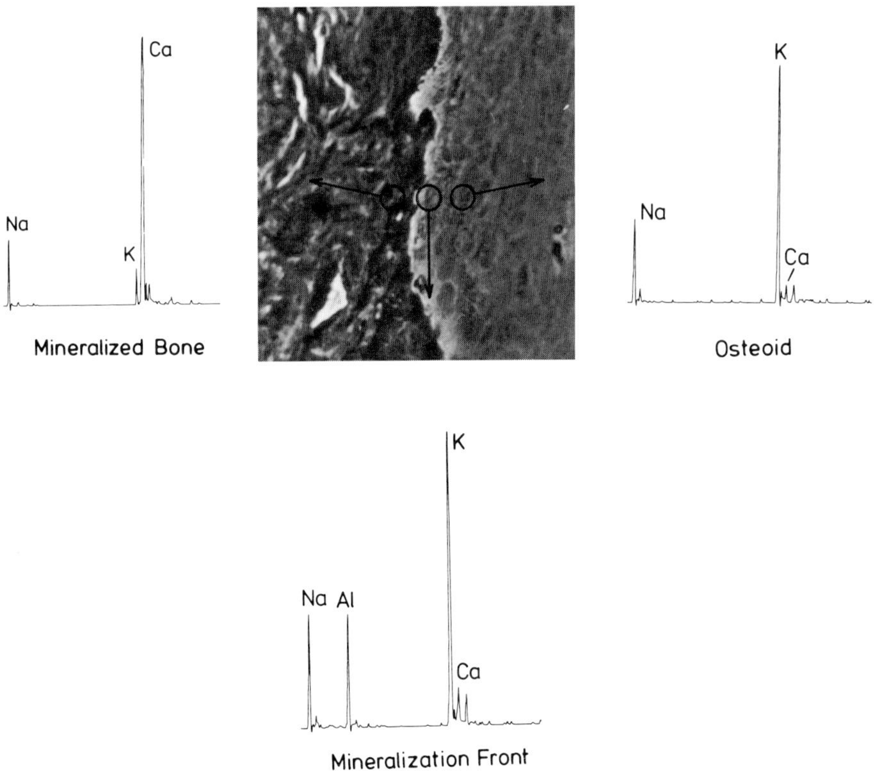

Fig. 2 Aluminum localization between mineralized and nonmineralized bone matrix

the mineralization front of calcified tissue and the osteoid similar to results of histochemical staining techniques (9) and results, obtained by ion microprobe measurements (10).

Additionally LAMMA could detect an increased aluminum content in osteocytes of patients with dialysis osteomalacia. Fig. 3 shows mass spectra from osteocytes of a patient suffering from dialysis osteomalacia and from an osteocyte of a patient with normal kidney functions and without treatment with aluminum containing drugs. These spectra demonstrate that osteocytes in dialysis osteomalacia - especially osteocytes near the boundary between the calcified tissue and the osteoid-show a very

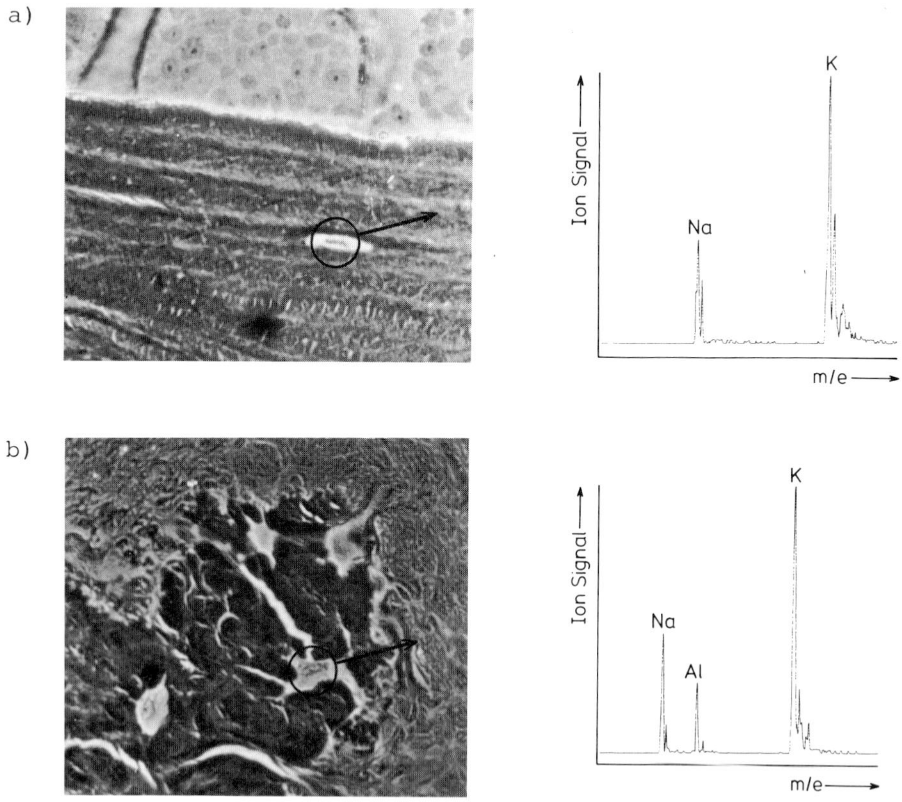

Fig. 3 Al-content in osteocytes of patients without (a) and with (b) dialysis osteomalacia

high peak of aluminum. Also in single fibroblasts of "fibrotic" osteoid an elevated aluminum level could be detected.

These results support the hypothesis of a direct toxic effect of aluminum on bone cells with consecutive development of dialysis osteomalacia.

These investigations show that AAS provides quantitative information on the aluminum concentration and can give an indication of an aluminum intoxication while LAMMA offers the possibility to estimate the local aluminum distribution with an extremely low limit of detection (10^{-19}g) and a high sensitivity (ppm). In opposition to other detection procedures LAMMA enables to localize aluminum also in low concentrations, which is a presupposition for the determination of intracellular aluminum content.

To localize aluminum in bone tissue usually histochemical staining methods are used. But when the bone aluminum content was in the range of 20 to 50 ppm no stain could be seen (2). Also the analysis of aluminum using x-ray microanalysis is only possible if aluminum is accumulated in a high concentration deposit: the sensitivity of this method is in the range of 800 ppm, the detection limit is about 10^{-16} g.

The sensitivity and the spatial resolution of the ion microprobe is comparable to LAMMA but the use of this method is restricted by the light microscopical imaging system (reflexion mode).

In general by the determination of the aluminum concentration in the mineralization front and in cells of bone tissue from patients with different severe osteomalacic lesions it will be possible to evaluate if aluminum interferes with mineralization and what aluminum concentration causes defects in the mineralization activity of bone cells and prevents the mineralization process of the organic matrix.

Acknowledgement

The technical assistance of Mrs. H. Schwantge is gratefully acknowledged.

References

1. Alfrey, A.C., Le Gendre, G.R., Kaehny, W.D.: New Engl. J. Med. 294, 184-188 (1976).

2. Zumkley, H., Bertram, H.P., Lison, A., Knoll, O., Losse, M.H.: Clin. Nephrol. 12, 18-21 (1979).

3. Delling, G., Schäfer, H.-J., Krumme, H., Schulz, W.: Nieren- und Hochdruckkrankheiten 12, 207-213 (1983).

4. Drüeke, T.: Nephron 26, 207-210 (1980).

5. Zumkley, H., Schmidt, P.F., Bertram, H.P., Lison, A., Winterberg, B., Spieker, K., Losse, H.: Verhandlungen der Deutschen Gesellschaft für innere Medizin 89, 1227-1235 (1983).

6. Kaufmann, R., Hillenkamp, F., Wechsung, R.: Med. Progr. Technol. 6, 109-121 (1979).

7. Schmidt, P.F.: Lokalisation von Spurenelementen mittels eines Laser Mikrosonden Massen Analysators (LAMMA). In: "Spurenelemente". Ed. H. Zumkley, Georg Thieme Verlag Stuttgart, 12-24 (1983).

8. Schmidt, P.F., Ilsemann, K.: Scanning Electron Microsc./ 1984/I, 77-85.

9. Ott, S.M., Maloney, N.A., Coburn, J.W., Alfrey, A.C., Sherrand, D.J.: New Engl. J. Med. 307, 709-713 (1982).

10. Cournot-Witmer, G., Zingraff, J., Plachot, J.J., Escaig, F., Lefevre, R., Boumati, P., Bourdeau, A., Garabedian, M., Galle, P., Bourdon, R., Drüeke, T., Balsan, S.: Kidney International 20, 375-385 (1981).

INTERACTION OF PLATINUM WITH SELENIUM AND OTHER TRACE ELEMENTS IN MICE TREATED WITH THE ANTITUMOR AGENT CISPLATIN

P.S.Tjioe, C.J.A. Van den Hamer, J.J.M. De Goeij
Interuniversity Reactor Institute, 2629 JB Delft, The Netherlands

J.G. McVie
Cancer Institute, 1066 CX Amsterdam, The Netherlands

Introduction

The antitumor drug cisplatin (cis-diamminedichloro-Pt(II)) is effective in the treatment of cancer of the testis and ovaries. However, serious toxic side effects, particularly renal damage, have been reported (1). The side effects may be directly due to cisplatin or secondary to its interaction with other trace elements (TE). An increased level of Mg, Zn, and Cu in the urine and a decreased plasma concentration of Mg, K, and Ca were observed (2). A protective effect of Se (as selenite) against the toxicity of heavy metals such as Hg (3), Cd (4), and Pb (5) has been reported. From animal studies it was concluded that selenite can also interact with Pt, reducing the toxic effects of cisplatin (6).

Materials and Methods

Female Swiss random-bred mice of approximately 20 g were used throughout the study. Pt and the other TE's were determined by neutron activation analysis (NAA) as previously described (7). [^{195m}Pt]-cisplatin was synthesized following an adaptation of the procedure of Dhara (8), using ^{195m}Pt ($t_{\frac{1}{2}}$= 4.0 d; principle γ-radiation=99 keV) produced by the reactor of our Institute by irradiation of enriched (80%) ^{194}Pt with thermal neutrons (flux 1x 10^{13}n. $cm^{-2}.s^{-1}$) during 100 h, yielding an activity of

Trace Element - Analytical Chemistry in Medicine and Biology, Vol. 3

40 µCi/mg Pt at the end of the irradiation. Non-radioactive cisplatin was synthesized by the same method. Its chemical purity was checked spectrophotometrically and its antitumor activity using the leukaemia cell line L1210 (REPGO-TNO, Rijswijk, The Netherlands). [^{75}Se]-selenite was obtained from the Radiochemical Center (Amersham, U.K.).

In single isotope experiments whole body radioactivity of mice was determined using two 3x3" NaI(Tl) crystals with 1¼x2" wells and placed opposite to each other, forming a cavity for 4π counting (9). In double isotope (^{195m}Pt + ^{75}Se) experiments both radioactivities in mice and tissue samples were determined using a 60 cm^3 flattop Ge(Li) detector. The biological $t_{\frac{1}{2}}$'s of the retention curves were estimated using the data points from 5-12 days.

Distribution of radionuclides over soluble cytosolic proteins of kidney and liver was determined as follows: the tissue was homogenized with 3 volumes of Tris-acetate buffer 0.05M, pH 7.4. After centrifugation for 45 min at 100,000xg a sample of the supernatant was passed through a 1.4x56 cm column of Sephacryl S-300, equilibrated and eluted with the same buffer.

Results and Discussion

Results of the analysis of kidney and liver of mice, which were treated repeatedly with cisplatin, are shown in Table 1. Cr is known to be part of the glucose tolerance factor.It was decreased in kidney and in liver. Treatment of rats with cisplatin was shown to impair glucose tolerance (11). Co is the central atom in methylcobalamin. This vitamin B_{12} auxotroph is reported to methylate Pt-complexes (12), which may account for the observed increase of Co in kidney and liver. The increase of Se in kidney and liver may be the result of a redistribution of this element caused by cisplatin. A redistribution of Se has been described as a result of its interaction with Hg or Cd.

	Kidney Controls (n=6)		Kidney Pt-treated (n=10)	Liver Controls (n=4)		Liver Pt-treated (n=10)
As	0.003 ± 0.001		0.003 ± 0.001	0.002 ± 0.001		0.002 ± 0.001
Br	3.5 ± 0.6	(*)	6.3 ± 0.8	1.3 ± 0.1	(*)	1.9 ± 0.3
Cr	0.52 ± 0.31	(*)	0.07 ± 0.05	3.6 ± 2.8	(**)	1.5 ± 1.1
Hg	0.006 ± 0.001		0.005 ± 0.002	0.002 ± 0.001	(*)	0.004 ± 0.001
Sb	0.005 ± 0.003		0.006 ± 0.001	0.004 ± 0.001	(*)	0.006 ± 0.001
Se	2.7 ± 0.8	(*)	4.0 ± 0.4	1.8 ± 0.1	(*)	2.4 ± 0.1
Au	0.004 ± 0.001	(*)	0.001 ± 0.001	0.006 ± 0.001	(*)	0.003 ± 0.001
Pt	0		5.4 ± 1.5	0		6.3 ± 1.7
Rb	0.8 ± 0.2	(*)	1.2 ± 0.1	1.13 ± 0.24	(*)	1.6 ± 0.2
Cd	0.28 ± 0.27	(**)	0.10 ± 0.02	0.04 ± 0.01		0.03 ± 0.01
Co	0.25 ± 0.006	(*)	0.74 ± 0.19	0.20 ± 0.07	(*)	0.82 ± 0.14
Cu	6.3 ± 3.3	(**)	8.3 ± 0.7	8.0 ± 1.9	(*)	4.7 ± 1.4
Fe	158 ± 54	(*)	239 ± 48	211 ± 36	(**)	345 ± 89
Mo	0.74 ± 0.13		0.79 ± 0.08	0.92 ± 0.1	(**)	1.20 ± 0.20
Zn	34 ± 4	(*)	45 ± 3	43 ± 4	(*)	49 ± 3

Table 1.: Trace element content of kidney and liver of mice treated with cisplatin (schedule of 6x 10 μmol/kg two-weekly intervals) and of controls. Values in μg/g (wet weight) ± stand. dev. Significancy: (*) $p<0.01$ (**) $0.01<p<0.1$ Food given as described before (10).

Organ	[^{195m}Pt]-cisplatin t=1h	t=6h	t=24h	[^{195m}Pt]-$PtCl_6^{2-}$ t=1h	t=6h	t=24h
pancreas	0.93	0.85	0.80	1.55	1.68	1.49
spleen	0.92	0.78	1.00	1.32	1.43	1.32
stomach	0.80	1.03	0.92	1.56	2.63	2.25
intestine	1.72	2.30	4.17	2.67	3.24	3.38
liver	2.92	2.41	2.15	3.24	3.52	3.05
kidney	4.00	2.90	2.92	6.87	6.83	6.00
lung	1.26	1.04	0.94	3.06	2.93	2.35
skin	1.17	1.14	1.25	2.46	2.59	2.31
muscle	0.44	0.44	0.41	1.19	1.30	1.12
bone	0.69	1.58	1.34	2.53	2.64	2.07
brain	0.10	0.08	0.08	0.14	0.13	0.07
heart	0.56	0.44	0.43	1.83	1.74	1.42
carcass	0.94	0.99	1.03	2.07	2.02	1.97

Table 2.: Pt content of tissues of mice in μg/g (wet weight). Each value is the average of 3 animals. Mice were treated with either [^{195m}Pt]-cisplatin or [^{195m}Pt]-$PtCl_6^{2-}$. (dose: 10 μmol/kg).

The observed changes in TE content obtained by NAA are within easy reach of experimental verification with the use of radiolabeled compounds. The distribution of ^{195m}Pt over the tissues of mice after an i.v. injection of 10 µmol [^{195m}Pt]-cisplatin is given in Table 2 (left).
In most tissues, the ^{195m}Pt remained fairly constant between 1 and 24 h post dose. It decreased in the liver and increased in the intestine as function of time, both being probably the result of an excretion of ^{195m}Pt via the bile as has also been observed in man (13) and dog (14). Because $PtCl_6^{2-}$ may be a contaminant or a metabolite of cisplatin, and more toxic than cisplatin, the distribution of ^{195m}Pt was also studied after an i.v. administration of 10 µmol [^{195m}Pt]-$PtCl_6^{2-}$ per kg. The tissue concentrations of the Pt from $PtCl_6^{2-}$ were about twice of those after [^{195m}Pt]-cisplatin injection (Table 2 - right), probably because much of the injected cisplatin is rapidly lost in the urine. The distribution over the tissues was quite similar for both labeled compounds, suggesting similar metabolic pathways.

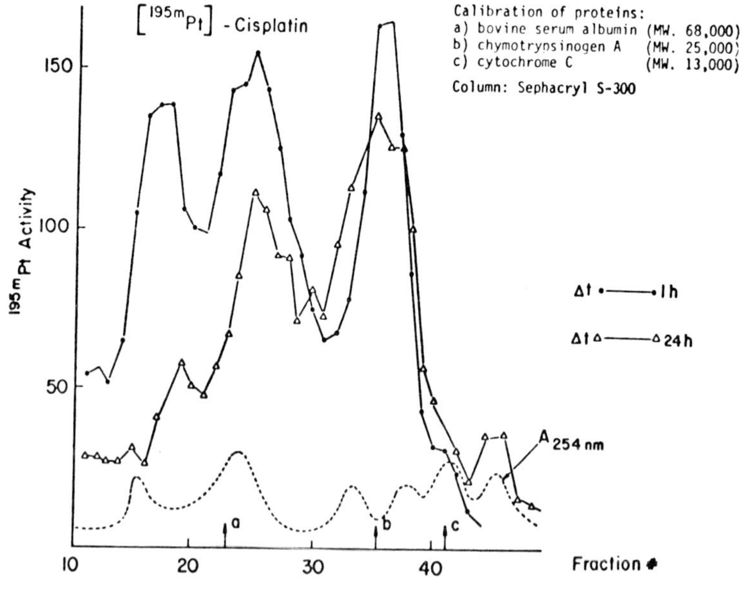

Fig. 1: ^{195m}Pt distribution from cisplatin as function of time post dose over the cytosolic proteins of the mouse kidney.

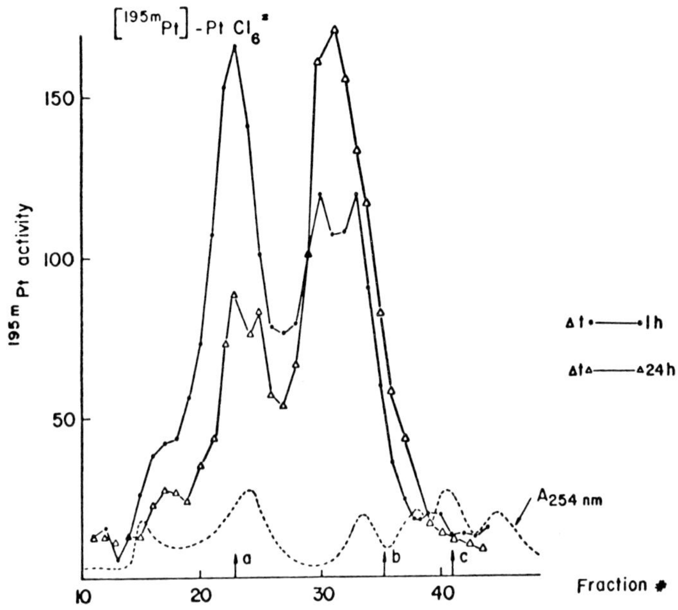

Fig. 2: ^{195m}Pt distribution from $PtCl_6^{2-}$ (details as in Fig. 1)

Preliminary results indicate a shift in distribution of Pt over the cytosolic proteins from the kidney (Fig. 2) and liver (not shown) as function of time after injection of $PtCl_6^{2-}$ towards proteins of lower molecular weight. A corresponding shift was not seen after injection of [^{195m}Pt]-cisplatin (Fig. 1). Further work along these lines leads to better understanding of the intracellular metabolism of cisplatin.

The total body retention in mice of subcutaneously administered [^{195m}Pt]-cisplatin (20 μmol/kg) showed an apparent retention (R_a) of 19% (Fig. 3). Neither this R_a nor the biological $t_{\frac{1}{2}}$ (8 - 9 d) of ^{195m}Pt was much influenced by simultaneous injection of selenite (6, 12 and 25 μmol/kg). The total body retention of s.c. administered [^{75}Se]-selenite (12 μmol/kg) showed a R_a of 9% (Fig. 4). This R_a increased to 14, 21 and 29% when cisplatin (10, 20 and 40 μmol/kg, resp.) was given simultaneously; the biological $t_{\frac{1}{2}}$ of ^{75}Se remained fairly constant, varying between 7 and 9 days.

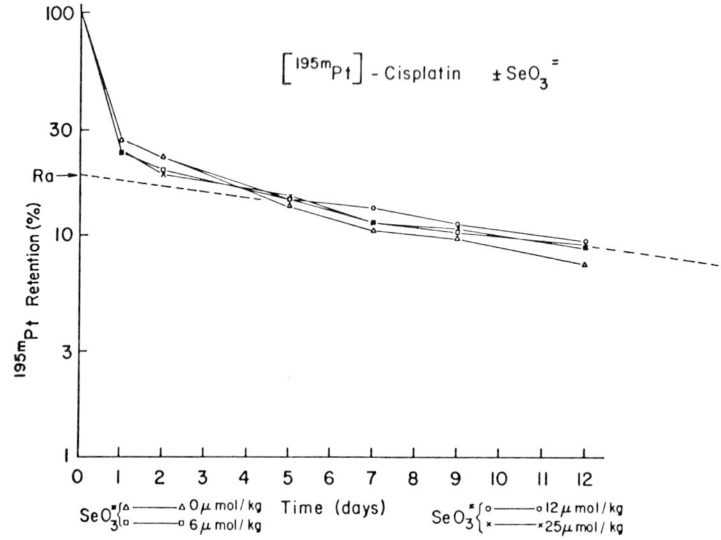

Fig. 3: The whole body retention of ^{195m}Pt from [^{195m}Pt]-cisplatin was not influenced by simultaneous administration of selenite.

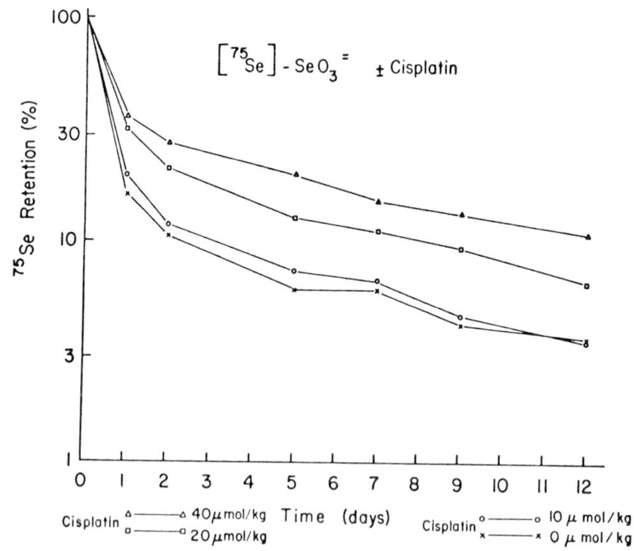

Fig. 4: The whole body retention of ^{75}Se from selenite is linearly increased by simultaneous administration of increasing amounts of cisplatin.

The increasing R_a of ^{75}Se (Fig. 4) shows that more Se is retained when cisplatin was given (an effect taking place in the first day post dose), suggesting an interaction. Only part of the injected Se dose is involved in this interaction because the R_a increases from 9% to 29%. Using the values given in the previous paragraph one can make the following estimation: adding 20 µmol cisplatin to [^{75}Se]-selenite resulted in an extra ^{75}Se retention of 12% (Fig. 4) or 1.4 µmol ^{75}Se. Conversely, when one looks at the ^{195m}Pt under the same conditions and assumes an equimolar interaction, one would expect also an extra retention of ^{195m}Pt of 1.4 µmol or about 7% of the cisplatin dose of 20 µmol. The curves in Fig.3 would have been further apart unless Pt is involved that also in the absence of selenite would have been retained, though in a different chemical form.

In these experiments 40 µmol cisplatin per kg was found to be lethal in 5 days unless 12 or 25 µmol selenite per kg had been administered simultaneously, clearly demonstrating protection by selenite.

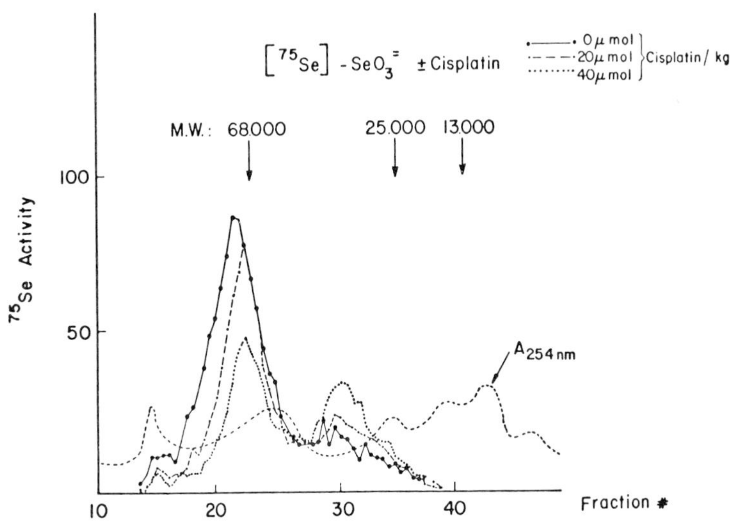

Fig. 5; The ^{75}Se distribution over the cytosolic proteins of the kidney when simultaneously cisplatin was administered.

The observed increase in whole body retention of ^{75}Se when cisplatin had been administered simultaneously points, like the influence of selenite on the lethality of cisplatin, to an interaction between cisplatin (or a metabolite) and selenite. The results shown in Fig. 5 are in accordance with this assumption: the distribution of ^{75}Se over the soluble cytosolic proteins of the kidney after injection of [^{75}Se]-selenite was changed when cisplatin had been given simultaneously. The nature of these Se-binding proteins is still unknown.

References

1. Weiner, M.W., Jacobs, C.: Fed. Proc. 42, 2974-2978 (1983).
2. Zumkley, H., Bertram, H.P., Preusser, P., Kellinghaus, H., Straub, C., Vetter, H.: Clin. Nephr. 17, 254-257 (1982).
3. Naganuma, A., Imura, N.: Pharmacol. Biochem. Behav. 13, 537-544 (1980) 15, 449-454 (1981).
4. Chen, R.W., Whanger, P.D., Weswig, P.H.: Bioinorg. Chem. 4, 125-133 (1975).
5. Flora, S.J.S., Singh, S., Tandon, S.K.: Acta Pharmacol. Toxicol. 53, 28-32 (1983).
6. Naganuma, A., Satoh, M., Yokoyama, M., Imura, N.: Res.Commun. Chem. Pathol. Pharmacol. 42, 127-134 (1983).
7. Tjioe, P.S., Volkers, K.J., Kroon, J.J., De Goeij, J.J.M., The, S.K.: 1st Workshop on Carcinogenic and/or Mutagenic Metals, Geneva (1983), in press.
8. Hoeschele, J.D., Butler, T.A., Roberts, J.A., Guyer, C.E.: Radiochim. Acta 31, 27-36 (1982).
9. Van Barneveld, A.A., Van den Hamer, C.J.A.: Nutr. Rev. Int. 29, 173-182 (1984).
10. Van Barneveld, A.A., Morse, H.: Proc. 1st Meeting of FELASA, 1981
11. Goldstein, R.S., Mayor, G.H., Gingerich, R.L., Hook, J.B., Rosenbaum, R.W., Bond, J.T.: Toxic. Appl. Pharmacol. 69, 432-441 (1983).
12. Fenchang, Y.T., Ridley, W.P., Wood, J.M.: J. Amer. Chem. Soc. 101, 1442-1447 (1979).
13. Vermorken, J.B., Van der Vijgh, W.J.F., Pinedo, H.M.: Res. Commun. Chem. Pathol. Pharmacol. 28, 319-330 (1980).
14. LeRoy, A.F., Lutz, R.J., Dedrick, R.L., Litterst, C.L., Guarino, A.M.: Cancer Treatm. Rep. 63, 59-71 (1979).

D i s c u s s i o n Session VIII - 2

Subjects of the discussion

- LAMMA-technique
- origin of Al in case of dialysis
- uptake of Al into different tissues
- Al/Zn interaction
- Al in the mineralization process
- side effects of antitumor reagents

Contributors: Brätter, Gardiner, Tölg

Session IX

ANALYSIS OF TRACE ELEMENTS IN BIOLOGICAL MATERIALS

Plenary Lecture given by: M. Stoeppler (FRG)

Chairman:

G. Tölg (FRG)

Poster Reports given by: G. Knapp (Austria)

J.M. Ottaway (UK)

VANADIUM IN HUMAN SERUM BY NEUTRON ACTIVATION ANALYSIS

M. Simonoff, Y. Llabador, G.N. Simonoff
E.R.A. 144 - C.E.N.B.G. - Le Haut Vigneau - 33170 GRADIGNAN - France

C. Beraud, P. Couzigou
Service de Gastro-entérologie - Hôpital du Haut Lévêque -
33604 PESSAC Cedex - France

C. Conri, B. Fleury
Service de Médecine Interne - Hôpital du Haut Lévêque -
33604 PESSAC Cedex - France

Introduction

Vanadium has been shown to be an essential trace element in animals (1) and appears to have several physiological functions (2, 3, 4, 5). Several studies have also shown vanadium to be a powerful inhibitor of Na-K-ATPase at physiological concentrations (6, 7). This ability has aroused considerable interest for analytical methods permitting the precise determination of vanadium in human tissues and plasma.

The literature data concerning vanadium concentrations in biological fluids, particularly plasma, show little agreement. Published values for plasma over the last 5 years vary from some picograms/mL (neutron activation analysis) (8) to several nanograms/mL (atomic absorption) (9, 10). The assumption that the lower values are the more accurate suggests that such low and widely dispersed values are due to environmental pollution and that vanadium may not be an essential trace element for man. The answer to these question requires that further measurements be carried out by the method giving the lowest values (neutron activation analysis), but taking advantage of recent technological advances, particularly in the detector field.

Trace Element - Analytical Chemistry in Medicine and Biology, Vol. 3

We have therefore measured serum vanadium in 23 subjects by neutron activation analysis, employing rapid chemical separation after irradiation. ^{51}V accounts for 99.75 % of natural vanadium. The cross section of the $^{51}V(n, \gamma)^{52}V$ reaction is large (4.8 barns) and ^{52}V gives rise to an easily-detectable photon of 1.434 MeV. The period is short (3.76 min), however, particularly since it is indispensable to eliminate the strong activities of ^{24}Na, ^{38}Cl, ^{80}Br and ^{66}Cu, which are always present in biological materials and could saturate the very sensitive detector employed.

Experimental Methods

1) Decontamination of sample containers : The 70 mL pyrex flasks to be used for blood collection and the quartz tubes for plasma irradiation were immersed in boiling nitric acid (6 mol/L) ("Suprapur", Merck), then rinsed several times with triple-distilled water from a quartz still. The same operation was repeated with 3 mol/L, "Suprapur" sulfuric acid and triple-distilled water. The foregoing treatment with the two acids was carried out a second time and the containers were then immersed in boiling triple-distilled water for 3 hours, then rinsed several times. Finally, the containers were heated in an oven at 100°C for 4 hours, then at 500°C for 8 hours. Each container was then kept in a sealed polyethylene envelope until use.

2) Collection and preparation of blood samples : Blood samples were collected from healthy volunteers (mainly hospital staff) by means of a polypropylene-over-the needle catheter (Vygon). To minimize contamination risks, the first 10 mL was rejected. The succeeding 60 mL was retained and centrifuged (3000 xg) in the cleaned pyrex flask (see above) to separate the serum. The latter was decanted into a second cleaned pyrex flask and dried for 24 hours at 60°C. The dried serum was then ashed for 8 hours at 300°C, then for a further 8 hours at 500°C. The product was dissolved in three times 1 mL of 3 mol/L HCl ("Suprapur") and the solution poured into a cleaned quartz tube (height, 70 mm ; diam, 12 mm) for irradiation. The tube was then placed in the oven for 24 hours at 100°C, followed by 3 hours at 200°C. After cooling to room temperature, the tube

was sealed with a polyethylene capsule and placed in a dessicator until irradiated.

3) Irradiation : The quartz tubes containing the dried serum were placed in sealed aluminium containers and irradiated for 1 min in a neutron flux of 2×10^{14} $n.cm^{-2}.s^{-1}$ in the OSIRIS reactor of the Commissariat à l'Energie Atomique at Saclay. In addition to the serum sample, each irradiated container enclosed a second quartz tube containing 50 ng of vanadium for normalization purposes. Irradiation of empty, cleaned, quartz tubes in presence of a second tube containing 50 ng of vanadium showed that no detectable vanadium passed from one tube to the other (see Fig. 1a).

4) Separation of vanadium after irradiation : Immediately after irradiation, the sample was dissolved in 2 mL of 3 mol/L hydrochloric acid. The irradiated tube was then rinsed with two more 2 mL portions of the same acid, the three portions being combined in a separation funnel containing 15 mL of water, 1 mg of vanadium carrier (as NH_4VO_3) and a sufficient quantity of ^{48}V for the calculation of the extraction yield ($^{48}VO_2Cl$, obtained from Amersham, U.K.). After addition of 2.5 mL of 6 % cupferron and 10 mL trichloromethane, the funnel was vigorously agitated for 30 s and the organic phase (containing the ^{48}V and ^{52}V) separated into a counting vial. The time elapsed between the end of the irradiation and the start of the measurement was about 6 or 7 min.

5) Activity measurements : The radioactivity in the organic phase was measured for 4 min, using a Ge(Li) detector (130 cm^3, 29.6 % relative efficiency, 1.69 keV resolution) at the Pierre Süe laboratory at Saclay. The first few samples were counted two or three times in order to check that the period of the 1.434 MeV peak corresponded to that of ^{52}V, thus indicating the absence of "parasite" peaks at that energy. Among the undesired reactions which could lead to ^{52}V production in a neutron flux are $^{52}Cr(n, p)^{52}V$ and $^{55}Mn(n, \alpha)^{52}V$. These two reactions are mainly induced by fast neutrons and have cross sections 50 to 100 times smaller that of the main reaction with thermal neutrons. The flux of the latter, in addition, was considerably higher than that of fast neutrons at the site of irradiation. The small quantities of chromium and manganese in plasma

(11, 12) should thus not effect the present results. Sample pollution with those elements, on the other hand, could falsify the results, so that all precautions were taking during collection and preparation of sample.

Results

Figure 1a shows the spectrum obtained with a quartz "blank", which had been irradiated in company with a standard (50 ng V in a quartz tube). No contamination by transfer can be observed. Fig. 1b shows another blank corresponding to the plasma ashing step. The pyrex flasks containing $\sim$ 30 mL of serum were placed in the ashing oven in groups of 5 and a sixth empty flask was placed in the oven with each group, in order to detect possible contamination during the procedures carried out in the oven. The mean value obtained with 5 "ashing procedure" blanks was 0.029 ± 0.010 ng V/mL.

Figure 1c shows the spectrum obtained with one sample of serum. The yields of the post-irradiation chemical separation are shown by the ^{48}V activity to be generally > 80 %. By adding ^{48}V to a serum sample, we have also determined that no loss of vanadium occured before irradiation (final recovery in the quartz tube represented 98 ± 1 % of the initial activity, provided the pyrex flask was rinsed 3 times with 1 mL of 3 mol/L HCl). Serum from one subject was divided into two parts, one of which was ashed in a quartz flask and the other in pyrex. The final results for serum vanadium differed by only 5 %, indicating that ashing in pyrex or in quartz led to no significant difference in the final result.

The analytical method described was tested by analysing ten samples of "standard" water obtained from the National Bureau of Standards (USA) (fig. 1d). The certified vanadium content was 53 ± 3 ng/mL. The present method gave 52.9 ± 0.8 ng/mL for five 1 mL samples and 53 ± 2.8 ng/mL for five other samples made up from 5 mL of water containing 2 mL of plasma.

The following Table shows the results and the corresponding error obtained for the 23 subjects tested. No correlation with sex or age is apparent. The mean value is 0.48 ± 0.01 ng/mL.

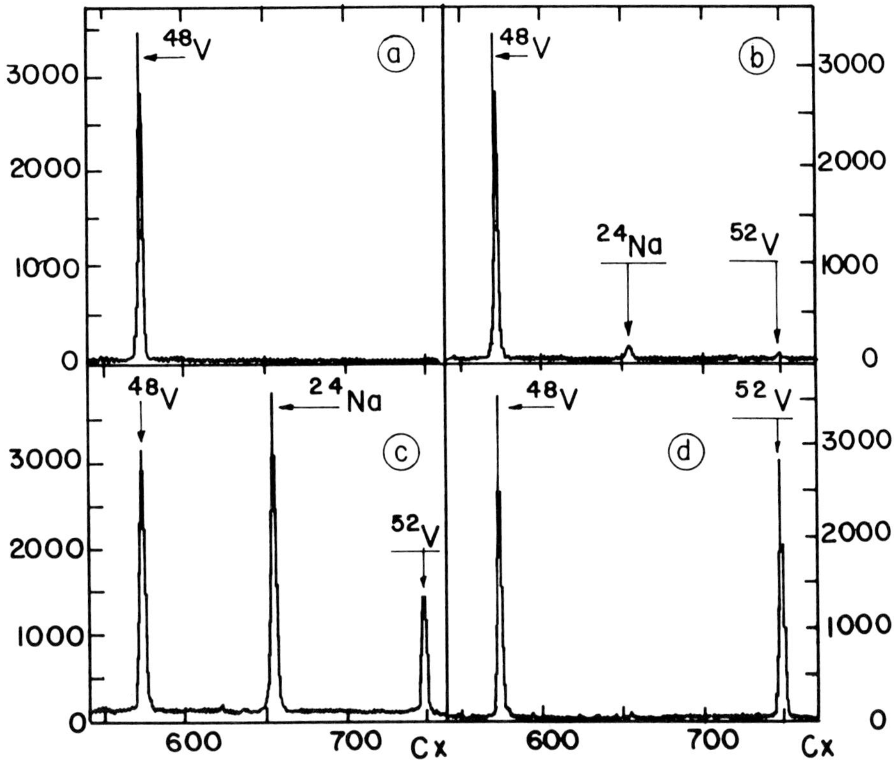

Figure 1 : γ spectra in the energy range of the ^{48}V, ^{24}Na and ^{52}V peaks

a) Quartz blank (see text)

b) Blank for ashing in pyrex, followed by final drying in quartz tube

c) Spectrum for 30 mL of serum from a healthy subject

d) Spectrum for 1 mL of standard (NBS) water

Discussion

We note, first of all, that the present results lie within a fairly narrow range (0.26-1.30 ng/mL), lying above the values (0.024 to 0.939) obtained by Cornelis et al. (1981) (13) by the same method. From our results, it appears that serum vanadium values are fairly constant in healthy subjects.

All published values for serum vanadium have been obtained by atomic absorption spectroscopy (AAS) or by neutron activation analysis (NAA). The

Table 1

V concentration in serum (ng/mL)							
N°	Sex	Age	ng V/mL	N°	Sex	Age	ng V/mL
1	M	30	0.77 ± 0.04	13	F	45	1.14 ± 0.10
2	F	49	0.58 ± 0.03	14	F	25	0.80 ± 0.06
3	M	32	1.22 ± 0.08	15	M	32	0.61 ± 0.06
4	M	56	0.55 ± 0.04	16	M	60	0.46 ± 0.04
5	M	45	1.30 ± 0.11	17	M	54	0.69 ± 0.05
6	M	25	1.30 ± 0.12	18	M	41	0.40 ± 0.03
7	F	37	0.98 ± 0.05	19	F	--	0.37 ± 0.03
8	M	34	0.27 ± 0.03	20	M	71	0.27 ± 0.02
9	M	45	0.81 ± 0.05	21	M	27	0.34 ± 0.03
10	M	57	0.29 ± 0.03	22	M	29	0.52 ± 0.03
11	M	36	0.82 ± 0.04	23	F	54	0.69 ± 0.06
12	M	32	0.26 ± 0.03				

following discussion compares these earlier results with those reported here.

1) Vanadium by AAS : Published values by AAS fall between 2 and several hundred ng/mL (10, 14). The sensitivity of the AAS method was considerably improved by Stroop et al. (1982) (10) who obtained 2 to 4 ng V/mL for 10 samples of serum. Prost et al. (1979) (15) obtained values of about 5 ng/mL for whole blood, and Dick et al. (1982) (9) gave an average value of 7.5 ng/mL for normal subjects. Although considerably lower than earlier values obtained by AAS, the foregoing are considerably higher than those reported in Table 1. Since we have shown experimentally that no loss of vanadium occured during analysis, we conclude that the present results are to be preferred over the higher AAS values.

2) Vanadium by NAA : Literature values for serum vanadium by NAA vary from several pg/mL (13) to 16 ng/mL (16).

The chief difficulty encountered in NAA resided in the short period of ^{52}V and the need to separate it from ^{24}Na and ^{36}Cl, in particular. Avoiding this problem by carrying out a chemical separation before irradiation involves the risk of sample contamination, which perhaps accounts for the rather high values (3.8 to 11.3 ng/mL) obtained by Sabbioni et al. (17). Values observed when chemical separation is carried out after irradiation are much lower. Byrne and Kosta (18) found a value less than 0.3-0.5 ng/mL for whole blood. Cornelis et al. (13) obtained a mean value of 0.031 ± 0.010 ng/mL and Allen and Steinnes found 0.8 ng/mL (19).

Since the values obtained with pre-irradiation chemical separation were often in agreement with those obtained by AAS, it has been suggested (10) that the rapid post-irradiation chemical extraction resulted in poor extraction yields. We have shown here, however, that the extraction yields were excellent, and the results obtained for standard (NBS) water samples were in perfect agreement with the certified values, even in presence of added plasma.

With regard to the values obtained by Cornelis et al. (13), which were much lower than ours and relatively more dispersed, it may be noted that our experimental conditions (flux intensity, detector characteristics, plasma weight, etc...) resulted in a higher sensitivity by a factor of ∿ 50. If we divided the activity of our ^{52}V peaks by that factor, the results would be statistically much less reliable.

In spite of the difficulties inherent in NAA with post-irradiation chemical separation, this would appear to be the most sensitive and accurate method available. The role of vanadium in human physiology is only beginning to be studied and these investigations will depend on the availability of numerous accurate analytical measurements, particularly in blood.

Conclusion

Vanadium concentrations have been measured in the blood of 23 healthy subjects by neutron activation analysis, with post-irradiation chemical separation. The values obtained fall in a fairly narrow range (0.26 - 1.3 ng/mL, mean value 0.48 ± 0.01 ng/mL), which suggests a physiological role for this element.

The method was checked by analysis of standard (NBS) water samples and serum samples spiked with ^{48}V. Excellent agreement was obtained in both cases and the reported results for human serum are considered to be more reliable than those obtained by atomic absorption spectroscopy.

Acknowledgments

The authors are grateful for the help and support of Drs G. Revel and O. Stulzaft from the P. Süe Nuclear Activation Analysis Laboratory (CNRS-CEA, Saclay) and from the department of Biology (Service Hospitalier Frédéric Joliot) for their expert assistance in many crucial ways in the course of this work.

They wish to thank Dr. A. Mackenzie Peers for help with the english text and M. C. Hamon for his care in glass decontamination.

References

1. Strasia, C.A. : Ph. D. Thesis, University Microfilms, Ann. Arbor (1971)
2. Macara, I.G. : Trends Biochem. Sci. 5, 92-94 (1980)
3. Borchord, U., Fox, A.A.L., Greef, K., Schlicper, P. : Nature 279, 339-341 (1979)
4. Werden, K., Beuriedel, G., Bozsik, M. et al : Biochem. Biophys. Acta 577, 364-383 (1980)
5. Komar, A., Corder, C.N. : J. Pharmacol. Exp. Ther. 213, 85-90 (1980)
6. Cantley, L.C., Resh, M.D., Guidotti, G. : Nature 272, 552-554 (1978)
7. Glynn, I.M., Grantham, J.J. : Am. J. Physiol. 236, F530-F535 (1979)
8. Cornelis, R., Mees, L., Hoste, J., Ryckebush, R., Versieck, J., Barbier, F. : Symposium Nuclear activation technique in the life Sci. IAEA-SM-227/25, 165-177 (1978)
9. Dick, D.A.T., Naylor, G.J., Dick, E.G. : Psychological Medicine 12, 533-537 (1982)
10. Stroop, S.D., Helinek, G., Greene, H.L. : Clin. Chem. 28/1, 79-82 (1982)
11. Simonoff, M., Llabador, Y., Simonoff, G.N., Besse, P., Conri, C. : Conference Proceedings will be published in Nucl. Inst. and Meth. 231 (1984)

12. Casey, C.E., Hambidge, K.M. : Adv. Nutr. Res. 3, 23-63 (1980)
13. Cornelis, R., Versieck, J., Mees, L., Hoste, J., Barbier, F. : Biological Trace Element Research 3, 257-263 (1981)
14. Bello-Reuss, E., Grady, T., Mazumdar, D. : Ann. Inter. Med. 91, 743 (1979)
15. Prost, R.L., Hunt, D.P., Walderhaug, M.D., Perkins, R.C., Park, J.H., Beth, A.H. : In Na, K-ATPase : Structure and kinetics, ed. Skou, J.C. and Norby, J.G., p. 389-401, Academic Press, London
16. Ward, N.I., Ryan, D.E. : Anal. Chem. Acta 105, 185-197 (1979)
17. Sabbioni, E., Marafante, E., Pietra, R., Goetz, L., Girardi, F., Orvini, E. : Proc. Int. Symp. Nucl. Act. Techn. in the Life Sci. - IAEA-SM-227/115, 179-192 (1979)
18. Byrne, A.R., Kosta, L. : Sci. Total Environ. 13, 87-90 (1979)
19. Allen, R.D., Steinnes, E. : Anal. Chem. 50, 1553 (1978)

A SIMPLE RADIOCHEMICAL PROCEDURE FOR ANALYSIS OF MOLYBDENUM AND TUNGSTEN IN BIOLOGICAL MATERIALS BY RADIOCHEMICAL ACTIVATION ANALYSIS

A.R.Byrne
"J.Stefan" Institute, "E.Kardelj" University,
61000 Ljubljana, Yugoslavia

1. Introduction

The essential role of molybdenum is related to its presence in numerous metallo-flavin and porphyrin enzymes. Molybdenum also interacts with copper antagonistically, a relationship complicated by interactions with sulphur[1]. Though increasing amounts of information are becoming available on the low natural levels of molybdenum in tissues, e.g. human blood and serum concentrations now appear to be around 1 $ng.ml^{-1}$ or somewhat below[2,3], analyses for molybdenum are hampered by the virtual absence of standard reference materials (SRMs) with known low level contents.

The present work describes development of a simple radiochemical neutron activation analysis (RNAA) technique for determination of nanogram levels of molybdenum, and its application to a number of widely used SRMs.

The method also allows simultaneous determination of tungsten, an element whose functions are little studied, but which interestingly enough is known to have antagonistic interactions with molybdenum mechanism[4,5,6].

2. Experimental

In outline, the method consists of neutron irradiation of a sample with Mo and W standards, wet ashing with carriers, followed by solvent extraction with N-benzoylphenylhydroxylamine (BPHA) from mixed $HCl-H_2SO_4$ medium.

Trace Element - Analytical Chemistry in Medicine and Biology, Vol. 3

2.1. Irradiation

Samples were irradiated in the Institute's TRIGA Mark II reactor at a flux of $2x10^{12} n.cm^{-2}sec^{-1}$ for about 20 hours and cooled for about 24 hours before processing. Polythene vials, precleaned with nitric acid, were heat sealed, and further encapsulated in polythene foil.

2.2. Standards

Stock solutions of 0.5 $mg.g^{-1}$ of Mo and W were prepared from MoO_3 and WO_3 in 2% ammonia. Dilution produced working standards of 100 and 10 $ug.g^{-1}$ for Mo and W, respectively, also in 2% NH_3. Standards were prepared by sealing weighed amounts in small bore polyethylene tubes. After irradiation the contents were quantitatively transferred to a counting vial with ammoniacal carrier solution. The tubes were leached with this carrier for some time to ensure complete recovery from the walls.

2.3. Procedure

Cooled samples were transferred to 100 ml long-necked quartz Kjeldahl flasks containing 2 ml conc. H_2SO_4, 200 ug V(V), 50 ug Mo and 50 ug W carriers. The samples were ashed by heating with repeated additions of HNO_3 until a pale solution was obtained which did not darken on heating. Complete destruction and removal of traces of HNO_3 and halogens was ensured by addition of 1 ml conc. $HClO_4$ which was fumed off. The solution was cooled, diluted with 2-3 ml H_2O, and transferred to a 60 ml separatory funnel with 10 ml 10M HCl. Water was added to about 19 ml (rinsing the Kjeldahl flask), and 1 ml of 0.1N $KMnO_4$, the funnel stoppered and shaken for about 10 secs.Then 10 ml 0.2% BPHA in toluene was added, and the funnel shaken for 30-40 secs. The violet organic phase(V complex) was allowed to separate, the aqueous phase discarded, and the organic phase washed once briefly with 10 ml 5M HCl. 9.5 ml of the organic phase was pipetted into a 25 ml flask.

The vanadium carrier catalyses wet destruction, and the coloured BPHA complex gives visual indication of the extraction and phase separation. (Vanadium may also be analysed simultaneously by this procedure, as discussed in 3.)

2.4. Counting

Solutions were counted on an intrinsic Ge detector, connected to a 4000-channel analyser, in optimum geometry, i.e. directly on the detector in a flat bottomed flask. For tungsten, the 479 and 686 keV peaks of ^{187}W (23.8 hours) were recorded, while for molybdenum, after waiting for re-establishment of radiochemical equilibrium between ^{99}Mo (66 hours) and ^{99m}Tc (6 hours), i.e. at least 36 hours, the 140 keV peak of ^{99m}Tc was used.

2.5. Chemical yields

As described in the next section, yields were established by separate tracer experiments on various matrices. As they were found to be quantitative for Mo and nearly so for W, they did not require re-measurement in the irradiation procedure.

3. Results and Discussion

The starting point of the technique was a method developed for RNAA of vanadium in biological materials(7), which was later developed to allow analysis of molybdenum(8) (with arsenic if desired). In the present work, the co-extraction of tungsten was investigated, and the chemical yields of molybdenum and tungsten in the whole procedure examined carefully, since it was desired to use the method to help establish values for these elements (especially Mo) in SRMs.

Table 1 below shows chemical yields for Mo and W found by tracer experiments using non-irradiated biological materials.

Table 1: Chemical yields of procedure by tracer experiments

Material	Weight used (mg)	Mo % recovery	W % recovery
Milk powder			
	250	101.6, 101.2	
	500		97.0
Bovine Liver	200	102.2	98.4
	400		100.0
Wheat flour	300	100.5	97.9
Bowen's Kale	50		97.5
	100		98.9
	150		96.3^{x}
	200	99.3	92.7^{xx}
	400	100.0	91.4^{xx}, 93.3^{xx}
		mean: 100.8 ± 1.1(6)	mean: 98.4±0.9(6) ^{x}excluded (^{xx}mean: 92.5±1.0(3) $CaSO_4$ formation)

It is clear that yields of molybdenum are quantitative, and that the yield of tungsten is generally almost so, at 98.4 ± 0.9 % for 6 trials of various substances. However, from the results given for Bowen's Kale, the yield drops slightly as the sample weight increases, coinciding with visual observation of cloudiness or a precipitate of $CaSO_4$, which appears to adsorb some tungsten. Thus 200-400 mg portions gave a yield of 92.5 ± 1.0 % (3), increasing to 96 % for 150 mg, and reaching its normal value of 98 % for aliquots of 100 mg or less. Hence for very accurate work with plant materials or other samples where $CaSO_4$ is deposited on dilution of the aqueous phase, the chemical yield of W should be checked for the aliquot size used.

Using the method developed, the Mo and W contents of a number of SRMs and CRMs were determined, as shown in Table 2

Table 2: Mo and W contents of some SRMs by proposed method

Material	Mo ug/g	W ng/g
NBS Bovine Liver SRM 1577	3.4, 3.6	3.6, 4.1
Certificate	(3.4)	
NBS Citrus Leaves SRM 1572	0.153 ± 0.016 (8)	8.1 ± 0.5 (5)
Certificate	0.17 ± 0.09	
NBS Milk Powder SRM 1549	0.342 ± 0.010 (6)	0.43 ± 0.03 (6)
Certificate	(0.34)	
Bowen's Kale	2.35 ± 0.12 (6)	50 ± 1.4 (6)
recommended	2.3	60
IAEA Milk Powder A-11	0.089 ± 0.003 (5)	1.3 ± 0.1 (3)
IAEA Animal Muscle H-4	0.042 ± 0.003 (5)	≤ 0.2 (5)

Bowen's Kale appears to be to only SRM of known tungsten content, while there are virtually no materials with known sub-ppm levels of molybdenum. Two IAEA materials, Milk Powder A-11 and Animal Muscle H-4 are currently the subject of an IAEA co-ordinated follow-up study designed to improve the status of these materials with respect to trace element content. In the case of A-11, since the original Intercomparison[9] doubts have been expressed about some of the values recommended, particularly for Cd, Cr, Cu and Mn[10,11]. In this follow-up study, a small number of selected laboratories are participating, and some of the results recently obtained[12] for Mo are shown in Table 3.

Table 3: Mo values obtained for IAEA A-11 and H-4 by participants in a current follow-up study(12)

Participant	Method	A-11 ng/g	H-4 ng/g
This work, Ljubljana	RNAA	89 ± 3 (5)	42 ± 3 (5)
de Goeij, Delft	RNAA	93 ± 10 (8)	41 ± 4 (4)
Iyengar, Jülich	RNAA	101 ± 8 (6)	45 ± 5 (6)
Parr, Vienna	RNAA	79 ± 5 (15)	38 ± 2 (6)

From the values presented, it is clear that the method developed gives results which are in excellent agreement with those of other workers and accepted values, and offers an attractive approach to low-level molybdenum and tungsten determination.

It should also be noted that vanadium can be analysed simultaneously with Mo and W by a short re-irradiation immediately prior to radiochemical processing, a technique we have termed(8) LICSIR, from the initial letters of long irradiation, cooling, short irradiation, radiochemistry.

Finally, possible interference from ^{99}Mo produced from fission of uranium can be discounted in view of the very low levels of uranium in most biological materials. In certain plants where uranium can be present at the 10-100 ng g^{-1} level, molybdenum is generally at the ppm level.

Acknowledgements

We thank the International Atomic Energy Agency, Vienna, and the National Bureau of Standards, Washington, for assistance through contracts 3393/TC and NBS(G)-251, respectively. We are also indebted to the Slovene Research Community for support.

References

1. E.J.Underwood, Trace Elements in Human and Animal Nutrition, Academic Press, 4th Edition, 1977.
2. J.Versieck, J.Hoste, F.Barbier, L.Vanbattenburghe, J.deRudder, R.Cornelis, Clin.Chim.Acta, 87 (1978) 135.
3. J.Versieck, R.Cornelis, Anal.Chim.Acta 116 (1980) 217.
4. R.E.Davies, B.L.Reid, A.A.Kurnich, J.R.Couch, J.Nutr. 70 (1960) 193.
5. E.S.Higgins, D.Richert, W.W.Westerfield, J.Nutr. 59 (1956) 539.
6. J.L.Johnson, K.V.Rajagopalan,H.J.Cohen, J.Biol.Chem. 249 (1974) 859. and in "Molybdenum in the Environment" (W.Chappell, K.Peterson, eds.) p.201, Dekker, New York, 1979.
7. A.R.Byrne, L.Kosta, J.Radioanal.Chem. 44 (1978) 247.
8. A.R.Byrne, Radiochem.Radioanal.Letters, 52 (1982) 99.
9. R.Dybczinski, A.Veglia, O.Suschny, Report on Intercomparison Run A-11 for the Determination of Inorganic Constituents of Milk Powder, IAEA/RL 68, IAEA, Vienna, 20 July 1980.
10. L.Kosta, A.R.Byrne, J.Radioanal.Chem. 69 (1982) 117.
11. J.J.M. de Goeij, L.Kosta, A.R.Byrne, J.Kučera, Anal.Chim.Acta 146 (1983) 161.
12. Personal communication, R.M.Parr and individual analysts, March 1984.

RADIOCHEMICAL NEUTRON ACTIVATION ANALYSIS OF ZINC FOR STABLE ISOTOPE TRACER STUDIES IN HUMAN NUTRITION

İnci G. Gökmen, Namık K. Aras
Department of Chemistry, Middle East Technical University, Ankara, Turkey

Glen E. Gordon
Department of Chemistry, University of Maryland, College Park, MD, 20742, USA

Robert, I. Henkin
Laboratory for Molecular Nutrition and Sensory Disorders, Georgetown University, Medical Center, Washington, D.C. 20007, USA

Introduction

Since Zn is one of the most important trace element nutrients measurement of Zn absorption will be very valuable for assessment of Zn requirement. Radioactive tracers have been used to explore Zn metabolism and measure its absorption (1-4). Stable isotope tracers do not expose subjects to radiation as in case of radiotracers, so they will be very useful for nutrition studies of pregnant women and children under 18. However, the method should be validated on adults against radioactive tracers before applying to above mentioned groups.

In stable isotope tracer technique, stable element with one of the isotopes enriched (usually the least abundant isotope) is used as a tracer. One of the very important requirements for this technique is the measurement of two different Zn isotopes (5-7). Of the two suitable measurement techniques (8, 9) mass spectrometry (MS) and neutron activation analysis (NAA), the latter was chosen for this study.

Trace Element - Analytical Chemistry in Medicine and Biology, Vol. 3

Experimental

As can be seen on Table I, natural abundance of ^{70}Zn is very low, administration of enriched ^{70}Zn will result in better sensitivity (5). However, ^{70}Zn has low neutron cross section (0.008) and the NAA product ^{71m}Zn has relatively short half life (3.96 h) and a gamma ray at the lower energy region of the spectrum (386.7 keV) which can be covered by other interfering activities that have better nuclear properties and/or present in larger quantities. In this work elimination of some major elements, Na, K, Cl, Br, and some trace elements, Mn and Cu, were necassary to detect short-lived Zn isotopes, ^{69m}Zn and ^{71m}Zn.

Table 1. Some Nuclear Properties of Zinc Isotopes

Isotope	Abundance %		σ(barns)[b]	Product	
	Natural[a]	Enriched[b]		$(t_{1/2})$[b]	γ(KeV)[b]
64	48.6	15.84	0.78	244 d	1115
66	27.9	8.95	0.85	-	-
67	4.10	1.32	6.9	-	-
68	18.8	8.38	0.072 0.81	13.8 h 56 m	438 v.weak
70	0.62	65.51	0.008 0.09	3.9 h	386

[a]Ref. 10

[b]Oak Ridge National Laboratories, Certificate of Analysis (1981).

The tracers which contain both radioactive ^{65}Zn and enriched ^{70}Zn were administered to four adult subjects and monitored by collection of blood, urine, feces samples, and whole body counting of ^{65}Zn at known intervals (7, 11). Samples were assayed for ^{65}Zn after collection and at the end of separation to obtain yields.

Portion of freeze-dried, homogenized (12) feces samples, plasma and red blood cells were digested in acid digestion bombs with 2.0 mL high purity conc. HNO_3 at 130^0 C for 1 hr. Urine subsamples (10 mL) and digested feces, plasma, red blood cells were passed through pre-cleaned. Chelex columns (7, 13). At the end of pre-irradiation separation, which is summarized on Figure 1, Na, K, Br, Cl were eluted from the column but Zn, Mn and Cu were kept in Chelex resin. Urine, plasma and red blood cells were irradiated for 3 hours and feces samples for 30 min to 1 hr at a flux of 5×10^{13} n/cm^2-sec, together with two standards and a Co flux monitor.

In order to detect ^{71m}Zn with reasonable statistics Chelex separation was not sufficient. It was necessary to eliminate ^{56}Mn and ^{64}Cu. Double precipitation of MnO_2 and extraction of Cu with dithizone at pH $\leq$ 1 was sufficient to eliminate all Mn and Cu. Then, to obtain a better counting geometry Zn was precipitated as $ZnHg(SCN)_4$. Post-irradiation separation scheme is summarized on Figure 2.

Results

Spectra were collected using large volume (50-85 cm^3) Lithium-drifted germanium (Ge(Li)) detectors which have full width at half maximum of 1.8 to 2.0 keV for 1332.5 keV γ ray of ^{60}Co. Two spectra were collected for pre-irradiation separation yield determinations using 1115 keV γ-ray of 244 d ^{65}Zn. After the irradiations 3 spectra were collected. From the first spectrum taken at 25 cm away from the detector, 438 keV γ-ray of 13.8 h ^{69m}Zn was observed. After the separation of Mn and Cu, samples were counted closer to the detector (10 cm) to

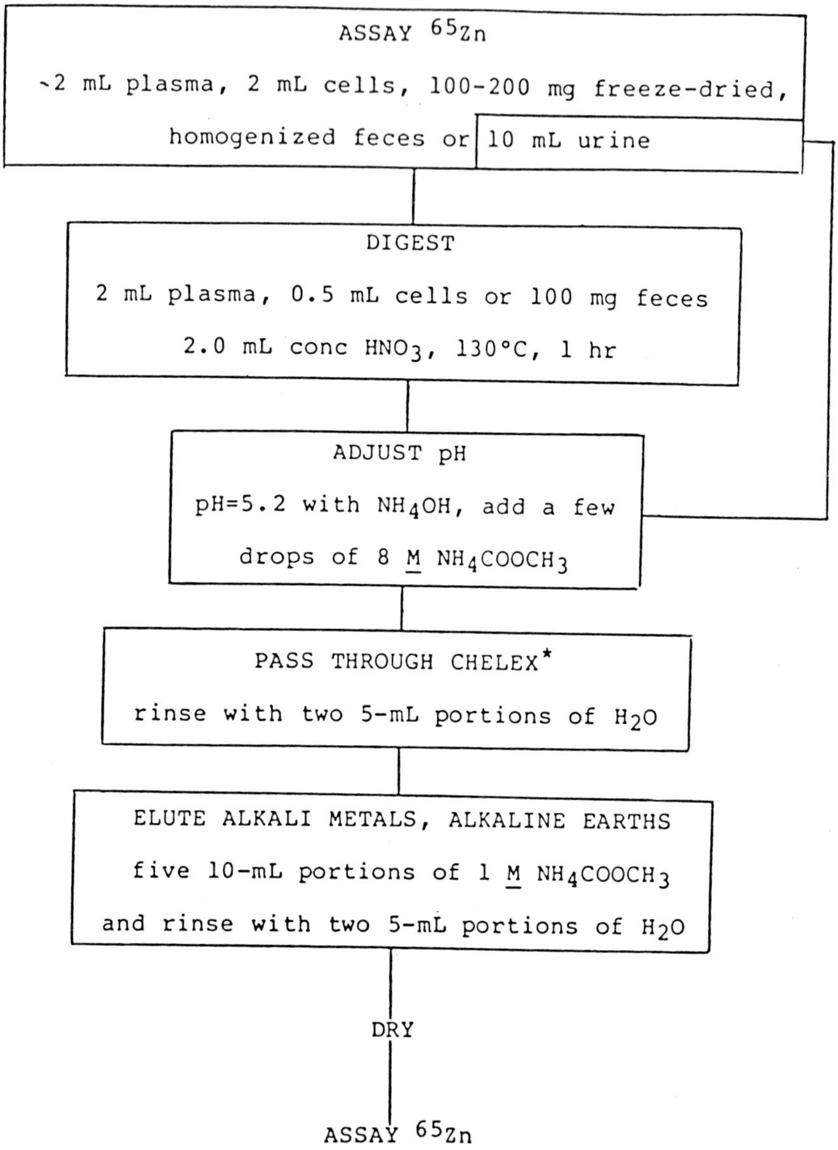

Figure 1. Chelex Separation Scheme

*400 mg Chelex cleaned with three 5-mL volumes of 2.5 M HNO_3, rinsed with two 5-mL portions of H_2O, converted to NH_4^+ form with two 5-mL portions of 2.0 M NH_4OH, and rinsed with two 5-mL portions of H_2O.

ELUTE ZINC

with three 5-mL portions of 2.5 M HNO_3

ADD CARRIERS

10 mg Mn^{++}, 1 mg Cu^{2+}, 1 mg Zn^{2+}

add ~100 mg $KBrO_3$

PRECIPITATE MnO_2

heat solution until dark brown,

centrifuge and discard precipitate

add 10 mg Mn^{2+}, 100 mg $KBrO_3$, repeat precipitation

EXTRACT Cu

with dithizone, discard dithozone

EXTRACT Zn

make remaining solution basic with conc NH_4OH,

extract Zn into dithizone phase

(until dithizone phase becomes green)

BACK-EXTRACT Zn

into 30 mL 1 M H_2SO_4, discard dithizone phase

PRECIPITATE Zn

add 10 mg Zn^{2+} carrier, 5-mL NH_4HgSCN, stir,

cool, filter the precipitate

Figure 2. Post-Irradiation Separation Scheme

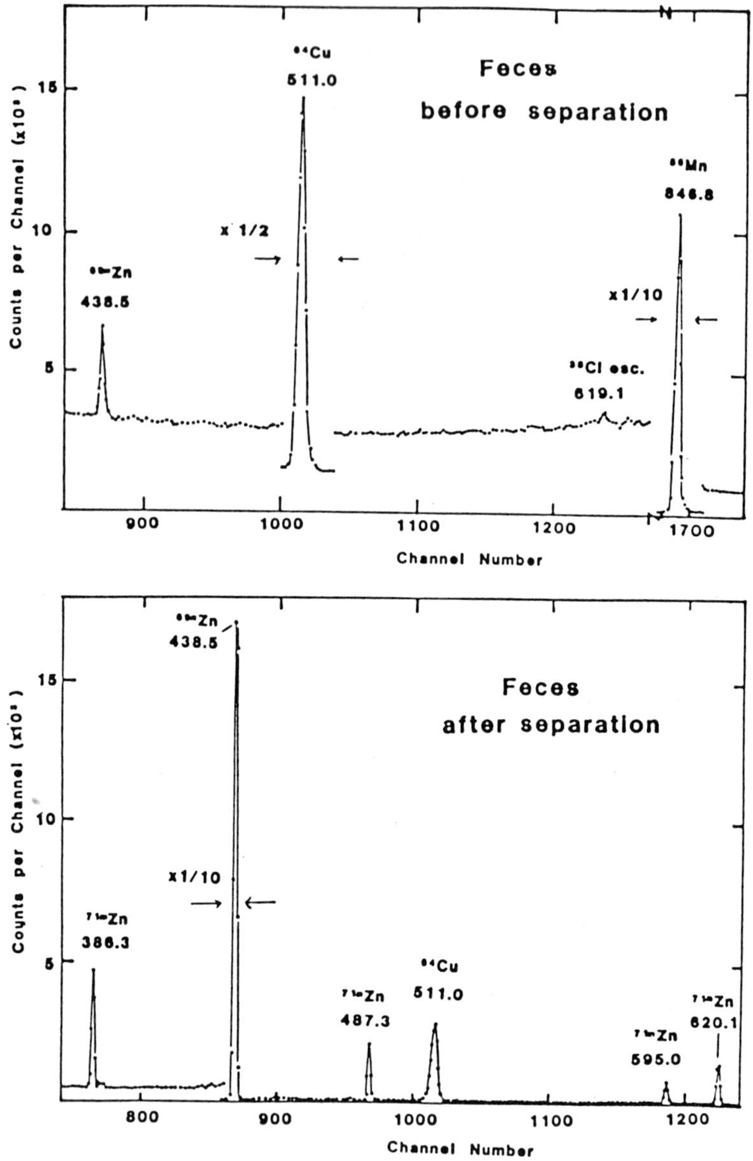

Figure 3. Portion of Gamma-Ray Spectrum of Freeze-Dried Chelex Separated Feces

a. Observed 23 min after 1-hr neutron irradiation and 2-hr cooling

b. Observed 1 hr after 1-hr neutron irradiation and post-irradiation separation.

increase the sensitivity, Gamma rays of ^{65}Zn and ^{71m}Zn at 1115, 438 and 386 keVs, respectively, were observed during this count. Portions of the gamma ray spectra of freeze dried, Chelex separated and post-irradiation separated feces sample is given on Figure 3a and 3b, respectively. Then the samples were placed at 25 cm from the detectors and a third spectra were taken. From the comparison of 438 keV γ-ray of ^{69m}Zn, obtained from the first and third spectra, post irradiation separation yields were found. The yields for pre and post-irradiation separation for 90 samples were 86±16 % and 70±13 %, respectively.

References

1. Aamodt, R.L., Rumble, W.F., Johnston, G.S., Foster, D., Henkin, R.I., Am.J.Clin. Nutr., 32, 559 (1979).
2. Aamodt, R.L., Rumble, W.F., Johnston, G.S., Markley. E.J., Henkin, R.I., Am.J.Clin.Nutr., 34, 2684 (1981).
3. Aamodt, R.L., Rumble, W.F., Babcok, A.K., Foster, D.M., Henkin, R.I., Metabolism 31, 326 (1982)
4. Babcokc, A.K., Henkin, R.I., Aamodt, R.L., Foster, D.M., Berman, M., Metabolism, 31, 335 (1982).
5. Gordon, G.E., Gökmen, İ., Aras, N.K, Henkin, R.I., J. Radional. Chem., 70, 146 (1982).
6. Janghorbani, M, Young, V.R., in Advances in Nutritional Res., H. Draper edn., Plenum Pub., New York, Vol.3, p.127 (1980).
7. Gökmen, İ.G., Ph.D. Thesis, Dept. of Chemistry, University of Maryland (1983).
8. Janghorbani, M., Young, V.R., Gramlich, J.W., Machlan, L.A., Clin. Chim. Acta 114, 163 (1981).
9. Solomons, N.W., Am.J. Clin.Nutr., 35 , 1048 (1982).
10. Lederer, C.M., Shirley, V.S., Browne, E., Dairiki, J.M., Doebler, R.E., Table of Isotopes, Wiley Interscience New York, 1978, 7 th ed.
11. Gökmen, I.G., Gordon, G.E., Aras, N.K. Fitzgerald, J., to be published in J.Clin. Nutr., (1984).
12. Zeisler, R., Langland, J.K., Harrison, S.H., Anal.Chem., 55, 2431 (1983).
13. Kingston, H.M., "Quantitative Ultratrace Trace Metal Analysis of High Salinity Waters Utilizing Chelating Resin Separations", Interagency Energy-Environment R & D Program Report, NBS, EPA, EPA-600/7-79-174 (1979).

VOLTAMMETRIC DETERMINATION OF NICKEL IN URINE AND BLOOD BY CHELATE ADSORPTION.

Bente Gammelgaard, Jan Rud Andersen
Royal Danish School of Pharmacy, Department of Chemistry AD
DK-2100 Copenhagen Ø

Asmus Poulsen
Finsensinstitutet
DK-2100 Copenhagen Ø

Introduction

The fact that nickel can cause allergy and in some of its compounds is carcinogenic has lead to the comprehension that monitoring nickel levels in biological materials are of great importance [1]. In cases of occupational exposure, health screening programs are often necessary to detect the fraction of the environmental concentration absorbed, and hence to control an eventual increased absorption in order to avoid hazardous body levels. For this reason a routine analysis method capable of handling organic matrices is desirable.

In our laboratory, we have tentatively adopted the electrochemical procedure recently described by Pihlar *et al.* [2]. These authors recommended the method due to its high sensitivity and reliability, but were less than specific about the necessary pretreatment of biological samples high in dry matter such as human blood.

Consequently, the method was subjected to a "field test" consisting in analyzing a series of blood and urine samples. They originated from a female nickel-sensitive patient and were collected during her treatment with a nickel "scavenger".

The patient, who had no history of nickel hypersensitivity when hospitalized, had a stainless steel implant placed in the leg. After some time she developed an eczema in the skin of the leg. Patch testing showed hypersensivity to nickel at this stage. The patient was treated with Antabuse (tetraethylthiuramdisulfide) which in vivo acts as a nickel chelating agent [1]. As hoped, the eczema disappeared, and the metal implant, which was suspected to have provoked the allergy, was removed. The Antabuse treatment was subsequently continued for three months. During these periods, the blood and 24-hours urine samples were collected in pre-rinsed polyethylene containers and were kept deep frozen until analyzed.

Experimental

The voltammetric method described by Pihlar et al. involves a preconcentration step of the analyte at the electrode-solution interface before the actual determination, hence the unsurpassed sensitivity [2]. Our version, equipment and instrument settings are summarized below.
After pretreatment of the sample - vide infra - dimethylglyoxime (DMG) is added to form a neutral chelate with the nickel ion - $Ni(DMG)_2$. An ammoniabuffer is used as supporting electrolyte.
By applying a negative potential of 0.7 V with stirring, the chelate is accumulated at the surface of a hanging mercury drop electrode. The solution is equilibrated and a cathodic scan is subsequently applied in the differential pulse mode. The peak due to reduction of $Ni(DMG)_2$ appears at ÷ 1.05 V (vs. Ag/AgCl). The sample is purged with nitrogen prior to analysis to eliminate oxygen.
The analytical equipment is a PAR Model 303 Standard Mercury Drop Electrode, a PAR Model 264 Polarographic Analyzer and a Kipp & Zonen Model BD 90 X-Y recorder. Instrument Settings: Purge time - 4 min., drop size - medium, pulse amplitude - 50 mV,

clock time -0,5 s, scan rate - 5 mV/s, adsorption time - 120 s, equilibration time - 30 s, buffer concentration - 0,1 M, DMG-concentration - 10^{-4}M.

All analyses were performed in a normal laboratory environment. The electrode was situated in a Laminar Air Flow (LAF)- bench and as many manipulations as possible were carried out in this. All reagens used were of suprapure quality (Merck or BDH), and water was purified with a Milli-Q^{TM} water system. Glass-ware, plastic containers, utensils, pipette tips etc. were soaked in 7 M nitric acid for several days and then rinsed in Milli-Q water.

Urine. 2 ml of thawed sample was pipetted into a 30 ml Kjeldahl flask, 0,2 ml H_2SO_4 (98%) and 1 ml HNO_3 (65%) were added, and the flask was heated on a sand bath to 160°C. After addition of 1 ml H_2O_2 (30%), the flask was heated to the point of thermal decomposition of the sulfuric acid as seen by the white fumes expelled. The sample was cooled, neutralized with NH_3 (25%) and made to a volume of 20 ml. To a 9 ml aliquot of this solution was added 1 ml ammoniabuffer (1 M), 10 µl DMG (0.1 M), and the nickel content was subsequently quantized by means of two standard additions.

Blood. Blood samples were freeze dried before decomposition. The method described above for wet ashing of urine was not sufficient for a total mineralization of blood samples. The treatment time had to be extended and larger reagent volumes were required. Due to the rougher treatment, the blanks became disproportionately high and problems with missing recovery of known added amounts arose. The latter problem, however, seemed to be overcome by adding a large excess of zinc ions to compete for the available adsorption sites, but as the blanks were still too high, the wet ashing approach was abandoned.

Dry ashing in a muffle furnace in various kinds of vessels was highly uncontrolled as manifested by high and varying blanks. Pressure decomposition with nitric acid was rather unsuccesfully attempted using a Uniseal^{TM} decomposition bomb made of stainless steel with an inner vessel of teflon. Even after extended

reaction times at high temperatures, mineralization was incomplete causing severe disturbance of the analytical signal, thus rendering this technique useless.
Preliminary experiments with Low Temperature Ashing (LTA) have shown promising results. LTA consist in combustion in a microwave induced oxygen plasma, and the high reactivity of this reagent makes complete decomposition possible at relatively low temperatures i.e. of the order of a few hundred degrees centigrade.
The equipment used was a LTA Model 302 from LFE Corporation using containers of quarts. The fact that the decomposition takes place in a closed, clean system, and that additional handling of the ashing residues can be carried out under controlled conditions in a LAF-bench makes the blank values acceptable.

Results and Discussion

The urine nickel levels and the total daily excretion are shown in Fig. 1. The concentrations varies between 4 μg/l and 30 μg/l with standard deviations of 0.5 μg/l and 3 μg/l, respectively. The blank value is 0.16 μg/l with a standard deviation of 0.04 μg/l.
As is seen from Fig. 1, the urinary nickel excretion increases abruptly shortly after the ingestion of Antabuse. Until the implant is removed, the excretion remains at the elevated level, whereupon it declines towards the initial value after removal. It is not clear whether the lowered excretion is caused by the removal of the implant or the withdrawal of Antabuse, but the latter explanation appears more likely, as the excretion before ingestion of the chelating agent is low.
Fig. 2 displays the preliminary results for the nickel contents of the blood samples, and they are in fairly good agreement with the results for urine. A similar behavior has recently been observed in another nickel sensitive patient treated with

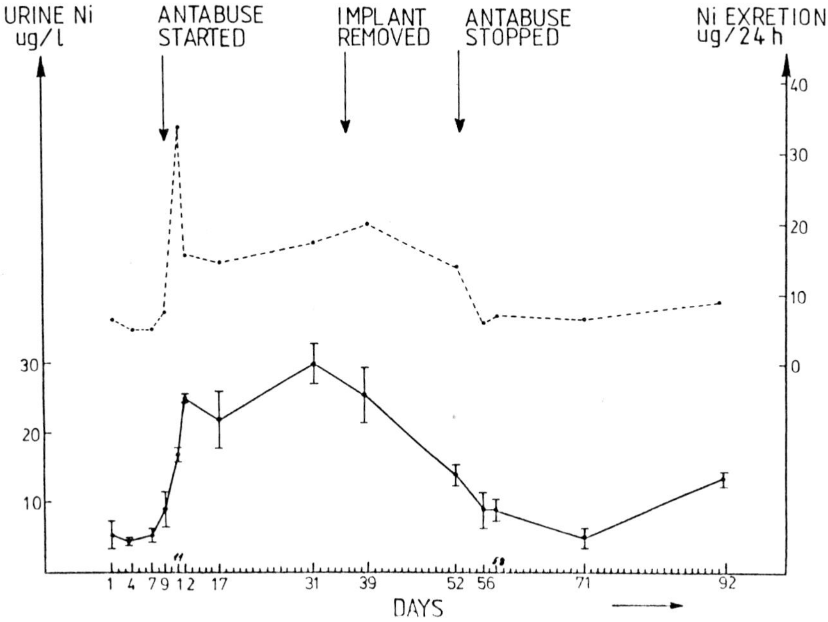

Fig. 1: Urine nickel levels from a nickel sensitive patient. During treatment with Antabuse - urinary nickel levels µg/l ---- total urinary excretion µg/24 hours.

Antabuse [3].
From our results it is impossible to demonstrate a correlation between the presence of the stainless steel implant and the body fluid levels before the administration of the Antabuse. The initial levels are not significantly higher than the levels found by us and others [1] in unexposed persons <u>i.e.</u> 1-5 µg/l. However, our results do show that once the Antabuse is administered, it acts very effectively as a nickel scavenger.

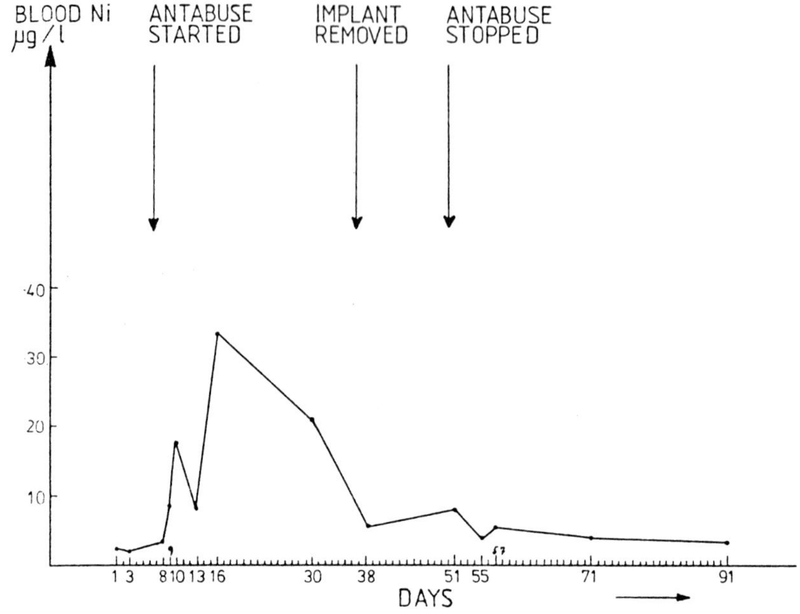

Fig. 2: Blood levels in a nickel sensitive patient during treatment with Antabuse.
Preliminary results. Each point represents a single determination.

As to the voltammetric procedure used, our experiences are that it is indeed sensitive and reliable so far as the detection is concerned. The sample pretreatment is critical in order to obtain meaningful results, and for complex biological matrices, non-trivial ashing techniques have to be resorted to.

Acknowledgement

The valuable technical assistance of mr Niels Vinther is gratefully appreciated.

References

1. Nriagu, J.O. (Ed.) : Nickel in the Environment, John Wiley & Sons, New York 1980.

2. Pihlar, B., Valenta, P., Nürnberg, H.W. : Fresenius Z. Anal. Chem. 307, 337-346 (1981).

3. Kaaber, K., Menne, T., Tjell, J.C., Veien, N. : Contact Dermatitis 5, 221-228 (1979)

SIMULTANEOUS DETERMINATION OF TRACE METALS IN VEGETABLE MATERIALS BY ALTERNATING CURRENT ANODIC STRIPPING VOLTAMMETRY AND ATOMIC ABSORPTION SPECTROSCOPY

Clinio Locatelli, Francesco Fagioli, Corrado Bighi, Luca Scanavini
Laboratory of Analytical Chemistry, Department of Chemistry
Via L. Borsari 46 44100 FERRARA (Italy)

Introduction

Some metals have been determined in vegetable materials (Standard Reference Material n. 1571, Orchard Leaves of the National Bureau of Standards) at trace levels by employing either polarographic[1-5] or spectroscopic techniques[6]; a few of these metals are essential for plant mineral nutrition (i.e. zinc, copper, iron, manganese), whereas others are likely to represent a pollution problem (i.e. cadmium and lead).

Sample preparation

The samples (Orchard Leaves, Standard Reference Material n. 1571 of the National Bureau of Standards) were prepared employing two procedures.

Method A

The samples were prepared by dry ashing partially following the method proposed by 'Official Methods of Analysis'[7].

About 0.50 g of the sample were accurately weighed into a platinum crucible, ashed two hours at 500 °C and let cool. Successively the sample ashes were wet with ten drops of H_2O and then three ml of HNO_3 (1:1) were carefully added. The excess HNO_3 was evaporated on a hot plate set at 100-120

Trace Element - Analytical Chemistry in Medicine and Biology, Vol. 3

°C. The crucible was put again on the furnace and further ashed for one hour at 500 °C and then left to cool. Finally two ml of $HClO_4$ 1M were added to the ashes and then transferred to a 50 ml volumetric flask.

Method B

Every sample, weighing about 0.25 g, was introduced into a thick-walled pirex digestion tube having a calibration mark at 25 ml with two ml of concentrated sulphuric acid. The tubes were agitated by Vortex, then introduced into a block digestor, heated at a heating rate of 3-4 °C/min up to 350 °C and maintained at this temperature for ten minutes. After cooling, the tubes were made up to the mark with deionized water. Blanks were always run in the same manner.

Polarographic technique

Very few papers have dealt with the determination of elements in vegetable material by polarographic technique[1-5].

Since the concentration of trace metals in samples of special interest (e.g. natural waters, industrial effluents, biological matters) is frequently in the μg/L-ng/L range, anodic stripping voltammetry (ASV) with stationary or semistationary mercury electrode has often been chosen for its high sensitivity.

In the present work zinc, cadmium, lead and copper are analysed by alternating current anodic stripping voltammetry (ACASV), iron and manganese by alternating current voltammetry (ACV) (fundamental harmonic).

A.C. current-potential curves were recorded with an AMEL mod. 471 multipolarograph equipped with a stirrer AMEL mod. 291/LF and a timer AMEL mod. 452/T.

A three-electrode system was used. The working electrode was a new type of semistationary mercury electrode: the long-lasting sessile-drop mercury electrode (LLSDME)[8,9] and potential was measured against an aqueous saturated calomel reference electrode (SCE) with auxiliary platinum electrode. The voltammetric cell was thermostated at 25.0±0.5 °C.

The solutions were deaerated with pure nitrogen prior to analysis and, during the experiments, a nitrogen atmosphere was maintained above the solutions at all times. During the electrolysis time, the solutions were stirred with a magnetic stirring bar. All chemicals were 'Suprapur' grade.

The best conditions for carrying out the analytical measurements were chosen on the basis of a critical analysis of a different parameters:

a) Zn, Cd, Pb, Cu		b) Fe, Mn		
E_d=	-1.15	E_i=	0 (Fe)	-1.0 (Mn)
t_e=	120 (240 for Cd)	t_d=	60 (Fe)	240 (Mn)
t_r=	30	dE/dt=	20	
dE/dt=	20	f=	100	
f=	100	ΔE=	10	
ΔE=	10	φ=	+7.8 (Fe)	+7.4 (Mn)
φ=	0			
u=	800			

where:

E_i=Initial potential (V/SCE); E_d=Electrodeposition potential (V/SCE); t_e=Electrolysis time (s); t_r=Rest time between the end of electrodeposition and the start of recording (s); t_d=Delay time before the start of recording (s); dE/dt=Potential scan rate (mV/s); f=Frequency (Hz); ΔE=Amplitude of the alternating current voltage (mV); φ=Demodulation phase angle (degree); u=Stirring rate (r.p.m.).

Once established these optimal experimental conditions, a 25 ml aliquot of sample, prepared according to method A, was pipetted into a teflon polarographic cell, diluted with 25 ml of $NaClO_4$ 0.2M and deaerated with pure nitrogen prior to analysis. As soon as the analysis of zinc, cadmium, lead and copper was accomplished, two 20 ml aliquots of sample solution were respectively diluted with 20 ml of sodium oxalate 0.2M (for the determination of iron) and 20 ml of sodium cyanide 2.0M (for the determination of manganese). In all cases, the standard addition method was used. The analytical results are reported in Table 1.

TABLE 1

Analytical results obtained by polarographic technique (Zn, Cd, Pb, Cu by ACASV; Fe, Mn by ACV)

Elements	N. of samples	$\bar{c}$ NBS (μg/g) certified data	$\bar{c}$ NBS (μg/g) experimental data	e%	s_r%
Zn	6	25±3 (*)	23.9±1.5 (*)	-4.4	2.37
Cd	6	0.11±0.01	0.12±0.01	+9.1	2.98
Pb	6	45±3	45.7±1.3	+1.6	1.14
Cu	6	12±1	12.6±0.7	+5.0	2.22
Fe	6	300±20	298.9±8.1	-0.4	1.06
Mn	6	91±4	90.9±4.7	-0.1	2.03

where e% is the relative error and s_r% is the relative standard deviation. (*) 95% confidence limit.

Spectroscopic technique

A Perkin-Elmer mod. 603 atomic absorption spectrophotometer equipped with deuterium background corrector and mod. 056 recorder was used for all measurements. Single-element Intensitron hollow cathode lamps were used for all measurements. A premix design, 10 cm, titanium, single-slot burner head was used. The fuel was acetylene and the oxidant was air. The instrument settings were those recommended by the manufacturer[10]. A scale expansion was employed in the case of copper. An integration time of two seconds was used for all the elements determined by flame atomization, i.e. zinc, lead, copper, iron and manganese. For the determination of cadmium, an HGA-76B graphite furnace with conventional graphite tubes was used.

Absorbance was read in the peak height mode.
Solutions were injected into the graphite-tube (20 μl) with an AS-1 sampler. The instrumental settings, checked through the charring/atomization curves, was the following: drying at 100 °C for 30 s (heating rate: 15 °C/s), charring at 350 °C for 30 s (heating rate: 77 °C/s), atomization at 1500 °C for 5 s. During atomization, the flow of the internal purge gas (argon, 280 ml/min) was set automatically in the mini-flow mode (60 ml/min).
A block digestor system was used for the digestion of the samples. All dilutions were performed by sampling through Gilson pipetman and Eppendorf micropipettes with disponsable plastic tips.
The atomic absorption measurements were made on the samples prepared following either method A or method B.
The results obtained through the analysis of samples prepared according to method A are reported in Table 2.

TABLE 2

Analytical results obtained by spectroscopic technique (Method A)

Elements	N. of samples	$\bar{c}$ NBS (μg/g) certified data	$\bar{c}$ NBS (μg/g) experimental data	e%	s_r%
Zn	6	25±3(*)	26.9±1.1(*)	+7.6	4.0
Cd	6	0.11±0.01	0.12±0.01	+9.1	2.9
Pb	6	45±3	45.9±1.2	+2.0	2.4
Cu	6	12±1	12.4±1.4	+3.3	5.3
Fe	6	300±20	282.3±9.4	-5.9	4.1
Mn	6	91±4	91.0±0.8	0.0	0.8

where e% is the relative error and s_r% is the relative standard deviation. (*) 95% confidence limit.

Alternatively to classical dry and wet digestion methods of vegetable samples, a new method (method B) was developed, consisting in a partial wet ashing of the samples with concentrated sulphuric acid. The carbonaceous material so obtained was diluted to 25 ml with deionized water and directly sampled as a slurry either into a flame (determination of zinc, lead, copper, iron and manganese) or into a graphite-tube (determination of cadmium) of an atomic absorption spectrophotometer.
The results reported in Table 3 are obtained by means of analytical calibration curve for the determination of zinc, lead, copper, iron and manganese, and by analyte addition technique in the case of cadmium.
Deuterium background corrector did not completely remove on the flame the absorbance value of the blank from that of carbonaceous slurry, and therefore each absorbance reading was corrected for blank background absorbance. In the case of cadmium, determined by graphite-tube furnace, the background corrector was not necessary.

TABLE 3

Analytical results obtained by spectroscopic technique (Method B)

Elements	N. of samples	$\bar{c}$ NBS (μg/g) certified data	$\bar{c}$ NBS (μg/g) experimental data	e%	s_r%
Zn	6	25±3 (*)	23.4±1.4 (*)	-6.4	8.3
Cd	6	0.11±0.01	0.12±0.01	+9.1	3.8
Pb	6	45±3	42.8±3.1	-4.9	5.9
Cu	6	12±1	12.5±0.5	+4.2	5.5
Fe	6	300±20	288.1±7.2	-4.0	3.6
Mn	6	91±4	89.4±1.4	-1.8	2.1

where e% is the relative error and s_r% is the relative standard deviation. (*) 95% confidence limit.

The comparison of the results obtained by spectroscopic technique utilizing the sampling of the carbonaceous slurry with voltammetric and spectroscopic results achieved with the samples prepared by dry ashing method reveals what follows: the accuracy and precision of spectroscopic measurements (method A and B) are quite similar, but the spectroscopic results achieved either by method A or method B are less accurate and precise than the results relative to voltammetric technique. However it seems evident that it is difficult to discriminate between voltammetric and spectroscopic technique as far as these parameters are concerned.
In fact, the analytical results (voltammetric and spectroscopic measurements) may be considered satisfactory for most practical purposes, even if the atomic absorption spectroscopy is perhaps a more rapid technique, but the polarographic technique is certainly much less expensive.

References

1) Jones G.B., Anal. Chim. Acta 7, 578 (1952).
2) Jones G.B., Anal. Chim. Acta 11, 88 (1954).
3) Sirois J.C., Analyst 87, 900 (1962).
4) Batley G.E., Florence T.M., J. Electroanal. Chem. 55, 23 (1974).
5) Wisser K., Wöhrle G., Microchim. Acta 1, 129 (1980).
6) Fagioli F., Landi S., Lucci G., Ann. Chim. (Rome) 72, 63 (1982) and their references.
7) Horwith W., A.O.A.C., Washington, USA (1981).
8) Andruzzi R., Trazza A., Talanta 28, 839 (1981).
9) Andruzzi R., Trazza A., Marrosu G., Talanta 29, 751 (1982).
10) "Analytical Methods for Atomic Absorption Spectrophotometry", The Perkin-Elmer Corp., Norwalk, Connecticut, USA (1982).

D i s c u s s i o n Session IX - 1

Subjects of the discussion

- differences in "normal values" for V in the literature
- stable isotopes studies
- sample preparation techniques for low element concentrations
- use of Standard Reference Materials for testing decomposition techniques

Contributors: Brätter, Knapp, Tölg

RECENT IMPROVEMENTS FOR NICKEL ANALYSIS IN BIOLOGICAL MATERIALS

Markus Stoeppler

Institute of Applied Physical Chemistry, Nuclear Research Centre (KFA) Juelich, P.O. Box 1913, 5170 Juelich, Federal Republic of Germany

1. Introduction

Nickel, atomic number 28, atomic weight 58,17 is the twenty-fourth most abundant element in the earth's crust. Nickel contents typically range from several mg/kg (sometimes even higher) in rocks, soils, plants and some food down to the low μg/kg level in most biological materials. The nickel content of open sea water usually is below 1 μg/l. Recent methodological progress for Voltammetry but also for GFAAS has resulted in increasingly reliable data for nickel contents in environmental and biological materials (see table 1) as well as in food (1-9).

This paper thus aims at describing the present state-of-the art of nickel analysis mainly in biological materials. Based on remarkable methodological progress and refinement achieved within the last couple of years it can be stated that even lowest nickel contents can be determined quite reliably in principle now. Thus, a progress also can be expected in the field of biomedical research on adverse and toxic effects of nickel due to an increased performance of the applied methodology. Subsequently starting with sampling and sample treatment the most promising methodological approaches will be described and their advantages and disadvantages compared so that a clear picture can be obtained of the present analytical potential.

Trace Element - Analytical Chemistry in Medicine and Biology, Vol. 3

Table 1

Averages and ranges for normal nickel contents in biological specimen types from recent studies (values in µg/kg F.W.)

Specimen	Average	Range	Remarks
Serum	<0.5	0.1 - 1.3	
Whole Blood	<1.5	0.6 - 1.8	
Urine	2	<1 - 4.0	µg/g Creatinine
Saliva		0.5 - 7	
liver, human	10.5	3 - 40	n = 15
liver, bovine		0.4 - 16	
kidney, bovine		2.5 - 13	limited numbers
meat, bovine		1.5 - 13	
nails (foot)		200 - 300	limited study
nails (hand)		700 -3000	contamination?

Sources: 6,8,10,11,12; bovine tissue included for comparison

Note: Nickel contents in body fluids of exposed persons (occupation, implants) are up to a factor of 50 higher

2. Importance of careful sampling and sample storage

Due to various nickel sources in the general and the laboratory environment careful sampling procedures are necessary to arrive at meaningful data. During all steps discussed here, rigid contamination precautions are mandatory. Thus the persons doing sampling and subsequently necessary manipulations play the most important role. They should be fully aware of all difficulties and should be skilled in this type of work and also have sufficient time, patience and motivation to perform all duties at this stage with utmost care. For instance, all analytical manipulations should be performed by wearing plastic gloves to avoid nickel contamination from sweet on the fingers of the staff members.

Collection of blood and blood serum by venepuncture for normal levels first requires a careful cleaning of the antecubital fossa of the arm with ethanol and drying by evaporation. A tourniquet is applied subsequently and a polyethylene cannula is inserted into an antecubital vein since stainless steel needles usually contaminate the blood by variable nickel amounts depending on the type and manufacturer of the needles applied (2,8). The stylus of the cannula is removed, and the cannula is flushed with 3 ml of blood, which is discarded. A polypropylene syringe is then used to collect the required volume (usually 5 - 10 ml) of blood (8). The bottles for urine collection should be acid-washed, i.e. first heated with warm (70 ^{o}C), dilute acids, then allowed to stand for several hours at ambient temperature, rinsed with ultrapure water and stored like other laboratory ware under contamination minimized conditions. Collection of urine has to be performed after having taken a shower or a bath in a rather clean environment and by no means at the working place.

If hair, nails, bones and tissues have to be cutted or dissected all instruments made from steel alloys have to be avoided and replaced by those from titanium, quartz or plastic. Cutting e.g. of finger and toe nails with stainless steel scissors can introduce mg/kg-amounts of nickel. On the other hand there is strong evidence that finger nails, predominantly of particular fingers, commonly are contaminated and thus do not reflect true endogenous levels which appears not to be the case for toe nails (6).

In general, an effective laboratory organization for nickel analysis should be based besides minimization of contamination from the staff by wearing plastic gloves and the prohibition of cigarette smoking and dust-free or dust-minimized conditions. The most successful approach is a filter system for the air inflow and the additional use of laminar-flow working places for all steps which may introduce contamination. Only these means can guarantee an adequate analytical performance if extremely sensitive determination methods are applied. The latter should be used continuously for blank control of all laboratory ware, water and chemicals needed (8,13,14). For medium and long-term storage cooling is recommendable also to minimize contamination and losses (15). For urine an acidification, preferably with hydrochloric acid, to approximately pH 2 has proven to be of advantage to minimize losses by adsorption on particulate matter (14).

3. General Analytical Aspects

The classical approach, due to severe matrix interferences observed in former - mainly AAS - methods commonly started with a more or less complete decomposition. This was followed by separation from the interfering matrix by solvent extraction. Because of limited detection power of the used instruments, the last step mainly included preconcentration as well (2,3,7). Despite carefully evaluated - also reference - methods (2,10,16) in routine application frequently contamination could not be controlled as required. Thus reliable analysis at the µg/l-level was comparatively rare (5).

In the meantime, i.e. in the last couple of years, however, more reliable methodological concepts have been developed and successfully applied. Moreover, a significant increase in detection power, particularly for voltammetry and compensation ability by introduction of Zeeman compensation instrumentation for atomic absorption spectroscopy is to note. This offers at present the potential for faster and more reliable nickel determinations down to the lowest levels in all appropriate biological specimen types and shall be addressed for Voltammetry, conventional and Zeeman-graphite furnace atomic absorption spectroscopy (GFAAS and ZGFAAS) and by description of a new radiochemical Neutron Activation method.

Finally efforts will be discussed to improve the situation for internal and external quality control. In table 2 the at present attainable determination limits for Voltammetry, ZGFAAS and Radiochemical Neutron Activation Analysis are given for comparison.

Table 2

Relative determination limits in μg/l or μg/kg for Voltammetry (Adsorption Pulse Voltammetry, ADPV) Zeeman Graphite Furnace Atomic Absorption Spectroscopy (ZGFAAS) and Radiochemical Neutron Activation Analysis (RNAA). The given concentrations commonly can be quantified with a relative error of approx. 30 %

Method	determination limit	remarks
1) ADPV	$\leq$ 0.002	non interfering analyte solution
	0.05 - 0.1	blood, urine, tissue after digestion
2) ZGFAAS	$\leq$ 0.2	non interfering analyte solution
	$\leq$ 0.5	directly in urine 20 μl dosage
	$\leq$ 1.0	1:1 diluted whole blood, 20 μl dosage
3) GFAAS	$\leq$ 1.0	analyte solution
3) RNAA	$\leq$10	sample amount 1g, ^{65}Ni

Notes:

1) The sample volume for ADPV always is $\geq$5 ml in the polarographic cell, determination limit in biological materials depends on blank levels of digestion

2) These values are based on an injected volume of 50 μl for analyte solutions

3) Due to the relatively low flux of 5 x 10^{11} n.cm^{-2}.s^{-1} for 3 h in the used SLOWPOKE reactor (17) it can be expected that under more favourable irradiation conditions a significantly lower determination limit well below 1 μg/kg may be achieved.

4. Voltammetry

Polarographic and voltammetric methods are of increasing significance for the determination of various ecotoxic trace metals. They generally combine due to their foundation on Faraday's law (1 mole of substance is equivalent to the enormous charge of n . 96.000 Coulomb) a high sensitivity with good precision and accuracy (18). For nickel and cobalt it has been observed that addition of dimethylglyoxime to the electrolyte significantly enhances the detection power (3,4, 19,20). In this case preconcentration is not performed by cathodic deposition of metal amalgams as is common in the normal stripping technique. The formed metal dimethylglyoximates, however, are adsorbed at the surface of the electrode at -0.7 V, before the potential is scanned from -0.7 to -1.2 V (versus a saturated Ag/AgCl electrode) the potential range of reduction of the nickel complex. At pH 9.2 and if all working adjustments have been carefully made, a determination limit of approximately 50 ng nickel per litre analyte solution could be easily achieved, if the differential pulse mode was applied (6). Meanwhile, however, progress both in analog and digital electronics have made accessible the square wave mode for commercial instruments. This approach makes analysis approximately 40 times more sensitive and 2 times faster compared to the beforementioned approach (see figure 1). In this case a determination limit (instrument: PAR 364 B) around 2 ng/l is attainable.

The determination limits in biological materials, however, are significantly poorer. This is due to the fact that the materials cannot be analyzed directly by voltammetry. Hence a complete wet digestion has to be performed prior to analysis. Commonly this is achieved by a $HNO_3/HClO_4$ digestion in quartz vessels, shown schematically together with the preparatory step for voltammetric determination in Figure 2.

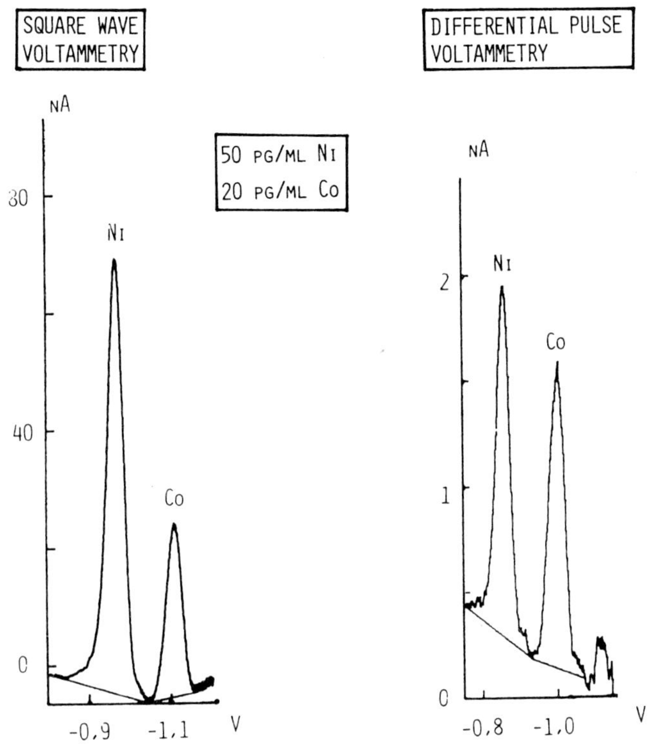

Figure 1 Voltammogrammes for nickel and cobalt - same concentration by square wave and differential pulse voltammetry

Since typical but very constant, nickel blanks are at the order of 0.1 ng/ml digestion solution due to the in-situ enrichment without any further contamination - GFAAS would in this case require an additional extraction or preconcentration step with possible additional contamination - under properly maintained conditions this is equivalent to a determination limit of 0.05 - 0.1 µg/l (µg/kg) in biological materials.

Obtainable precision ≥1 µg/l (µg/kg) commonly is ≤5 %. It has to be mentioned that, due to significantly lower blanks for cobalt (0.01 ng/ml digestion solution), the determination limit for this metal is ≤0.01 µg/l (µg/kg) and precision ≤3 %. Applying this procedure, very precise and accurate values for nickel and cobalt in biological specimens could be achieved (see Table 1). A further improvement for both elements by especially purified reagents and extremely clean fume hoods appears to be within realization.

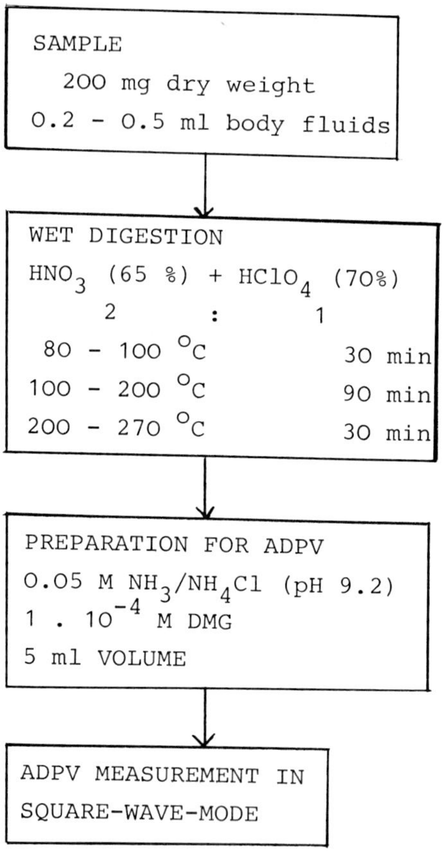

Figure 2 Scheme for digestion and sample preparation for ADPV determination of nickel and cobalt

5. Atomic Absorption Spectrophotometry

During the last couple of years the analytical potential of graphite furnace atomic absorption spectrophotometry (GFAAS) for nickel determination has been considerably improved. This is due to the commercial introduction of more sophisticated and powerful atomisers allowing rapid temperature raising during atomization, and completely new instrumental approaches as well. Together with pyrolytically coated graphite tubes of excellent performance absolute detection limits below $1 \cdot 10^{-11}$ g and analytical tube life times up to 400 atomizations (firings) can now be reached even in daily routine (23,24). A severe limitation of all hitherto used instruments, the poor ability to compensate for nonspecific background signals, could be terminated by the commercial introduction of the so-called Zeeman compensation. This principle uses the splitting of resonance lines under the influence of a strong magnetic field, hence enabling a very effective background compensation using a single element (hollow cathode or electrodeless discharge) lamp (24). This provides, in conjunction with matrix or element modification, a successful application of direct analysis in body fluids as well as solid sampling for tissues with somewhat higher contents (22,24). Another in general very promising improvement in GFAAS, the temperature stabilized (L'vov)platform, usually made from pyrocarbon, advantageous for elements with lower boiling points, however, appears not to be very promising for the analysis of nickel in body fluids and solutions from various digestion techniques (22). There are at present available two different types of instrumentation: Zeeman splitting at the atomizer (inverse Zeeman compensation) and splitting at the light source (direct Zeeman compensation) (24). Both systems have advantages and disadvantages. The direct method, for instance, is less sensitive for the instruments available at present but allows the introduction of solid samples up to

10 mg into the graphite furnace deposited on a platform made from regular graphite (24). This doubtless is a valuable tool for the fast characterization of materials with somewhat higher nickel levels, e.g. control and environmental materials, tissue and body fluids from persons which are occupationally exposed to nickel and nickel compounds.

Since all these improvements and new concepts allow a more sensitive and more reliable nickel determination in biological materials, subsequently a few examples will be given starting with whole blood and serum.

Since normal nickel contents in whole blood and serum are rather low, formerly used analytical procedures always required a digestion first and subsequently an extraction/preconcentration step with frequently less reliable results.

Just recently a ZGFAAS method has been proposed for the extremely sensitive determination of nickel in blood serum and whole blood after a simple matrix-modification/deproteinisation procedure and minimal dilution (8).

To 1 ml serum samples in centrifuge tubes during agitation (vortex mixer) 50 μl of concentrated nitric acid are added. Mixing is continued for 1 min after acidification. The tubes are stoppered and placed in a water bath at 70 °C for 5 min. The tubes are mixed for 10 s with a vortex mixer, and centrifuged at 900 g for 10 min. The protein-free supernatants are transferred into sample cups and capped until ZGFAAS determination. For whole blood a somewhat different procedure is proposed. 1 ml samples of heparinized blood are placed in centrifuge tubes. Ultrapure water (0.9 ml) is added to each tube. While the contents are mixed with a vortex mixer, 200 μl concentrated nitric acid is added directly into the sample. Deproteinization is performed as described for serum. The protein-free extracts are decanted and recentrifuged, if necessary to remove flecks of precipitated protein.

The heating step is necessary to free completely all nickel bound to blood cells and proteines.

Calibration is performed against aqueous standard solutions by peak area evaluation. The detection limits reported are 0.05 µg/l for serum and 0.1 µg/l for whole blood. Day-to-day precision is 8 % for an average concentration of 3.5 µg/l. With this method nickel contents in whole blood and urine could be found which very well agreed with values obtained with ADPV (see Table 1).

In Table 3 the instrumental settings are given for this method. Possibly a further improvement can be achieved if additionally a matrix modifier like $Mg(NO_3)_2$ is added: see next procedure.

Table 3

Instrumental settings for nickel in serum and whole blood after deproteinization with HNO_3 and heating, PE 5000 Z with HGA 500 and pyrotubes, sample volume 50 µl, wavelength 232.0 nm, spectral band-width 0.2 nm

Step	Temp (°C)	ramp-time (s)	hold-time (s)	purge gas (ml/min)
drying	100	1	1	300
	140	60	10	300
	190	30	5	300
charring	1200	80	45	300
	1200		5	30
atomization	2600	0	5	30
cleaning	2700	1	3	300

Notes: The atomization temperature of 2600 °C is calibrated each day by an optical pyrometer. The spectrophotometer is programmed to integrate the signal during the 5 s of atomization.

For higher contents in whole blood in occupational exposure we have developed recently a direct ZGAAS method without any treatment prone to contamination (25). Procedure: 1 Vol. whole blood is mixed with 0.5 Vol. 0.5 % Triton-X-100 solution and 0.5 Vol. of a 2 % $Mg(NO_3)_2$-solution. All reagent solutions are continuously controlled and nickel blanks minimized. Since a carbon black formation can be observed after several sample cups have been taken by the instrument, every 5th cup should contain solely 1 M HNO_3 to clean the tubes. This is sufficient to avoid any carbon black formation. The method reaches a determination limit around 1 µg/l and a day-to-day reproducibility $\leq$10 % at a moderately elevated level of 4 µg/l if a P.E. Zeeman 5000 is used. With the Hitachi-Zeeman 180/80 the same reproducibility can be reached at a nickel concentration of 8 µg/l. The instrumental settings are shown in Table 4 below; evaluation is performed by a matrix matched calibration graph. A typical print of the data system 3600 is shown in Figure 3 for approx 4 µg/l.

Table 4

Instrumental settings for nickel in whole blood after 1 + 1 dilution with Triton-X-100 and matrix modifier solution, PE 5000 Z, HGA 500 and pyrotubes, sample volume 20 µl; other settings see Table 3

Step	Temp (°C)	ramp-time (s)	hold-time	purge gas (ml/min)
drying	90	10	10	300
	150	20	10	300
charring	1300	10	20	300
atomization	2700	0	20	gas stop
cleaning	2800	1	1	300

The determination of nickel in urine can be performed either by solvent extraction from the matrix (26) or by direct injection of acidified urine (25). The advantage of the solvent

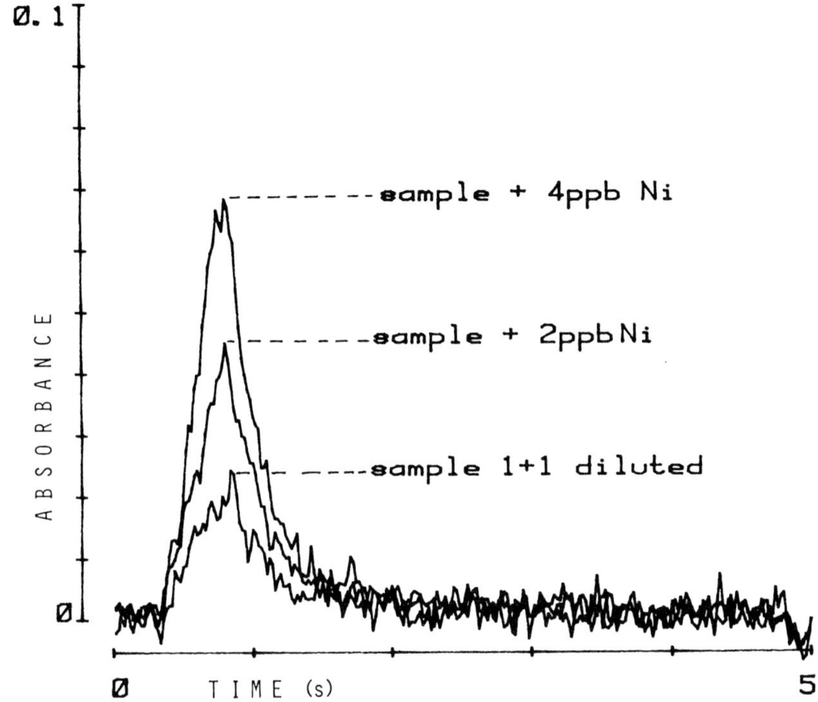

Figure 3 Nickel in whole blood, direct analysis with Perkin-Elmer 5000 Z and HGA 500; sample volume 20 µl; nickel content approx. 4 µg/l, P.E. data system 3600

extraction method is that due to the separation from the matrix no particular instrument is required, hence ZGFAAS is not necessary.

First the procedure of the extraction method shall be described. Nickel is complexed by addition of 0.5 ml of hexamethyleneammonium-hexamethylenedithiocarbamidate in Xylene-diisopropylketone to 4 ml urine and vigorous mixing. After centrifuging the determination of nickel in the organic layer is performed by GFAAS. Evaluation is performed against an aqueous standard solution treated in the same manner. The method attains, based on 1:8 enrichment a determination limit of 0.2 µg/l. Day-to-day reproducibility is 13 % at

3.5 µg/l and 5.1 % at 70 µg/l. The instrumental settings are given in Table 5.

Table 5

Instrumental settings for nickel in urine after 1:8 enrichment by HMA-HMDC-extraction, PE 4000 or 5000 with deuterium compensation; sample volume 50 µl

Step	Temp (^{O}C)	ramp-time (s)	hold-time (s)	purge gas (ml/min)
drying	110	10	15	300
charring	1000	20	25	300
atomization	2700	0	10	30
cleaning	2700	1	2	300

Zeeman GFAAS meanwhile also renders possible the direct injection of practically undiluted urine samples. This has been investigated together with the direct determination of nickel in whole blood. The method is rather simple: To 0.9 ml of urine 10 µl of 65% nitric acid and 90 µl H_2O is added. With 20 µl injections - instrumental settings are the same as already given in Table 4 - the determination limit is $\leqslant$0.5 µg/l with PE 5000 Z and HGA 500. Day-to-day reproducibility is <10 % at 2 µg/l (25). The method requires standard addition and is sufficient to distinguish between exposed and nonexposed persons. Compared to the P.E. 5000 the Hitachi 180/80 attains a determination limit of $\leqslant$1 µg/l and a 10 % day-to-day reproducibility at 4 µg/l (25). A typical print of the PE data system 3600 is shown in Figure 4. For solid materials with somewhat higher nickel contents solid sampling using the direct Zeeman effect can be successfully applied too (24).

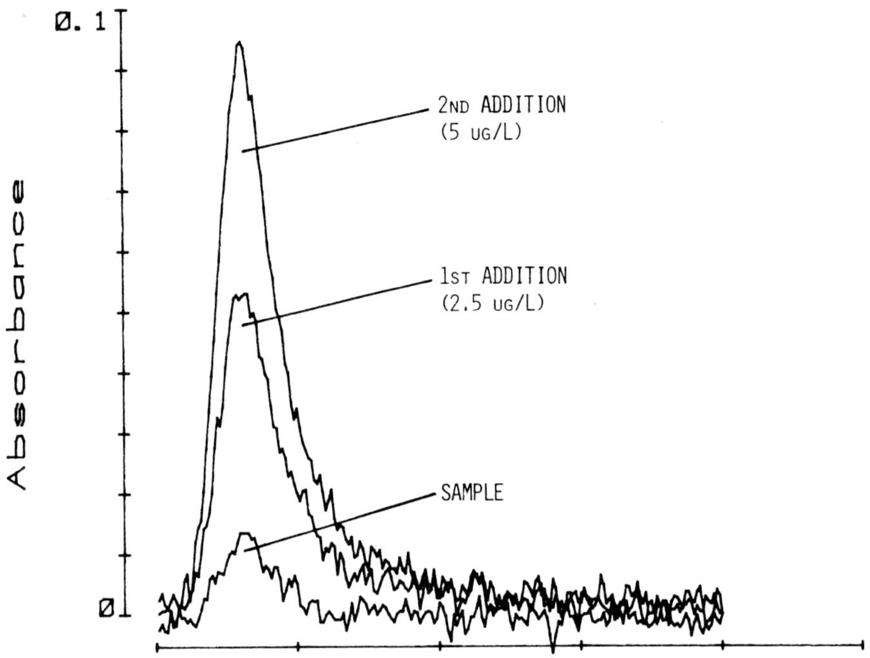

Figure 4 Nickel in urine, direct analysis with Perkin-Elmer 5000 Z and HGA 500, sample volume 20 μl, nickel content approx. 0.7 μg/L. P.E. data system 3600

6. Neutron activation analysis

Recently an improved method for radiochemical neutron activation analysis of nickel has been developed. Environmental and biological material is dry ashed, followed by HCl/HF treatment under addition of H_2O_2. After adjustment to approximately pH 9 nickel is preconcentrated by coprecipitation with α-benzildioxime. The precipitate is irradiated using a SLOWPOKE reactor at a thermal neutron flux of $5.10^{11} n.cm^{-2}.s^{-1}$ and subsequently dissolved in a mixture of 2 ml dimethylsulfoxide and 8 ml of chloroform. To avoid adsorption losses of the formed ^{65}Ni ($t_{1/2}$ = 2.52 h, mainly β-radiation) 0.1 ml of an aqueous solution containing 0.1 mg of inactive nickel carrier is added. Since during initial precipitation copper, cobalt, and manganese partially coprecipitate and trace amounts of chlorine and bromine are adsorbed onto the surface of the precipitate, these interferences are removed by washing the above described DMSO-chloroform solution with an aqueous solution of citric acid containing inactive hold-back-carriers of bromide, chloride, copper(II), manganese(II) and sodium ions. Finally nickel is stripped into the aqueous phase with 2 M hydrochloric acid. A new liquid scintillator, based on a Triton-X-100/ethanol mixture with high counting efficiency and which can tolerate large volumes of water is used for quantification (17).

Standard conditions: irradiation time 3 h, decay time 60 min, counting time 30 min. Under these conditions contents as low as 5 ng/g could be detected. An imprecision of less than 10 % was found for 100 ng/g. The method is probably useful for comparative analysis if more favourable irradiation conditions can be used.

7. Quality control

Quality control can be performed either by the use of physically or chemically different methods or appropriate control and reference materials. Therefore the prospects due to the availability of recently developed more reliable and more sensitive procedures can be regarded as promising for method's comparison.

However, despite some progress in biological reference materials - IUPAC now offers, besides already commercially available artifical urines with assigned nickel concentrations a new urine control material based on pooled natural urine (27) - the production and certification of appropriate materials for blood, blood serum and tissue still is a challenge. In general, there are now available in limited amounts reference materials of environmental origin - mainly plants - with known nickel contents (28).

8. Conclusions and future prospects

Recent progress in methodology has led to various promising methods also for normal nickel levels. It can be expected from very recent improvements in detection power that optical emission spectroscopy with plasma excitation additionally will be able to contribute significantly to the analysis of nickel in biological materials. This can be utilized in the near future to accomplish our knowledge on nickel distribution in hitherto not well characterized organs and secreta and to obtain more certain data e.g. for whole blood and serum. Since the newly developed methods are physically and chemically satisfactorily different also a stimulation for the production of more satisfactory biological and environmental reference materials - which are urgently required - can be forecasted. Doubtless, IUPAC actions in this field will play the most important role due to the efforts of the very active Nickel Subcommittee of the Clinical Chemistry Division of this international organization (15,16).

A further challenge for future research and method development doubtless is speciation of nickel - a growing field in which every activity should be encouraged in order to obtain a better insight into biochemical reactions. The progress in analytical potential certainly will constitute a good basis for particular studies in the near future, probably also in conjunction with environmental specimen banking and monitoring related to banking (13).

From the matter treated here, it is evident that a difficult task like trace and ultratrace analysis of nickel can be handled successfully on the long run only by a motivated and highly skilled staff together with modern equipment. Thus, training by further internal and external intercomparisons and organization of particular training programmes is highly necessary for a steady progress towards a better insight into facts and processes of metal biochemistry.

9. References

1. Bruland, K.W. and Franks, R.P.: Anal. Chim. Acta, 105, 233-245 (1979)

2. Sundermann, F.W., Jr.: Pure Appl. Chem. 52, 527-544 (1980)

3. Stoeppler, M., Analysis of nickel in biological materials and natural waters, in: Nriagu, J.O., ed., Nickel in the Environment, pp. 661-821, Wiley and Sons Inc (1980)

4. Pihlar, B., Valenta, P. and Nürnberg, H.W.: Fresenius Z. Anal. Chem. 307, 337-346 (1981)

5. Schaller, K.-H., Stoeppler, M. and Raithel, H.J.: Staub-Reinhalt Luft, 42, 137-140 (1982)

6. Ostapczuk, P., Valenta, P., Stoeppler, M. and Nürnberg, H.W., Voltammetric determination of nickel and cobalt in body fluids and other biological materials in: Brown S.S. and Savory, J., eds., Proc. IUPAC 2nd Int. Conf. Clin. Chem., Chem. Tox. of Metals, pp. 61-64, Academic Press, 1983

7. Stoeppler, M., Analytical chemistry of nickel in: Proc. WHO workshop, Nickel in the Human Environment, in press (1984)

8. Sunderman, F.W., Crisostomo, C., Reid, M.C., Hopfer, S.M. and Nomoto, S.: Ann. Clin. Lab. Sci, in press (1984)

9. Ostapczuk, P., Valenta, P., Nürnberg, H.W.: Ztschr. Lebensm. Unters. Forsch, in preparation

10. Ader, D. and Stoeppler, M.: J. Anal. Toxicol., 1, 252-260 (1977)

11. Schaller, K.H. and Zober, A.: Ärztl. Lab. 28, 209-214 (1982)

12. Narres, H.D., Valenta, P., Nürnberg, H.W.: Fresenius Z. Anal. Chem. 317, 484-485 (1984)

13. Stoeppler, M., Dürbeck, H.W., Nürnberg, H.W.: Talanta 29, 963-972 (1982)

14. Stoeppler, M., Analytical Aspects of Sample Collection, Sample Storage and Sample treatment in: Brätter, P., Schramel, P., eds., Trace Element - Analytical Chemistry in Medicine and Biologie, Vol. 2, pp. 909-928, Walter de Gruyter, Berlin-New York, 1983

15. Nieboer, E., Jusys, A., Cecutti, A.G., Collection Storage and Shipment of Blood-derived Fluids and Urine, IUPAC Nickel Subcommittee, to be published in Pure Appl. Chem.

16. Brown, S.S., Nomoto, S., Stoeppler, M and Sundermann, F.W., Jr.: Pure Appl Chem. 53, 773-781 (1981)

17. Bem, H., Holzbecher, J. and Ryan, D.E.: Anal. Chim. Acta 152, 247-255 (1983)

18. Nürnberg, H.W.: Pure Appl. Chem. 54, 853-876 (1982)

19. Flora, C.J. and Nieboer, E.: Anal. Chem. 52, 1013-1020 (1980)

20. Weinzierl, J. und Umland, F.: Fresenius Z. Anal. Chem. 312, 608-610 (1982)

21. Ostapczuk, P., Stoeppler, M., Nürnberg, H.W. in: Proc. 3rd Int. Conf. on Nickel Metabolism and Toxicology, in preparation

22. Stoeppler, M.: Spectrochim. Acta, 38B, 1539-1568 (1983)

23. Welz, B., Atomabsorptionsspektrometrie, 3. Auflage, Verlag Chemie, Weinheim (1983)

24. Kurfürst, U.: Fresenius Z. Anal. Chem. 313, 97-102 (1982)

25. Bagschik, U. and Stoeppler, M., in preparation

26. Angerer, J., Schaller, K.H., Fleischer, M., Nickel, Bestimmung in Harn, in: Analytische Methoden Band 2, Senatskommission zur Prüfung gesundheitsschädlicher Arbeitsstoffe der Deutschen Forschungsgemeinschaft - Arbeitsgruppe "Analytische Chemie", 7. Lieferung, 1983

27. Thomassen, Y., personal communication, 1984

28. Parr, R.M.: IAEA/RL/103, September 1983

D i s c u s s i o n Session IX - 2
Plenary Lecture

Subjects of the discussion

- detection limits for neutron activation analysis
- Ni concentration in blood in case of Ni exposure
- contamination problems during sampling

Contributors: Brätter, Schramel

Session X

STANDARD REFERENCE MATERIALS IN THE BIOMEDICAL FIELD

Plenary Lecture given by: H. Muntau (Italy)

Chairman:

P. Schramel (FRG)

Poster Reports given by: P. Schramel (FRG)

NEWER ESSENTIAL TRACE ELEMENTS AND ANALYTICAL RELIABILITY

H. Muntau
Joint Research Centre, Commission of the European Communities
I-21020 Ispra, Prov. Varese, Italy

Introduction

According to a recent statement of MERTZ [1] trace elements, including all of the naturally occuring elements expect the so-called bulk elements which make up the bulk of living matter such as carbon, hydrogen, oxygen, nitrogen, sulphur, sodium, phosphorus, potassium and chlorine, are best classified into two categories, those "whose essentiality has been established by accepted scientific standards and those for which proof of essentiality does not exist". The same author, at the occasion of the first International Workshop on Trace Metals in Medicine and Biology, defined a non-essential element simply as "an elements whose essentiality has not yet been proven".

As a matter of fact, an impressingly high number of new essential elements, namely Mo, Se, Cr, Ni, V, Si and As has been added to the list during the past three decades as definitively accepted and a number of further elements is subject to open controversies amongst researchers, and lithium, aluminium, cadmium and tin, to name a few [2], are good examples of these trace elements which have at least one common feature - they are problem cases, analytically seen, due either to their ubiquitous nature or the rather poorly developed analytical methods used for their determination.

Certainly it is not the aim of this paper to enter the discussion on trace element essentiality testing; nevertheless, a brief review of essentiality criteria will show to which extent analytical reliability influences

Trace Element - Analytical Chemistry in Medicine and Biology, Vol. 3

results and conclusions of such experiments.
The most recent set of essentiality criteria available today are those given by Henning presented at 4. Spurenelement-Symposium, Jena 1983 [3].

1. Disturbed growth and disturbed reproduction (transport via placenta).
2. Identification of the element as component of an enzyme, hormone or other biologically active substance.
3. Evident changes in enzyme or hormone status of immunoactivity.
4. Pathological changes.
5. Removal of reversible deficiency symptoms by supplementation with the element.
6. Dose-dependent effects.
7. Determination of status under pair-fed conditions.

Taking into account the aforementioned criteria, it is quite probable that a large number of controversies on essentiality of trace elements is bound to analytical problems and despite of creating the ultra-clean environment for essentiality testing of trace elements the analytical control of such environments, diets, beverages and air included as well as the "emission"-products, such as urin, feces and sweat, presents a considerable challenge even for the case of the much more abundant classical trace elements.

Figure 1 shows a schematic trace metal balance model essentiality testing.
A given organism A set in an ultra-clean environment B is fed under controlled conditions.

Parts of the trace element under test, introduced into the ultra-clean environment along with air, water and diet will escape to direct feeding and circulate along with some endigenous trace element originating from the inner surfaces of the test environment and some residues from incomplete removal of excreted trace elements (breath, sweat, etc.), before leaving the system or reaching again the test organism.

The total metal intake of the test organism is distributed throughout the soft and hard tissues and the body liquids and, depending upon nature of the element and organism, stored and/or released with excretions.

If, in ultimate analysis, the test organism would be a human, its consumption would be about 30 000 litres of air, 2 litres of water and between 0.1 and 1 kg of dry diet per day, the excreted mass depending upon its composition.

Taking further into account, that 50 µg/day were set as maximum recommended safe and adequate dietary intake for some elements such as chromium and selenium [1] and assuming that at least 50 % of safety margin were allowed setting the estimated critical threshold limit for full health requirements, a maximum total immission of 25 µg/day of the trace element into the ultra-clean environment, distributed between air, water and diet, must be controlled.

In the light of the results of numerous analytical interlaboratory exercises on trace elements in which the major European trace analysis laboratories took part [4,5,6], the analytical control of such an experiment seems a rather desperate venture, even for the more abundant trace elements and this impression is endorsed by the fact, that in the majority of published papers on essentiality tests, the analytical details are rather poorly described and very often even ignored.

Systematic measurement quality control is the exception rather than the rule and as a consequence the existence of controversies between different research groups can be attributed to some extent also to the different analytical

Fig. 1

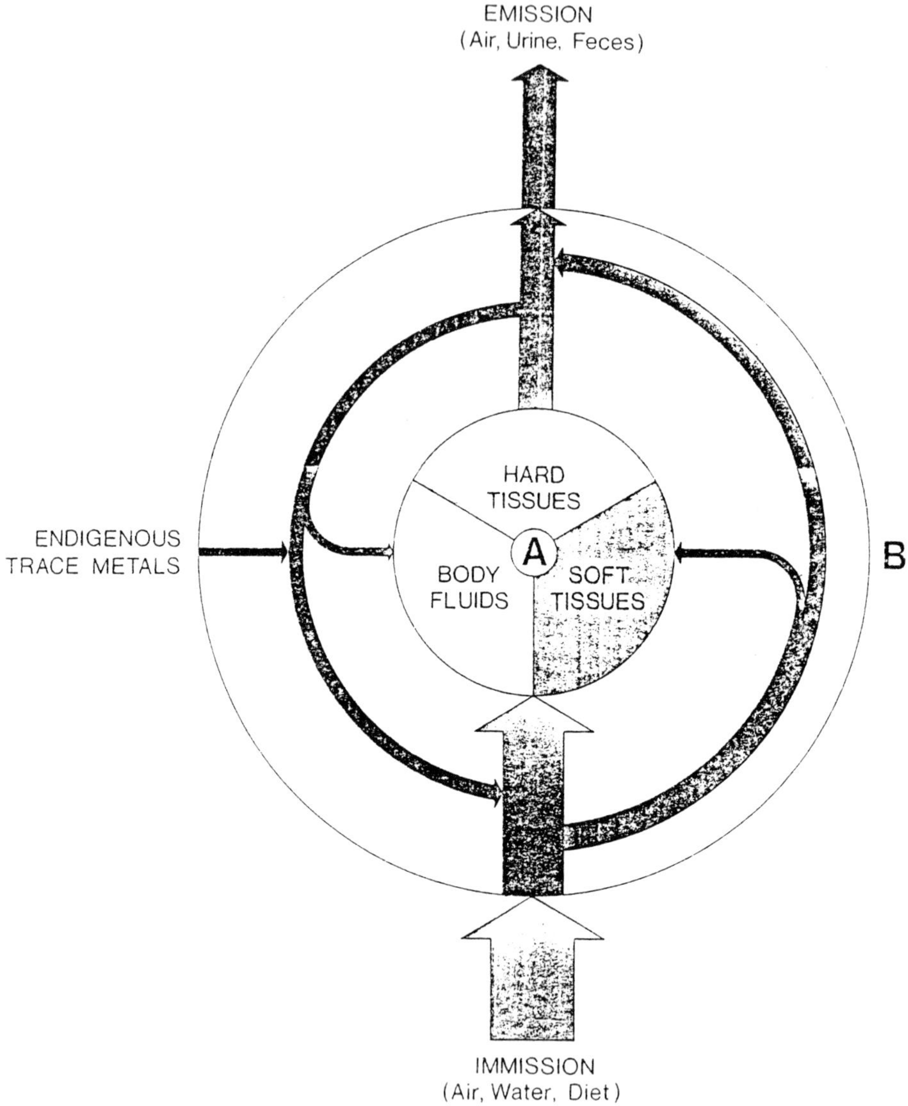

Figure 1. TRACE METAL BALANCE MODEL IN ESSENTIALITY TESTING.

systems used and the different systematic measurement errors encountered.

Three different analytical problem groups can be identified in such an experimental system:

- At the so-called zero-level, which represents in practice extremely low concentrations at or near to the detection limits of the methods used, systematic errors are particularly high and adequate reliability seems only obtainable, if at all, by relative measurements with standards.
- Low to extremely low concentrations are encountered in the immission control of water, diet and air of the pair-fed experiments and the emitted excretions.
- The analytical control of the test organism itself requires high reliability measurements usually at low concentration levels and on extremely small samples, such as biopsy material, for example.

Summarizing we conclude that, with all probability, numerous open controversies on presumed trace element essentiality have their roots in the analytical aspects of the matter and that progress in the establishment of high degrees of reliability in trace element analytical measurements will allow clear statements about the essential functions of a number of trace elements and a decision on the essentiality status of others.

The basis of analytical improvements

Trace element measurements, as required in the type of project described above, need to be accurate and precise, i.e., the expected variability of metal accumulation in different organs must not be masked by the variability of the measurements.

The overall reliability of analytical processes can be improved in two ways:

- Employment of definitive methods.
- Use of adequate certified reference materials.

Actually in the field of trace element determination in biological materials only isotope dilution mass spectrometry might be considered, with some reluctance, and for few trace elements only, as a definitive method. This is, for the time being, the unavoidable conclusion derived from numerous certification campaigns, conducted during the last five years. [7,8,9]. Consequently, the only available possibility of approaching the true values in trace analysis is the use of adequate certified reference materials.

By "adequate" we intend to classify materials, which besides of showing the generally accepted minimum requirements of certified reference materials, i.e., homogeneity, stability and data quality of the highest today obtainable state-of-the-art standards, match the single specimens to be analyzed as perfect as possible with respect to matrix composition and trace element concentrations. This is an easily made demand, its achivement, however, shall request the reactive power and the efforts of some generations of dedicated scientists.

Anyway, some progress has been achieved during the past decade and this paper aims at a short balance of debits and credit, of assets and liabilities in the field of biological certified reference materials.

It was the wish of the Scientific Committee of this Workshop, that the materials discussed in this paper be restricted to human and animal origin only and to the so-called "newer" and debated essential elements such as silicon, arsenic, nickel, vanadium, aluminium, lithium and cadmium since adequate certified reference materials will significantly promote the ongoing research on these elements.

Available CRM's of human and animal origin

In the terminology of modern reference material production the term "Certified Reference Material" should be of rather restricted use and in the context of this paper it is defined as a material certified by an organization of long and undiscussed reputation, ready to accept full warranty with respect to the material quality, whose properties were determined at the state-of-the-art level of reliability. Biological materials which correspond to these standards are available from very few organizations.
Unfortunately, the small number of such materials available is still more reduced with respect to the elements called here the "newer" essential trace elements. Table 1 shows, what is available today as certified reference materials of animal or human origin for the elements mentioned above. Materials claimed to be "in preparation" were not considered for abvious reasons.
Evidently cadmium does not present excessive difficulties to the attempt of certification. Similarly behaves arsenic, which was not envisaged for certification in the case of the BCR milkpowder project [10] for economical rather than for technical reasons. Tin, silicon and lithium were not considerd at all, and the reasons were certainly rooted in the extreme analytical difficulties (Sn, F and Si) but also in a certain lack of interest (Li) expressed by the buyers of the materials. A late addition to this small list is NBS-909 Human Serum, certified for its lithium content of 11.4 ppm.

There is no doubt that uncertified reference materials cannot replace their certified sisters, but they can help in many analytical situations whenever certified RM's are missing, provided they show adequate data quality.

Table 1

CERTIFIED RM's OF HUMAN AND ANIMAL ORIGIN

	NBS OYSTER TISSUE	NBS BOVINE LIVER	BCR No. 63 MILKPOWDER	BCR No. 150 MILKPOWDER	BCR No. 151 MILKPOWDER
Li	—	—	—	—	—
Al	—	(2)	(0.6)	—	—
Ni	1.03	—	(11.2)*	(61.5)*	(56)*
V	2.3	—	—	—	—
As	13.4	0.047	—	—	—
Cd	3.5	0.44	2.87*	21.8*	101.5*
Si	—	—	—	—	—
F	(5.2)	—	—	—	—
Sn	—	—	—	—	—

Values expressed as μg/g; except * (ng/g)

Table 2

NON-CERTIFIED RM's OF HUMAN AND ANIMAL ORIGIN
JOINT RESEARCH CENTRE ISPRA

CODE	MATERIAL	Al*	As**	Cd***	Ni***	V*
T 1	SPRAY DRIED ANIMAL BLOOD	26	<0.04	0.066	1.3	ND
T 2	SPRAY DRIED ANIMAL BLOOD	28	<0.04	0.029	7.2	ND
T 3	SPRAY DRIED ANIMAL BLOOD	18	<0.04	0.028	11.3	ND
T 4	HUMAN HAIR	228	0.10	0.48	—	ND
T 5	ANIMAL BONE	280	0.08	0.079	18.1	ND
T 6	FRESHWATER PLANKTON	1260	5.6	0.87	44.0	5.6
T 7	HUMAN HAIR	150	0.14	0.30	—	ND

Concentrations in μg/g

*ICP **HAAS ***DPASV

Table 2 (continued)

NON-CERTIFIED RM's OF HUMAN AND ANIMAL ORIGIN
JOINT RESEARCH CENTRE ISPRA

CODE	MATERIAL	Al*	As**	Cd***	Ni***	V*
T 8	PIG KIONEY	200	0.22	2.6	0.78	—
T 9	BOVINE LIVER	180	0.010	0.24	1.9	—
T 10	BOVINE MUSCLE	—	—	0.014	0.39	—
T 11	CONNECTIVE TISSUE	142	0.14	5.3	0.85	0.50
T 12	MUSSELTISSUE (CARDIUM)	582	5.2	0.12	27.1	2.8
T 13	MUSSELTISSUE (MYTILUS)	50	3.5	0.20	1.6	0.14

Concentrations in $\mu g/g$

*ICP **HAAS ***DPASV

Table 2 (continued)

NON-CERTIFIED RM's OF HUMAN AND ANIMAL ORIGIN
JOINT RESEARCH CENTRE ISPRA

CODE	MATERIAL	Al*	As**	Cd***	Ni***	V*
T 14	TUNA MUSCLE	3.6	0.60	0.042	0.27	ND
F 4	ALBUMENE	265	<0.04	0.065	0.23	ND
F 5	EGG YOLK	53	<0.04	0.016	0.050	ND
F 6	WHOLE EGG	48	<0.04	0.012	0.17	ND
F 9	SPIKED MILKPOWDER	3.7	0.16	0.062	5.0	2.6

Concentrations in $\mu g/g$

*ICP **HAAS ***DPASV

The Ispra Environmental Reference Material Bank contains a number of materials of animal and human origin, which cover for some of the elements discussed here relatively wide ranges of concentration (Table 2).

There are doubts concerning the aluminium values of T8, T9, T10 and T11, since strong contamination during freeze drying might well have been occured. Minor contamination is possible for the other materials due to the ubiquitous nature of aluminium.

In qualitative terms, the spiked milkpowder F9, will approach the status of a certified material since the well known

Table 3

Lithium-Concentrations Measured in Ispra-RM's by DCP*

MATERIAL	Li-CONCENTRATION [µg/g]
T 1 Animal dry blood	< 0.01
T 2 Animal dry blood	0.34
T 3 Animal dry blood	< 0.01
T 4 Human hair	2.0
T 5 Animal bone	0.82
T 6 Plankton	1.3
T 11 Connective tissue	< 0.01
T 12 Cardium mussel	7.6
T 13 Mytilus mussel	0.47
T 14 Tunafish muscle	< 0.01
F 4 Albumene (egg)	0.17
F 5 Egg yolk	0.31

* P. Schramel and S. Hasse

spikes have been, with the exception of mercury, confirmed by an interlaboratory exercise amongst highly specialized milk laboratories [11].

Some elements, however, are missing. Fluorine seems relatively uninteresting and the limited sensitivity of ICP for vanadium requires the employment of NAA for its lower concentrations. Lithium was not considered in the early stage of our material evaluation and its determination does not present excessive difficulties.

Actually the lithium study on the Ispra material series is in progress and first data have been measured by DCP (Table 3).

Silicon, however, is a more challenging problem. First of all it is extremely difficult to avoid silicon contamination during reference material processing, when the material is not made on purpose, which was never the case in our laboratory, and its storage in glass bottles will upset the trace analyst. Contamination will furthermore differ considerably in chemical form with respect to the endigenous silicon of the tissue.

The analytical methods applied were gravimetry where feasible and the spectrophotometric molybdate procedure which produces acceptable values, when extreme care is observed with respect to contamination.

Laser microprobe mass analysis (LAMMA) has been used in clinical research, especially for the localization of silicon in extremely small tissue sample areas around 1 μm^2, but it is hardly a certification quality method.

On the other hand, silicon seems to play an important role in cartilage ageing and bone development and the on-going research could be promoted by the improvement of the analytical measurement accuracy and precision and adequate RM's are highly desirable.

Table 4 presents very rough data for the Ispra tissue series and one should take into account that virtually none of these materials is free from contamination. Further analytical evaluation will show whether at least a few of the materials can be developed further for their silicon content.

The spiked milkpowder Ispra-F9 and the "normal" milkpowder IAEA-A11 are probably the only two available tissue materials which show reliable fluorine and tin values.

The lithium spike of F9 corresponds to 0.83 µg/g which fits satisfactorily to the "normal" value of the IAEA-powder [12] which is also the only available material quoting a silicon value. Though 34 µg/g in probably still one order of

Table 4

SILICON-CONCENTRATION OF SOME NON-CERTIFIED RM's

JOINT RESEARCH CENTRE ISPRA

CODE	MATERIAL	SILICON [µg/g]
T 1	DRY ANIMAL BLOOD	280
T 2	DRY ANIMAL BLOOD	188
T 3	DRY ANIMAL BLOOD	230
T 4	HUMAN HAIR	1269
T 5	ANIMAL BONE	423
T 6	PLANKTON	18600
T 7	HUMAN HAIR	1020
T 8	PIG KIDNEY	210
T 9	BOVINE LIVER	235

Table 4 (continued)

SILICON-CONCENTRATION OF SOME NON-CERTIFIED RM's

JOINT RESEARCH CENTRE ISPRA

CODE	MATERIAL	SILICON [μg/g]
T 10	BOVINE MUSCLE	120
T 11	CONNECTIVE TISSUE	7050
T 12	MUSSEL TISSUE (CARDIUM)	4935
T 13	MUSSEL TISSUE (MYTILUS)	329
T 14	TUNA MUSCLE	795
F 4	ALBUMENE	190
F 5	EGG YOLK	47
F 6	WHOLE EGG	47
F 9	SPIKED MILKPOWDER	86

Table 5

NON-CERTIFIED DATA FOR SOME RARE ELEMENTS

ISPRA F 9 SPIKED MILKPOWDER 1.37 μg/g Li, 5.35 μg/g F, 15 μg/g Sn
IAEA A 11 MILKPOWDER 34 μg/g Si, 0.54 μg/g Li, 122 ng/g Sn, 0.26 μg/g F

magnitude above the endigenous silicon level, taking human milk as a reference [13]. IAEA-11 is, with all probability, the best compromise in view of realistic silicon tissue concentrations until the at-purpose development of a true soft tissue material, be it animal or human.

Summarizing, the following conclusions can be drawn:

- Certified tissue RM's for the elements Sn, F, Li, Al, Si are inexisting and near to inexisting for V, Ni and As.

- The ubiquitous elements Si and Al require special RM-production techniques which have to be developed, unless it is shown that minor contamination in a different chemical form can be tolerated.
- The measurement methodology in view of the development of future CRM's is promising in the case of Li, Ni, V, As, Cd and Al. For the elements silicon, fluorine and tin the analytical evaluation of candidate materials to certification is not feasible at the time being due to the

Table 6

HIGH PRIORITY CERTIFIED REFERENCE MATERIALS FOR ESSENTIAL ELEMENT RESEARCH

HUMAN MATERIALS
URINE, BLOOD, SERUM, PLACENTA, HAIR
ANIMAL MATERIALS
KIDNEY, LIVER, BONE, BRAIN, LUNG, HEART

Table 7

RECENT ACHIEVEMENTS AT THE JRC ISPRA

HUMAN PLACENTA
LIQUID HUMAN URINE - FREEZE DRIED HUMAN URINE - DIGESTED HUMAN URINE

lack of more than one reliable technique. Consequently, for these elements method development must be given priority, and the existence of some uncertified RM's, such as the Ispra RM's, will be of help in this process.

- Materials of top priority, considering also the availability of large quantities of raw material, are human urine, blood serum, placenta and hair and animal kidney, liver, bone, brain, lung and heart which represent the major target organisms in trace element metabolism.

Recent RM-developments at the JRC Ispra

The most recent RM-developments at the JRC Ispra in the biochemical field, again in cooperation with P. Schramel, include, besides of human placenta, a series of human urine materials.

Fig. 2

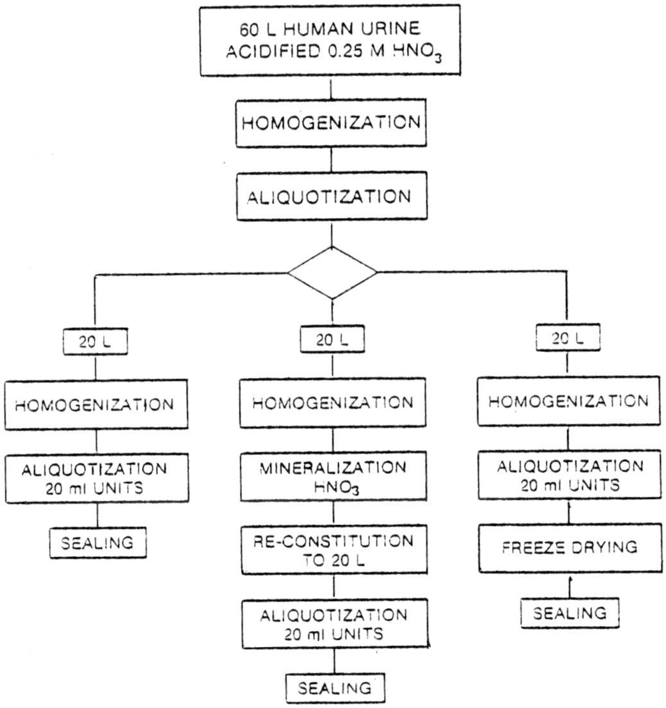

Figure 2.

Table 8

Lithium and Aluminium Concentrations in Freeze-dried and Liquid Human Urine RM's T 15 and T 16

MATERIAL	Al [$ng \cdot ml^{-1}$]	Li [$ng \cdot ml^{-1}$]
T 15 Liquid Human Urine	76 n = 5, s = 6	117 n = 5, s = 5
T 16 Freeze-dried Human Urine	91 n = 5, s = 17	106 n = 5, s = 9

DCP - Measurements by P. Schramel and S. Hasse

The analysis of human urine, for heavy metal traces in particular, is of eminent importance in the field of occupational health. The numerous problems associated with material stability and homogeneity led to the design of a small experiment aiming at the exploration and definition of the most suitable processing of such RM's.
To this end 60 l of human urine (24 h-samples, both sexes), collected at the JRC Ispra were subdivided after appropriate mixing and stabilization into three identical subfractions of 20 l each.
The single fractions were further processed to yield a liquid and acid-stabilized, a freeze-dried and an acid-digested material (Fig. 2).

First analytical data for the liquid and the freeze-dried material indicate fair homogeneity for both of the materials for the elements Li, P, Zn, Ca, and Cu (Table 8 and 9).

Table 9

Some Elements in Freeze-Dried and Liquid Human Urine RM's T 15 and T 16

MATERIAL	P [μg/ml]	Zn [ng/ml]	Ca [μg/ml]	Mn [ng/ml]	Fe [ng/ml]	Cu [ng/ml]
T 15 Liquid Human Urine	848 n = 5, s = 0.5	552 n = 5, s = 1.6	172 n = 5, s = 0.9	1.3 n = 5, = 0.2	86 n = 5, s = 7	8.5 n = 5, s = 1.2
T 16 Freeze - dried Human Urine	845 n = 5, s = 0.5	524 n = 5, s = 2.6	169 n = 5, s = 0.4	2.0 n = 5, s = 1.1	112 n = 5, s = 44	8.5 n = 5, s = 0.2

ICP - Measurements (P. Schramel and S. Hasse)

Al and Fe, homogeneously distributed in the liquid material, show scattering of the results in the freeze-dried material

Table 10

Nickel, Cobalt and Lead Concentrations in FREEZE-DRIED HUMAN URINE*

	NICKEL	COBALT	LEAD
T 16	2.6 n = 5, s = 1.1	0.49 n = 5, s = 0.08	20 n = 5, s = 2.7

* ng/ml

accompanied by an increase of the concentrations and data for Mn show the same trend, indicating a slight contamination by structure materials during freeze-drying. Lead and cobalt (Table 10) do not appear altered by contamination or inhomogenous distribution (DPASV-measurements).
Further measurements are in progress and a comprehensive report on the experiment shall be presented at the 4. Workshop on Trace Elements in Medicine and Biology, 1986.

References

1. Mertz, W., Science, Vol. 213 (1981), 1332
2. Anke, M., Proc. 4. Spurenelement-Symposium, Leipzig 1983, p. 380
3. Hennig, A., Proc. 4. Spurenelement-Symposium, Leipzig 1983, p. 6
4. Muntau, H., Leschber, R., Korresp. Abwasser, Vol. 28 (1981), 200
5. Muntau, H., Suess, A., Rosopulo, A., Europ. Symposium Sewage Sludge, Cadarache, 1979
6. Muntau, H., Schramel, P., Brätter, P., Knapp, G., Proc. 2. Int. Symp. Trace Elements Medicine and Biology, Neuherberg (1982), p. 819
7. Griepink, B., Muntau, H., Colinet, E., Fres. Z. Anal. Chem., Vol. 315 (1983), 193
8. Griepink, B., Colinet, E., Haemers, L., Muntau, H., Fres. Z. Anal. Chem., Vol. 315 (1983), 20
9. Muntau, H., Griepink, B., Paper presented to Int. Symp. Microanalysis, Amsterdam 1983
10. Griepink, B., Colinet, E., Muntau, H., Fres. Z. Anal. Chem. (In print)
11. Muntau, H., Int. Dairy Fed. Bull., Vol. 153 (1982), 35
12. Anonymus (IAEA), AQCS-Programme, Vienna (1983)
13. Iyengar, G., Kollmer, W., Bowen, H.J.M., The Elemental Composition of Human Tissues and Body Fluids, Verlag Chemie, Weinheim 1978

D i s c u s s i o n Session X - 1
Plenary Lecture

Subjects of the discussion

- silicon analysis
- preparation of an urine standard material
- speciation in standard materials

Contributors: Brätter, Müller, Schramel

RELIABILITY OF TRACE ANALYSIS OF BIOLOGICAL MATERIALS IN THE LIGHT OF IAEA's INTERLABORATORY COMPARISONS

Leon Pszonicki

International Atomic Energy Agency
A-1400 Vienna, Austria

Introduction

For many years the International Atomic Energy Agency has organized interlaboratory comparisons of analytical results. The main purpose of this activity is the promotion of nuclear analytical methods and the support for laboratories engaged in their developments. Any method, however, cannot be checked by itself and therefore all laboratories active in the field of trace analysis, using also others than nuclear methods, were invited to participate in these exercises. It seems that the contribution of various analytical methods to the results provided for individual intercomparisons corresponds to the actual frequency of their common application. Usually more than 90% of results are obtained by different variants of four methods: neutron activation analysis, atomic absorption, atomic emission and X-ray fluorescence. The intercomparisons are organized on a voluntary basis and any laboratory interested in the co-operation is accepted. The number of participants varies from 25 to 55 depending on the type of intercomparison. Most of the results orginate from industrial countries, however, a significant part of them is provided by laboratories all over the world and it may be assumed that the quality of results represents the average level of the analytical chemistry of trace elements.

Many of the materials used as the objects of IAEA's intercomparisons are typical biological materials, e.g. milk powder [1], animal muscle [2], animal bone [3], rye flour [4], cotton cellulose [5] and freeze-dried

Trace Element - Analytical Chemistry in Medicine and Biology, Vol. 3

animal blood [6]. The large number of analytical results received for these materials represent well the abilities of the present-day chemical analysis in the field of trace elements.

Results

The number of minor and trace elements determined, and for which at least five laboratories provided the results, varied from 20 to 30 depending on the type of intercomparison. All participants had been requested to make six independent individual determinations of each element. These determinations were used to calculate laboratory means and their confidence intervals representing the intralaboratory dispersion of results. The laboratory means found were the basis for calculation of overall means and their confidence limits which in turn were used as the best estimate of "true values" [1,2,3,4]. The confidence interval of the overall mean represents the interlaboratory dispersion of results. The subsequent investigation of artificially prepared materials containing known concentrations of the elements to be determined [7,8] showed, however, that the overall medians and their confidence limits are more robust estimates of the true values and therefore they were used in the latest intercomparisons instead of overall means [5,6].

The results of the statistical evaluation were selected using certain test criteria [6]. In general, these criteria were based on the following data: reasonable confidence interval of the median for the concentration level concerned, a sufficient number of results and of different analytical methods (techniques) used, reasonable agreement between the results obtained by various methods, and a limited number of outlying results. Only results fulfilling the test criteria were classified as consensus (certified) values.

The collected experimental material is very large and its complete presentation in a short paper is impossible even in a condensed form. Below only some typical examples showing the most general features of the collected data are given.

Table 1

Determination of minor and trace element concentrations in IAEA/A-13, Animal Blood

Conc. interval	No.	Element	Units	Overall median	Confidence interval (0.05)	Classification of results: certified	Classification of results: non-certified
above 100 µg/g	1	Ca	mg/g	0.29	0.23 - 0.33	+	
	2	Cl		13.0	10.6 - 17.6		+
	3	Fe		2.4	2.2 - 2.5	+	
	4	K		2.5	2.1 - 2.8	+	
	5	Na		12.6	11.6 - 13.5	+	
	6	P		0.94	6.68 - 1.12		+
	7	S		6.5	5.9 - 7.0	+	
1 - 100 µg/g	8	Br	µg/g	22	19 - 24	+	
	9	Cu		4.3	3.7 - 4.8	+	
	10	Mg		99	81 - 139		+
	11	NI		1	0.6 - 1.4		+
	12	Rb		2.3	1.7 - 3.1	+	
	13	Zn		13	12 - 14	+	
below 1 µg/g	14	Cd	ng/g	13	1 - 30		+
	15	Co		160	5 - 2500		+
	16	Cr		300	60 - 640		+
	17	Mn		50	1 - 65		+
	18	Pb		180	150 - 290		+
	19	Se		240	150 - 310	+	
	20	Sr		690	170 - 1100		+

Table 1 presents the results of determinations of inorganic components in freeze-dried animal blood [6]. Only those elements are listed for which the results were provided by at least five laboratories. They are divided into three groups according to their concentration: minor elements with a concentration above 100 µg/g, and two groups of trace elements with a concentration 1 to 100 µg/g and below 1 µg/g. The confidence limits were established at a significance level of 0.05.

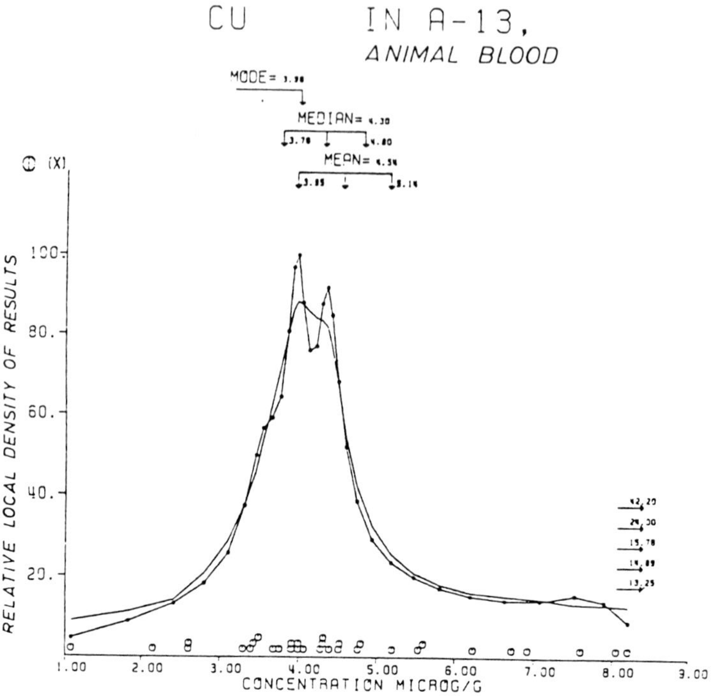

Fig. 1 Distribution of results of copper determination in animal blood.

The distribution of results of copper determinations in animal blood is shown in Fig. 1 as a function of their local density. The local density is defined as the number of results per one concentration unit in the vicinity of each data point. It is calculated as a reciprocal value of the mean distance between the given point and two neighbouring points. This function is plotted as a line with small black circles. The line without circles represents the same function in transformed form after elimination of local maxima resulting from the random distribution of data. The values of the overall mean and median accompanied by their confidence limits and the value of the mode are indicated by vertical arrows placed in the upper part of the figure. The values of results rejected as outliers with horizontal arrows indicating their position are listed on the right side of the plot.

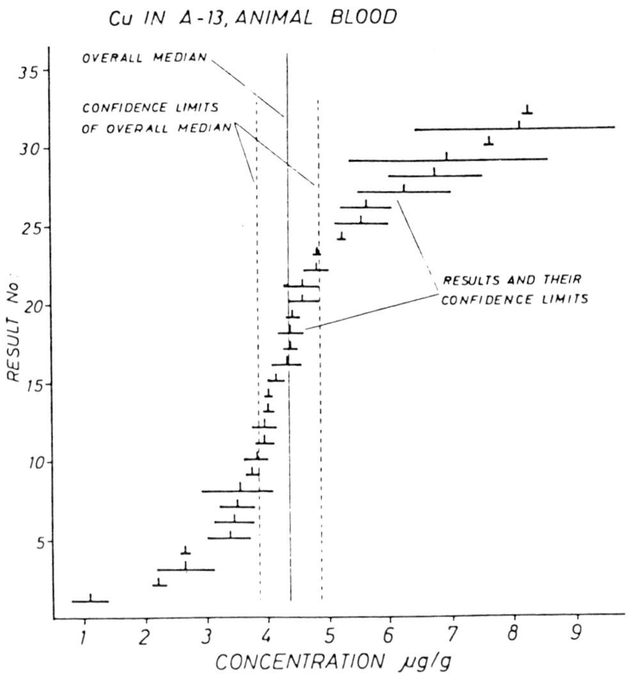

Fig. 2 Position of overall median, laboratory means and their confidence intervals for copper determination in animal blood.

The position of results (laboratory means) provided by individual participants and their confidence intervals in relation to the position of accepted value (overall median) and its confidence interval are shown in Fig. 2.

Discussion

Table 1 shows that the results for only 10 of the 20 elements determined in IAEA/A-13, Animal Blood, passed the test criteria and their medians could be certified. One observes also that the quality of the results decreases with the decreasing concentration. In general, in our intercomparisons of biological materials about 80% of all values for minor elements (concentration above 100 ug/g), 50% to 70% for trace elements in the concentration range from 1 to 100 ug/g, and below 30% for concentrations less then 1 ug/g could be accepted. This observation refers only to elements for which results from at least 5 laboratories were available. For many "more difficult" elements, e.g. 9 of 29 elements the determination of which was requested in animal blood, the number of received results varied from zero to four and they could not be evaluated at all.

The above considerations refered to sets of results provided by laboratories for individual elements. Let us now consider the quality of individual results (laboratory means). It is possible only for the results which belong to the set of data the central values of which were accepted for certification. Only then the adequate qualification criteria could be formulated.

The dispersion of results (laboratory means) of copper determination in animal blood is presented in Fig. 1. 5 of 37 results received were recognized as outliers by a statistical test and excluded from further evaluation. The remaining results are dispersed in a concentration interval extended over almost one range of magnitude. Their main cluster is found in the interval from 3 to 5 ug/g and it is contained in the confidence interval of the median significantly better than in that of mean. Fig. 2 shows that the confidence intervals of 16 results overlap at least partially the confidence interval of the median. This suggests that the dispersion of these results is random and they are free from systematic errors. The remaining 16 (7 lower and 9 upper) results contain a systematic error which in some cases is quite large. This group together with the outliers make up 57% of all results. This explains the commonly known fact that interlaboratory results are usually significantly more dispersed than the intralaboratory results. It also shows that a considerable part of all laboratories produces results of trace element determinations with systematic errors.

Conclusions

The experimental materials collected in the IAEA's interlaboratory comparisons of trace element determination in biological materials suggest the following conclusions:

1. A significant part of all results (often more than 50%) carries systematic errors.

2. The quality of results decreases with decreasing concentration of the elements to be determined. At lower ppm (µg/g) and at ppb (ng/g) concentration ranges it must be still considered as unsatisfactory.

3. The fact that a considerable number of outlying results is produced also by "good" laboratories suggests that the bad quality of results is due to a significant degree to the present, still unsatisfactory state of art of the methods used in trace analysis.

These conclusions should be kept in mind by analysts as well as by users of analytical results. For the analysts they can be a stimulus for further improvement of analytical methods. For the users they are an information that the results of trace element determiantions should be regarded with some caution. The results obtained by one laboratory and by one method may be useful when relative changes of concentration, e.g. at a purification process, are investigated. However, if the knowledge of an exact absolute concentration is required, the results of determinations should be verified carefully by more than one analytical method and, better, by more than one laboratory.

References

1. Dybczynski, R., Veglia, A., Suschny, O.: Report No. IAEA/RL/68, July 1980.

2. Parr, R.M.: Report No. IAEA/RL/69, October 1980

3. Parr, R.M.: Intercomparison of Minor and Trace Elements in IAEA Animal Bone (H-5), International Atomic Energy Agency, January 1981.

4. Pszonicki, L., Veglia, A., Dybczynski, R., Suschny, O.: Report No. IAEA/RL/91, January 1982.

5. Pszonicki, L., Hanna, A.N., Suschny, O.: Report No. IAEA/RL/97, March 1982.

6. Pszonicki, L., Hanna, A.N., Suschny, O.: Report No. IAEA/RL/98, March 1983.

7. Pszonicki, L., Veglia, A., Suschny, O.: Report No. IAEA/RL/94, May 1982.

8. Pszonicki, L., Veglia, A., Suschny, O.: Report No. IAEA/RL/95, June 1982.

HOMOGENEITY STUDIES IN BIOLOGICAL REFERENCE AND CONTROL MATERIALS WITH SOLID SAMPLING AND DIRECT ZEEMAN-AAS

U. Kurfürst
Grün-Optik Wetzlar GmbH
D-6330 Wetzlar

H.H. Grobecker
Institut für Pflanzenökologie der Universität Gießen
D-6300 Gießen

M. Stoeppler
Institut für Angew. Phys. Chemie der KFA Jülich
D-6170 Jülich

Introduction

The classical approach for elemental analysis in solids consists in the transformation into an analyte solution. For some materials this is feasible by dissolving subsamples in water or other appropriate solvents. For biological and environmental materials a more or less elaborate digestion is required (1). From this approach and the results achieved for e.g. some NBS Standard Reference Materials (9) it has been concluded that even in well prepared solids of environmental and biological origin subsample amounts of 100-200 mg (at least) should be taken for digestion due to heterogeneities within the materials. From practical analytical experience and with no particular metal heterogeneity present the level of homogeneity appeared to be around 5 % from day-to-day reproducibility estimates, in samples of approximately 100 mg.

If these estimates would be correct, subsamples of considerably lower amounts would lead to a drastic increase in imprecision, e.g. $>$50 % for a subsample weight of 1 mg. This would make solid sampling AAS due to the necessary low sample amounts only a semiquantitative approach.

Trace Element - Analytical Chemistry in Medicine and Biology, Vol. 3

The commercial introduction of Zeeman compensated AAS instruments in the meantime has rendered possible solid sampling with the aid of the L'vov platform in numerous biological and environmental materials with sample weights typically around and even below 1 mg.

In contrast to former estimates and predictions, however, numerous analytical results gained within the last couple of years using this technique showed a surprisingly low imprecision ranging from 5 to 15 % in well homogenized, mainly biotic materials (2). From this it appeared that earlier estimates of heterogeneity probably would be erroneous.

For the Zeeman/L'vov solid sampling technique the main analytical error is caused by heterogeneity thus it is possible to determine the homogeneity factor which is independent from sample weight. Hence subsequently the theoretical basis and a series of carefully performed investigations have been described which should contribute to a clearer picture of the situation in this field.

Theory

Wilson has given an exact expression for the sampling errors which are introduced into the chemical analysis, when samples taken from their heterogeneous powders are examined (3). The sampling variance for an element E in a mixture of several components is

$$S_{INH}^2 = \frac{1}{2} \sum_{i=1}^{m} \sum_{j=1}^{m} \left[(c_i - c) d_i - (c_j - c) d_j \right]^2 \cdot \frac{W_i \, W_j}{d_i \, d_j} \cdot \frac{1}{n} \qquad \text{(i)}$$

where $W_{i,j}$ are the proportions of the component i and j resp.

$c_{i,j}$ are the concentrations of E in these components

$d_{i,j}$ are the particles densities of the components

n is the number of particles

Klockenkämper has illustrated this equation for the spectrochemical analysis in some simplified forms, discussing the influence of the different variables (4).
Solberg and Langmyhr have applied this equation to estimate the sampling error for mercury sulphide distributed in a matrix of iron sulphide (5,6). In this worse case the sampling error may be serious und combination of unfavourable conditions (large particle size, low analyte content). But such extreme heterogenous distribution does not exist for biological samples.
Mostly the sampling error cannot be calculated from eq. (i) because the characteristic values of real samples are not known. But the known sample weight is inversely proportional to the sampling error, so that eq. (i) can be simplified to

$$S_{INH} = h_E \frac{1}{\sqrt{w}} \qquad \text{(ii)}$$

where $h_E = h_E$ (W,c,d,N) is the homogeneity factor for the element E,
N is the number of particles per unit weight so that $n = N \times w$ and
w is the weight of samples taken.

Using S_{INH} in % (RSD), h_E can be expressed as

$$h_E = H_E \cdot c/100 \qquad \text{(iii)}$$

where H is the relative homogeneity factor and
c is the mean concentration of the sample.

The homogeneity factor is a characteristic criterion of a powdered solid sample and depends on the original material (distribution of E) and the sample preparation (grinding). h can be calculated from the standard deviation of the determination of E in a narrow weight range, if other errors than S_{IHN} are small or well known.

That does not hold true for many analytical techniques of trace elements, because the errors are not known (e.g. digestion) or the sample amounts are too large so that S_{IHN} becomes very small. The observed error S_{OBS} of the solid sampling technique can be expressed as

$$S_{OBS} = |S_R^2 + S_W^2 + S_{INH}^2|^{1/2} \qquad \text{(IV)}$$

where S_R is related to the repeatability of the analytical method and S_W is the weighing error.

Because $S_R < 3\%$ if $w < 3$ mg and
$S_W < 1\%$ if $w > 0.1$ mg we have from eq. (iv):
$$S_{OBS} \sim S_{INH} \text{ if } S_{OBS} > 3\ \% \qquad \text{(V)}$$

Eq. (v) gives us the chance to determine the homogeneity factor with this method.

Instrumentation and Materials

- Direct Zeeman-AA-Spectrometer SM 1 (Grün-Optik, Wetzlar) with a Massmann graphite furnace and a graphite boat, acting as a L'vov platform (7,8)
- Microbalance 4503 Mikro MP 6 (Sartorius, Göttingen)
- Scanning - Foto - Sedimentograf analysette 20 (Fritsch, Idar-Oberstein) for particle size distribution
- Materials: SRM National Bureau of Standards (9)
  KFA Control Materials of the German
  Environmental Speciman Bank (10)

Results

a) Determination of the homogeneity factor

Tab. 1 shows accuracy and precision of the determination of cadmium in solid samples of two biological materials for different ranges of weight. Fig. 1 shows the RSD over the

Table 1 Examples for the determination of h for cadmium with the solid sampling technique

Material Ref.-conc. (µg/g)	Range of weight (mg)	Concentration solid sampling (µg/g)	RSD (%)	n
KFA Grass I 0.099	0.03 - 0.07	0.108 ± 0.020	18.5	10
	0.07 - 0.10	0.113 ± 0.012	10.2	8
	0.10 - 0.15	0.116 ± 0.009	7.6	7
	0.20 - 0.40	0.102 ± 0.008	7.5	9
NBS Spinach 1.5	0.04 - 0.10	1.20 ± 0.29	24.0	9
	0.10 - 0.20	1.26 ± 0.21	16.3	8
	0.2 - 0.3	1.15 ± 0.11	9.3	7
	0.3 - 0.5	1.19 ± 0.09	7.2	7

Table 2 Relative homogeneity factor H of some SRMs and control materials for lead and cadmium

Sample	H_{Cd} (% $\sqrt{}$ mg)	H_{Pb} (% $\sqrt{}$ mg)	mass of particles <50 µm (%)
KFA Grass I	3.3	11.0	62
KFA Poplar Leaves I	2.9	3.3	80
NBS Orchard Leaves	5.4	4.7	22
NBS Bovine Liver (12)	5.0	8.9	25
NBS Tomato Leaves	6.8		
NBS Spinach	5.5		

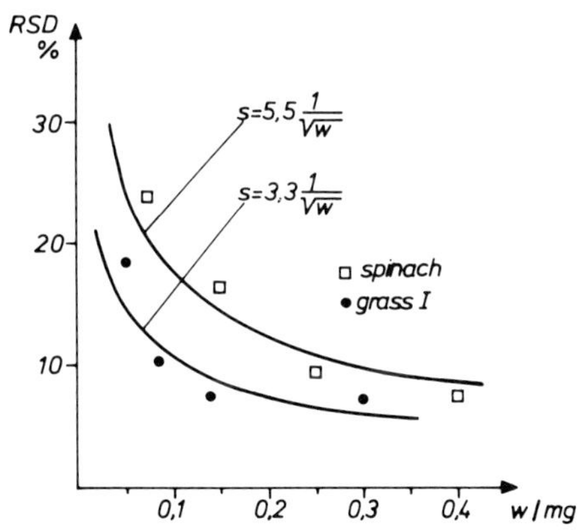

Figure 1 Graphic display of the examples in Tab. 1 and the regression curves calculated from these measurements

Fig. 2 shows the particle size histograms for this sample in comparison with the well homogenized KFA control material poplar leaves I. There is a correlation between the homogeneity factor H_{Cd} and the mass of particles smaller than 50 µm (Tab. 2). The Cd-value of the samples tomato leaves and spinach are not certified because of insufficient homogeneity (9). But Tab. 2 shows that we have found also a low homogeneity factor for these samples. With a 300 mg sample of tomato leaves (certification report) one can expect only $S_{INH} < 0.5\,\%$ (see Fig. 3).

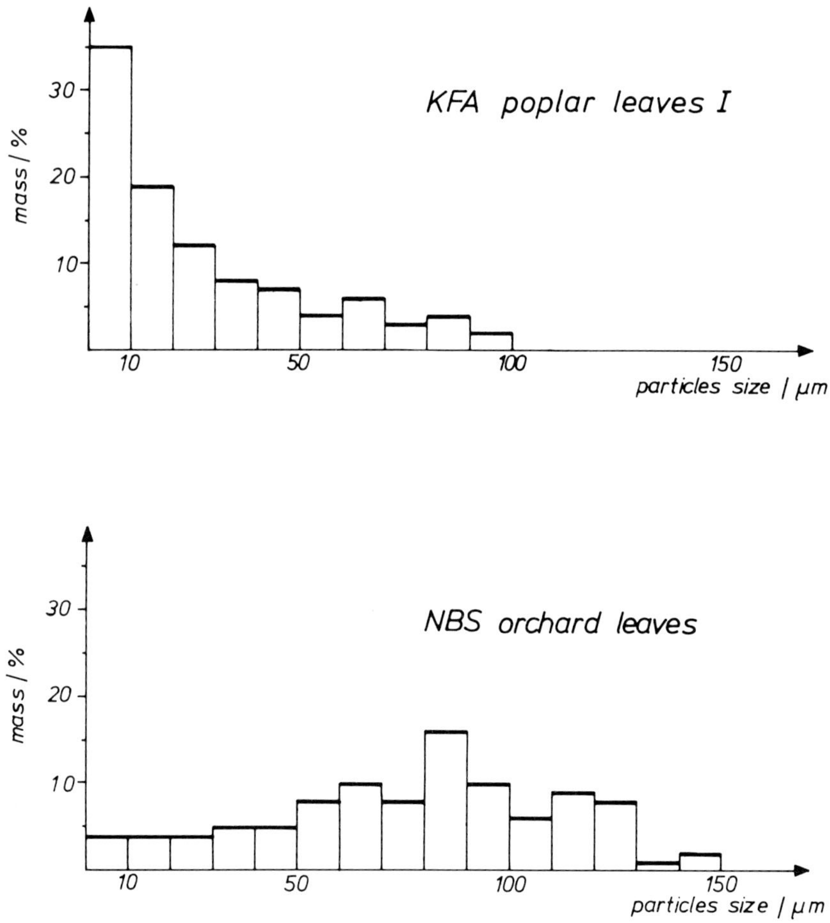

Fig. 2 Particle size distributions

L'VOV has given examples of the error caused by heterogeneous distribution of trace elements in solids using mg - samples (11). The determination of impurities (Bi, Sn, Cd, Pb, Sb) in zirconium dioxide leads also to relative homogeneity factors of about 9% $\sqrt{mg}$.

mean of the weight. The curves are calculated by linear regression after a transformation of the ordinates ($w \rightarrow w^{1/2}$). There is a significant relationship between S and w in the form of eq. (ii). The constant factor of these equations is the relative homogeneity factor H.

Tab. 2 shows the homogeneity factors for cadmium and lead of some biological SRMs and control materials.

H is typically below 10 (10 % RSD at the 1 mg weight level), even for the not properly homogenized NBS sample orchard leaves.

Table 3
Solid sampling analysis of manganese in NBS tomato leaves

Measurement No	Sample weight (mg)	Concentration (μg/g)
1	0.044	202
2	6	215
3	19	243
4	27	265
5	20	228
6	44	207
7	17	211
8	15	217
9	38	183
10	20	180
	0.025 ± 0.013	215 ± 27 (RSD 13%)
		238 ± 7 (certified)

Table 4 Relative homogeneity factors for manganese obtained with different analytical methods

methode	ZAAS/solid	AAS flame	XRF	INAA
sample weight (mg)	0.025	100	500	10
H (% $\sqrt{}$ mg)	2.1	8.1	18	5.1/4.0
sample	NBS tomato leaves			BCR aquatic plants
Ref.	this work	(13)	(13)	(14)

b) Homogeneity Factor H_{Mn} of NBS tomato leaves
For NBS tomato leaves a degree of homogeneity of S_{INH}=0.8 % was found by XRF analysis with a sample amount of 500 mg (13). This would lead to H_{Mn} = 18 (Fig. 3).
Tab. 3 shows the result of our solid sample analysis of the same material. From the RSD of 13% and the mean of the sample weight of 0.025 mg can be calculated H_{Mn} = 2.1.

Tab. 4 shows the homogeneity factors obtained by different methods. It appears that only with very small sample amounts the true homogeneity factor can be determined.
We have to mention, however, that these solid sampling measurements only give informations about "within bottle" homogeneity. The measurements in (13) and the NBS declaration about suspected heterogeneity are related to the "between-bottle" homogeneity. You sure can determine the between-bottle homogeneity by means of the solid sampling technique.
c) Calculation of the homogeneity factor H_{Cd}, expected for homogenized wheat
Tab. 5 shows the concentration of cadmium in the fractions of wheat grains and the percentage of these fractions in the whole grain (15,16).
In the homogenized wheat (N = 2500 particles/mg) are found 1.2±0.01 µg Cd/g (concentration calculated from Tab. 5: 1.20 µg Cd/g).

Fraction	Part of the grain (%)	Concentration solid sampling (µg/g)
Perikarp	3	1.9
Embryo	3	8.2
Endosperm (flour body)	84	0.52
Rest (Aleuron, Testa, etc.)	10	4.0

Table 5
Concentration of cadmium in different fractions of the wheat grain

With the values of Tab. 5 and $g_i = g_j$ it is possible to calculate the homogeneity factor. Eq. (i), (ii), (iii) give $H_{Cd} = 4.4\ \%\ \sqrt{mg}$.
Even the very heterogeneous distribution of cadmium in the natural biological sample gives (statistically and practically) a good degree of homogeneity in the mg-range.

For a 0.2 mg sample H = 4.4 gives a RSD of 9.8 % (Fig. 3). This has to be compared with the measured RSD of 7.8 % for the homogenized wheat sample.

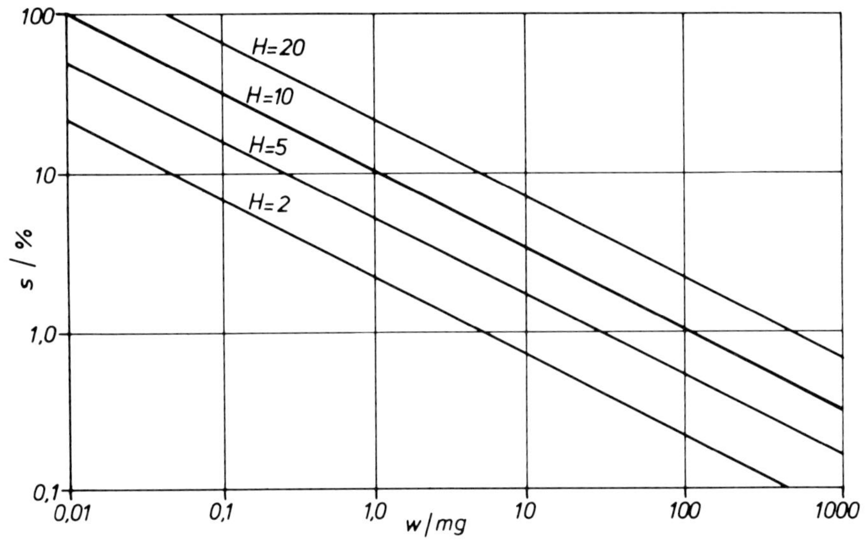

Figure 3 Sampling error S_{INH} over sample weight, with different homogeneity factors H

Discussion and conclusion

The presented results show good agreement of observed with calculated homogeneity factors. Therefore it is evident that, regardless of the analytical method used, if the material investigated is properly homogenized, for sample amounts ≥ 100 mg the sample heterogeneity should contribute theoretically less than 1 % to the whole analytical imprecision (see Fig. 3), i.e. should in fact not affect day-to-day precision.

Strong deviations observed for numerous analytical determinations in solid materials predominantly attributed to sample heterogeneity from these results must be due partly to not properly controlled or recognized contamination sources during sample treatment. Moreover, from own experience, for example for lead in difficult matrices, there have been strong differences observed for the analytical signal in various digests of the same homogenized sample. Until now this behaviour cannot be explained satisfactorily but is by no means attributable to a heterogeneous distribution of lead within the materials investigated (17).

From our findings the following conclusions can be drawn:

- Solid sampling with Zeeman-AAS and the L'vov platform can be regarded as a very useful and economic technique for homogeneity investigations that are of paramount importance in all cases where control and reference materials have to be produced and checked for homgeneity.
- In conjunction with this method significant improvements for digestion procedures can be expected for all types of materials as well.
- Furthermore, possible improvements for solid sampling-Zeeman-AAS with L'vov platform by true peak area integration and optimized furnace conditions can be expected to make this method more and more accurate and increaslingly recommendable for monitoring and reference duties in many routine laboratories, since at least all error sources from decomposition procedures can be omitted which, together with a good cost-benefit ratio, more than balances a somewhat higher RSD compared to precesion liquid sample analysis.

Acknowledgement: The authors wish to thank Mr. E.D. Korn for valuable practical help and Mr. U. Gerber for the determination of particle size distribution

References

1. Stoeppler, M., Analytical Aspects of Sample Collection, Sample Storage and Sample Treatment, in: Brätter, P., Schramel, P., eds., Trace-Element-Analytical Chemistry in Medicine and Biology, Vol. 2, pp. 909-228, Walter de Gruyter, Berlin-New YOrk, 1983
2. Rosopulo, A., Grobecker, K.-H., Kurfürst, U., Fresenius Z. Anal. Chem., in press
3. Wilson, A.D., Analyst 89, 18-30 (1964)
4. Klockenkämper, R., Z. Anal. Chem. 285, 345-352 (1977)
5. Langmyhr, F.J., Talanta 24, 277-282 (1977)
6. Solberg, Thesis, University of Oslo
7. Kurfürst, U., Z. Anal. Chem. 315, 304-320 (1983)
8. Kurfürst, U., Z. Anal. Chem. 316, 1-7 (1983)
9. National Bureau of Standards, Washington DC., Certificates: SRM 1571 orchard leaves, SRM 1577 bovine liver, SRM 1573 tomato leaves, SRM 1570 spinach
10. Stoeppler, M., Dürbeck, H.W., Nürnberg, H.W., Talanta 29, 963-972 (1982)
11. L'vov B.V., Atomic Absorption Spectrochemical Analysis, Adam Hilger, London 1970
12. Klüßendorf, B., graduation work, in preparation, Gießen (1984)
13. Guzzi, G., Colombo, A., Girardi, F., Pietra, R., Rossi, G., Toussaint, N., Rad. J. Chem. 39, 263-276 (1977)
14. Colinet, E., Griepink, B., Muntau, H., Certification report BCR 16 (1982)
15. Pieczonka, K., Aufnahme, Verteilung und chemische Bindung der Elemente Cadmium, Chrom, Kupfer, Nickel und Zink in den Karyopsen verschiedener Getreidearten, Thesis, Techn. University of München (1984)
16. Grün-Optik, Wetzlar, ZAAS-Applications Nr. 12 (1980)
17. Mohl, C., Stoeppler, M., Fresenius Z. Anal. Chem., in preparation

D i s c u s s i o n Session X - 2

Subjects of the discussion

- preparation of standard reference materials
- laboratory intercomparison
- solid sampling technique in flameless AAS

Contributors: Muntau, Schramel, Stoeppler, Weigert

Session XI

ON THE SIGNIFICANCE OF TRACE ELEMENTS IN MEDICINE

Plenary Lecture given by: J. Kiem, L.E. Feinendegen (FRG)

Chairman:

L.S. Hurley (USA)

M. Anke (GDR)

Poster Reports given by: O. Plantin (Sweden)

U. Weser (FRG)

T. Westermark (Finland)

SIGNIFICANCE OF THE NEWER ESSENTIAL TRACE ELEMENTS FOR MEDICINE

Josef Kiem and Ludwig Emil Feinendegen
Institute of Medicine, Nuclear Research Center Jülich GmbH, D-5170 Jülich, F.R.Germany

Among the 15 trace elements which are considered to be essential for humans or animals, physicians are familiar only with a few, despite the remarkable accumulation of data in this field during the past years (1 - 89). This is partly due to a lack of knowledge regarding physiological, clinical and dietetic aspects of these elements, and partly due to lack of commercial supply of sufficiently useful pharmaceuticals containing proper doses of such elements.

Representative Number of Selected Oral Drugs (Rote Liste 83, F.R.G.)

Fig. 1

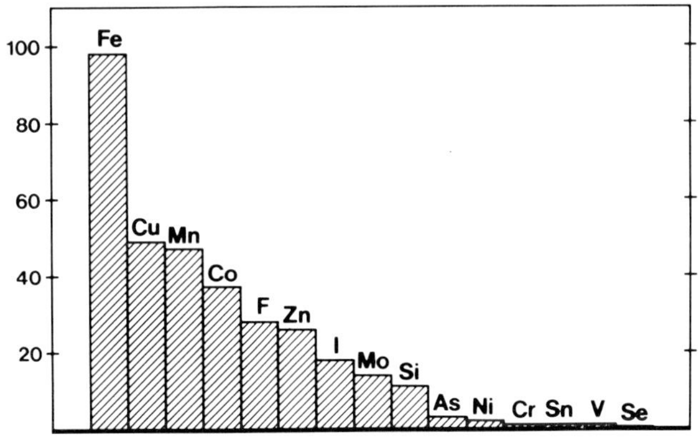

The pharmaceuticals manufactured in the Federal Republic of Germany are registered in a physicians reference list, the so-called "Rote Liste". Fig. 1 shows the frequency-distribution of the oral drugs from the "Rote Liste" of 1983 which contain essential trace elements. They are listed in Table 1. Not even one selenium-containing drug is registered. Measured

Trace Element - Analytical Chemistry in Medicine and Biology, Vol. 3

on their essentiality and probability of dietary deficiency some of these elements are clearly underrepresented from a therapeutical view-point. Fig. 2 and 3 show the compound frequency of the listed oral medications for 9 trace elements.

Fig. 2

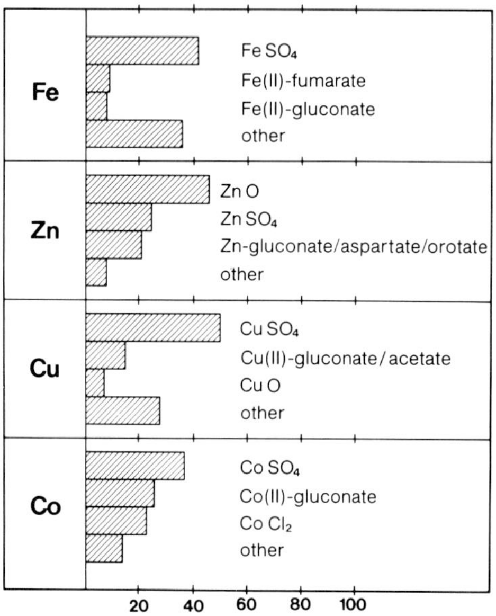

It is obvious that the various drugs contain the individual trace elements in widely varying form. The chemical form in which a trace element is orally supplied determines its absorption. These elements are also supplied in various forms of oral preparation (Fig. 4). The preparation type is less important than the chemical form, but still it may influence the site of absorption, and therefore to some extent its degree, since pharynx, oesophagus, stomach, duodenum, jejunum, ileum and colon have different absorption rates for various elements. For instance, capsules protect their filling from mixing with intestinal contents until they are digested, whereas solutions provide immediate contact of the elements with the mucosa of the gastro-intestinal tract from the pha-

Fig. 3

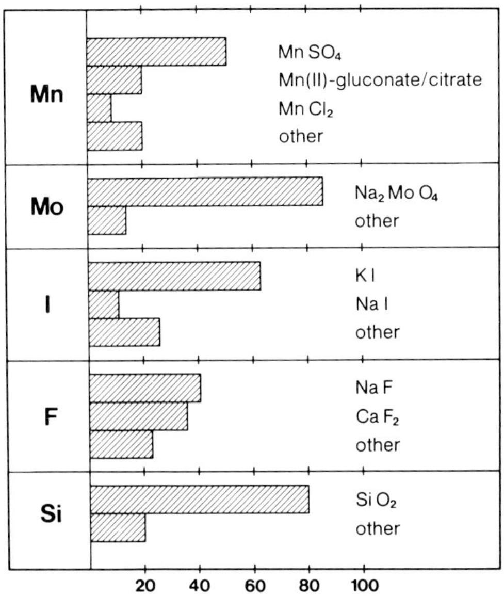

Compound frequency |%| of oral drugs
(Rote Liste 83, F.R.G.)

rynx onwards. There is obviously a considerable uncertainty of how a trace element should be given to the patient regarding chemical form and preparation.

Fig. 4

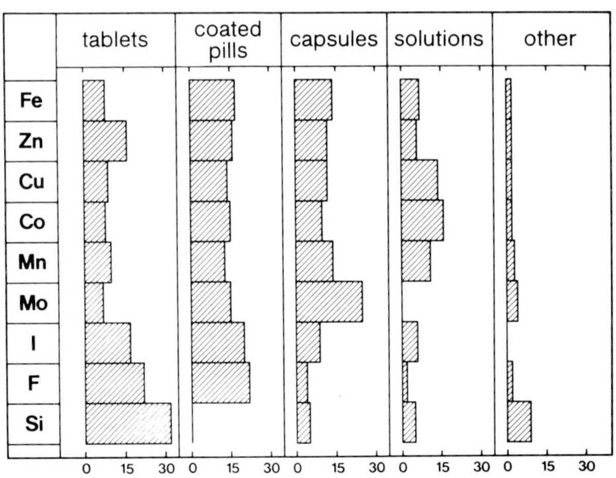

Preparation frequency |%| of oral drugs (Rote Liste 83, F.R.G.)

Thus it is useful to summarize what practical knowledge is presently available to physicians regarding the essential trace elements. Therefore, we wish to consider the physiological, clinical and dietary aspects and refer to the already introduced drugs which are at the disposal of physicians. We will discuss these drugs regarding their adequacy in element content and comment on factors influencing element absorption and utilization including their mutual antagonism. Fig. 5 recalls some parameters for the so-called reference-man of 70 kg. Deviations due to sex and variations in body size and weight are derived by simple conversions. As emphasized on the right, this reference-man daily drinks 2 liters of water and eats 500 g of food dry weight. Therefore, 1 ppm of a given element in drinking water is equivalent to 2 mg and in dry food to 0.5 mg daily.

Fig. 5

Reference - Man

	body content |kg|	daily intake |g|
wet weight	70	3500
total water	42	3000
food water		1000
drinking water		**2000**
dry weight	28	**500**
carbohydrates	1	285
lipids	10	100
proteins etc.	14	100
minerals	3	15

Fig. 6

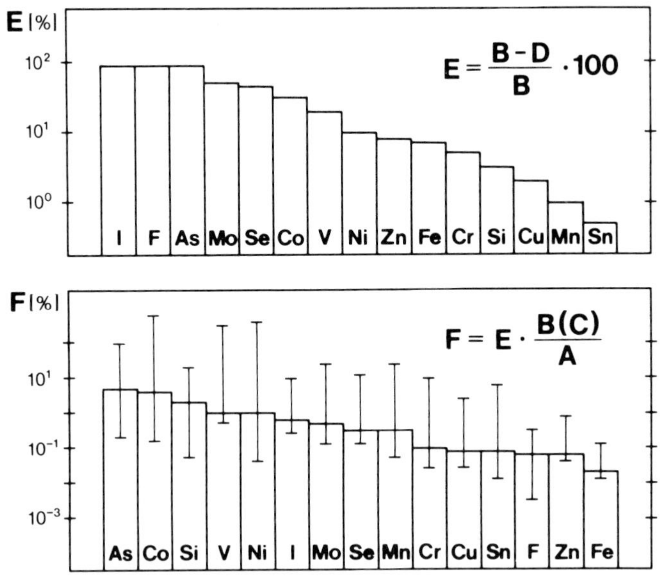

A = recommended body content [g]
B = recommended daily intake [g]
C = tolerated daily intake [g]
D = recommended daily fecal excretion [g]
E = effective absorption [%]
F = recommended (tolerated) utilization [%]

As shown on bottom of Fig. 6 the expressions A, B, C, D, E and F are introduced which are based on personal interpretation of the examined literature. A stands for the recommended body content of a given element in gramm, B for the recommended daily intake in gramm by solid food and liquids, C for the tolerated daily intake in gramm regarding upper and lower limits on a long term basis, D for the recommended daily fecal excretion in gramm either from non-absorption or intestinal secretion. E is the effective absorption in %, and F the recommended and tolerated utilization in %. The right upper corner defines E. Therefore, what we call here the effective absorption E is not identical with the commonly considered true absorption, but resembles the net absorption of an element, i.e. its oral intake minus its fecal excretion. This difference between true and effective absorption is especially important for elements such as copper which has a relatively high true absorption, but a very low effective absorption as can be seen from its late position on the third place from right. This is due to the relatively large amount of copper which is daily secreted through the bile. The average effective absorptions for the different trace elements shown in the upper pannel of Fig. 6 would be only valid if there is no special dietary interference and if the daily intakes of the elements are near the recommended values B, with no critical interactions among them. Yet, as will be mentioned later, type of diet and elemental composition are of practical importance and they may alter effective absorption either by interference or interelemental antagonism. The middle pannel gives the definition of F, and its values for the 15 essential trace elements. F should be directly correlated to the turnover rate of an element provided the body stores are not significantly bypassed towards excretion pathways such as urine, sweat, etc. . Whereas the recommended values for F are shown as horizontal bars in this figure, the ranges of corresponding tolerated values are given by vertical lines. In fact, it is well known that iron has a very slow turnover rate and is efficiently retained by the body and reutilized. Only three elements surpass the 100% level for F with their upper ranges of tolerated values, namely cobalt, vanadium and nickel. This means that, on a short-term basis, these elements can be ingested in relatively high quantities without toxic effects, so that their daily effectively absorbed fraction may exceed their total body content. In the following discussion, the individual elements will be treated with regard to A, B and C extracted from the literature (1 - 89) with special consideration of Underwoods review (52).

Fig. 7

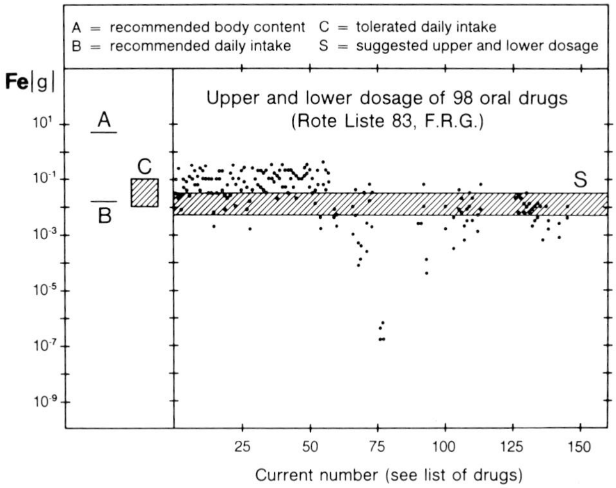

Fig. 7 lists the values for iron. In this as well as in the following figures A, B and C are shown on the left side. Obviously B is only slightly above the deficiency border of C. This partially explains why iron deficiency is the most prevalent deficiency state among trace elements. This dilemma is common knowledge to every physician. The range S on the right side of this and the following figures represents the dosage which we would like to suggest for drug preparations. It ranges around the value of B. Therefore the lower border of S corresponds to one third of B and the upper border of S to twice the value of B which is six times the lower border of S. The dots represent the actual dosages of the trace element within the mentioned oral medications on the market for physicians in the F.R.G. . If, for instance, a manufacturer recommends to take one of his tablets or capsules etc. daily, this dosage is shown on this figure as one dot. Each drug is easily identified according to a current number given in the abscissa of the figure and listed in Table 1. If a manufacturer recommends to take between 1 and 8 capsules daily, two dots corresponding to the elemental content of 1 and 8 capsules, have been inserted in the figure. Clearly, the bulk of

Table 1: List of drugs containing trace elements (Rote Liste 1983, Bundesverband der Pharmazeutischen Industrie e.V., 6000 Frankfurt, F.R.G.)

current number	registered number	registered name
1	08001	Ascofer
2	08002	Ce-Ferro
3	08003	Eisen Chelat-Tabletten
4	08004	Eisensulfat 150 Stada
5	08006	Ferro 66 DL
6	08007	Ferroglukonat-ratiopharm
7	08008	Ferrokapsul
8	08009	ferrolande
9	08010	Ferrophor
10	08011	ferro sanol
11	08012	ferro sanol duodenal
12	08013	Ferrum Hausmann
13	08020	Aegrosan
14	08021	Aktiferrin F
15	08022	Aleukon
16	08024	Ferlixir
17	08025	Ferlixir triplex
18	08026	Ferro-Folgamma
19	08027	Ferronovin
20	08029	Hämatopan
21	08030	Hepanovin
22	08031	Indovert
23	08032	Leukobosin
24	08033	Parkevit-Fe
25	08034	Resoferix fol B
26	08035	Rulofer
27	08036	Selectafer B12
28	08038	Berubi-compositum
29	08039	Ferroinfant
30	08041	Kendural-Plus
31	08042	Kobalt-Ferrlecit
32	08044	Ferro Cytofol
33	08045	Ferro-Folsäure-Vicotrat
34	08046	Ferro-Folsan
35	08048	Kendural-Fol-500
36	08049	Plastulen
37	08050	Tardyferon-Fol
38	08051	Athensa-ferro-Saft
39	08052	Ce-Ferro forte
40	08053	Eisen-Indovert
41	08054	Eryfer
42	08055	Ferrlecit
43	08056	Ferrlecit forte

Table 1 (continued)

44	08057	Ferrodux
45	08059	ferro sanol B
46	08060	Kendural C
47	08061	Liquifer
48	08062	Multibionta-Eisen
49	08063	Rulofer
50	08064	Tufemin
51	08065	Aktiferrin
52	08067	Ferro-C-Calcium
53	08068	Ferrum Klinge
54	08069	Ferrum-Strath
55	08071	Presselin Olin 2
56	08072	Resoferix
57	08073	Tardyferon
58	31296	Solvezink
59	42030	Vivioptal
60	52028	Gomban
61	53001	Fluoretten 1 mg
62	53002	Zymafluor 1 mg
63	53003	Calceno "F"
64	53004	Calciduran-F 20
65	53013	Zymafluor-Calcium
66	61020	Condossan
67	61022	Hevert-Vitan
68	61023	Imogenil
69	61024	Kavigeba
70	61026	Natabec F
71	61030	Osspulvit
72	61031	Presselin Telagut
73	61037	Calcium cum...
74	61046	Flucalmavit
75	61047	Oscevitin
76	61049	Calcium-Fides
77	61050	Pasivital
78	61080	Chemifluor 40
79	61082	Kupferorotat
80	61084	Mikroplex Fluor
81	61085	Mikroplex Kobalt
82	61086	Mikroplex Kupfer
83	61088	Mikroplex Mangan
84	61092	Mikroplex Zink
85	61093	Natriumfluorid
86	61094	Natriumfluorid 25 Baer
87	61095	Ossin
88	61096	Zink-aspartat
89	61097	Zinkorotat 40
90	61098	Biometalle II-Heyl
91	61099	Biometalle III-Heyl
92	61105	Inzelloval
93	61108	Membravit

Table 1 (continued)

94	61110	Mikroplex Mangan-Kobalt
95	61111	Mikroplex Mangan-Kupfer
96	61112	Mikroplex Mangan-Kupfer-Kobalt
97	61113	Mikroplex Nickel-Kobalt
98	61114	Mikroplex Zink-Kupfer
99	61115	Mikroplex Zink-Nickel-Kobalt
100	61116	Multi-Mineral Chelat-Tabletten
101	61118	Ossiplex retard
102	61123	Tridin
103	72013	Repursan masc.
104	72015	zirkulinol Lecithin
105	72020	Floradix Kräuterblut-Dragees
106	72023	Herz-Punkt
107	72028	Phosvitanon
108	72041	Bio-Nektarin
109	72049	Optonicum
110	72052	Promonta Nervennahrung
111	72053	Protopanyl
112	72055	Siozwo Tonicum
113	72074	Hevert-nerv
114	73014	Astrumin
115	73039	Presselin Thyre
116	73041	Rö-Strumal
117	73044	Thyreo-Pasc
118	73045	Jodetten
119	73046	Jodid-Tabletten
120	73047	Mikroplex Jod
121	73048	Destrumin
122	73050	Strumedical
123	73051	Strumex
124	81141	Eviprostat
125	83120	Fluor-Vigantoletten 500
126	83153	Cobidec
127	83155	Combionta
128	83156	Eunova forte
129	83157	Fortiplex-Multivitamin
130	83161	Lohprevit
131	83165	Obron
132	83166	Polyvital-SE
133	83167	Poly-Vitamin-Saar
134	83169	Prenatal
135	83171	Supradyn
136	83172	Taxofit Multivitamine
137	83173	Vi-Magna forte
138	83174	Vita Truw
139	83175	Viterra
140	83176	Addivit 12 "Fischer" compositum
141	83179	F M 7 für Erwachsene
142	83182	Mediovit
143	83183	Mulgatol
144	83187	Multi-Vitamin "Hormosan"
145	83188	Multivitamin Phytopharma

Fig. 8

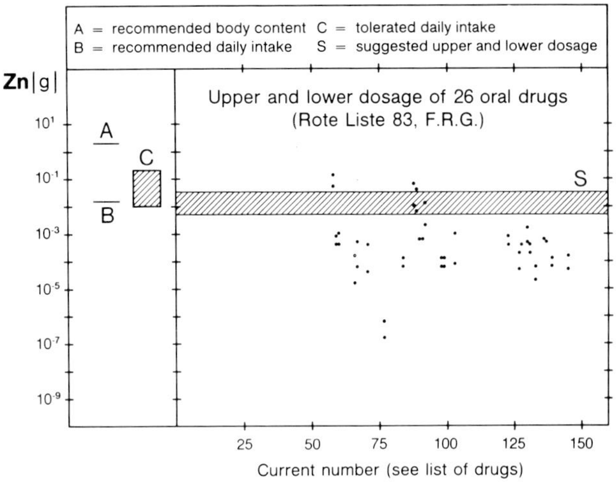

the iron-containing drugs lies above the range S and often even above the toxicity-border of C. This has two reasons: first, the metabolism of iron is well known and the dosage appears acceptable, and second, only on a long-term basis high dosages would become toxic, according to the definition of the range C.

Fig. 9

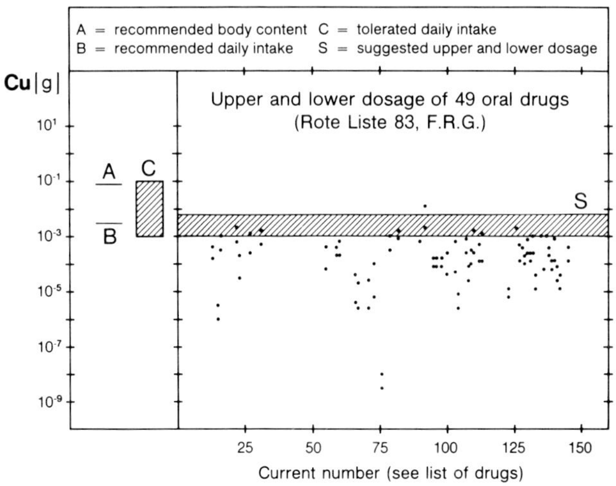

Fig. 8 shows the values for zinc. Also for zinc, B is very near to the deficiency-border of C. Therefore one might expect that zinc-deficiency is likely to be more common than actually diagnosed. One reason for the difficulty in detecting borderline states of zinc deficiency might be the difficulty of establishing normal ranges of the zinc concentrations when using plasma. For instance, in our laboratory we presently use a normal range for plasma zinc between 0.65 and 1.25 µg/ml. Yet, as will be discussed later, smaller normal ranges are probably available when using other tissues such as platelets. At any rate, this figure indicates that most of the drugs have daily dosages which are too low by a factor of about 100.

Fig. 10

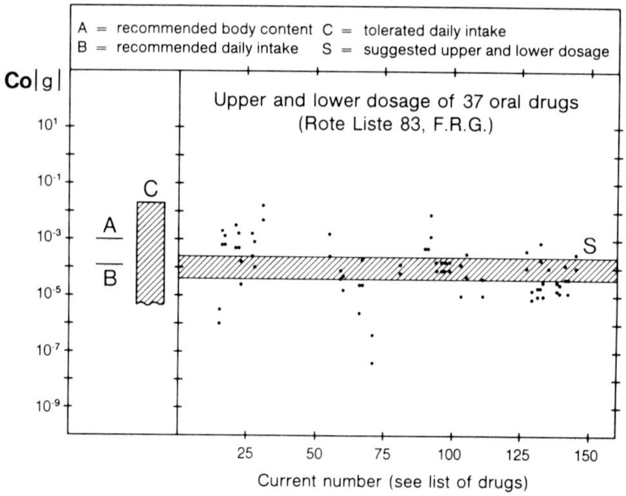

Comparing B and C in Fig. 9, copper-deficiency is likely to develop less readily than zinc-deficiency. Again, most of the available drugs show too low dosages.

Although the lower border for C of cobalt appears uncertain (Fig. 10), it is unlikely that cobalt-deficiency develops in man provided sufficient quantities of the cobalt-containing vitamin B12 are supplied. This is to be stressed, since humans are not able to synthesize sufficient vitamin B12 from

Fig. 11

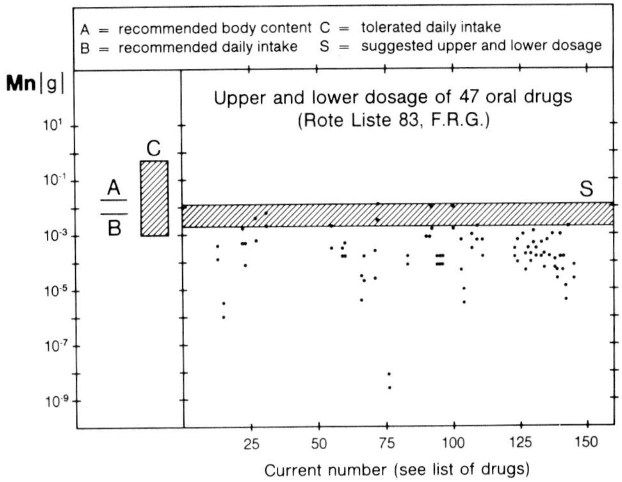

cobalt intakes with the help of the gastrointestinal flora. Therefore a sufficient supply of cobalt, unless with vitamin B12, is rather unimportant for man. It is astonishing that, nevertheless, the dosage of non-vitamin-B12 cobalt is rather high in many of the available oral drugs.

Fig. 12

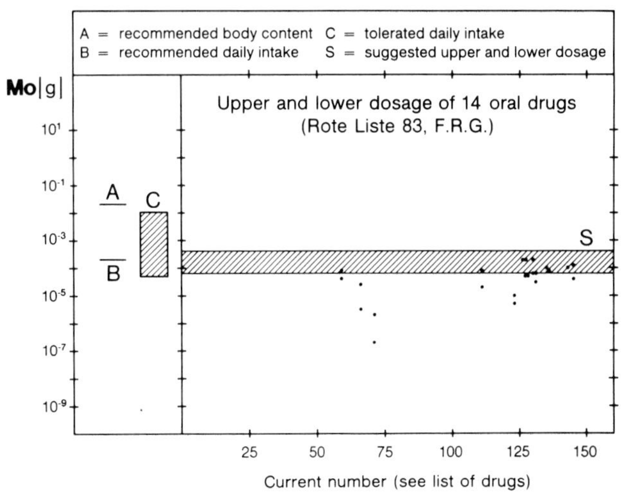

Fig. 11 shows the data for manganese. Comparing B and C, it again appears that manganese deficiency states are not likely to develop very often in man, and most of the available commercial preparations show dosages too low for treating deficiences.

A similar comment can be given for molybdenum (Fig. 12), although a higher fraction of the available drugs appears properly dosed.

Fig. 13

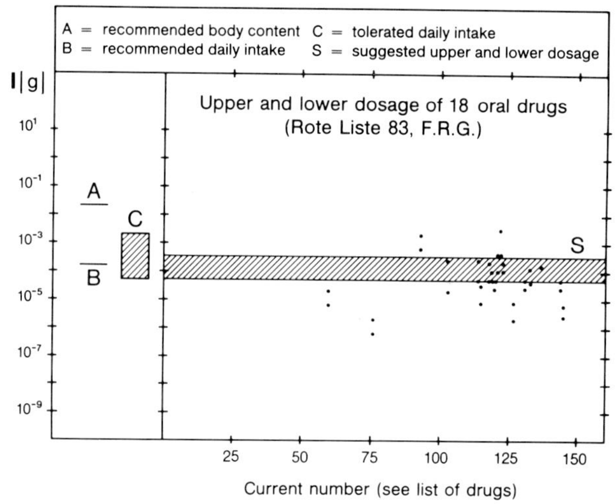

Iodine (Fig. 13) is, like iron, well known to physicians. Obviously, the dosage of the available drugs appears in most instances satisfying. Iodine deficiency results in the development of goiter. In medical practice this is usually treated with the iodine containing thyroid hormones. Yet prophylactic substitution with non-hormone iodine is practiced in many countries.

Regarding fluorine (Fig. 14), comparing B and C lets one expect that fluorine deficiency should not be frequent. But it actually is, since fluorine is mainly consumed with water, apart from the intake of seafood and tea, and water fluorine concentrations vary widely among different regions. This fi-

Fig. 14

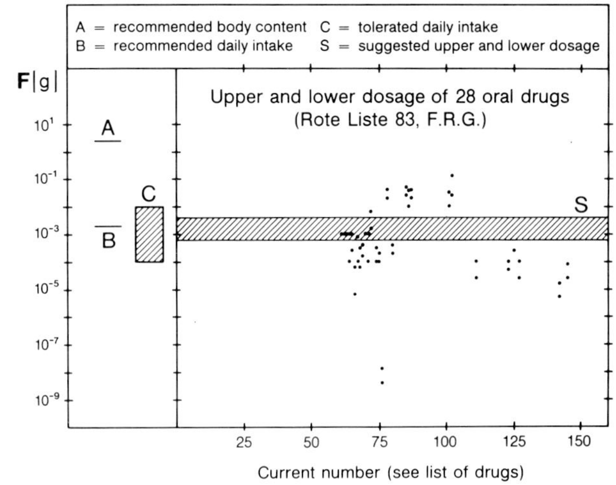

gure shows three groupings of dosages: a group with an insufficient dosage, a correctly dosed group, and this comprises mainly caries-prophylactic medications for long-term substitution, and a third group using dosages which would be toxic on a long-term basis. This group gives fluorine for treating osteoporosis, and it is not recommended for long-term substitution.

Fig. 15

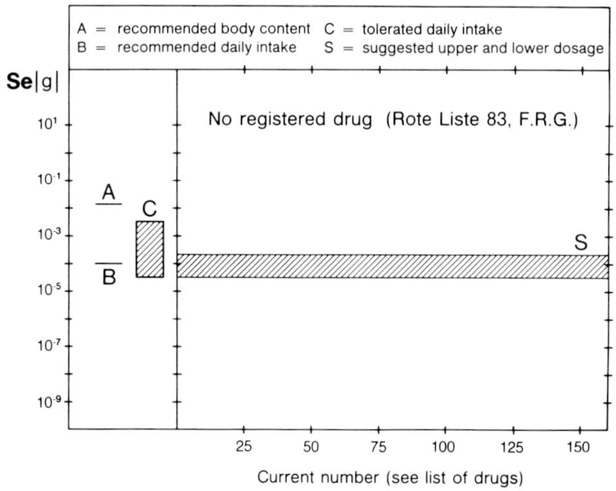

No registered drug containing selenium is presently available (Fig. 15), although some preparations are commercially supplied from imports. This is surprising in view of the well established importance of this element for human health, as again discussed during this workshop.

Fig. 16

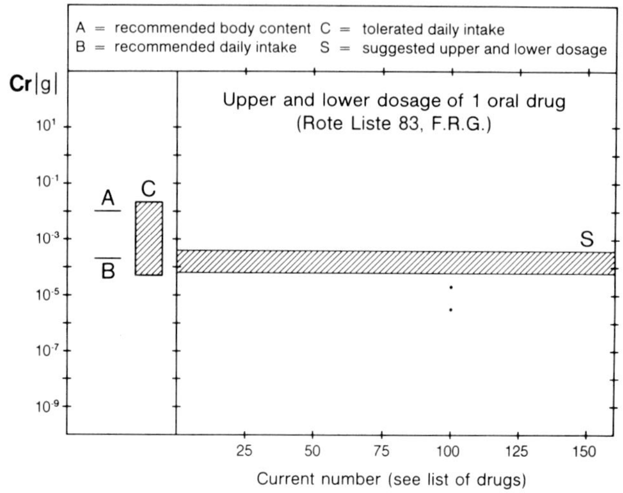

With respect to chromium (Fig. 16), only one oral drug is registered in the physicians reference list. This is again surprising since it is well established, but perhaps less well known, that chromium deficiency may cause late onset diabetes mellitus with decreased insulin-responsive glucose utilization. Comparing B and C it seems that chromium deficiency is not likely to develop easily. Nevertheless, chromium absorption may be low and thus deficiency may develop. Such decreased effective absorption of chromium may be caused by increasing the ingestion of phytate, zinc, vanadium and trivalent rather than hexavalent chromium. We will summarize here, for the purpose of being concise, that effective absorption of all trace elements can deviate from the values of E shown in Fig. 6. For instance, there is a mutual antagonism in effective absorption between iron, zinc, copper, cobalt, manganese and molybdenum so that increasing the fraction of one element decreases those of the others. As said before, physicians who want to confidently make use of trace elements for cure and prophylaxis of deficiency states, should have a profound knowledge of such interferences and interactions.

Fig. 17

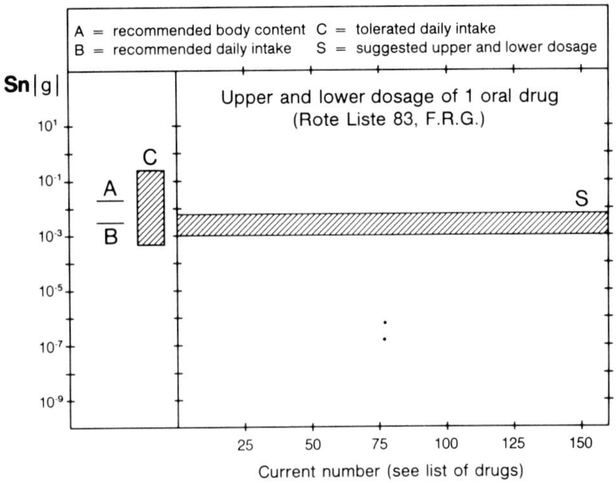

Also tin is an element that attracts little attention on the pharmaceutical market (Fig. 17). There is only one drug that contains tin. Tin toxicity is rare as illustrated by the values for B and C. The reason for the high upper limit of C is the poor gastrointestinal absorption of tin. Therefore, the danger deriving from tin-coated cans should not be overestimated.

Also silicon appears underrated in pharmaceuticals (Fig. 18). Most of the dosages for silicon are several decades below recommended daily intakes. Again, looking at B and C, a deficiency of this element should hardly develop. Deficiency may affect bone, teeth, cartilage, tendons and other connective tissues. A and B are close together which means that nearly a quantity corresponding to the total body content of silicon, is daily ingested. But the effective absorption is poor as mentioned earlier. Silicon is especially supplied with

Fig. 18

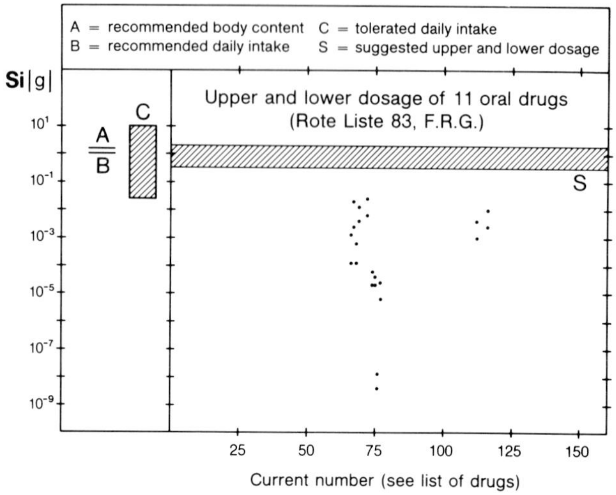

leafy vegetables. Such vegetables are simultaneously rich in iron, manganese, nickel and vanadium. Regarding toxicity there is no danger for humans to develop silica urolithiasis as for pasture animals, whereas silicosis of the lungs due to inhalation of silica particles is well known to physicians.

Fig. 19

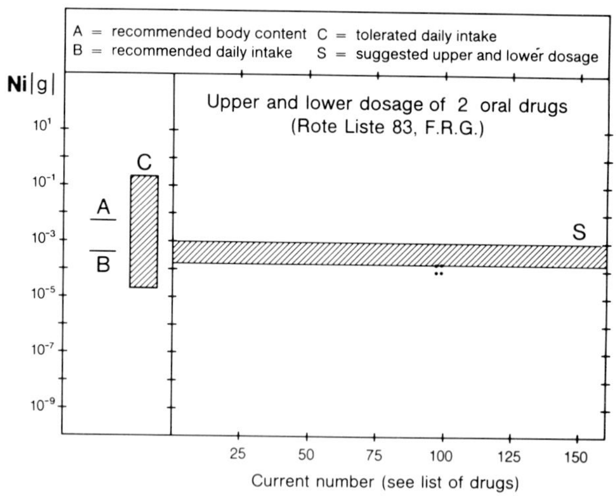

Fig. 19 shows the data for nickel. Neither nickel deficiency nor toxicity is likely to develop because of wide overlap of C with A and B. It seems that the body tissues including the intestinal epithelial cells can actively regulate the utilization of this element so that even daily intakes of more than ten times the total body nickel content do not lead to toxic symptoms. The two available drugs are acceptably dosed.

Fig. 20

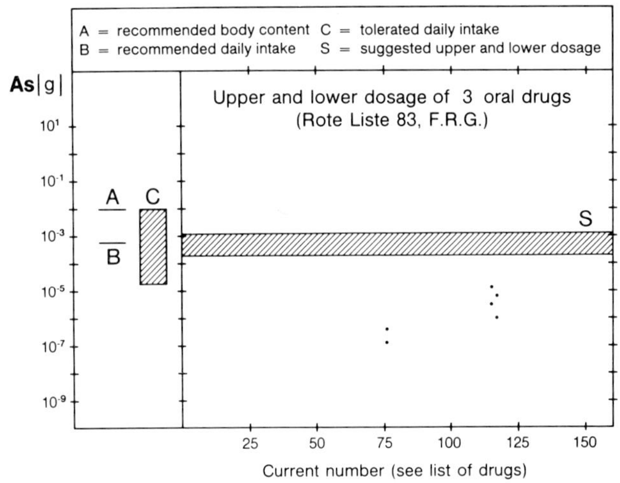

Arsenic (Fig. 20) shows a constellation of B and C that makes deficiency unlikely. In particular, seafood provides larger quantities of this element together with iodine and fluorine. Arsenic is among the elements with the highest effective absorption E and also has the highest F-value (Fig. 6), so that normally an amount corresponding to more than 5% of the total body arsenic is daily effectively absorbed. Toxicity symptoms on a long-term basis such as weakness, muscular pain, headache, drowsiness and convulsions are already encountered when the daily intake of arsenic exceeds the recommended intake by a factor of about 15. The drugs in the physicians reference list can be neglected because of their low dosages.

Vanadium-deficiencies are obviously much more likely to develop than chronic toxicity-states (Fig. 21). This element is known to influence the lipid metabolism whereas its influence on dental caries seems questionable. There is only one drug containing vanadium and its dosage is close to the recommended daily intake.

Fig. 21

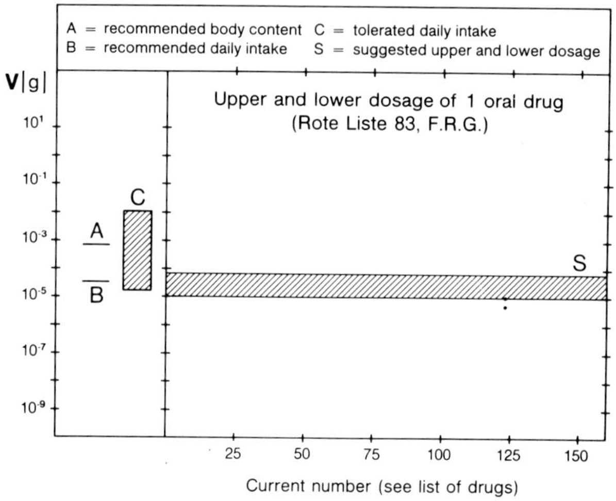

Among the elements which are possibly important or even essential for human life, rubidium is a candidate. Using techniques as described elsewhere (90), we investigated the kinetics of rubidium and compared it with its alkali analogs, cesium and potassium, in human platelets. First we describe cesium (Fig. 22). Platelets from 7 healthy human males were incubated for 29 hours at 37°C in their own blood plasma to which was added cesium-134-chloride. The ratio (mean, n=7) between the intracellular and extracellular concentration for stable cesium and added radioactive cesium, as shown by the expression i/e on the ordinate scale, was then determined, on a wet weight basis, at different times of incubation. The concentration of total cesium in platelets was about 29 times higher than in plasma, as determined by instrumental neutron activation analysis (INAA). The same i/e-ratio was reached for cesium-134 after 25 hours of incubation, as indicated by

Fig. 22

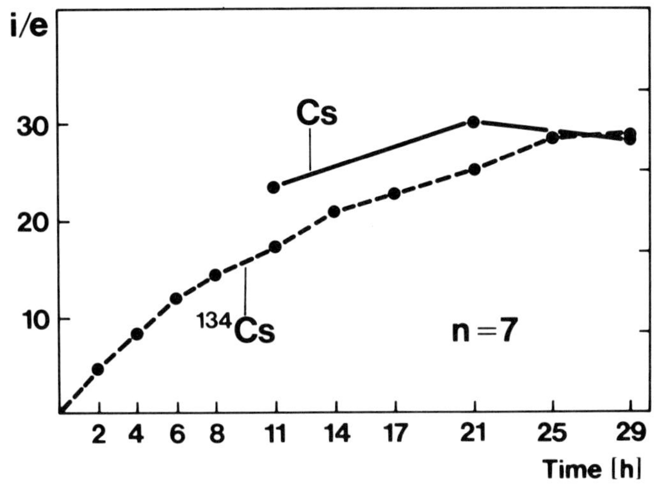

the dashed line. Hence this element is completely exchangeable in normal human platelets.

Fig. 23

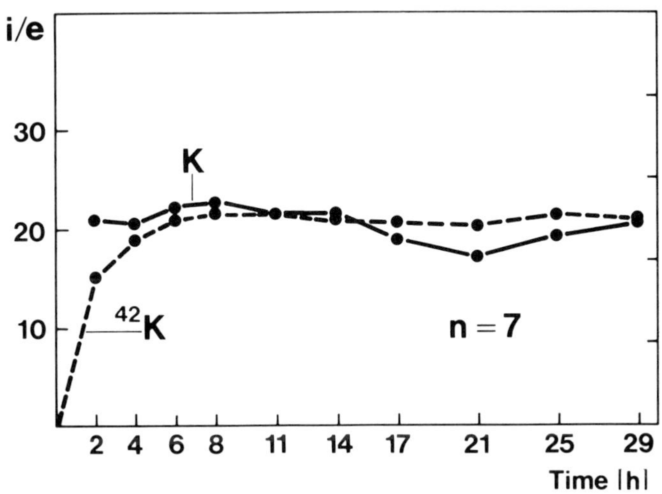

Fig. 23 shows that a similar pattern is observed for potassium. This element is 21 times more concentrated in platelets than in plasma as determined by AAS. Also potassium is, although at a rate faster than for cesium, within six to eight hours completely exchanged in these cells.

The same principal pattern emerges from studies with rubidium (Fig. 24). The concentration of rubidium in platelets is close to 40 times that in plasma, as determined by INAA. On the other hand, using rubidium-86-chloride as a tracer, the ratio comes to a plateau of 29 meaning that the added rubidium did not totally exchange with the intracellular rubidium. In fact, the data show that about 25% of the intracellular rubidium is non-exchangeable, corresponding to 1.5 µg per gramm wet weight. Future research may clarify whether this non-exchangeable rubidium pool constitutes a special fraction of non-ionized rubidium compounds, and which functions they may have.

Fig. 24

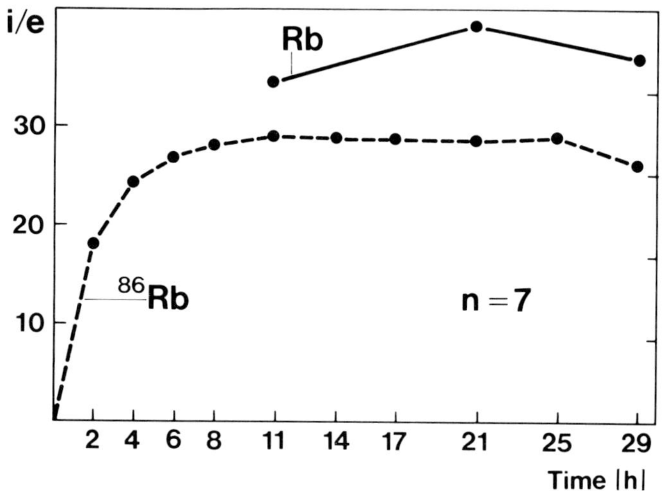

A final comment now relates to the need of improvement of precision in trace element analysis for clinical diagnostic purposes. We recently published the concentrations of three elements in normal human platelets (90). These were, in µg per gramm wet weight, 630 $\pm$ 33 for calcium, 198 $\pm$ 10 for magnesium and 29.4 $\pm$ 1.4 for zinc (mean $\pm$ SD). Therefore, what we called biological variation coefficients between different subjects, were only 5% for all three elements. This is far lower than the variation coefficients of at least 10% which are generally reported for biological materials. This means that most reported variations in biological tissues of different subjects are likely due to methodological inadequacies. Such inadequacies may arise from overlooking the fact that every biological tissue with a given overall elemental concentration and an overall weight, is always a mixture of several biological components with individual elemental concentrations and partial weights. Because variations of elemental concentrations may be smaller in specified tissue components they should preferentially be analysed instead of the whole tissue as examplified by these data for platelets.

Conclusion

Clinical practice demands reliable values with small normal ranges so that practical conclusions for therapy may not be misleading. Especially subclinical or borderline states of element deficiency may be easily missed, and these are, in view of public health, indeed quite important and may even be most important. Our own results on elemental composition of platelets indicate that it is principally possible to achieve narrow normal ranges and thus should be achievable also for other tissues and elements. Especially it is hoped that also the pharmaceutical industry directs its attention to:

1) the desirability of reliable trace element analysis for diagnostic purposes;
2) the possibility that this is in principle achievable;
3) the need of proper adjustment of trace element dosages in the commercial preparations.

References

1. Monier-Williams, G.W.: Trace Elements in Foods. Chapman and Hall, London (1949).
2. Symposium on Copper Metabolism (W.D.McElroy and B.Glass, eds.). Johns Hopkins, Baltimore (1950).
3. Trace Analysis (J.H.Yoe and H.J.Koch, eds.). Wiley, New York (1957).
4. Trace Elements (C.A.Lamb et al., eds.). Academic Press, New York (1958).
5. Metal-Binding in Medicine (M.J.Seven and L.A.Johnson, eds.). Lippincott, Philadelphia (1960).
6. Browning, E.: Toxicology of Industrial Metals. Butterworth, London (1961).
7. Bothwell, T.H., Finch, C.A.: Iron Metabolism. Churchill, London (1962).
8. Bersin, T.: Biochemie der Mineral- und Spurenelemente. Akademische Verlagsgesellschaft, Frankfurt (1963).
9. Rosenfeld, I., Beath, O.A.: Selenium. Academic Press, New York (1964).
10. Wayne, E.J., Kontras, D.A., Alexander, W.D.: Clinical Aspects of Iodine Metabolism. Blackwell, Oxford (1964).
11. Fluorine Chemistry, Vol.4 (J.H.Simon, ed.). Academic Press, New York (1965).
12. Bowen, H.J.M.: Trace Elements in Biochemistry. Academic Press, New York (1966).
13. Zinc Metabolism (A.S.Prasad, ed.). Thomas, Springfield (1966).
14. Nuclear Activation Techniques in the Life Sciences - 1. IAEA, Vienna (1967).
15. Selenium in Biomedicine (O.H.Muth et al., eds.). AVI, Westport (1967).
16. Activation Analysis in the Study of Mineral Metabolism in Man. IAEA, Vienna (1968).
17. Browning, E.: Toxicity of Industrial Metals. Butterworth, London (1969).
18. Chemical Fallout (M.W.Miller and G.C.Berg, eds.). Thomas, Springfield (1969).
19. Mineral Metabolism, Vol.1-3 (C.L.Comar and F.Bronner, eds.). Academic Press, New York (1960-1969).
20. Mineral Metabolism in Paediatrics (D.Barltrop and W.J. Burland, eds.). Blackwell, Oxford (1969).

21. Iron Deficiency: Pathogenesis, Clinical Aspects, Therapy (L.Hallberg et al., eds.). Academic Press, New York (1970).

22. Trace Element Metabolism in Animals - 1 (C.F.Mills et al., eds.). Livingstone, London (1970).

23. Intestinal Absorption of Metal Ions, Trace Elements and Radionuclides (S.C.Skoryna and Waldron-Edward, eds.). Pergamon, Oxford (1971).

24. Newer Trace Elements in Nutrition (W.Mertz and W.E.Cornatzer, eds.). Dekker, New York (1971).

25. Selenium in Nutrition. National Academy of Sciences, Washington (1971).

26. Davies, I.J.T.: The Clinical Significance of the Essential Biological Metals. William Heinemann, London (1972).

27. Epstein, E.: Mineral Nutrition of Plants: Principles and Perspectives. Wiley, New York (1972).

28. Metallic Contaminants and Human Health (D.H.K.Lee, ed.). Academic Press, New York (1972).

29. Metal Ions in Biological Systems (S.K.Dhar, ed.). Plenum, New York (1973).

30. Nuclear Activation Techniques in the Life Sciences - 2. IAEA, Vienna (1973).

31. Organic Selenium Compounds: Their Chemistry and Biology (D.L.Klayman and W.H.H.Günther, eds.). Wiley, New York (1973).

32. Koch, O.G., Koch-Dedic, G.A.: Handbuch der Spurenanalyse, Vol.1-2. Springer, Berlin (1974).

33. Lang, K.: Wasser, Mineralstoffe, Spurenelemente. Steinkopff, Darmstadt (1974).

34. Chromium. National Academy of Sciences, Washington (1974).

35. Elektrolyte und Spurenelemente in der Intensivmedizin (U.Henneberg et al., eds.). De Gruyter, Berlin (1974).

36. Glutathione (L.Flohe et al., eds.). Thieme, Stuttgart (1974).

37. Protein-Metal Interactions (M.Friedman, ed.). Plenum, New York (1974).

38. Selenium (R.A.Zingaro and W.C.Cooper, eds.). Van Nostrand Reinhold, New York (1974).

39. Trace Element Metabolism in Animals - 2 (W.G.Hoekstra et al., eds.). Univ. Park Press, Baltimore (1974).

40. Schröder, H.A.: Elements in Living Systems. Plenum, New York (1975).

41. Heavy Metals in the Environment (F.W.Oehme, ed.). Dekker, New York (1975).

42. Mineral Resources and the Environment. National Academy of Sciences, Washington (1975).

43. Nickel. National Academy of Sciences, Washington (1975).

44. Spurenelemente in der Entwicklung von Mensch und Tier (K.Betke and F.Bidlingmaier, eds.). Urban und Schwarzenberg, München (1975).

45. Trace Elements in Soil-Plant-Animal Systems (D.J.D.Nicholas and A.R.Egan, eds.). Academic Press, New York (1975).

46. Trace Subst. Environ. Health, Vol.1-8. Proc. Univ. Mo. Annu. Conf. (1968-1975).

47. Fluoride, Teeth and Health: a Report of the Royal College of Physicians. Pitman, London (1976).

48. Molybdenum in the Environment, Vol.1-2 (W.Chappell and K.Peterson, eds.). Dekker, New York (1976).

49. Nuclear Techniques in Animal Production and Health. IAEA, Vienna (1976).

50. Selenium. National Academy of Sciences, Washington (1976).

51. Trace Elements in Human Health and Disease, Vol.1-2 (A.S. Prasad and D.Oberleas, eds.). Academic Press, New York (1976).

52. Underwood, E.J.: Trace Elements in Human and Animal Nutrition. Academic Press, New York (1977).

53. Arsenic. National Academy of Sciences, Washington (1977).

54. Copper. National Academy of Sciences, Washington (1977).

55. Iron-Sulfur Proteins, Vol.1-3 (W.Lovenberg, ed.). Academic Press, New York (1973-1977).

56. Zinc Metabolism (G.J.Brewer and A.S.Prasad, eds.). Liss, New York (1977).

57. Iyengar, G.V., Kollmer, W.E., Bowen, H.J.M.: The Elemental Composition of Human Tissues and Body Fluids. Verlag Chemie, Weinheim (1978).

58. Luckey, T.D., Venugopal, B.: Metal Toxicity in Mammals, Vol.1-2. Plenum, New York (1977-1978).

59. Prasad, A.S.: Trace Elements and Iron in Human Metabolism. Wiley, New York (1978).

60. New Trends in Bio-inorganic Chemistry (R.J.P.Williams and J.R.R.F.Da Silva, eds.). Academic Press, New York (1978).

61. Nutritional Disorders, Vol.3 (M.Rechcigl Jr., ed.). CRC Press, Cleveland (1978).

62. Trace Element Metabolism in Man and Animals - 3 (M.Kirchgessner et al., eds.). ATW, Freising-Weihenstephan (1978).

63. Zinc and Copper in Clinical Medicine (K.M.Hambidge and B.L.Nichols Jr., eds.). Spectrum Publications, New York (1978).

64. Bowen, H.J.M.: Environmental Chemistry of the Elements. Academic Press, New York (1979).

65. Hamilton, E.I.: The Chemical Elements and Man. Thomas, Springfield (1979).

66. Young, R.S.: Cobalt in Biology and Biochemistry. Academic Press, New York (1979).

67. Chromium in Nutrition and Metabolism (D.Shapcott and J. Hubert, eds.). Elsevier/North-Holland, Amsterdam (1979).

68. Nuclear Activation Techniques in the Life Sciences - 3. IAEA, Vienna (1979).

69. Spurenelemente: Analytik, Umsatz, Bedarf, Mangel und Toxikologie (E.Gladtke et al., eds.). Thieme, Stuttgart (1979).

70. Trace Metals in Health and Disease (N.Kharasch, ed.). Raven Press, New York (1979).

71. Naumann, D.: Fluor und Fluorverbindungen. Steinkopff, Darmstadt (1980).

72. Saner, G.: Chromium in Nutrition and Disease. Liss, New York (1980).

73. Scholz, H.: Mineralstoffe und Spurenelemente. Paracelsus, Stuttgart (1980).

74. Iron in Biochemistry and Medicine, Vol.1-2 (A.Jacobs and M.Worwood, eds.). Academic Press, New York (1974-1980).

75. Mineral Tolerance of Domestic Animals. National Academy of Sciences, Washington (1980).

76. Molybdenum and Molybdenum-containing Enzymes (M.P.Coughlan, ed.). Pergamon, Oxford (1980).

77. Molybdenum Chemistry of Biological Significance (W.E.Newton and S.Otsuka, eds.). Plenum, New York (1980).

78. Nickel Toxicology (S.S.Brown and F.W.Sunderman Jr., eds.). Academic Press, New York (1980).

79. Trace Element Analytical Chemistry in Medicine and Biology - 1 (P.Brätter and P.Schramel, eds.). De Gruyter, Berlin (1980).

80. Dollwet, H.H.A.: The Copper Bracelet and Arthritis. Vantage, New York (1981).

81. Owen, C.A., Jr.: Copper Deficiency and Toxicity. Noyes, Park Ridge (1981).

82. Copper in Soils and Plants (J.F.Loneragan et al., eds.). Academic Press, New York (1981).

83. Copper Proteins (T.G.Spiro, ed.). Wiley, New York (1981).

84. Mineral Elements '80, Vol.1-2. Academy of Finland, Helsinki (1981).

85. Selenium in Biology and Medicine (J.E.Spallholz et al., eds.). AVI, Westport (1981).

86. Iron-Sulfur Proteins (T.G.Spiro, ed.). Wiley, New York (1982).

87. Trace Element Metabolism in Man and Animals - 4 (J.M. Gawthorne et al., eds.). Springer, Berlin (1982).

88. Shamberger, R.J.: Biochemistry of Selenium. Plenum, New York (1983).

89. Trace Element Analytical Chemistry in Medicine and Biology - 2 (P.Brätter and P.Schramel, eds.). De Gruyter, Berlin (1983).

90. Kiem, J., Weese, H., Feinendegen, L.E.: Fresenius Z. Anal. Chem. 317, 438-440 (1984).

D i s c u s s i o n Session XI - 1
Plenary Lecture

Subjects of the discussion

- validity of data used in a shown equation
- the need for getting more information on trace elements before supplyng them as drugs
- danger of the shown equations and formulas in connection with supplementation
- essentiality of Rb

Contributors: Binnerts, Brätter, Hurley, Kirchgessner, White

TRACE ELEMENTS IN RENAL INSUFFICIENCY. THE 5/6-NEPHRECTOMIZED RAT AS A MODEL OF THE URAEMIC STATE.

Dieter Gawlik, Dietrich Behne
Hahn-Meitner-Institute for Nuclear Research Berlin

Dieter Kraft, Gerd Offermann
Department of Internal Medicine Klinikum Steglitz,
Freie Universität Berlin

Introduction

There is still some interest in the effect of renal insufficiency on trace element metabolism. Into the data base MEDLARS 2 of the Deutsches Institut für medizinische Dokumentation und Information during the last 4 years a total of 252 articles on this subject were entered. The majority dealt with the elements iron and zinc and especially with aluminium. In merely a smaller part of the publications (approximately 60) the consequences of renal insufficiency itself are described, whereas in a far greater part - particularly in most articles on the aluminium problem - the adverse effects of haemodialysis and medication are discussed.

The multiple function of the kidney however may have impacts on the trace element metabolism long before the start of the dialysis treatment. The anaemia accompanying most renal diseases may also be the cause of trace element changes. Two contributions to renal anaemia are of interest here: 1. haemolysis, shortening the lifetime of the red cells and 2. a diminished production of red cells caused by a deficiency of the stimulating hormone erythropoietin. Further down we shall demonstrate, that the shortening of the life-time of erythrocytes may lead to a higher mean zinc content of these cells.

Moreover, we would like to report on studies on a group of non dialyzed

Trace Element - Analytical Chemistry in Medicine and Biology, Vol. 3

uremic patients with the intention to prove if the element contents of plasma or red cells might be used as an indicator for the progress of the disease.

Secondly we would like to discuss, to what extent the trace elements contents of plasma and erythrocytes in this particular disease reflect the element changes in particular organs or in the total body and could therefore be used for the diagnosis of enrichment or deficiency.

Finally we want to compare the trace element changes in chronically uraemic rats with those occuring in man to test the significance of animal experiments for investigations in this field.

We used neutron activation analysis as tool for element determination. Doing so, we were able to detect calcium, phosphorus, sodium, potassium, aluminium and fluorine in bone after a short irradiation of 20 seconds. After an irradiation of 10 days we have measured the electrolyte elements rubidium and caesium and the mainly protein-bound elements zinc and selenium in plasma, erythrocytes, organs and other soft tissues and in bone.

Patients: 24 (Interstitial nephritis, cystonephrosis)
No dietary restrictions.

Controls: 11 (Sex and age matched).

	Controls	Patients	Patients/Controls
Plasma creatinine (μmole·l^{-1})	< 8o	1oo-1ooo 1/	1,25-12,5
Hematocrit (%)	48,1	36,o	o,75
Relative dry mass, blood plasma (%)	9,3o	9,27	1,oo

1/ 3 groups: < 25o
25o-5oo
> 5oo

Fig. 1: Clinical data of patients and controls

Figure 1 presents some clinical data of the group of patients investigated by us. In the column at the right the relative changes in comparison with the controls are shown.

The easily obtainable creatinine or urea content of plasma is commonly used in clinics as a measure of the progress of the disease. Therefore we chose these parameters in preference to the creatinine clearence being more significant for the glomerular filtration but less easy to perform. Because of the unevitable scatter of the plasma creatinine content caused by influences like physical activity or muscle mass we kept to groups with a relatively wide margin of creatinine values during the investigation of changes in the trace element content with the progress of the disease. The upper limit of 1000 μmole $\cdot l^{-1}$ corresponds to the twelvefold normal value of plasma creatinine.

The renal anaemia becomes apparent in the haematocrit being decreased to 75 % of the normal value.

The relative dry mass of the blood plasma as a not very accurate measure of the protein content is unchanged.

We would like to mention, that the patients had non dietary restrictions.

5/6 nephrectomy:

It is the aim of the 5/6 or more generally of the subtotal nephrectomy to produce over a larger period an unsufficient renal function in the animals thus causing a chronical uraemia for about 4 till 6 weeks. The mass of kidney to be removed must range between one kidney and the total ectomy of both. In the latter case the animals survive owing to the then developing acute uraemia a couple of days only.

In order to prevent the regeneration of the tissue in the partly removed kidney it is necessary to irradiate the organ after nephrectomy with 400 rad X-rays.

Figure 2 shows the clinical data of rats treated that way four weeks after the nephrectomy. Again the right column presents the relative changes in comparison to the controls.

The nephrectomized animals having a normal body weight show a plasma urea increased by a factor of 3,8.

The formation of the renal anaemia can be realized by the haematocrit

being 12 % lower than in the controls.
The animals were pair fed and as with the uraemic patients they have an unchanged relative dry mass of the blood plasma indicating an unchanged total plasma proteine.

	Controls	5/6 nephr.	5/6 nephr./Contr.
Plasma urea ($mmole \cdot l^{-1}$)	5,9	22,5	3,8 ($p < o,oo1$)
Body weight (g)	221	231	1,o5
Hematocrit (%)	44,4	39,1	o,88 ($p < o,oo2$)
Relative dry mass, blood plasma (%)	7,48	7,64	1,o2

Figure 2: Clinical data of 5/6 nephrectomized rats 4 weeks after nephrectomy.

The composition of the commercial maintenance diet was tested by us. In the middle column of figure 3 you will find the contents of the trace elements which have been investigated in the rats. The selenium and the zinc content of the diet is surely sufficient.

The rubidium content equals approximately that of human food. From the potassium content of 1 % results a rubidium/potassium ratio of $0,4 \cdot 10^{-3}$. Besides we want to point out, that in the diet used by us which is rather frequently used in Germany and in another well known german product we have found aluminium contents of 150 till 250 $mg \cdot kg^{-1}$. This corresponds

Rat diet

Altromin 132o 1/

Na	o,31 %	Rb	4,o	$mg \cdot kg^{-1}$	Co	18o	$\mu g \cdot kg^{-1}$
K	1,o %	Cs	32	$\mu g \cdot kg^{-1}$	Cr	1,o	$mg \cdot kg^{-1}$
P	1,3 %	Zn	73	$mg \cdot kg^{-1}$	Al	155-25o	$mg \cdot kg^{-1}$
Ca	o,81 %	Se	o,56	$mg \cdot kg^{-1}$			
		F	1o,o	$mg \cdot kg^{-1}$			

1/ Altromin Spezialfutterwerke GmbH
Lange Straße 4o
D-4937 Lage
Deutschland

Figure 3: Element content of rat diet (dry weight base)

to the three to tenfold content of human food. Therefore it has to be taken into account in the case of experiments with oral supplementation of aluminium.

Results and discussion

In the following presentation of the results the relative changes of the contents compared with the controls are given. Thus a ratio of element content of 1 means no change caused by the disease.

Figure 4 and 5 present the element content of plasma and erythrocytes of undialyzed patients in an early stage of renal insufficiency. Within the triplets of columns each left column corresponds to the lowest plasma creatinine content.
Apparently the elements rubidium, zinc and selenium are decreased while the caesium content is elevated.
With the observation of a 15 till 20 % lower zinc content we are in accordance with approximately 30 other authors.
As we do, Alfrey et al. (1) find decreased values for rubidium and Cornelis et al. (2) find similar changes as we for all four elements, but that in a group of dialyzed patients.

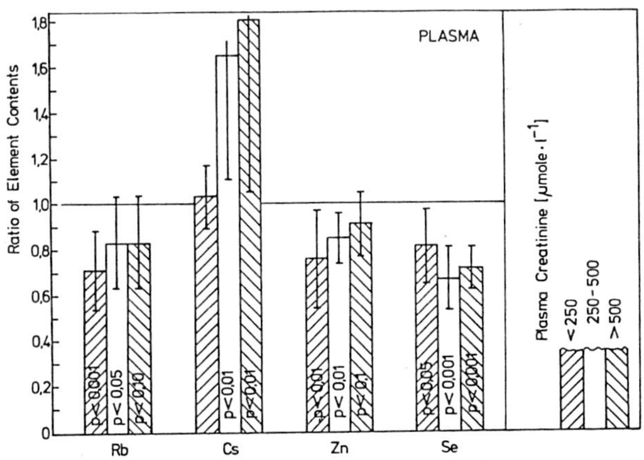

Figure 4 : Element contents of blood plasma in renal insufficiency compared with healthy controls. Grouped according to plasma creatinine content. (Bars: Single standard deviation).

Obviously all element changes besides plasma caesium do not depend on the plasma creatinine content.

For the same group of patients figure 5 shows the changes of the element contents in the erythrocytes.
Again rubidium is decreased and the caesium is elevated but the protein bound elements selenium and zinc behave differently as in the plasma: While selenium remains in the normal range, zinc in erythrocytes is elevated from the beginning.

Keeping to the question if the change in element content in blood may be used as a diagnostic aid, so one has to deny this concerning the range of creatinine contents investigated. All changes of element contents besides caesium are independent of the creatinine content of the plasma and consequently of the degree of damage to the kidney once this damage is so far in progress, that the creatinine content is significantly elevated. Moreover the underlying causes of the element changes are unsufficiently known for most elements.

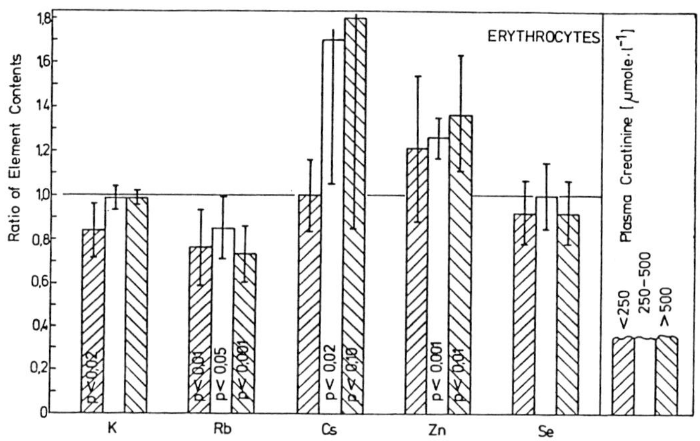

Figure 5: Element contents of erythrocytes in renal insufficiency compared with healthy controls. Grouped according to plasma creatinine content. (Bars: Single standard deviation).

Rubidium likewise, suggested by Alfrey et al. as diagnostic aid, seems not to be suitable. The diminution of the plasma rubidium is not that specific for renal insufficiency and can also be found for example in pregnancy (3) . Anyway we shall discuss possible experiments to clarify the role of rubidium later together with the rat experiment further down.

To continue with the second part of the report, namely the comparison of plasma and erythrocyte element contents with those in organs and other tissues: On the left side of figure 6 one can see the changes of rubidium (the hatched columns) and of zinc (the open columns) in plasma and in erythrocytes measured by us and already mentioned above. (A decrease of rubidium and zinc in plasma and of rubidium in erythrocytes and an elevation if zinc in the erythrocytes).

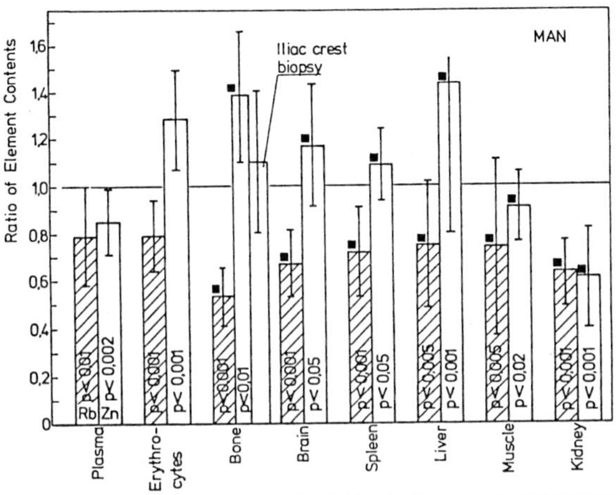

Figure 6 : Rubidium and zinc content of blood plasma and various organs in renal insufficiency compared with healthy controls.
(Filled squares: A.C. Alfrey, W.R. Smythe, "Trace Element Abnormalities in Chronic Uremia", Rep. AK-4-2217, from autopsies.
Bars: Single standard deviation).

For comparison with the changes in tissues of non dialyzed human uraemics we are dependant on the comprehensive work of Alfrey and Smythe (1). We come to know, that rubidium is decreased in all tissues by app. 30 %, while zinc, decreased in plasma, turns out to be significantly elevated in a number of tissues. The exceptions with a decreased zinc content are muscle and the damaged kidney. In contrast to Alfrey we have measured in iliac crest biopsies of uraemic patients normal zinc values.

We would like to include into the discussion the results of corresponding measurements on nephrectomized rats.

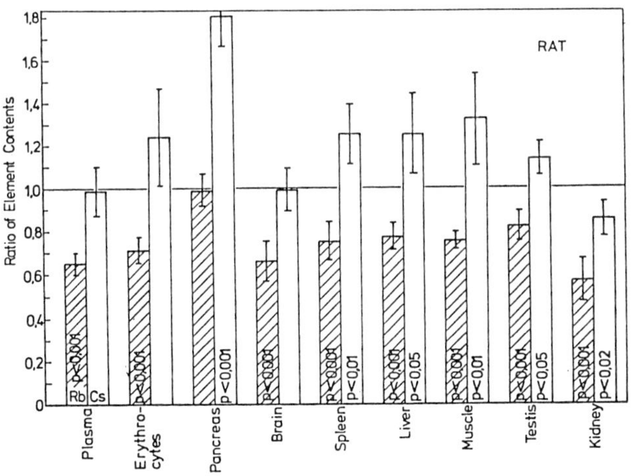

Figure 7 : Rubidium and caesium content of plasma and various organs of 5/6 nephrectomized rats compared with healthy controls.
(Bars: Single standard deviation).

Rubidium (the hatched columns) is decreased in all tissues, as well as in plasma and in the erythrocytes, while caesium (the open columns) here in rats in contrast to uraemic man remains unchanged in the plasma but is

elevated in erythrocytes and in some organs. The kidney is the only organ that shows a decrease of the caesium content.
As already mentioned above the potassium content of all tissues remains unchanged after the nephrectomy, a fact which is not shown here.

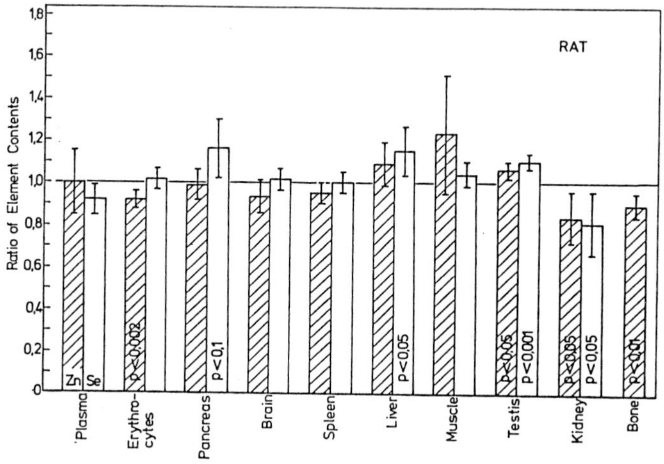

Figure 8: Zinc and selenium content of plasma and various organs of 5/6 nephrectomized rats compared with healthy controls.
(Bars: Single standard deviation).

The elements zinc and selenium (fig.8) are generally not changed in the blood and in the organs of the uraemic rats (the exceptions are again the kidney and the elevated contents for both elements in the testis).
The significant decrease of the zinc content in the erythrocytes will be discussed lateron. We remember, with human uraemics we had altogether observed elevated zinc contents in the erythrocytes.
Answering the question, to what extent the contents in the blood components reflect changes in organs, other tissues or in the whole body we have obviously to classify the investigated trace elements into three groups:
1. The electrolyte element rubidium is decreased in plasma, in the erythrocytes and also in all of the measured organs. Here the plasma content seems to reflect the total body status.
2. This is not the case with the other alcaline element caesium. This element is unchanged in the plasma of the uraemic rats but elevated in some of the organs.

In contrast to rubidium and caesium, the potassium level in the plasma and in the erythrocytes of our group of uraemic patients and in all investigated organs of the rats is normal. So these two elements behave differently as potassium.

This fact should be considered in the interpretation of tracer experiments using ^{86}Rb for the investigation of the potassium metabolism.
The drastic loss of rubidium of the whole body still requires an explanation. While the tubular backresorption of potassium is still sufficient despite the diminished number of nephrons this may not be true for rubidium .

3. A third group consists of the elements zinc and selenium which are decreased in the plasma of man, but unchanged in that of the rats. Here the plasma values seem to be least significant for the contents of the organs or the whole body
With the exception of the decreased values in the kidney there is no indication for a deficiency of these elements in any organ, a fact already mentioned for zinc by C.J. Condon et al. in 1970 (4). Alfrey even estimated an elevation of total body zinc in uraemia (1).

It is possible, that the slight decrease of the plasma selenium and zinc content depends on nutrition. The fact, that with pair fed uraemic rats the zinc and selenium plasma levels do not decrease compared to controls is a clue to the influence of nutrition. Unfortunately we did not use ad lib. fed controls besides the pair fed animals to confirm this assumption.

We have already compared in our report the results for renal insufficient patients with those for uraemic rats and we want to comprise this for plasma, erythrocytes and bone (fig. 9).

All hatched columns are the values for man and the open

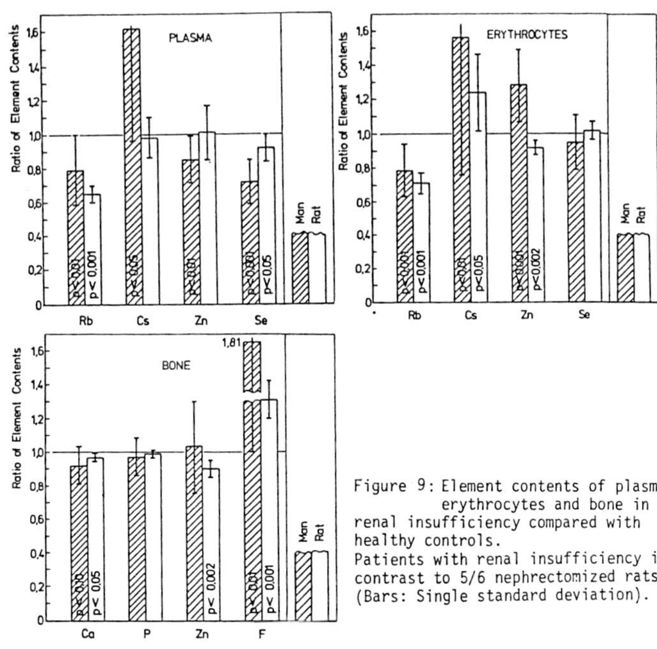

Figure 9: Element contents of plasma, erythrocytes and bone in renal insufficiency compared with healthy controls.
Patients with renal insufficiency in contrast to 5/6 nephrectomized rats.
(Bars: Single standard deviation).

columns those for the rats.
The element rubidium behaves in plasma, in erythrocytes and, as we know, in all other tissues investigated by us in man as in rat the same.
Caesium, the other investigated electrolyte element is elevated in the erythrocytes of man and rat but there is a contrast in the plasma values, which are only elevated in man and not in the rat.

The diminished capability of excretion of the kidney for fluorine both in man and rats results in a significant elevation of the bone fluorine contents. This is demonstrated by the comparison of rat femuri with human iliac crest biopsies on the bottom left of the picture.

It was mentioned above, that the different behaviour of the proteine bound elements zinc and selenium in plasma with man and rats depends possibly on nutrition. So it cannot be excluded, that the rat might serve as a model for the investigation of the metabolism of these elements in uraemia too.

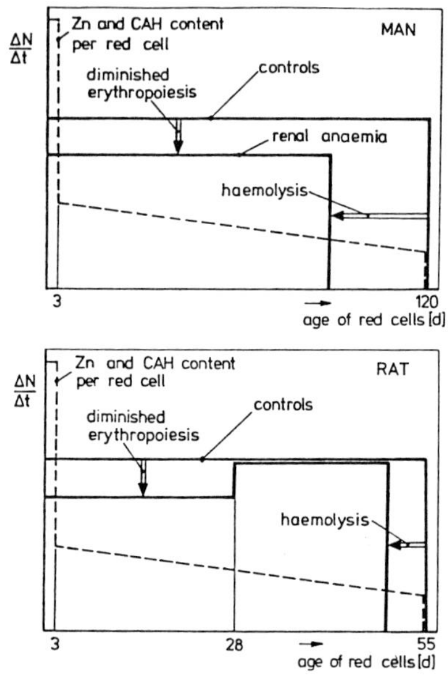

Figure 10: Number of red cells per life span interval N/ t (solid lines) and zinc and carbonic anhydrase (CAH) content per cell (broken lines) versus age of cells. (Schematic)

Top: Effect of renal anaemia in man.

Bottom: Rats 28 days after 5/6 nephrectomy.

There has yet an explanation to come for the different changes of erythrocyte zinc in man and nephrectomized rats. We think, the underlying cause is a change in the age spectrum of the red cells. (figure 10).

The upper part of the picture is showing the facts for man. The solid lines show the number of the red cells Δ N per life span interval Δ t in dependance of the age of the cells, the age spectrum.

Normally in man this covers a period of 120 days, the lifetime of the red cells. Anaemia can reduce either the lifetime by haemolysis or the number of the cells produced per time by diminished erythropoiesis.
Consequently in the case of haemolysis there are on the average more young erythrocytes in the blood.

We found, that the content of carbonic anhydrase and therefore the zinc content also is higher in reticulocytes and young erythrocytes than in old ones. The age dependance of the carbonic anhydrase activity is schematically illustrated in the picture by the broken line with the peak value for reticulocytes during the first three days of the lifetime.
So the haemolytic part of the renal anaemia causes an increase of the average zinc content of the red cells.
We presume hereby, that during a time comparable with the lifetime of the red cells only minor changes in the production rate have occured, in other words, that a quasi-stationary state has developed.

This stationary state does not exist in the nephrectomized rats as illustrated in the lower part of the picture. Because the experiment ended 28 days after the nephrectomy the production rate is diminished not longer than this time and on the average there is a greater number of old erythrocytes resulting in a lower zinc content than in the controls.
The contrast between the decreased zinc contents in red cells of chronically uraemic rats and the elevated values in man seems to us a good example for a possible wrong interpretation of an animal experiment.

On the other hand the parallel behaviour of rubidium in rat and man asks for more experiments with rats to clarify the reason of the total body loss of this element. The fact, that we did not find a dependance between the rubidium content of plasma and erythrocytes and the plasma creatinine in the range of 100 till 1000 $\mu mole \cdot l^{-1}$ does not exclude the possibility, that plasma rubidium might be changed by an even less damaged kidney. One must keep in mind, that the plasma creatinine rises measurable not before approximately half of the kidney parenchyma is inactive.

References

1. Alfrey, A. C., Smythe, W. R.: Rep. AK-4-4-2217 (1981)

2. Cornelis, R., Ringoir, S., Mees, L., Lameire, N., Wallaeys, B., Hoste, A.: Tema 4, Perth, Australia (1981)

3. Rösick, U.: Personal communication

4. Condon, C. J., Freeman, R. M.: Annals of Internal Medicine, 73, 531 - 536 (1970).

Acknowledgement

We are indebted to Mrs. Gerda Röhlicke and Mr. Wolfram Gatschke for their skilful performance of the neutron activation analysis.

KINETICS OF ^{109}CADMIUM IN CEREBRAL STRUCTURES AND BLOOD OF RATS SUPPLEMENTED WITH ZINC

Ana María de la Torre, Samed M. Katarneh, Isabel Pérez Escolano, Asunción Arranz and Bartolomé Ribas.

Instituto de Bioquímica, Centro Mixto C.S.I.C.-Universidad Complutense, Madrid-3, España.

Introduction

The biological interest of cadmium in the substitution of zinc in proteins, invites to study its pharmacokinetics in cerebral structures, both metals probably inducing the synthesis of metalloproteins in brain (1, 2). Cadmium has a slow turnover (3) but it can go through the blood brain barrier (4) in both directions from and to the cerebrospinal fluid (5), presenting a symptomatology of cephalea, vertigo and insomnia (6) with workers of electric storage batteries, showing direct effects in the CNS (7). It is evident therefore that Cd is an etiological agent in pathology. The toxicity of Cd is greater in young growing than in adult animals, inducing hemorrhagical lesions principally in encephalon and cerebellum, selective lesions of sensory ganglia (8) and bone marrow changes in cases of chronic cadmium poisoning in rabbits (9).

The administration of Cd-109 by intracerebroventricular way to the rat increases the motor activity, the serotonin and hydroxyindolacetic acid in the brain (10, 11) and the activity of some brain enzymes (11). Neurotoxic effects of Cd are more effective in young rats than in adult ones (12). After subcutaneous administration of Cd the toxicological effects are principally localized in the peripheral nervous system, in the sensitive ganglia (8) with demielinization of the peripherical nerve (13) and in the cortex of new born rats (12). The lesion

Trace Element - Analytical Chemistry in Medicine and Biology, Vol. 3

of the nucleus caudatus which normally inhibits the motor activity, increases this activity (7).

The kinetics of Cd in blood and other tissues is suggested to be peculiar (12, 14). Cd compared to other bivalent trace elements as Zn, Se, Cr and Hg is signalized to have a lower turnover and a kinetic sigmoid profile in fish (14, 15), certainly under the influence of the metallothionein biosynthesis in the presence of those metals.

The clearance of Cd in blood in a period of 120 min has three phases, the first of which has rapid kinetics (14, 16).

Material and Methods

To male Wistar rats of 250 g was administered 2 μCi $^{109}Cd\ Cl_2$ through intracerebroventricular way in both lateral ventricles each group previously treated or not with 5 doses of zinc. Other rat groups with the same conditions are used for the blood kinetics after endovenous $^{109}Cd\ Cl_2$ administration through the suprahepatic vein, each group previously treated with 5 or 10 doses of zinc by intraperitoneal way.

As scheduled, animals were sacrified, for the kinetics in cerebral structures, in function of time and hypothalamus, hypocampus, cortex, striatum, cerebellum, mesencephalon and medulla oblongata were dissected, weighed and cpm determined. For the experiments in blood kinetics, animals were anaesthetized with ether and after laparotomy the radionucleic ^{109}Cd, was injected with a single dose of 50μCi ^{109}Cd through the suprahepatic way. Blood samples were taken through the same way in function of time, and IV gouts weighed and cpm determined. The cpm of the samples of each experiment were processed with the HP-85 computer program Elsfit 1-2.

Results

In table 1 are expressed the pharmacokinetic parameters in blood of controls with 5 to 10 doses of zinc. These data are also plotted in the figures 1, 2 and 3.

PARAMETER	CONTROL	ZINC TREATMENT (5 dosis)	ZINC TREATMENT (10 dosis)
$\alpha (h^{-1})$	10.142	15.150	13.512
$\beta (h^{-1})$	0.119	0.152	0.080
$t\frac{1}{2}\ \alpha$ (h)	0.062	0.045	0.051
$t\frac{1}{2}\ \beta$ (h)	5.823	4.523	8.630
$k_{21} (h^{-1})$	1.047	1.072	1.994
$k_{12} (h^{-1})$	8.947	12.062	11.054
$k_{13} (h^{-1})$	1.266	2.164	0.544

Table 1. Pharmacokinetic parameters of ^{109}Cd in blood of control rats and treated ones with 5 and 10 doses of zinc sulphate processed with the Elsfit 1-2 program in a HP-85 computer. (α = velocity constant of the α phase; β = velocity constant of β phase; $t\frac{1}{2}\ \alpha$ = half-life of the α phase; $t\frac{1}{2}\ \beta$ = half-life of the β phase; k_{12}, k_{13} are constants of pharmacotransference).

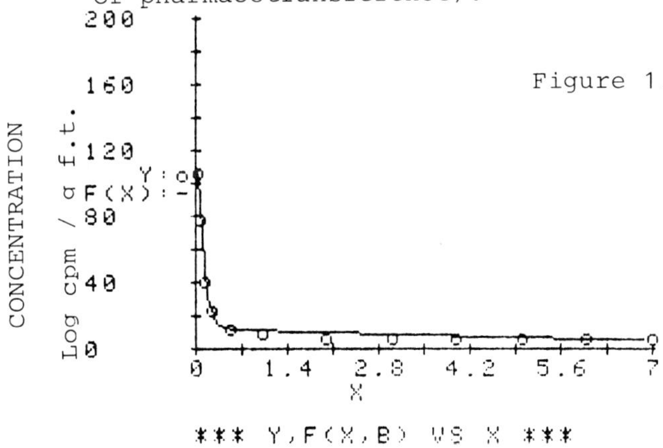

Figure 1. Kinetics of ^{109}Cd in control rats.

In tables 2 and 3 are observed the kinetic parameters in several cerebral structures of the control rats with doses of 1 mg zinc/kg respectively.

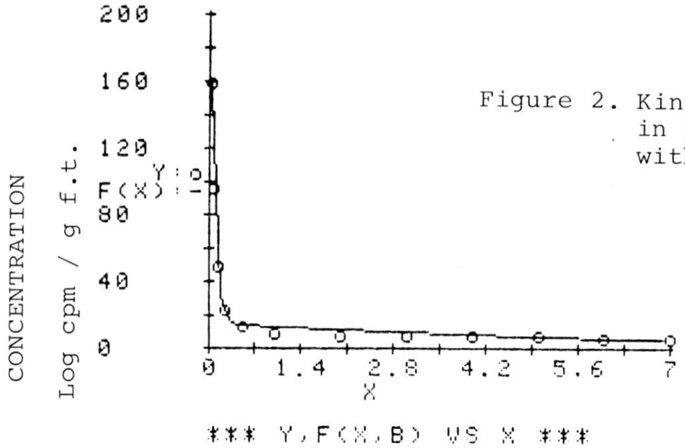

Figure 2. Kinetics of ^{109}Cd in rats supplemented with 5 doses of Zn.

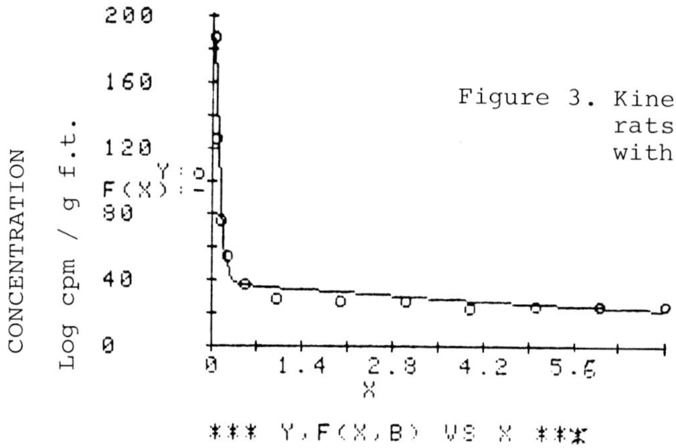

Figure 3. Kinetics of ^{109}Cd in rats supplemented with 10 doses of Zn.

Discussion

The peculiar kinetics of Cd in different animal species and tissues, suggested in various studies (12, 14, 15) is the consequence of the low Cd turnover (3) and the synthesis of several metalloproteins (1) induced by the physicochemical and biological characteristics of cadmium.

In blood are considered two pharmacokinetic phases, the first or inicial one being the incorporation of Cd into the hetero-

Table 2. Pharmacokinetic parameters of ^{109}Cd in cerebral structures of control rats, calculated with the method of Wagner and Nelson. The signification of α and β are the same as in table 1; Ka= absorption constant; k_{12}, k_{21} and k_{13} are the microscopic constants of pharmacotransference, AUC_{o}^{120} is the area under the curve. Values of the relative area taking the hypothalamus of zinc treated rats as reference unit, are also expressed.

CEREBRAL STRUCTURES	α (h^{-1})	β (h^{-1})	$t_{\frac{1}{2}}$ (d)	Ka (h^{-1})	K_{12} (h^{-1})	K_{21} (h^{-1})	K_{13} (h^{-1})	AUC_{o}^{120}	RELATIVE AREA
M. OBLONGATA	0.694	0.0059	4.86	1.7779	0.3742	0.3125	0.0132	1 608 479	0.457
CORTEX	0.460	0.0059	4.83	1.8641	0.2655	0.1857	0.0147	1 593 951	0.453
CEREBELLUM	0.411	0.0055	5.25	2.1600	0.1856	0.2207	0.0102	2 012 745	0.572
MESENCEFALON	0.420	0.0029	9.72	1.8258	0.2066	0.2109	0.0059	688 066	0.195
HYPOTHALAMUS	0.135	0.0059	4.83	0.3066	0.0469	0.0845	0.0095	2 097 854	0.596
HYPOCAMPUS	0.028	0.0040	7.16	0.0326	0.0047	0.0222	0.0050	2 480 421	0.705
STRIATUM	0.294	0.0040	7.12	0.4900	0.0967	0.1952	0.0061	1 901 891	0.541

Table 3. Pharmacokinetic parameters of ^{109}Cd in cerebral structures of rats treated with 10 doses of $ZnSO_4$, computed with the method of Wagner and Nelson. The signification of α, β, $t\frac{1}{2}\alpha$, $t\frac{1}{2}\beta$ are the same as in table 1. Ka, K_{12}, K_{21}, K_{13} and AUC_o^{120} are also expressed in table 2. Values of the relative area taking hypothalamus as reference unit are also signalized.

CEREBRAL STRUCTURES	α (h^{-1})	β (h^{-1})	$t\frac{1}{2}$ (d)	Ka (h^{-1})	K_{12} (h^{-1})	K_{21} (h^{-1})	K_{13} (h^{-1})	AUC_o^{120}	RELATIVE AREA
M. OBLONGATA	0.027	0.0045	6.37	0.0709	0.0084	0.0156	0.0079	1 858 295	0.528
CORTEX	0.444	0.0037	7.25	1.1258	0.2906	0.1457	0.0113	1 527 640	0.434
CEREBELLUM	0.399	0.0047	6.12	2.5410	0.2827	0.0908	0.0207	805 609	0.229
MESENCEFALON	0.851	0.0048	5.97	1.7559	0.5716	0.2842	0.0041	678 121	0.192
HYPOTHALAMUS	0.513	0.0093	3.08	0.6648	0.0740	0.4373	0.0109	3 515 408	1.0
HYPOCAMPUS	0.017	0.0037	7.74	0.0254	0.0042	0.0103	0.0061	3 374 488	0.959
STRIATUM	0.485	0.0029	9.87	0.8050	0.2954	0.1848	0.0076	2 151 702	0.612

genic cellularity of blood and systemic organs, and the second one being the biosynthesis of metalloproteins with the specificity of each organ. The first phase shows in blood a rapid biexponential kinetics as already known (17, 18), but others proposed three exponential components for the clearance of Cd in the first 120 minutes (18). The half-life of β phase in control rats is 5.82 h and in rats treated with 10 doses of zinc in 8.63 h. In rats supplemented with zinc, ^{109}Cd has a higher biological half-life, because the association of Cd to sulphur proteins and certain metalloproteins induced by zinc. At the same time the half life of the α phase is slower in rats treated with zinc ($t\frac{1}{2}$ α with 5 doses = 0.045 h and with 10 doses 0.051 h and for the controls 0.062 h) than for the control animals. The distribution constant k_{12} increased with the supplement of zinc from 8.947 h^{-1} to 11.054 h^{-1}.

In cerebral structures, cadmium is regularly incorporated,the contrary as in systemic organs (17) with a great differential specificity. Nevertheless, hypothalamus and hypocampus structures with greater neurochemical and behavioural signification the first being philogenetically older, are the two of higher ^{109}Cd incorporation.

These results are corroborated by previous works, increasing Cd the synthesis of 5-HT (10) and the rat motor activity (10, 11). Intracerebroventricular supplementation increases the accumulation of ^{109}Cd in cerebral structures as shows the areas under the curves. In hypothalamus, the absorption constant of supplemented animals is increased 0.664 h^{-1} compared to controls 0.306 h^{-1}. The half-life of ^{109}Cd in β phases in control rats oscilated between 4.83 and 9.72 days, whereas with 10 doses supplemented animals oscilates between 3.08 and 9.87 days. The high biological half-life of cadmium signalizes the presence of proteins with thiolic groups, its strong fixation, lower turnover and catabolism, and the induction of the synthesis of certain metalloproteins or metallothionein(1, 2).

References

1.- Metallothionein Proceedings lst. Intern. Meeting, Edts.: Kägi J.H.R. and Nordberg M., Birkhausen Verlag, Basel, Boston, Stuttgart (1979).

2.- Ribas B.: Trace Elements Anal.Chem.Med.Biol., vol. 2,Edts.: P. Brätter, P.Schramel, Walter de Gruyter Co., Berlin-New York, 585-597, (1983).

3.- Cotzias G.C., Borg D.C., Selleck B.: Am.J.Physiol., 201, 927-930, (1961).

4.- Berlin M., Ullberg S.: Arch.Environm.Health 3,176-184(1963)

5.- Katarneh S.M.: Tesina Facultad de Farmacia, Univ. Compl. Madrid: Farmacocinética del Cadmio-109 en estructuras cerebrales y sangre de ratas suplementadas con zinc (1983).

6.- Vorobjeva R.S.: Arch. Pathol., 8, 25-32 (1957).

7.- Villablanca J.R., Olmstead C.E., Levine M.S., Markus R.I.: Exp. Neurol. 61, 615-634 (1978).

8.- Gabbiani G., Gregory A., Baic D.: J. Neuropathol. Exp. Neurol. 26, 498-506 (1967).

9.- Berlin M., Fredricson B., Linge, G.: Arch. Environm.Health 3, 176-184 (1961).

10.- Ribas B., Ochoa M.C. and Santos Ruiz A.: Trace Elements Metabolism in Humans and Animals, Edts.: Hoekstra W.G., Suttie J.W., Ganther H.E. and Mertz W., University Park Press, Baltimore, 476-478 (1974).

11.- Rastogi R.B., Merali Z., Singhal R.L.: J.Neurochem. 28, 789-794 (1977).

12.- Wong K.L., Klassen C.D.: Toxicol.Appl.Pharmacol, 63, 330-337 (1982).

13.- Sato F., Iwamasa, T., Tsuru, T., Takeuschi T.: Acta Neuropathol.: 41, 185-190 (1978).

14.- Matsubara, J.: Kadmium Symposium, Friedrich Schiller Universität Jena (1978).

15.- Jones R.H., Williams R.L., Jones A.M.: Proc.Soc.Exp.Biol. Med. 137, 1321 (1971).

16.- Boisset M., Girard F., Godin J., Boudène Cl.: C.R.Acad.Sci. Paris, Serie D 287, 61-64 (1978).

17.- Klaassen C.D., Kotsonis F.N.: Toxicol. Appl. Pharmacol. 41, 101-112 (1977).

18.- Frazier J.M., Kingsley B.S.: Toxicol. Appl. Pharmacol, 38, 583-593 (1976).

Acknowledgement: We like to thank the W.H.O. for the help to the fellow Samed M. Katarneh, Ph.M. from Ministery of Health, Jordan.

PLASMA SOMATOMEDIN-C AND ZINC STATUS AS AFFECTED BY INTERACTIONS BETWEEN VARYING LEVELS OF DIETARY PROTEIN FED IN COMBINATIONS WITH VARYING LEVELS OF ZINC SUPPLEMENT

Zafrallah T. Cossack

Wayne State University School of Medicine, Dept. of Internal Medicine and the Veterans Administration Medical Center, Medical Research (151), Allan Park, Michigan 48101, U.S.A.

Introduction

Somatomedins are peptides of low molecular weight which are regulated by growth hormone (1), nutrition (2,3,4,5) and insulin (6). It has been documented that growth hormone (GH) produces growth indirectly through generation of a serum factor(s), somatomedin, which acts directly to produce cartilage proliferation and linear skeletal growth (7,8). However, low levels of somatomedin with associated growth failure despite elevation of GH was noted in laron-type dwarfism (9) and in protein-energy malnutrition (10,11). On the other hand, a normal level of somatomedin associated with low levels of GH was observed in children who were hyperphagic and hyperinsulinemic after hypothalamic surgery (12).

Recently, it has become apparent that nutritional status is the most important factor influencing growth and development. Changes in body weight in response to dietary manipulation has been used to identify the nutrients essential for growth. The recent establishment of a method(s) for measuring the activity of somatomedin (13,14) has prompted further investigation of the role of some factors in growth. However, studies on the effects of several nutritional factors on the activity of so-

Trace Element - Analytical Chemistry in Medicine and Biology, Vol. 3

matomedin in human and experimental animals have not been conclusive. An inverse relationship between dietary fat and plasma somatomedin was observed in rats (15). Moreover, the level of somatomedin increased with increasing levels of protein (3,15) and no change was observed in the levels (conc.) of growth hormone. Most recently, we have reported (4) a significant decrease in the level of somatomedin-C (SM-C) in zinc deficient rats when they were compared to pair-fed, continuously-pair-fed and ad libitum-fed rats. In the same study we observed that plasma SM-C correlated with zinc concentration in the tibia ($r=0.79$, $P < .005$) and with body wt. gain ($r=0.96$, $P < .001$). We concluded that growth retardation associated with zinc deficiency could be related to the adverse effect of zinc deficiency on the generation of SM-C,and zinc supplementation would reverse this effect (4).

Based on data from numerous research studies, optimal levels of certain nutrients (e.g. protein and zinc) should be provided in the diet in order to maintain optimal growth and normal body function. The availability of zinc is dependent upon a number of dietary factors, including the level and source of protein (16,17,18). On the other hand, protein metabolism in human and animal is influenced by the status of zinc (19,20). Poor quality sources of protein are needed at a higher level (compared to high quality sources) in the diet to promote optimal growth rate (21). In addition, supplementation of a poor quality source of protein (bean diet) with zinc has resulted in a modest increase in growth rate (22) compared to an unsupplemented diet.

The increased growth rate in the world population may severely limit the supply of protein. Thus, the establishment of a dietary regimen, where maximum utilization of nutrients could be achieved, is highly recommended. Malnourished populations have often been supplemented with grain sources of protein and no considerable attention has been paid to the potential role of zinc supplementation. Thus, the objective of the present study is to investigate the changes in growth, SM-C, trace elements status and other parameters, in rats, in response to varying levels of dietary zinc and protein fed in combinations.

Materials and Methods

Thirty six white male growing rats of the Holtzman strain (Holtzman company, Madison, Wis.) were fed 6 test diets(table 1) based on combinations of 2 levels of protein (7.5% and 20%) and 3 levels of zinc supplements (0.9ppm, 55 ppm and 110 ppm). Table 2 shows a typical composition of the diets used in this study. The animals were housed in individual stainless steel wire buttom cages under 12-hour light-dark periods in a controlled room temperature (22°C). Initial body weight of the rats ranged from 95 to 100 g. The rats were fed a standard diet (purina rat chow, St.Louis, MO) for 4 days afterwhich they were randomly allotted into the 6 experimental diets (table 1). In a separate 3-weeks preliminary study we observed that among all groups of rats fed the experimental diets (ad libitum), the group of rats fed the ZD-Lpr (7.5% protein and 0.9 ppm Zn) diet have consumed the lowest amount of food (g/rat/day). Based on this observation, the amount of food given to all groups (in the present study) matched the intake of their counterparts of the group fed ZD-Lpr diet. The rats were given free access to distilled- deionized water throughout the experimental period (21 days) afterwhich they were killed by decapitation. Rats were mildly anesthesized with ether and blood drawn by heart puncture. The blood was collected (using EDTA as anticoagulant) in plastic tubes free of zinc, immediately centrifuged (2,000 g) for 10 min. and plasma was separated and kept at -60°C. Tibia and liver were isolated, cleaned from debris with distilled- deionized water, lyophilized, weighed and digested with nitric acid for the

Table 1. Experimental Design

Protein (%)	Zinc Supplement (PPM) 0	55	110
7.5	ZD-Lpr[1]	MZn-Lpr[2]	HZn-Lpr[3]
20	ZD-Hpr[4]	MZn-Hpr[5]	HZn-Hpr[6]

1 Zinc deficient (0.9 ppm), low protein diet.
2 Medium zinc, low protein diet.
3 High zinc, low protein diet.
4 Zinc deficient, high protein diet.
5 Medium zinc, high protein diet.
6 High zinc, high protein diet.

Table 2. Composition of Basal Diet[1]

Ingredients	g/Kg diet
Egg white solids, spray dried	varied[2]
Glucose Monohydrate (Dextrose)	varied[2]
Fiber, non-nutritive	30.00
Oil, corn	100.00
Mineral Mix[3]	25.6902
Vitamin Mix[3]	10.00
Biotin	0.004
Zinc Sulfate ($ZnSO_4$)	varied[2]

1 contained 0.9 ppm of zinc by analysis.
2 depending upon protein and zinc levels in the diets.
3 as recommended by the AIN for growing rat.

determination of zinc using a Perkin Elmer flame atomic absorption spectrophotometer (Perkin Elmer 406, Norwalk, Conn.).

Zinc in plasma was determined according to an established procedure (23). Plasma somatomedin-C was assayed using radioimmunoassay method (reagents including SM standards and controls were supplied by Nichols Institute, CA, USA, and the method is decribed by Furlanetto and co-workers, ref.13).

Results

Table 3 shows the changes in body wt. gain, concentrations of zinc in tibia, plasma and liver and the levels of SM-C, in response to dietary treatments in rats. Body wt. gains in the groups of rats fed a zinc deficient diet were significantly ($P<.001$) lower when compared to those of zinc supplemented groups regardless of the level of protein in the diet. Increasing the level of protein in the zinc deficient diet did not result in a significant increase in body wt. (ZD-Lpr vs. ZD-Hpr). However, body weight gain increased significantly ($P<0.005$) when the level of protein was increased from 7.5 to 20% in combination with moderate or high zinc supplement (MZn-Lpr vs. MZn-Hpr and HZn-Lpr vs. HZn-Hpr). In zinc deficient rats, increasing the level of dietary protein did not result in an increase in the conc. of zinc in the tibia (116 vs. 120 ug/g dry wt.). However, the conc. of zinc in this tissue increased significantly ($P<.005$) in response to an increased level of dietary protein in rats supplemented with moderate or high levels of zinc (Table 3). Liver zinc conc. increased significantly($P<.0001$)in response to zinc supplementation (ZD-Lpr vs. MZn-Lpr) and also in respose to increased levels of dietary protein (ZD-Lpr vs. ZD-Hpr ; $P<.002$). Maximum conc. of liver zinc occured when both protein and zinc were increased in the diet (MZn-Hpr). There was no effect of high (20%) dietary protein on plasma zinc conc. when rats were fed a zinc deficient diet. However, a significant ($P<.0000$) increase in plasma zinc conc. was observed in response to higher level of protein in combination with moderate or high levels

Table 3

Body Wt. Gain, Concentration of Zinc in Tissues and Plasma Somatomedin-C (means ± SD)[1]

	Dietary Protein %	Zinc Supplement (ppm)		
		0	55	110
Body Wt.gain (g/rat/day)	7.5	0.9 ± 0.1[a]	1.7 ± 0.1[b]	1.6 ± 0.3[b]
	20	1.2 ± 0.1[a]	2.2 ± 0.3[c]	2.3 ± 0.2[c]
Tibia zinc (ug/g dry wt)	7.5	116 ± 8[a]	213 ± 9[b]	217 ± 11[b]
	20	120 ± 15[a]	236 ± 15[c]	250 ± 19[c]
Plasma zinc (ug/100 ml)	7.5	54 ± 6[a]	141 ± 20[b]	147 ± 12[b]
	20	54 ± 10[a]	181 ± 25[c]	179 ± 26[c]
Liver zinc (ug/g dry wt)	7.5	92 ± 19[a]	136 ± 14[b]	136 ± 16[b]
	20	112 ± 10[c]	150 ± 15[d]	140 ± 11[bd]
Somatomedin-C (unit/ml plasma)	7.5	0.9 ± 0.1[a]	1.24 ± 0.3[b]	1.22 ± 0.1[b]
	20	0.89 ± 0.1[a]	1.37 ± 0.1[b]	1.20 ± 0.1[b]

[1] means not having common superscripts are significantly different.

of zinc supplement. Regardless of the level of protein intake, rats supplemented with 55 or 110 ppm of zinc exhibited significantly higher levels of plasma SM-C compared to zinc deficient rats (Table 3). Maximum levels of SM-C were observed in rats fed diets supplemented with 55 ppm of zinc. In addition, further increase in zinc supplement (110 ppm) did not result in any increase in the levels of SM-C. However, plasma SM-C conc. correlated ($r=0.82$, $P<.0005$) with the conc.

of zinc in the tibia (Fig.1), and with plasma zinc conc. ($r = 0.84$, $P < .0005$; Fig.2).

Discussion

The purpose of the present investigation was to study the effect of 3 levels of zinc supplement, fed in combinations with 2 levels of dietary protein, on the level of plasma SM-C and zinc status in rats. It has been documented that plasma SM-C is dependent upon, in addition to other factors, the level of zinc in the diet (4) and it correlated with zinc status in rats. In addition, the utilization of both nutrients (zinc and protein) are interdependent in animal (17,18,20) and man (16,24). The utilization of protein was impaired in zinc deficient states. On the other hand, the availability of zinc was shown to be dependent upon the source and level of dietary protein (16,17,18,20,24).

In the present study, the growth of rats (expressed as body wt. gain) increased significantly when the diet was supplemented with 55 ppm of zinc regardless of the level of protein (7.5% or 20%). However, increasing the level of dietary protein in zinc deficient rats did not improve the status of zinc in these rats as assessed by the conc. of zinc in tibia, plasma, SM-C or body wt. gain. This finding contradicts results obtained by Wallwork et al (17) and Pederson and Eggum (25). This discrepancy may be due to the differences in the levels of zinc and protein used. However, our results are supported by data from Magee and Grainger (18) where similar levels of protein and zinc have been used. The present data and data from other investigations (17,18,25) suggest that the synergism between zinc and protein occurs when marginal (not deficient) or moderate level from each nutrient is supplied in the diet. Plasma zinc conc. in rats fed ZD-Lpr diet did not change in response to a higher level of protein (20%). This

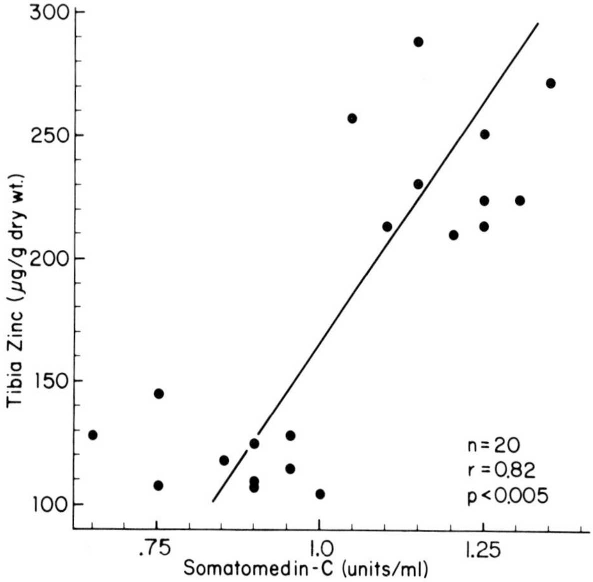

Fig. 1 Correlation between plasma SM-C and tibia zinc conc.

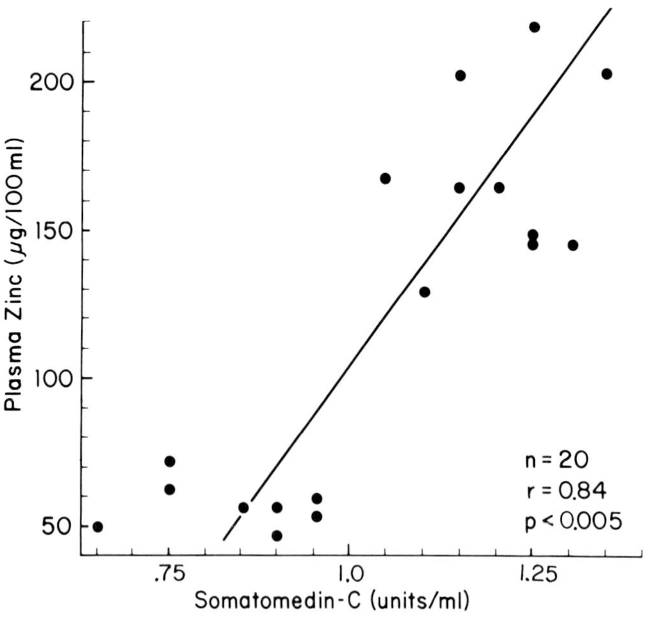

Fig. 2 Correlation between plasma SM-C and plasma zinc conc.

is consistent with previous reports (26) using casein or isolated soybean. Tibia and liver zinc conc. followed a similar trend with regard to the effects of zinc supplementation and/ or level of protein. The effect of dietary treatments was more pronounced in the tibia.

It has been well documented that dietary composition has a direct effect on plasma SM activity (2,3,4,5,10). Improvement in growth in response to zinc supplementation has been reported in children (27,28,29) and experimental animals (4). In the present study, supplementation of zinc deficient rats with 55 ppm of zinc resulted in an increase in the levels of SM-C regardless of the levels of dietary protein (Table 3). However, increasing the levels of protein did not affect the level of SM-C in zinc deficient rats and caused a slight increase in rats supplemented with 55 ppm of zinc. Reports on the effect of protein on somatomedin are inconsistent. Yeh and co-workers (30) showed an increased level of somatomedin in fasted rats when refed a protein diet. In contrast, Phillips and Young (2) found that somatomedin activity is increased most by refeeding fat and least by refeeding protein when the calories consumed were identical. The decrease in plasma SM-C in response to dietary protein, as shown in the present study, is remarkable and deserves further investigation. One possible explanation for this finding is that the calorie intake, in rats fed high protein diets, was reduced. In addition, the decrease in circulating SM-C could be a result of an increase in cartilage responsiveness to somatomedin activity when a high protein diet was fed (2).

It is noteworthy to indicate that the decline in serum GH during underfeeding in rats (31) is in marked contrast to the rise in blood GH levels observed in human subjects (32). Thus, further studies are needed to elucidate the zinc-protein interaction with regards to its effect on SM-C in human subjects.

References

1. Salmon, W.D., Jr., Daughaday, W.H.: J. Lab. Clin. Med. 49, 825-836 (1957).
2. Phillips, L.S., Young, H.S.: Endocrinology 99, 304-314 (1976).
3. Switzer, B.R., Prewitt, T.E., D'Ercole, A.T.: Fed. Proc. 40, 3974 (1981).
4. Cossack, Z.T.: Experientia (in press, 1984).
5. Phillips, L.S., Orawski, A.T., Belosky, D.C.: Endocrinology 103, 212-127 (1978).
6. Phillips, L.S., Orawski, A.T.: Diabetes 26, 864-869 (1977).
7. Daughaday, W.H., Hall, K., Raben, M.S., Salmon, W.D., van den Brande, J.L., Van Wijke, J.J.: Nature 235, 107 (1972).
8. Daughaday, W.H., Phillips, L.S., Herington, A.C.: Methods Enzymol. 378, 93-100 (1975).
9. Laron, Z., Pertzelan, A., Mannheimer, S.: Israel J. Med. Sci.:2, 152-155 (1966).
10. Hintz, R.L., Suskind, R., Amatayakul, K., Thanagkul, O., Olson, R.: J. Pediat. 29, 153-156 (1978).
11. Tanner, J.M., Whitehouse, R.H., Takaishi, M.: Arch. Dis. Childh. 41, 545-551 (1966).
12. Costin, G., Kogut, M.D., Phillips, L.S., Daughaday, W.H.: J. Clin. Endoc. Met. 42, 370-375 (1976).
13. Furlanetto, R.W., Underwood, L.E., Van Wijke, J.J., D'Ercole, A.J.: J. Clin. Invest. 60, 684-692 (1977).
14. van den Brande. J.L., Du Caju, M.V.L.: Acta Endocr. (Kbh). 75, 233-242 (1974).
15. Reeves, R.D., Dickinson, L., Lee, J., Kilgore, B., Branham, B., Elders, M.J.: J. Nutr. 109, 613-620 (1979).
16. Cossack, Z.T., Prasad, A.S.: Nutr. Res. 3, 23-31 (1983).
17. Wellwork, J.C., Johnson, L.K., Milne, D.B., Sandstead, H.H.: J. Nutr. 113, 1307-1320 (1983).
18. Magee, A.C., Grainger, F.P.: Nutr. Rep. Int. 20, 771-776 (1979).
19. Sandstead, H.H.: Am. J. Clin. Nutr. 26, 790-793 (1973).
20. Hsue, J.M., Anthony, W.L.: J. Nutr. 105, 26-30 (1975).
21. Cossack, Z.T., Weber, C.W.: Nutr. Rep. Int. 28, 203-217 (1983).
22. Boloorforooshan, M., Markakis, P.: J. Food Sci. 42, 1671-1672 (1977).

23. Whitehouse, R.C., Prasad, A.S., Rabbani, P.I., Cossack, Z.T.: Clin. Chem. 28, 475-480 (1982).

24. Sandstrom, B., Cederblad, A.: Am. J. Clin. Nutr. 33, 1778-1783 (1980).

25. Pederson, B., Eggum, B.O.: Nutr. Rep. Int. 27, 441-453 (1983).

26. Hardie-Muncy, D.A., Rasmussen, A.I.: J. Nutr. 109, 321-329 (1979).

27. Collip, P.J., Castro-Managa, M.C., Petrovic, M., Thomas, J., Cheruvansky, T., Yao Chen, G., Sussman, H.: Ann. Nutr. Met. 26, 287-290 (1982).

28. Prasad, A.S., Cossack, Z.T.: Ann. Int. Med. 100, 367-371 (1984).

29. Walravens, P.A., Hambidge, K.M.: Am. J. Clin. Nutr. 29, 1114-1121 (1976).

30. Yeh, J., Aloia, J.F.: Metabolism 27, 507-509 (1978).

31. Campbell, G.A., Kurcz, M., Marshall, S., Meites, J.: Endocrinology 100, 580-587 (1977).

32. Glick, S.M., Roth, J., Yallow, R.S., Berson, S.A.: Recent Prog. in Horm. Res. 21, 241-248 (1965).

DETERMINATION OF METALLOTHIONEIN AS A TOOL FOR "BIOLOGICAL SPECIATION" OF ZINC

Herbert E. Heilmaier[1], Peter Schramel[2] and Karl H. Summer[1]

Gesellschaft für Strahlen- und Umweltforschung München, Abteilung für Toxikologie (1) und Physikalisch-Technische Abteilung (2), D-8042 Neuherberg, F.R.G.

Introduction

Metallothionein (MT) is a low molecular weight sulfur-rich protein with high affinity for zinc (Zn) and other metals. It is present in many mammalian tissues and efficiently controls Zn homoestasis and its bioavailability. Studying Zn metabolism therefore requires the identification of Zn bound to specific ligands like MT (biological speciation).

Presently available methods to detect MT are either laborious or relatively insensitive. We now describe a modified assay for MT with high sensitivity, which is based on the cadmium (Cd) saturation method of Onosaka (1). The method takes advantage of the quantitative binding of Cd to MT, the heat stability of MT and the definite molar ratio of 7 mol Cd per mol MT (2). This test permits the accurate measurement of 160 ng MT.

The assay was applied to differentiate between values of Zn bound to MT and total tissue Zn in various organs of untreated and Zn-pretreated rats.

Material and methods

Animals and treatments. Male Wistar rats (inbred strain Neuherberg) weighing 180-200 g were fed a standard laboratory

Trace Element - Analytical Chemistry in Medicine and Biology, Vol. 3

diet (Altromin, Lage, FRG) ad libitum and had free access to tap water. The Zn content of the diet was 75 mg/kg. For Zn-pretreatment, animals were i.p. injected with zinc sulfate (20 mg Zn/kg in saline). Control animals received saline (2 ml/kg, i.p.). Eighteen hours after injection, rats were decapitated under light ether anesthesia. Organs were removed, immediately frozen and stored at -80°C. For measurements, organs were homogenized in 9 volumes of ice cold TRIS-HCl buffer (30 mM, pH 7.4).

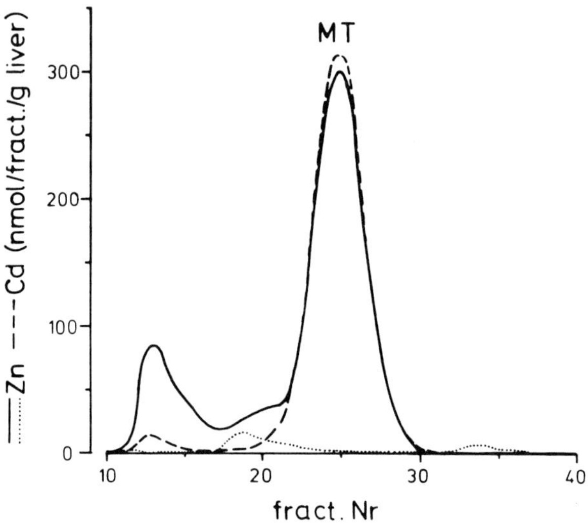

Fig. 1:

Cadmium saturation of metallothionein. Metal content of hepatic cytosol corresponding to 0.3 g of liver wet weight of Zn-pretreated (30 mg Zn/kg, i.p.) rats after separation on Sephadex G-75. One aliquot of cytosol was directly applied to the column and measured for zinc (——); another aliquot was incubated with cadmium, heated and zentrifuged. Heat stable supernatant was applied to the column and measured for zinc (……) and cadmium (- - -).

Gel chromatography. Hepatic cytosol (100000xg supernatant) from Zn-pretreated rats (30 mg Zn/kg) referring to 0.3 g wet wt. of liver was applied to a column (1.6 x 84 cm) packed with Sephadex G-75 and equilibrated with TRIS-HCl buffer.

Five ml fractions were collected at a flow rate of 18 ml/h. All chromatographic procedures were accomplished at 4°C. One aliquot of the cytosol was directly chromatographied and analysed for Zn, another aliquot has been incubated with Cd and heat labile proteins were precipitated by heating for 3 min at 100 °C and centrifugation for 5 min at 8000xg. Subsequently Zn and Cd was determined after chromatography.
Fig. 1 demonstrates the quantitative replacement of MT-zinc by cadmium and the stability of cadmium-MT under the heat denaturation conditions of the MT assay.

Assays. Zn and Cd were detected by atomic emission spectroscopy with inductively coupled plasma excitation (ICP-AES) using a JY 38 P sequence spectrometer from Instruments SA (3). Tissue homogenates were analysed after wet ashing with nitric acid at 140°C (4).
For routine analysis, metallothionein was determined in 18000xg supernatants of tissue homogenates (S18). A sample of S18 was diluted with TRIS-HCl buffer (30 mM, pH 7.4) to 1.1 ml and incubated at room temperature for 5 min with cadmium chloride (1000 ng Cd). To precipitate non MT-bound Cd, hemoglobin (100 μl of a 4% solution, w/v) was added and the sample was heated for 3 min at 100°C and centrifuged for 5 min at 8000xg. Addition of hemoglobin, heating and centrifugation of the samples was repeated three times. Resulting supernatant was measured for Cd.
Under our conditions test results were corrected for a small amount of non hemoglobin-precipitable Cd of 2 ng/ml. This experimental blank value could neither be decreased by increasing hemoglobin or decreasing Cd concentrations (results not shown). The interaction of Cd with low molecular weight SH-compounds like glutathione (GSH) particularly in liver samples will increase the experimental blank of non hemoglobin-precipitable Cd. Such samples therefore should be

properly diluted not to exceed a GSH concentration of 0.5 mM in the test. For GSH-rich tissues like mouse or rat liver a dilution factor of 20 will meet this requirement.

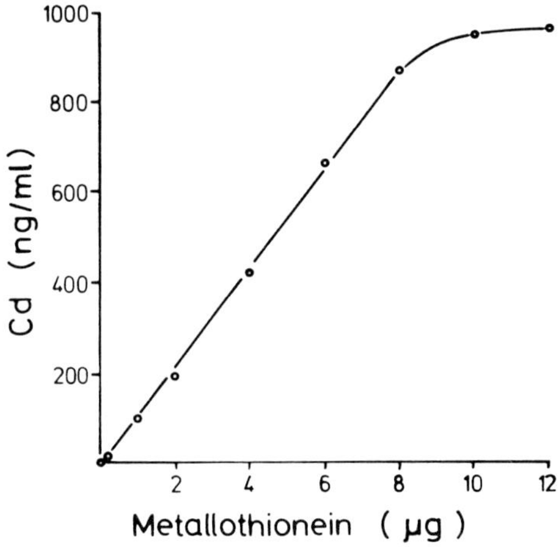

Fig. 2:

Linearity of the assay

The assay shows a linear relationship between purified zinc-MT quantified by a metal-independent thiolate-group analysis with 5,5'-dithio-bis(2-nitrobenzoic acid) according to (5) and measured Cd up to 8 μg MT corresponding to 870 ng Cd/ml (Fig. 2).

Results and discussion

The total Zn content in 11 rat tissues was 20-35 μg/g (Fig. 3, upper panel). Eighteen hours after a single i.p. injection of 20 mg Zn/kg tissue Zn content increased about 10fold in pancreas, 4fold in liver, twice in kidney and small intestine and marginally in spleen and stomach. MT content of control animals varied between 2 and 52 μg/g tissue wet wt., being highest in kidney and brain and lowest in lung and muscle. Injection of zinc sulfate increased MT contents in heart and those tissues which showed increased total Zn contents being most prominent in liver, pancreas, kidney and small intestine (Fig. 3, lower panel).

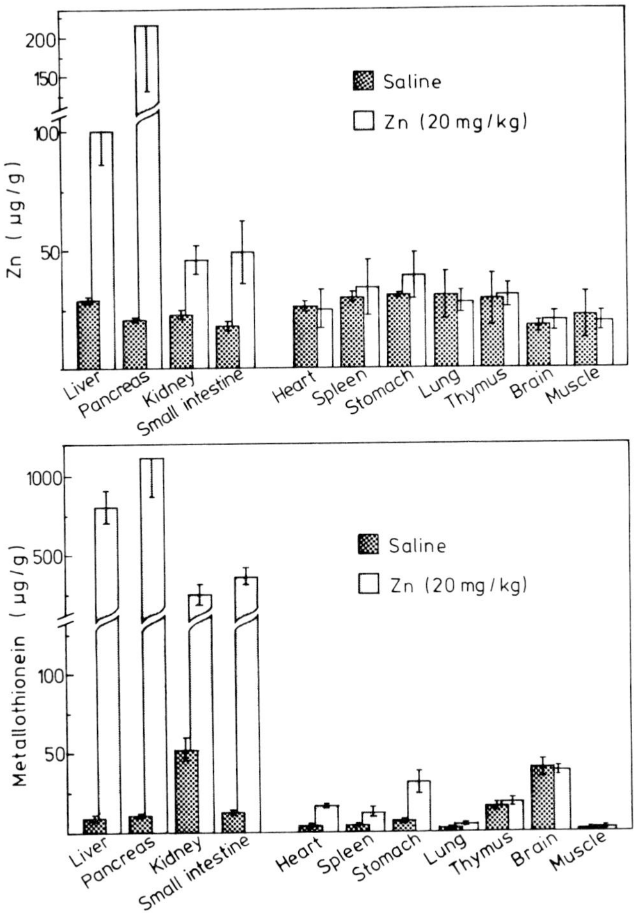

Fig. 3:

Content of zinc and metallothionein in various tissues of the rat 18 hours after i.p. injection of saline or zinc. Mean ± S.D. of 3 rats.

Zinc stimulates MT synthesis at the transcriptional level and is rapidly incorporated into newly synthesized MT (6). In rat liver a threshold concentration of 30 /ug Zn/g wet weight has been reported for the induction of zinc-MT (7). Among the tissues with highly induced MT, elevated MT levels in liver, kidney and small intestine do account for the major part of the increased tissue Zn level after Zn injection.

The ratio of MT bound Zn to total tissue Zn is strongly increased in liver, pancreas, kidney and small intestine and moderately in heart and stomach (Fig. 4). In liver for example MT-bound Zn under these conditions represents 56 % of total tissue Zn; MT then represents about 1 % ot total cytosolic protein.

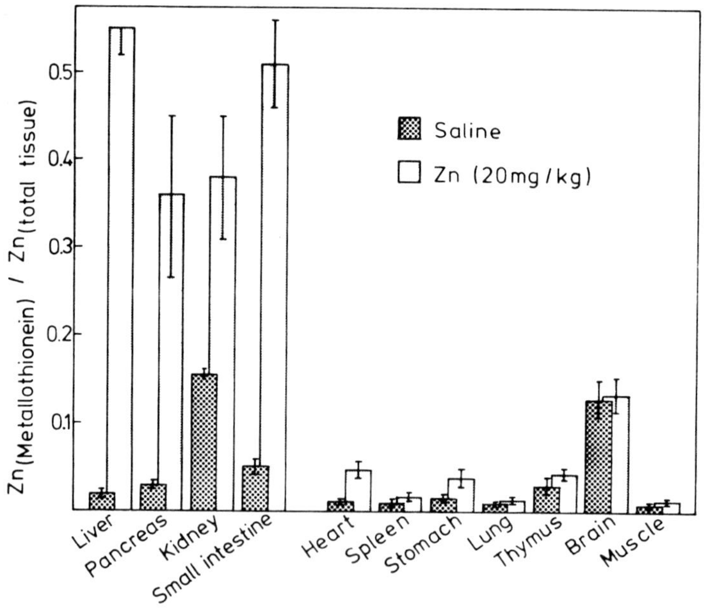

Fig. 4:

Ratio of MT-bound zinc to total tissue zinc in various tissues of the rat 18 hours after i.p. injection of saline or zinc. Mean ± S.D. of 3 rats.

Besides Zn and possibly Cd, copper is normally only a minor component of mammalian metallothionein. However, it binds to MT more firmly than Zn or Cd (8). Therefore, the applicability of the MT assay will be limited when copper becomes the major metallic component of MT e.g. in patients with Wilson's disease (9) or primary biliary cirrhosis (10).

In conclusion, studying the effects of metals on living systems require information on the chemical basis of metal interaction with intracellular targets (biological speciation). The determination of metallothionein as a major and inducible binding site for zinc is considered to be a first approach to a better understanding of the balance between essentiality and toxicity of this metal.

References

1. Onosaka, S. and Cherian, M.G.: Toxicol. Appl. Pharmacol. 63, 270-274 (1982).

2. Winge, D.R. and Miklossy, K.A. : J. Biol. Chem. 257, 3471-3476 (1982).

3. Schramel, P., Klose, B.J. and Hasse, S.: Fresenius Z. Anal. Chem. 310, 209-216 (1982).

4. Schramel, P., Wolf, A., Seif, R. and Klose, B.J.: Fresenius Z. Anal. Chem. 302, 62-64 (1980).

5. Sedlak, J and Lindsay, R.H.: Anal. Biochem. 25, 192-205 (1968).

6. Cousins, R.J.: J. Inher. Met. Dis. 6, 15-21 (1983).

7. Bremner, J., Davies, N.T. and Mills, C.F.: Biochem. Soc. Trans. 1, 982-985 (1973).

8. Weser, U. and Rupp, H.: in Metallothionein, J.H.R. Kägi and M. Nordberg, Eds., Birkhäuser, Basel, 221-230 (1979).

9. Evans, G.W., Dubois, R.S. and Hambridge, K.M.: Science 181, 1175-1176 (1973).

10. Janssens, A.R., Bosman, F.T., Ruiter, D.J. and Van den Hamer, C.J.A.: Liver 4, 139-147 (1984).

SIGNIFICANCE AND ROLE OF ZINC IN THE ANTIBACTERIAL /GROUP-B STREPTOCOCCI/ ACTION OF THE AMNIOTIC FLUID

László Salgó, Attila Pál, Károly Moholi, Kálmán Gyurkovits, László Kovács

Department of Paediatrics and Department of Gynecology, Medical University of Szeged, and Hospital of Szeged, 6701 Szeged, Hungary

Introduction

Intra-amniotic bacterial infection, either following rupture of the fetal membrane or without membrane rupture, presents a very significant problem in both mother and fetus. Potentially pathogenic bacteria have been isolated from the amniotic fluid in the presence of intact membranes /10/. Several researchers have reported that human amniotic fluid samples obtained near term inhibit bacterial growth /2,5,9, 14/. Other investigators have claimed that the amniotic fluid not only does not inhibit bacterial growth, but actually supports it /11,18/.

The bacterial inhibitory system contains two important components: the zinc cation and a low molecular weight peptide /13,14,15/. Other inhibitory substances, e.g. lysozyme, transferrin and immunoglobulins, may play a role in the host defenses against microbial invasion /8,9,12/.

The aim of the present investigation was to establish the inhibitory activity of the amniotic fluid against group B streptococci, and the change in this capacity in vitro when zinc was added to the liquor. The group B streptococci have a very important role as perinatal pathogenic microorganisms, and are now recognized as a most significant cause of infection in the neonatal period.

Trace Element - Analytical Chemistry in Medicine and Biology, Vol. 3

Material and Methods

105 amniotic fluid samples were obtained at different stages of gestation /15 to 42 weeks/ by transabdominal amniocentesis or cesarean section, or by puncture of the amniotic sac before term delivery, using disposable needles and plastic sterile syringes. All samples of amniotic fluid contaminated with meconium and/or blood were discarded. Patients receiving antibiotics were not included. The gestation age in weeks was calculated from the date of the last normal menstrual period. Each fluid was centrifuged at 10000 g for 30 minutes at 4 °C to remove cells and debris. All supernatant fluids were sterilized.

Biochemical assay

Zinc was determined by atomic absorption spectrophotometry; total protein by the biuret method; phosphorus with the molybdate reagent /SKI/Chemie Linz Test/; and lysozyme by turbidimetry at 405 nm /Testomar-Lysozym Test, Behringwerke/.

Organism

Group B streptococci were obtained from the WHO Collaborating Laboratory /Prague/ and served as the test organisms. The rate of growth was measured by recording the turbidimetric changes /spectrophotometrically at 405 nm, Eppendorf/ of the incubated fluid samples /both the controls and tests were examined round the clock at 2 , 4 , 6 , 8 , 10 , 12 and 24-hour intervals/ and viable plate counts /4 , 8 and 24-hour intervals/. After incubation, the initial borth culture was diluted to $1:15^5$ with Todd-Hewitt medium. From this final dilution 0.1 ml was added to each of the following media:

A. 0.9 ml amniotic fluid;

B. 0.9 ml amniotic fluid + 0.1 ml 125 µmol/l $ZnSO_4$;

C. 0.9 ml Tris buffer /pH 8.1/;

D. 0.9 ml Todd-Hewitt broth.

/The signs are the same on the Figures./

The standard Zn solution contained 125 µmol/l $ZnSO_4$. Zn was

added to all amniotic fluid samples.
Statistical analysis was performed by means of Student's t test.

Results and Discussion

The presence of antimicrobial activity in the amniotic fluid was first reported by Cattaneo /3/ in 1949, the presence of lysozyme being suggested.
Figure 1 shows the lysozyme and total protein levels in the amniotic fluid during pregnancy.

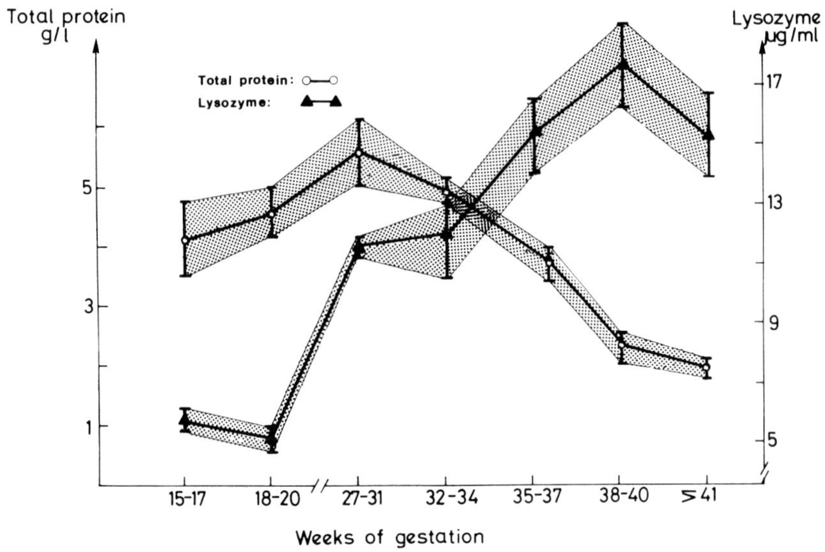

Fig. 1. Lysozyme and total protein levels in the amniotic fluid during the 15th to the 42nd weeks of gestation

The lysozyme level increased continuously during pregnancy, but with a decrease after term, while the total protein level increased up to the 31st week of gestation and thereafter

decreased. Our results are similar to those of Cherry /4/, Ford /7/. In our laboratory investigation, we have studied the possible correlation of the presence of lysozyme and the bacteriostatic properties of the amniotic fluid.
Figure 2 illustrates the zinc and phosphate levels in the amniotic fluid.

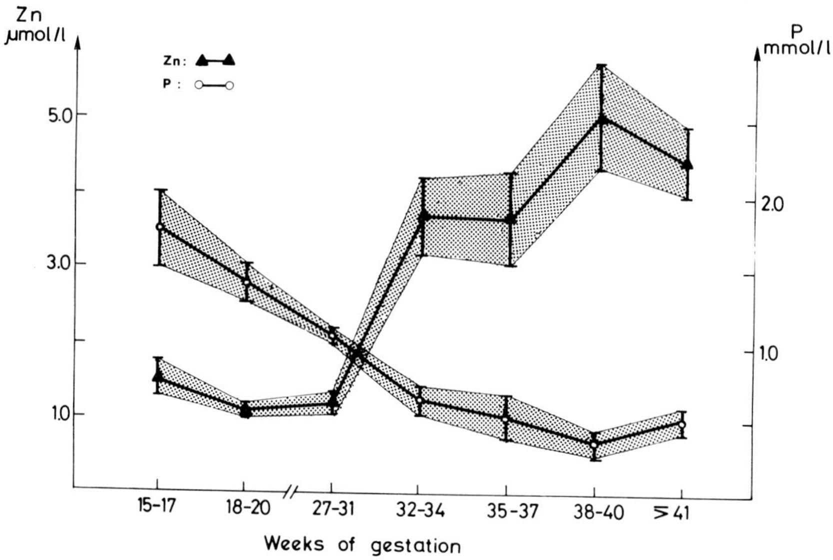

Fig.2. Changes in the amniotic fluid zinc and phosphate levels as a function of the duration of gestation

The zinc level increased after the 31st week of gestation, with a decrease after term, while the phosphate level decreased continuously during pregnancy. There were significant differences in the zinc levels of the amniotic fluid between the inhibitory and non-inhibitory liquors of women in the 37th to 40th weeks of gestation /$P < 0.01$/. The results of Tafari /16/ indicate that the low antimicrobial activity of the amniotic fluid of women in Addis Abeba may be related to the women's poor nutrition. The majority of such fluids were non-inhibitory. The zinc level in the amniotic fluid of Ethiopian women was very low. The results of Schlievert /15/ suggested that the dietary zinc supplementation of pregnant women may increase the amniotic fluid zinc content and thereby enhance the amniotic fluid antibacterial /against E. coli/ activity. Appelbaum /1/ demonstrated the effect of the in vitro addition of zinc on the growth-supporting property of non-inhibitory amniotic fluids in African women. Figure 3 depicts the group B streptococci growth curves, with the growth expressed in 1 x 10^{log}/ml of CFU on ordinate and time on the abscissa.

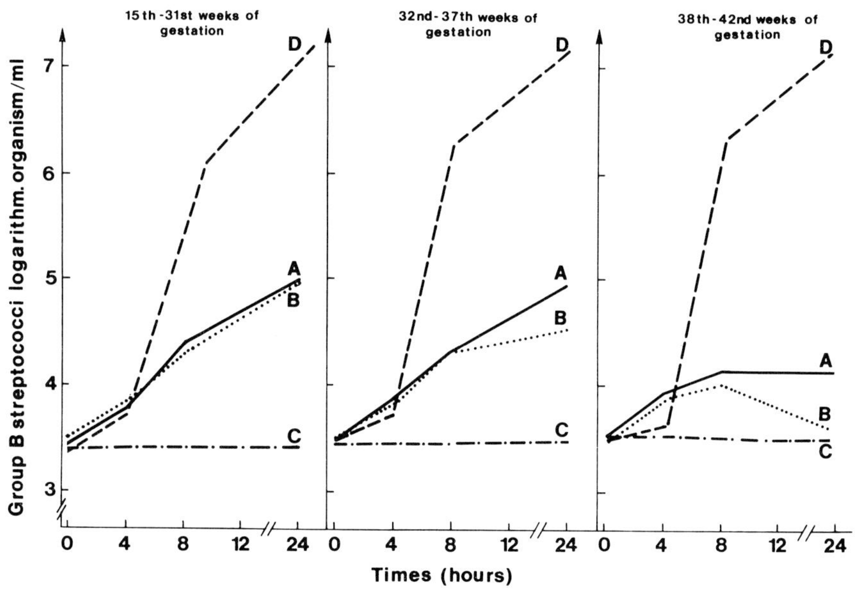

Fig. 3. Variability in antibacterial capacities

The antibacterial activity increased with the duration of gestation, the highest activity being observed between the 37th and 40th weeks of gestation.

The growth of group B streptococci in amniotic fluid was

measured spectrophotometrically. The results can be seen in Fig. 4.

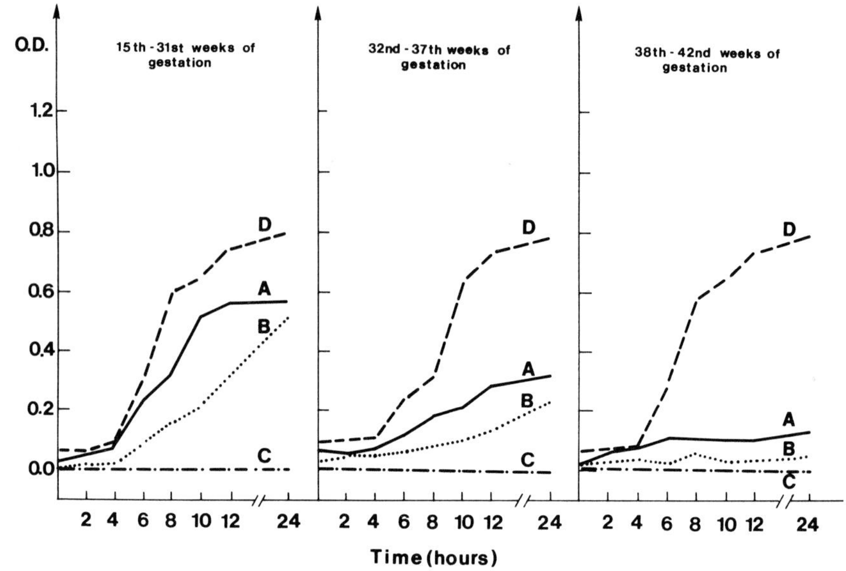

Fig. 4. Growth of group B streptococci in fluids. Growth measured spectrophotometrically.

The results presented showed that the amniotic fluid is not bactericidal between the 15th and 32nd weeks of gestation; from the 32nd to the 37th week of gestation the antimicrobial activity is minimal; subsequently, bactericidal effects can be observed.

The inhibitory or non-inhibitory capacities of fluid samples are summarized in Table I.

All samples obtained before the 32nd week of gestation are seen to be non-inhibitory. In the period between the 32nd and 40th weeks of gestation, there is an increase in the number of bacteriostatic fluid samples. However, the percent age of bactericidal fluid samples is zero/except between the 38th weeks of gestation/.

Table I. Correlation of gestation stage and percentages of non-inhibitory, bacteriostatic and bactericidal amniotic fluid samples

Weeks of gestation	No.	Non-inhibitory No.	%	Bacteriostatic No.	%	Bactericidal No.	%
15 - 17	5	5	100	0	0	0	0
18 - 20	6	6	100	0	0	0	0
27 - 31	2	2	100	0	0	0	0
32 - 34	7	5	71	2	29	0	0
35 - 37	17	13	76	4	24	0	0
38 - 40	60	30	50	26	43	4	7
- 41	8	5	62	3	38	0	0

The antibacterial activity was higher when 1.25 nmol $ZnSO_4$ was added to the amniotic fluid. The results are to be seen in Table II.

Table II. Correlation of gestation stage and percentages of non-inhibitory, bacteriostatic and bactericidal amniotic fluid samples after in vitro addition of $ZnSO_4$

Weeks of gestation	No.	Non-inhibitory No.	%	Bacteriostatic No.	%	Bactericidal No.	%
15 - 17	5	5	100	0	0	0	0
18 - 20	6	6	100	0	0	0	0
27 - 31	2	2	100	0	0	0	0
32 - 34	7	4	57	3	43	0	0
35 - 37	17	8	47	8	47	1	6
38 - 40	60	20	33	30	50	10	17
- 41	8	2	25	5	63	1	12

After Zn was added, the antibacterial activity of the amniotic fluid was increased.

Evans /6/, Galash /8/ and Thadepalli /17/ found that amniotic fluids inhibited Gram-positive, Gram-negative and anaerobic bacteria.

Conclusions

1. The amniotic fluid has antimicrobial /anti-group B streptococci/ activity, but the inhibition is highest between the 37th and the 40th weeks of gestation.
2. The inhibitory activity is bacteriostatic.
3. The antibacterial activity is higher when Zn is added to the amniotic fluid.
4. The zinc and lysozyme levels correlate well with the gestational age and with the observed increase in inhibitory activity.
5. The phosphate and total protein values decrease during pregnancy.

References

1. Appelbaum, P.C., Ross, S.M., Dhupelia, I., Naeye, R.L.: Am. J. Obstet. Gynecol. 135, 82 /1979/
2. Blanco, J.D., Gibbs, R.S., Krebs, L.F.: Am. J. Obstet. Gynecol. 147, 247 /1983/
3. Cattaneo, P.: Clin. Obstet. Gynecol. 51, 60 /1949/
4. Cherry, S.H., Filler, M., Harvey, H.: Am. J. Obstet. Gynecol. 116, 639 /1973/
5. Evaldson, G., Nord, C.E.: Med. Microbiol. Immunol. 170, 11 /1981/
6. Evans, H.E., Levy, E., Glass, L.: Obstet. Gynecol. 49, 35 /1977/
7. Ford, L.C., De Lange, R.J., Lebhara, Th.B.: Am. J. Obstet. Gynecol. 127, 788 /1972/
8. Galash, R.P., Snyder, I.S.: Am. J. Obstet. Gynecol. 102, 949 /1968/
9. Larsen, B., Snyder, I.S., Galash, R.P.: Am. J. Obstet. Gynecol. 119, 492 /1974/
10. Miller, J.M., Pupkin, M.J., Hill, G.B.: Am. J. Obstet. Gynecol. 136, 796 /1980/

11. Sarkany, I., Gaylarde, C.C.: Br. J. Dermatol. 80, 241 /1968/

12. Schlievert, P., Larsen, B., Johnson, W., Galash, R.P.: Am. J. Obstet. Gynecol. 122, 809 /1975/

13. Schlievert, P., Johnson, W., Galash, R.P.: Am. J. Obstet. Gynecol. 125, 899 /1976/

14. Schlievert, P., Johnson, W., Galash, R.P.: Infect. Immun. 14, 1156 /1976/

15. Schlievert, P., Johnson, W., Galash, R.P.: Am. J. Obstet. Gynecol. 127, 603 /1977/

16. Tafari, N., Ross, S.M., Naeye, R.L., Galash, R.P., Zaar, B.: Am. J. Obstet. Gynecol. 128, 187 /1971/

17. Thadepalli, H., Appleman, M.D., Maidman, J.E., Arce, J.J., Davidson, E.C.: Am. J. Obstet. Gynecol. 127, 250 /1977/

18. Walsh, H., Hildebrandt, R.J., Prystowsky, H.: Am. J. Obstet. Gynecol. 93, 590 /1965/

Key-words

Group B streptococci, amniotic fluid, antimicrobial activity, zinc

THE CHANGES OF Zn, Cu AND Fe CONTENTS IN THE ORGANISM'S FLUIDS AFTER GOITRE RESECTION

Wojciech Kwiatek, Józef Kajfosz, Stanisław Szymczyk
Institute of Nuclear Physics, Cracow, Poland

Zbigniew Sznapka, Maria K. Walas
Medical Academy, 2. Cathedra of Surgery, Cracow, Poland

Introduction

The significance of the trace elements in the phenomenon of metabolism is essential though not yet well known. As contained in many different active enzymes, coenzymes, hormones etc. they essentially influence the metabolic process. After the operation shock the metabolic processes are especially intensive and in a relatively short time they pass from the dominance of decay over synthesis, through the equilibrium phase, to the dominance of synthesis. Therefore it is very interesting to study changes of the trace element contents in cells and beyond the cellular space that occur during that time. In this paper the changes of Zn, Cu and Fe contents in the organism's fluids after goitre resection are presented. PIXE measurements of concentrations in blood, plasma and urine during 5 consecutive days were performed.

Material Studied and Method of Measurements

The studies were performed for 13 women operated because of simple goitre /9 cases/ and toxic goitre /2 cases - Graves Basedov disease, 2 cases - toxic nodular goitre/. The age of patients was from 19 to 63 years, the mean age being 37 years.

Trace Element - Analytical Chemistry in Medicine and Biology, Vol. 3

All women were identically anaesthetized and operated by the same surgeon. The operation consisted in extensive resection of thyroid slices after previously tying up the thyroid arteries and cutting the isthmus. During the operation and within 24 hours after it 2500 - 3000 ml of crystalloids, with 500 ml of dextrin were transfused through the veins and 200 - 300 mg of cortisol were administered. On the first and second days after the resection the patients received drinks orally and since the third day normal food at will. No complications were observed during the postoperation treatment. Plasma, full blood and urine were taken for measurements. The measurements consisted of four stages: 1. Sample collecting; 2. Target preparation; 3. Target exposition on a proton beam and spectra acquisition; 4. Analysis of results.

1. Sample collecting

The samples were collected before the operation /index - BR/, immediately after the operation and anaesthetization /index - AR/, and during the next four days /indexes - 1, 2, 3, 4/. Full blood was collected directly from the vein of the forearm to testtubes containing two drops of heparine and also to another clean tube for plasma extraction. The plasma was separated with a separate pipette for each sample. The urine samples were taken from the twenty-four hour urine collection. The test-glasses and pipettes had been boiled in nitrogen acid and then rinsed five times in threefold distilled water obtained from a quartz distiller.

2. Target preparation

The collected fluids were mixed with the water solution of $Y/NO_3/_2$ in a ratio 1:1. This water solution contained 200 ppm of yttrium and was used as an internal standard. One drop of thus prepared liquid /ca 0.03 ml/ was put on a thin formvar layer /ca 100 ug/cm^2/ in an aluminium frame. The targets, mounted on a support, were dried in an evacuated measuring chamber.

3. Target exposition

For PIXE analysis a proton beam of a small C-48 cyclotron in Cracow was used. The analyses were carried out with proton energy of 2.6 MeV, beam intensity of 40 nA, beam diameter of 5 mm, exposition time of 10 minutes. For the elimination of a large bremsstrahlung background caused by charging effects a technique of an intermediate air pressure /ca 10 mm Hg/ in the target chamber was used. A Si/Li/ detector of 200 eV resolution at 5.9 keV, CAMAC electronics and an ICA-70 multichannel amplitude analyser with a paper tape puncher for X-ray spectra registration were used.

4. Analysis of results

The processing of final results, which involved the least-squares fitting of individual peak areas to Gaussian functions, was carried out on an ODRA-1305 computer. The absolute elements concentrations in full blood, plasma and urine were determined by the comparison of X-ray yields with the X-ray yield of yttrium. The contents in blood corpuscules were calculated from the differences between the concentrations in full blood and plasma. Because of individual differences in concentrations the results were expressed in a logarithmic scale for further statistical analysis. The distribution of logarithms of the concentrations was proved to be normal. The confirmation of the significance of estimated differences was carried out by means of the Student´s test.

Results

In Table 1 and Figure 1 the mean values of obtained results /logarithms of concentrations in ppm/ with their standard deviations are presented.

1. Zinc

Full blood: until the end of the second day after surgery the

zinc content decreases; on the third day an increase, and on the fourth day a decrease, below the initial value, is observed.
Plasma: after surgery the zinc content increases until the end of the second day, then a decrease follows; the lowest value is reached on the fourth day.
Blood corpuscules: until the end of the second day the zinc content decreases. On the third day it exceeds slightly the initial value and on the fourth day it decreases slightly below.
Urine: during the whole period investigated the removal of zinc by urine is decreased.

Table 1. Results of the measurements

Specimen	Element	BR	AR	1	2	3	4
Full blood	Zn	1.09 .07	0.86 .09	0.84 .05	0.70 .10	1.15 .11	0.79 .12
	Cu	0.52 .20	0.70 .17	0.59 .17	0.73 .16	0.86 .19	0.57 .11
	Fe	2.86 .12	2.74 .07	2.76 .04	2.80 .08	2.99 .09	2.76 .12
Plasma	Zn	-.16 .15	0.29 .22	0.10 .20	0.21 .11	-.18 .21	-.30 .14
	Cu	0.74 .16	0.77 .14	0.70 .16	0.58 .12	0.73 .15	0.35 .11
	Fe	0.37 .16	0.49 .06	0.15 .16	0.22 .23	0.49 .27	-.01 .14
Blood corpuscules	Zn	1.25 .24	0.57 .22	0.74 .19	0.49 .23	1.33 .48	1.09 .23
	Cu	-.22 .23	-.07 .33	-.11 .25	0.15 .21	0.13 .20	0.22 .21
	Fe	2.49 .24	2.25 .11	2.61 .22	2.58 .17	2.50 .16	2.77 .18
Urine	Zn	0.83 .22		0.21 .22	0.52 .28	0.07 .21	0.12 .15
	Cu	0.88 .27		0.46 .32	0.61 .31	0.39 .12	0.36 .29
	Fe	0.52 .34		0.08 .22	-.08 .32	0.02 .18	0.41 .41

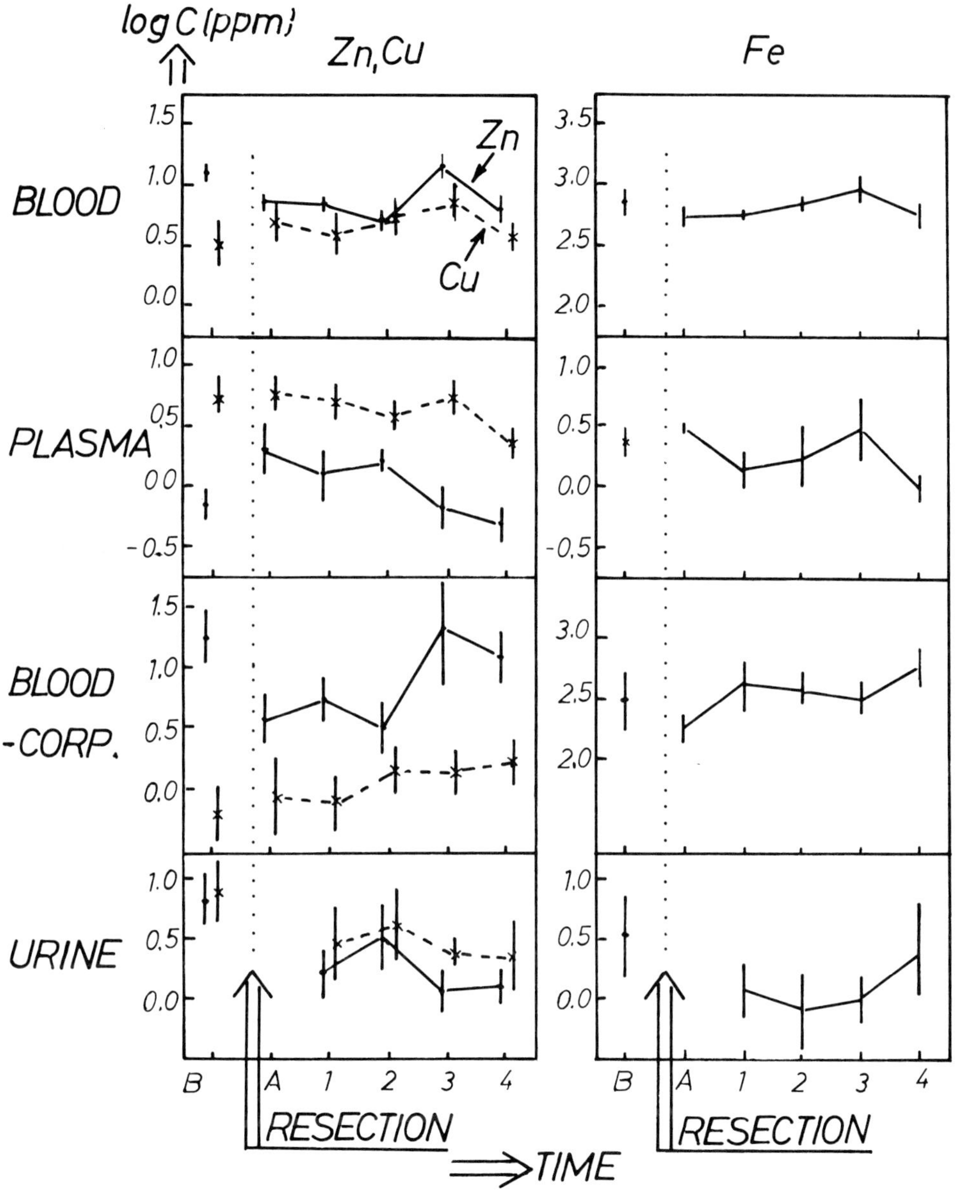

Fig. 1. Diagrams of changes of Zn, Cu and Fe contents in blood and urine. Time in days after resection. B - before resection A - immediately after resection

2. Copper

Full blood: until the end of third day the copper content slowly increases while on the fourth day it decreases roughly to the initial value.

Plasma: from the first day after surgery a decreasing tendency in copper content is observed.

Blood corpuscules: a steady increase of copper content in blood corpuscules was found.

Urine: the copper content decreases during the whole period under study.

3. Iron

Full blood: the iron content here immediately after surgery and in the first day is slightly lower than before. On the second day it starts to increase and on the third day exceeds the initial value. On the fourth day it does not differ significantly from the initial value.

Plasma: after an initial decrease, on the first day the iron content increases and on the third day it exceeds the initial value. Afterwards a remarkable decrease down to below the initial value occurs.

Blood corpuscules: immediately after surgery the iron content is slightly lower than before, afterwards it increases until the end of the period under study.

Urine: the iron content decreases until the end of the second day, then it increases but does not reach the initial value.

Discussion

During the whole period studied the organism after goitre resection retains zinc, copper and /especially in the first two days/ iron. Until the end of the second day zinc is being removed from blood corpuscules and in this time its content in plasma is increasing. This may indicate that zinc is being transferred from cells to other regions, probably to the wound-ed tissue. In the next two days the zinc content in blood corpuscules returns to its initial value and this effect is accompanied by a considerable decrease of zinc content in plasma. The nutrition, which the patients receive depending on their needs, prevents symptoms of zinc deficiency. Unlike zinc, copper is transferred to blood corpuscules all the time after surgery and its concentration in plasma slightly decreases. The decrease of iron content in blood corpuscules immediately after the operation, and its simultaneous increase in plasma are connected with a blood loss during the surgery and with the decay process as a result of catabolism. This phenomenon is not too strong and it confirms our earlier estimates /1, 2, 3/ of the intensity of the metabolic reaction after goitre resection. During the days after operation the iron content in blood corpuscules is increasing. In this period the iron concentration in plasma is changing but the majority of obtained values is less than the initial value. Generally, it indicates that a regeneration of erythrocytes in excess of the previous loss, takes place.

Conclusions

1. In the first four days after goitre resection the removal of zinc and copper by urine is diminished. A similar effect for iron is observed in the first two days only.
2. In the first two days after surgery the zinc content in blood corpuscules decreases and, simultaneously, an increase of its content in plasma occurs. During the two subsequent days the previous zinc content in corpuscules is restored. At the same time the zinc content in plasma decreases substantially below its value before the surgery.
3. In the whole period under study the copper content in blood corpuscules increases. At the same time its slight decrease in plasma is apparent.
4. After a short decrease of iron content in blood corpuscules

with a simultaneous increase of its content in plasma, a gradual increase of iron content in blood cells is observed. Simultaneously, a decrease in its content in plasma is observed.

References

1. Marczyńska, A., Sznapka, Z., Oszacki, J.: Polski Przegląd Chirurgiczny 50, 653 /1978/.

2. Sznapka, Z.: Polski Przegląd Chirurgiczny 48, 1451 /1976/.

3. Sznapka, Z., Kwiatek, W., Walas, M.K., Cholewa, M., Kajfosz, J., Szymczyk, S: Polski Przegląd Chirurgiczny / to be published /.

EFFECT OF LOW ZINC INTAKE ON PLASMA AND LEUKOCYTE ZINC CONCENTRATION IN PREGNANT EWES

George Everett and Jean Apgar

U.S. Plant, Soil and Nutrition Laboratory, Agricultural Research Service, USDA, Tower Road, Ithaca, N.Y. 14853 USA

Introduction

Indications that zinc (Zn) intake may be suboptimal (1) and that poor Zn status may be particularly deleterious for pregnant women (2) and for children (3) emphasize the need for methods of determining Zn status. The most widely used indicator of Zn status is serum or plasma Zn concentration (4). Although it is depressed in severe Zn deficiency, it may not be affected in cases of marginal deficiency (3). Changes in serum or plasma Zn caused by factors other than Zn status further complicate its use as an indicator (3). Leukocyte Zn concentration has been suggested as a better indicator of Zn status than serum or plasma Zn (5-10). We have measured leukocyte Zn in pregnant ewes given a low or adequate Zn intake. The method we used required only 2 ml of blood and permitted determination of DNA, protein, and phosphorus as well as Zn on the same sample.

Results and Discussion

Plasma Zn concentration decreased promptly after Zn intake was reduced (Fig. 1) and remained at approximately the same level during the rest of gestation and after parturition. Although excessive salivation appeared in the -Zn sheep 4-6 weeks after Zn intake was reduced, signs of severe deficiency such as marked wool loss and skin lesions did not appear in

Trace Element - Analytical Chemistry in Medicine and Biology, Vol. 3

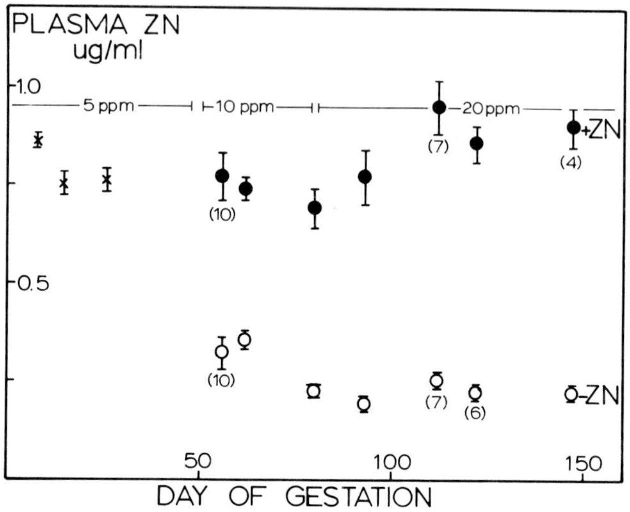

Fig. 1. Plasma Zn (mean ± SEM) in pregnant ewes given a low Zn diet with or without Zn in the drinking water. All ewes received 5 ppm Zn for first 50 days. Plasma Zn of unsupplemented ewes differed significantly (P <.01; one-way analysis of variance) from supplemented ewes at all times measured. Each point represents 10 animals except for the latter part of the experiment by which time some ewes had been killed for other measurements.

---

the ewes until they had received the low Zn intake for approximately 3 months. Fetuses/lambs of -Zn ewes weighed less than those of +Zn ewes of comparable gestational age after day 121 of gestation. Leukocyte Zn concentrations (Table 1) of the -Zn ewes, however, did not differ from those of the +Zn ewes. The values in Table 1 were obtained after 10 to 17 weeks of the low Zn intake and included the time during which several of the ewes had obvious signs of Zn deficiency. Zn concentrations were calculated on the basis of phosphorus, protein, and DNA rather than on the basis of cell number because the leukocytes were sticky by the end of

TABLE 1

ZINC IN LEUKOCYTES FROM PREGNANT SHEEP

WEEK OF LOW ZN INTAKE [1]	ZINC ADEQUATE	ZINC DEFICIENT
--------- Micrograms zinc per mg protein[2] ---------		
10	1.2 ± .1	1.1 ± .1
12	1.2 ± .1	1.2 ± .1
14	1.3 ± .1	1.2 ± .1
16	1.3 ± .1	1.2 ± .1
17	1.0 ± .1	.96 ± .1
--------- Nanograms zinc per μg phosphate[2] ---------		
10	59 ± 5	55 ± 5
12	58 ± 4	52 ± 4
14	72 ± 6	70 ± 8
16	67 ± 6	70 ± 5
17	61 ± 5	64 ± 5
--------- Nanograms zinc per μg DNA[2] ---------		
10	7.8 ± 0.7	8.9 ± 0.7
12	7.4 ± 0.7	7.8 ± 0.6
14	12 ± 1.1	15 ± 2.1
16	7.5 ± 0.5	8.6 ± 1.2
17	7.6 ± 0.7	9.9 ± 1.4

[1] Low Zn intake began day 50 of gestation. Week 10 of low Zn intake was therefore week 17 of gestation.

[2] Differences were not significant.

the isolation procedure making cell counts inaccurate due to clumping. Calculation of the various ratios (DNA/P, protein/P, etc) indicated that DNA values were more variable than the other values, possibly due to incomplete solubilization of the DNA. Agreement between duplicate measurements of the same sample in all four determinations was excellent (within 5 percent), but determinations in the same animal at different times varied as much as determinations in different animals. There was no indication of a difference between the +Zn and -Zn ewes, however, and no tendency for the concentration in the -Zn ewes to decrease with increasing duration of low Zn intake. The retention of Zn by the leukocytes of the -Zn ewes in spite of the much lower concentration of Zn in the plasma indicates that the leukocyte Zn was firmly bound. Changes in leukocyte Zn concentration in a -Zn animal would therefore probably have to take place during the development of the leukocyte.

Suggestions that changes in leukocyte Zn concentration reflect Zn status (5-10) have come from measurements in humans in which a larger proportion of the leukocyte population is neutrophils (11) than is the case in the ewe. Lymphocytes are a larger proportion of the leukocyte population in sheep (12). The shorter life-span of neutrophils (11-12 days (12)) compared to that of the lymphocytes (15-100 days (12)) could make neutrophils more sensitive to a change in Zn status. Experimentally-induced Zn deficiency in humans resulted in a 60 percent reduction in leukocyte Zn (5). Neutrophil chemotactic activity in patients with Acrodermatitis Enteropathica, a congenital Zn-responsive disorder, was impaired but was returned to normal after Zn therapy (13). However, phagocytosis by rat neutrophils has been reported not to be affected by Zn deficiency (14).

Changes in immune function of lymphocytes from Zn-deficient animals have been studied extensively (15). Since changes in lymphocyte function in Zn-deficient mice are well-documented (15), changes in Zn concentration of the lymphocyte might also be expected. Immune function in Zn-deficient sheep has not been studied as far as we know. If immune function of sheep lymphocytes is affected by Zn deficiency, it apparently is not due to a reduction in lymphocyte Zn concentration. It is possible, however, that Zn deficiency affects a subpopulation of lymphocytes which is too small to affect the Zn concentration of the whole population. Zn deficiency may also, of course, affect the function of lymphocytes without altering Zn concentration by affecting factors required by lymphocytes. Serum thymic factor (FTS), which is required for maturation of T-lymphocytes (16), and which was reduced in Zn-deficient mice (17), is one example of factors that could affect lymphocyte function without affecting lymphocyte Zn concentration.

Although leukocyte Zn concentration does not reflect Zn status in the ewe, it may be a useful measurement in humans. The method we have used is suitable for use with small amounts of blood and does not rely on cell counts for quantitation of the amount of material analyzed. Further studies of human populations may answer the question of the usefulness of leukocyte Zn concentration as an indicator of Zn status in the human.

Methods

Twenty Finn Dorset cross ewes were given throughout pregnancy a low Zn (<3 ppm) diet (18) supplemented three times per week with 100 g mixed grass hay (Zn:12 ppm). All ewes received deionized water supplemented with Zn for the first third (50 days) of gestation to insure normal implantation and to increase fetal survival. Depriving ewes of Zn throughout pregnancy in a previous experiment resulted in abortions, malformations, mummified fetuses, and poor survival of the lambs (19). The level of Zn in the water (5 ppm) was sufficient to maintain plasma Zn concentrations at the initial level. After 50 days ten ewes (+Zn) continued to receive Zn-supplemented water (10 ppm for 4 weeks; 20 ppm thereafter). The remaining ten ewes (-Zn) were unsupplemented. Blood samples with heparin (<1 ng Zn/ml blood from heparin) as anticoagulant were taken by jugular puncture at intervals throughout the experiment. Four -Zn and three +Zn ewes were killed day 110-121 for other measurements. The remaining ewes were sampled until 4-6 weeks after delivery.

Leukocytes were isolated from 2 ml of blood by the standard Ficoll-Pacque technique. The pellet was stored frozen in polypropylene tubes until assay. It was solubilized with 1 ml of 1 N NH_4OH-.02 percent Triton-X-100 (20). Polypropylene tubes were used because zinc was bound by glass tubes in the alkaline solution. Triplicate aliquots of a 10th volume were removed for Zn analysis and digested to dryness in a sand bath with 50 µl "ultrapure" HNO_3 in precleaned borosilicate tubes. The digested samples were taken up in 0.5 M "ultrapure" HCl for flameless atomic absorption on the Perkin Elmer 603. The standard was prepared in a matrix of low samples which was subtracted as the blank. The standard matrix curve could be reproduced consistently and had a lower slope than the standard prepared in 0.5 M HCl.

Plasma zinc was determined by aspiration of a 1:1 dilution of plasma with water into a Perkin Elmer 360 equipped for flame analysis. Standards in 12 percent sucrose were used to compensate for the viscosity of the plasma.

Duplicate 10th volume aliquots were analyzed for protein (21), DNA (22; modified), and total phosphate (23).

Summary and Conclusion

Better measures of Zn status are needed than plasma or serum Zn concentration. Leukocyte Zn concentration is one alternative that has been suggested. We have measured leukocyte Zn in 20 pregnant ewes given a diet low in Zn with or without supplemental Zn in the drinking water. Leukocytes were isolated and Zn as well as DNA, protein, and phosphorus were determined on the same sample. Expressing Zn concentration per unit DNA, protein, or phosphorus avoided the problem of determining cell number which was subject to inaccuracies due to clumping. Zn concentration of leukocytes from Zn-adequate and Zn-deficient ewes did not differ. Although leukocyte Zn did not reflect Zn status in the ewe, it may still be a useful measure in humans in which neutrophils are a larger fraction of the leukocyte population than in sheep. Further studies in the human will be necessary to determine the usefulness of leukocyte Zn as an indicator of Zn status. The methods used here permit quantitative measurement of Zn in leukocytes from 2 ml of blood.

References

1. Smith, Jr., J.C., Morris, E.R., Ellis, R.: Zinc Deficiency in Human Subjects (eds. A.S. Prasad, A.O. Cavdar, G.J. Brewer, P.J. Aggett), Alan R. Liss, Inc., New York (1983), p. 147-169.

2. Jameson, S.: Acta Med. Scand. Suppl. 593, 1-89 (1976).

3. Hambidge, K.M.: Phil. Trans. R. Soc. Lond. B 294, 129-144 (1981).

4. Solomons, N.W.: Am. J. Clin. Nutr. 32, 856-871 (1979).

5. Prasad, A.S., Rabbani, P., Abbasii, A., Bamersox, E., Spiney-Fox, M.R.: Ann. Intern. Med. 89, 483-490 (1978).

6. Meadows, N.J., Ruse, W., Smith, M.F., Day, J., Keeling, P.W.N., Scopes, J.W., Thompson, R.P.H.: Lancet 1, 1135-1136 (1981).

7. Meadows, N.J., Cunnane, S.C., Keeling, P.W.N., Thompson, R.P.H.: Zinc Deficiency in Human Subjects (eds. A.S. Prasad, A.O. Cavdar, G.J. Brewer, P.J. Aggett), Alan R. Liss, Inc., New York (1983), pp. 139-146.

8. Hinks, L.J., Colmsee, M., Delves, H.T.: Analyst 107, 815-823 (1982).

9. Hinks, L.J., Clayton, B.E., Lloyd, R.S.: J. Clin. Pathol. 36, 1016-1021 (1983).

10. Whitehouse, R.C., Prasad, A.S., Rabbani, P.I., Cossack, Z.T.: Clin. Chem. 28, 475-480 (1982).

11. Ganong, W.F.: Review of Medical Physiolgy, 7th Edition, Lange Med. Pub., Los Altos, California (1975), pp. 379.

12. Phillis, J.W.: Veterinary Physiology, W.B. Saunders Co. (1976), pp. 285-288.

13. Weston, W.L., Huff, J.C., Humbert, J.R., Hambidge, K.M., Neldner, K.H., Walravens, P.A.: Arch. Dermatol. 113, 422-425 (1977).

14. Davis, M.A., Wallwork, M.C. and Sandstead, H.H.: Fed. Proc. 43, 685 (1984).

15. Fraker, P.J.: Surv. Immunol. Res. 2, 155-163 (1983).

16. Good, R.A., Fernandes, G., Garofalo, J.A., Cunningham-Rundles, C., Iwata, T., West, A.: Clinical, Biochemical, and Nutritional Aspects of Trace Elements (ed. A. Prasad), Alan R. Liss, Inc., New York (1982), pp. 189-202.

17. Iwata, T., Incefy, G.S., Tanaka, T., Fernandes, G., Menendez-Botet, C.J., Pih, K., Good, R.A.: Cell. Immunol. 47, 100-105 (1979).

18. Apgar, J.: Proc. Cornell Nutr. Conf. for Feed Manuf., 9-13 (1983).

19. Apgar, J., Fitzgerald, J.A.: J. Am. Col. Nutr. 2, 285 (1983).

20. Fiszer-Szafarz, B., Szafarz, D, De Murillo, A.G.: Anal. Biochem. 110, 165-170 (1981).

21. Lowry, O.H., Rosebrough, M.J., Farr, A.L., Randall, A.J.: J. Biol Chem. 193, 265-275 (1951).

22. Puzas, J.E., Goodman, D.B.P.: Anal. Biochem. 86, 50-55 (1978).

23. Ames, B.N., Dubin, D.T.: J. Biol. Chem. 235, 769-775 (1960).

CADMIUM DETERMINATIONS IN PREGNANCY

László Salgó, Károly Moholi, Gábor Petheő, Lajos Bódis
Pediatric Department University of Szeged, Hospital of Szeged and Public Health Station of Szeged, 6701 Szeged, Hungary

Introduction

Cadmium is recognized as a toxic environmental contaminant. As a heavy metal, it can bind to a number of enzymes, thereby altering their physiological functions. Cadmium exhibits a significant affinity for -SH and -CH groups and N-containing bonds and accordingly blocks them.
The cadmium poisoning observed in the Japanese province of Toyama (Itai-itai or ouch-ouch disease), and manifested in gastrointestinal and renal lesions and osteomalatia, resulted from the industrial contamination of food and water supplies (8,13). Since then, considerable interest has been taken in the analysis of cadmium in human and animal biological materials. Long-term exposure to cadmium (which is toxic to virtually every human and animal body system, whether inhaled, injected or ingested) causes reproductive failures, fetal growth retardation, anemia (the common manifestation of chronic toxicity, due at least in part to its metabolic antagonism to copper and iron), renal lesions, osteomalatia, teratogenic effects, etc. (2,3,4). Injection of large doses of cadmium into pregnant mice or other animals causes fetal death or developmental anomalies. The data are contradictory as to whether the placenta does, or does not provide a "protective barrier" against cadmium (1,11,16).
The aim of our investigation was to establish the cadmium levels and, together with these, copper, zinc, iron and total protein levels in amniotic fluid and blood.

Trace Element - Analytical Chemistry in Medicine and Biology, Vol. 3

Material and Methods

In 1983-84 we analyzed 80 amniotic fluid samples and 62 maternal blood samples at different stages of gestation (15 to 42 weeks). The mothers ranged in age between 18 and 38 years. 30 blood samples were examined from healthy non-pregnant women (ages: 16-38 years, smokers and non-smokers) and 45 blood samples from the umbilical vein. Of the neonates, 21 were boys and 24 girls; with respect to the birth weight, 15 were premature (≤2500 g) and 30 mature.
The amniotic fluid samples were obtained by transabdominal amniocentesis or cesarean section or by puncture of the amniotic sac before term delivery. All samples of amniotic fluid contaminated with meconium and/or blood were excluded. Maternal blood samples were collected just before or during the onset of labor, whereas cord blood was taken immediately after delivery. The amniotic fluid and the sera were separated by centrifugation for 30 minutes at 10000x g in order to settle corpuscular and cellular components, and were kept at -20°C. Cadmium was determined by electrothermal atomization and atomic absorption; zinc, copper and iron by atomic absorption spectrophotometry; and total protein by the biuret method.

Results and Comments

The analytical results for cadmium and total protein levels in amniotic fluid are presented in Fig. 1.
The cadmium level was highest in the 15th to 17th weeks of gestation, and subsequently decreased. Between the 27th to 42nd weeks of gestation, the level fell stepwise. The total protein level increased up to the 31st week of gestation and later decreased.

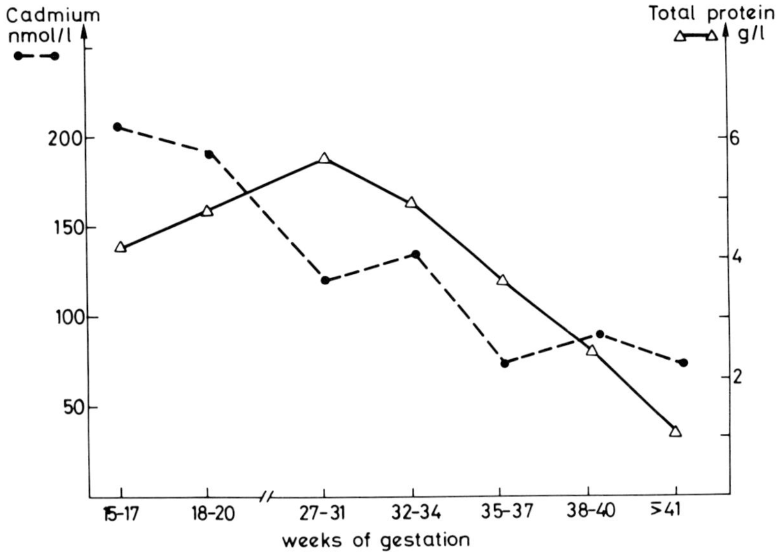

Fig. 1. Cadmium and total protein levels of amniotic fluid

Figure 2 shows the concentrations of copper, zinc and iron in amniotic fluid samples at different stages of gestation. It can be seen that the iron level increased continuously; the zinc level increased after the 31st week of gestation; while the copper level increased from the 15th up to the 34th week of gestation, and afterwards decreased.

The cadmium, copper, zinc, iron and total protein levels of blood samples from controls, newborns, and mothers in all three trimesters of pregnancy can be seen in Table I.

Since the cadmium levels in the umbilical cord blood were similar to those in the maternal blood, we consider that the placental passage of cadmium is *possible*. The blood copper levels in newborns were very low, but the iron, zinc and total protein concentrations were higher than in the mothers.

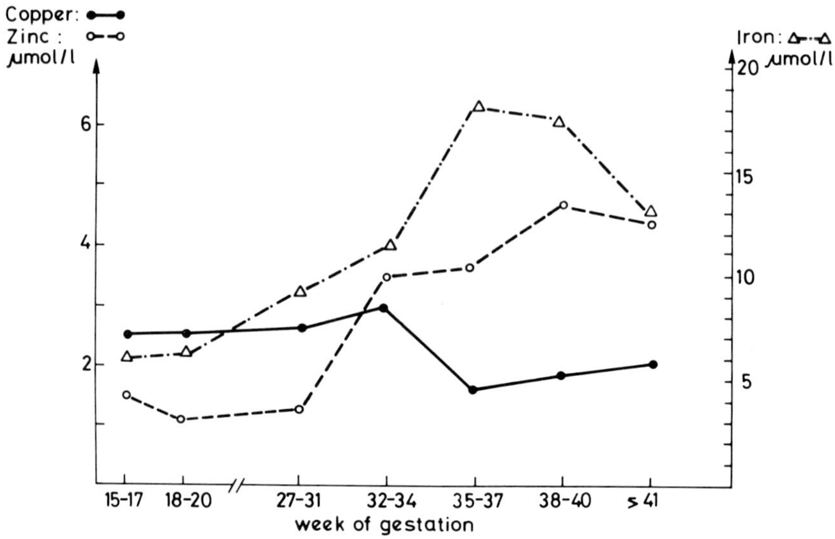

Fig. 2. Copper, zinc and iron levels of amniotic fluid

Table I. Cadmium, zinc, copper, iron and total protein levels in blood of mothers, newborns and controls

	Serum									
	Cd nmol/l		Cu mmol/l		Zn mmol/l		Fe mmol/l		total protein g/l	
	$\bar{x}$	SD	$\bar{x}$	SD	$\bar{x}$	SD	$\bar{x}$	SD	$\bar{x}$	SD
Controls	46.4	27	18.9	2.0	14.3	1.1	18	2.8	66.6	4.2
(No)	(30)		(30)		(30)		(30)		(30)	
Mothers										
trim.I	52.8	24	26.9	2.0	13.1	2.4	18.2	3.8	66.2	3.4
(No)	(14)		(20)		(20)		(20)		(20)	
trim.II	49.8	26	31.1	2.3	12.1	2.0	16.9	3.2	64.0	4.6
(No)	(14)		(20)		(20)		(20)		(20)	
trim.III	41.6	25	37.8	3.4	10.6	1.8	15.0	2.5	60.4	4.7
(No)	(13)		(22)		(22)		(22)		(22)	
Newborns	38.8	24	5.7	1.9	12.3	6.0	16.5	4.4	63.7	0.8
(No)	(25)		(45)		(45)		(45)		(45)	

The maternal blood cadmium value underwent practically no change (it decreased minimally) during pregnancy.

Discussion

The microelements may be divided into two groups: essential and non-essential (often toxic) elements.

A lower than normal intake of essential trace elements results in a decreased growth rate, and inhibition of fertility (changes in trace elements may also occur during the pathological condition of pregnancy). Some of the biological effects of non-essential, toxic elements are known, but the majority of these effects are still unknown.

Cadmium is a very toxic, non-essential element. The mean cadmium levels in human tissues fallow the sequence: renal cortex > renal medulla > liver > lung > brain > testis > ovary > lymph nodes > muscle (18). The salient features of cadmium metabolism are: a) lack of an effective homeostatic control mechanism; b) tendencious retention in the body with a very long half-life; and c) strong competition with other divalent metals (18).

The results reported by *Friberg* and *Vahter* (6)(from a study in which 10 countries participated) demonstrated that the exposure to cadmium varied substantially in the different areas: the geometric mean cadmium level in the blood ranged from 0.5 µg/l (in Stockholm and Jerusalem) to 1.2 µg/l (in Brussels and Tokyo). Smokers generally had considerably higher concentrations than non-smokers, while former smokers had intermediate values. Our cadmium levels (in controls and mothers) were similar to those given by *Friberg* and *Vahter* (6). The cadmium content of normal human blood is low and variable, but the cadmium levels of the amniotic fluid are higher than those of the blood (1,7,15).

The competition between cadmium and other metals has been studied previously (5,9,12). Zn antagonizes the testicular weight loss induced by cadmium (19). A mechanism for zinc-induced tolerance to cadmium toxicity was clarified by *Leber* and *Miya* (10), who showed that metallothionein induced by zinc served as a cadmium scavenger and contributed to the protec-

tion against cadmium-induced toxicity.
Cadmium is known to induce hypertension in animals, but the significance of the relationship between human hypertension and the cadmium levels of the blood (and/or urine) is uncertain and the results are contradictory (14,17).

Conclusions

From the data reported in this paper the following conclusions can be drawn:

1. The cadmium levels in the amniotic fluid are *higher* than those in maternal blood samples during pregnancy. The cadmium concentration is higher in the early period of pregnancy than later.
2. The cadmium levels in cord blood samples are *similar* to those in maternal blood. *We consider that the placenta does not provide a "protective barrier" against this element.*
3. The cadmium values of maternal blood samples are similar to those of control blood samples.
4. The iron and zinc levels *increase continuously,* while the copper level increases from the 15th to the 34th weeks of gestation, and afterwards decreases in the amniotic fluid.
5. The maternal blood copper concentration *increases* during pregnancy, while the zinc and iron levels *fall.*
6. The total protein level of the amniotic fluid is *very low* compared to that of the maternal blood.

References

1. Auermann, E., Heidel, G., Jacobi, J., Dietel, K., Meckel, U., Cumbrowsky, H.: Dt. Gesundh.-Wesen 37, 1161 (1982).
2. Chernoff, N.: Teratology 8, 29 (1973).
3. Duricic, J., Keserstankoviz, M., Duricic, Z., Zulic, I.: Jugosl. Physiol. Pharm. Acta 19, 35 (1983).

4. Ferm, V. H.: Biol. Neonat. 19, 101 (1971).

5. Francis, P. C., Petersen, R. L.: Bull. Environ. Contam. Toxicol. 30, 567 (1983).

6. Friberg, L., Vahter, M.: Environ. Research 30, 95 (1983).

7. Huber, J. C., Murana, R. F. L., Michalica, W., Murana, H., Wernisch, J.: Münch. med. Wschr. 125, 53 (1983).

8. Ishizaki, A., Fukushima, M., Sahamoto, M.: Japan J. Hyg. 25, 86 (1970).

9. Kojima, R., Tanaka, E.: Med. Biol. 86, 173 (1973).

10. Leber, A. P., Miya, T. S.: Toxicol. Appl. Pharmacol. 37, 403 (1976).

11. Lucis, J. O., Lucis, R., Shaikh, A.: Arch. Environm. Health. 25, 14 (1972).

12. Miyahara, T., Oh-E., Y., Takaine, E., Kozuha, H.: Toxicol. Appl. Pharmacol. 67, 41 (1983).

13. Murata, J., Hirono, T., Saeki, Y.: Bull. Soc. Intern. Cl. 1, 34 (1970).

14. Perry, H.M., Thind, G. S., Perry, E. F.: Med. Clin. North Am. 60, 759 (1976).

15. Rösick, E., Rösick, U., Brätter, P.: Trace Element-Anal. Chem. Med. Biol., Vol. 2., 1983, Walter de Gruter, Berlin-New York, pp. 464.

16. Szadkowski, D.: Med. Monatsschr. 26, 553 (1972).

17. Tulley, R. T., Lehmann, H. P.: Clin. Chim. Acta 122, 189 (1982).

18. Underwood, E. J.: Trace Elements in Human and Animal Nutrition. Acad. Press. New York - San Francisco - London 1977. pp. 243.

19. Webb, M.: Biochem. Pharmacol. 21, 2767 (1972).

HEMODIALYSIS INFLUENCE ON ZINC LEVELS IN SERUM, HAIRS AND DIALYSIS FLUIDS.

Alain Favier, Jean-Luc Lafond, Bertrand Desrousseaux, Josiane Arnaud.
Laboratoire de Biochimie C, Professeur LATURAZE, CHRUG, BP 217 X, 38043 GRENOBLE CEDEX (FRANCE).

Toufic Hachache, Robert Milongo.
A.G.D.U.C, CHRUG, BP 217 X, 38043 GRENOBLE CEDEX (FRANCE).

Trace elements during hemodialysis has been extensively studied especially zinc. But these various works show controversial results, some authors found that hemodialysis is responsible for a lower zinc level in serum (1 - 2), other found the contrary (3).

As hemodialysis led to an important change in blood content, we try to evaluate the variations of numerous factors know to be linked with zinc metabolism. This study was carried out to found the consequences of hemodialysis on zinc level in blood, in hairs, on total proteins, on specific proteins or enzymes, for eventually, giving more light on zinc status in this disease.

Material and methods

32 patients were studied during six months. Their blood was collected in polystyrene tube each month for trace element determination and for ionogram control, before and after hemodialysis session. Hemodialysis fluids were also withdrawn before, five minutes, two hours and four hours after the start of the session. Hairs of some patients was taken at the beginning and at the end of our study.

Trace Element - Analytical Chemistry in Medicine and Biology, Vol. 3

Zinc level was assayed by flameless AAS (Perkin Elmer HGA 500) : serum, after a 1/10 dilution (4), hairs after ashing with HNO_3 and dialysis fluid directly without dilution.

Total proteins were determined by Technicon SMA Analyser (Biuret method), α_2 Macroglobulin was determined by radial immunodiffusion (Behring), alkaline phosphatase and albumin by Dupont ACA II methodologies.

Vitamin A was determined by fluorimetric method of Brubacker and Vuilleumier.

Results and discussion

. Zinc status in hemodialysed patients

As shown in table 1 and in fig. 1, hemodialysis patients present a low serum zinc but we found a normal level in hairs of these patients. Indirect indicators of zinc deficiency are normals (α_2 Macroglobulin, albumin, alkaline phosphatase) except Vitamin A that is increased, and we found no relation between serum zinc with alkaline phosphatase and Vitamin A (Fig. 2). This paradoxal increase of retinol had yet been found by others (5).

	Zinc in serum µmol/l	Zinc in hairs µmol/l	Vit. A µmol/l	Alk. P. UI/dl	Albumin g/l	α_2M g/l
Hemodialysed Patients	11.53 ± 1.59 n = 33	2.79 ± 0.79	8.89 ± 2.55 n = 33	8.6 ± 3.1 n = 33	42.6 ± 3.1 n = 33	1.85 ± 0.60
Normals	16.6 ± 2.05 n = 30	2.87 ± 0.80 n = 19	3.2 ± 0.9 n = 16	6.1 ± 3.6	43.0 ± 5.0	1.5 to 4.0

Table 1

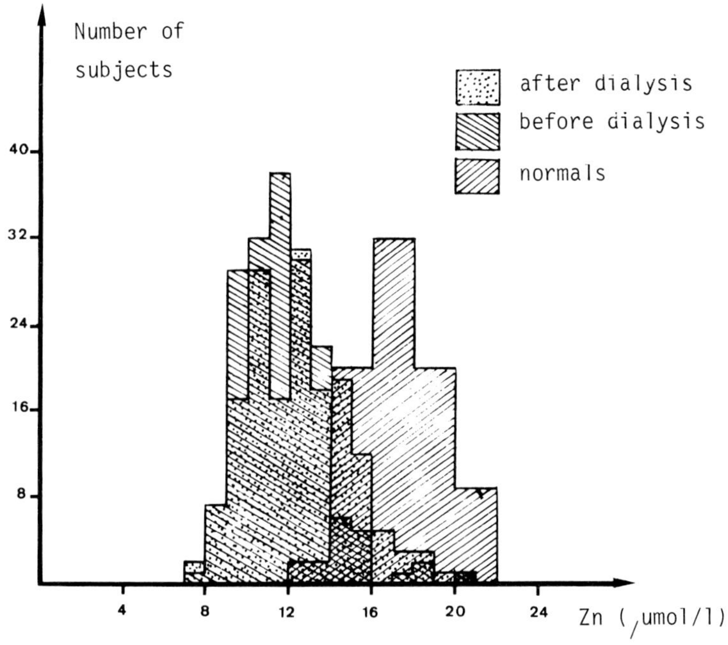

Figure 1 : Zinc level in serum.

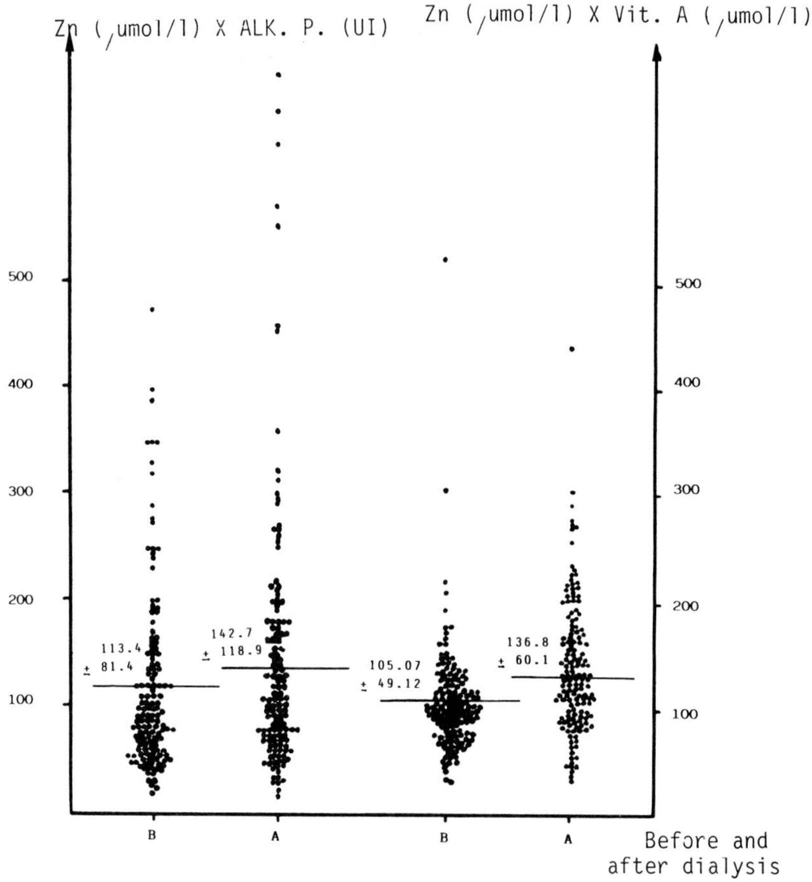

Zn (μmol/l) X ALK. P. (UI)
Zn (μmol/l) X Vit. A (μmol/l)
500
400
300
200
100
113.4
± 81.4
142.7
± 118.9
105.07
± 49.12
136.8
± 60.1
B
A
B
A
Before and after dialysis

Figure 2

The low level of zinc exists before the hemodialysis, newly dialysed patients present even a more low level than former (6).
In these patients (Table 2) hemodialysis seems to slightly increases level of zinc in serum, but this may be a consequency of changing in diet (more proteins) that start with hemodialysis treatment.

		Serum Zinc levels			
Units		μmol/l		μmol/g of proteins	
Dialysis		Before	After	Before	After
Former dialysed patients n = 26		11.47	13.26	0.168	0.171
Newly dialysed patients n = 7	before first session	10.06	11.02	0.142	0.141
Newly dialysed patients n = 7	after one month of dialysis	11.94	13.95	0.170	0.184
Newly dialysed patients n = 7	after 6 months of dialysis	11.70	12.66	0.164	0.169
Total patients n = 33		11.53 $\pm$ 1.59	13.14 $\pm$ 2.05	0.167 $\pm$0.022	0.171 $\pm$0.025

Table 2

. Influency of the hemodialysis session

During hemodialysis we found an increase of zinc level in serum as found by many other authors, but when we looked at the concentration of zinc expressed as μg per g of proteins, no variation can be found (fig. 2).
We found also identical values before and after the session, if zinc was expressed per g of specific protein as α_2 Macroglobulin and albumin, that are known binding proteins (fig. 3).

Fig. 3

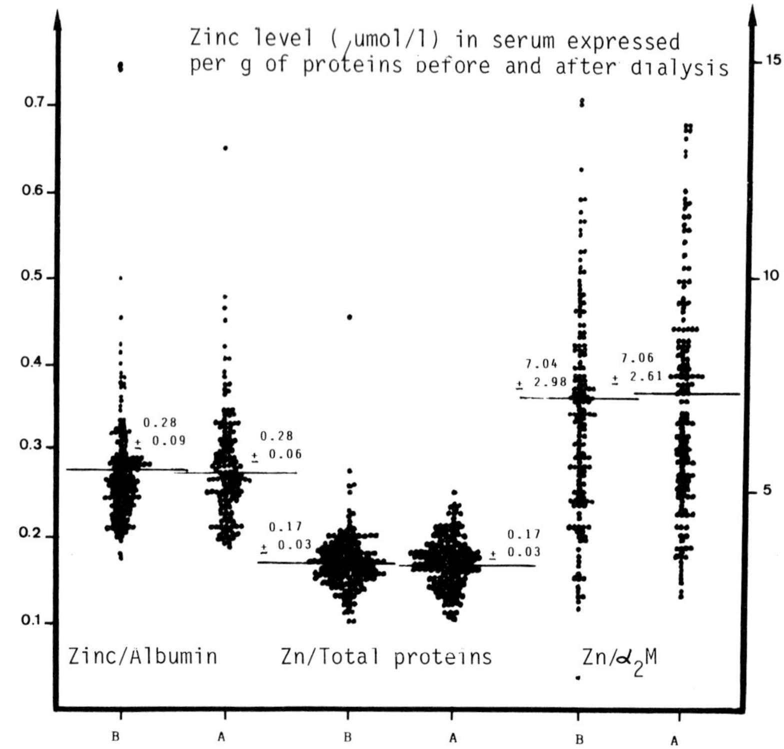

The level of zinc in dialysis fluid increase with a peak at 5 minutes during hemodialysis session, level before and two hours after the begining, remains identical (7). More hight is the level of zinc before hemodialysis more hight will be the peak at 5 minutes. So, we explain this increase of zinc by a release of the metal from the membrane (fig. 4). Dialysis membrane have been treated during the process of fabrication with zinc chlorid but, some have been not correctly washed and release more zinc. At the begining of the dialysis, small molecular weight compounds as amino acids, that are strong chelating agents, cross the membrane pick up its zinc, and clean up the membrane. That can explain the transient peak of zinc in dialysis fluid.

We survey serum zinc of our patients during a long period (up to six months) and during that time we found no change of this parameter. So, (table 2) dialysis has few long term consequency expect for newly dialysed as discussed before.

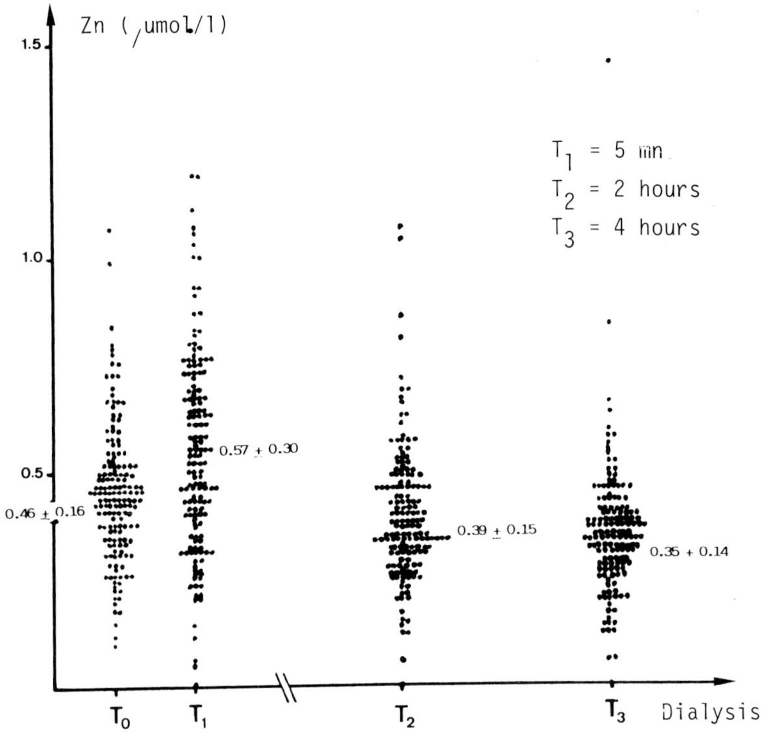

Figure 4 : Zinc in hemodialysis fluids.

Conclusion

In hemodialysis zinc status is low, and can cause a deficiency, but hemodialysis by itself has few consequency in zinc homeostasis and the apparent change in zinc level is only due to a concentration of proteins in blood of patients.

References

1. MOUNTOKALAKIS T., DAKANALIS D., BOUKIS D., VIRVIDAKIS K., VOUDIKLARIS S., KOUTSELINIS A.: Clinical Nephrology 12, 5, 206-209 (1979).

2. MAHAJAN S.K., GARDINER W.H., ABBASI A.A., BRIGGS W.A., PRASAD A.S., Mc DONALD F.D.: Trans. Am. Soc. Artif. Intern. Organs 24, 50-54 (1978).

3. BOGDEN J.D., OLESKE J.M., WEINER B., SMITH L.G., SMITH L.G Jr., NAJEM G.R.: Am. J. Clin. Nutr. 33, 1088-1095 (1980).

4. FAVIER A., RUFFIEUX D.: Biomedicine & Pharmacotherapy 37, 462-466 (1983).

5. CASEY C.E., MOSER M.C., HAMBIDGE K.M., LUM G.M.: J. Pediatr. 3, 434-435 (1981).

6. THOMPSON N.M., STEVENS B.J., HUMPHERY T.M., ATKINS R.C.: Kidney Int. 23, 9-14 (1983).

7. BOGDEN J.D., ZADZIELSKI E., WIENER B., OELSKE J.M., AVIV A.: Am. J. CLin. Nutr. 36, 403-409 (1982).

MANGANESE STATUS IN LIVER ALCOHOLIC DISEASES

Alain Favier, Josiane Arnaud, Laurence Dumolard.
Laboratoire de Biochimie C, Professeur LATURAZE, CHRUG, BP 217 X, 38043 GRENOBLE CEDEX (FRANCE)

Jean-Pierre Zarski.
Clinique d'Hépatogastroentérologie, Professeur RACHAIL, CHRUG, BP 217 X, 38043 GRENOBLE CEDEX (FRANCE)

Liver is a key tissue in manganese metabolism.

- Bile is the main way of endogenous manganese excretion in animals and enterohepatic circulation of bile has been considered as the principal regulatory mechanism of manganese homeostatis (1).
- Excessive manganese uptake by the rat liver can cause liver damage and cholestasis (2). The maximum capacity of liver to extract manganese and prevent toxicity is unknown.
- Manganese is the cofactor of many liver enzymes. Hepatic mitochondrial superoxide dismutase and pyruvate carboxylase are real manganese enzymes, glycogen synthetase and phosphoenolpyruvate carboxykinase are activated by manganese and sulphate (3).

  In one case of manganese deficiency, liver can't synthetise coagulation factors and prothrombin time is decreased (4).

In hepatic disease, manganese homeostasis is disturbed.

- VIERSECK and al (5,6) have found an increase of serum manganese and copper during liver metastases and hepatobiliary disease like acute hepatitis, chronic hepatitis, post hepatitis cirrhosis, cirrhosis and extrahepatic bi-

Trace Element - Analytical Chemistry in Medicine and Biology, Vol. 3

liary obstruction ;

These authors found also a significant positive correlation between serum albumin and serum manganese concentrations in acute, chronic and post necrotic cirrhosis (6) and between serum aminotransferase activities and serum manganese concentration, so the authors postulate that the increase value of serum manganese can be an index of liver cell damage.

- SULLIVAN and al (7) found an increase of serum manganese during cirrhosis. However, the difference was not significant.
- KANABROCKI and al (8) found an increase of serum manganese during liver cirrhosis.

Material and methods

Patients

The population studied consisted of fifteen alcoolic cirrhotic patients (13 males and 2 females, mean age : 60,3 years) (Group I), eleven chronic alcoholic (11 males, mean age : 51,8 years) (Group II) and thirty control subjects (Group III).

All patients of the group I had compensed cirrhosis. No patients presented clinical and/or biological hepatocell failure signs (PT $\geqslant$ 50 %). Acute alcoholic hepatitis patients were excluded. In all case, the diagnosis of liver disease was histologically proven. The chronic alcoholic patients had normal nutritional status (theorical weight, arm circumference, triceps skinfold, serum transferrin and urine creatinin are normals). Patients with histologically steatosis and without acute alcoholic hepatitis were included.

None of the patients received diuretic nor chelating drugs.

Laboratory investigations

15 ml of blood (tube 1) was collected in the morning on a polystyrene

free trace elements tube without additive.

5 ml of blood (tube 2) was collected on natrium citrate.

Urine was collected during 24 hours on a polypropylene free trace elements vial.

In serum (tube 1), we have determined :

- Manganese, copper, zinc and iron by flameless atomic absorption, after dilution 1/10 in deionised water (copper, zinc (9)), dilution 1/5 in triton X 100 % (manganese) (modification of our previous method (10)) or after deproteinization with hydrochloric acid for iron (11).

- Total proteins by Biuret's method (Technicon Microlyser), Albumin by radial immunodiffusion (Behring Nor partigen) and α_2 globulin by electrophoresis.
- Bilirubin by the method of Jendrassik (Technicon microlyser).
- Glutamyl transferase (γ GT) by Scaz's method, glutamic pyruvic transaminase (SGPT) by IFCC's method, Alkaline phosphatase (Alk P) by Bessey's method.

In total blood (tube 2), we have determined :

- Erythrocyte sedimentation rate (ESR)
- Prothrombin time (PT).

In urine, we have determined :

- Trace elements (manganese, copper, zinc, iron) by flameless atomic absorption after simple dilution in deionised water.
- Creatinin by Jaffe's method.

The instrument use for atomic absorption is a Perkin Elmer 560 with a HGA 500 furnace and a AS 40 automated sampler.

Statistical analysis

Means and standard deviation of each parameters were calculated for the three groups of subjects.
Regression line and correlation coefficients were calculated between some of the parameters in the same group of patients.
The Student's t test was used for statistical comparison of one parameter between the differents groups or between differents parameters in the same group.
P. value of less than 0.05 was considered to indicate a statistically significant difference between two means values.

Results

Our results are given in Table I.
In the cirrhosis group (group I), we observe a significant increase of serum manganese, ($P < 0.01$), and serum copper ($P < 0.001$), a significant decrease of serum albumin ($P < 0.001$) and urinary manganese ($P < 0.01$) versus the control group.
In the chronic alcoholic group (group II), we note only a significant decrease of urine manganese ($P < 0.05$) versus the control group.
The correlations between serum manganese and some others biological parameters give the results listed in table II.
In the alcoholic cirrhosis group, the increase of serum manganese is in relation with bilirubin, γ GT, serum copper, α_2 globulin and serum albumin.
There is no relationship between serum manganese, and prothrombin time, Alkaline phosphatase, transaminase, erythrocyte sedimentation rate, serum iron, total proteins, urine copper nor urine manganese.

In the chronic alcoholic group, we note only one significant relationship between serum manganese and α_2 globulin.

TABLE I : RESULTS EXPRESSED BY MEANS ± ONE STANDARD DEVIATION

	Control	Chronic Alcoholic n = 11	Alcoholic Cirrhosis n = 15
SERUM			
Mn nmol/l	20.5 ± 3.75	20.4 ± 9.1	27.8 ± 11.85
Zn µmol/l	16.6 ± 2.05	13.7 ± 3.0	10.9 ± 3.4
Cu µmol/l	21.2 ± 3.95	23.4 ± 2.6	26.6 ± 6.4
Fe µmol/l	22.35 ± 5.8	19.3 ± 5.2	20.3 ± 4.0
Bilirubin µmol/l	9 ± 5	28 ± 15	62 ± 40
SGPT nkat	280 ± 85	642 ± 278	1002 ± 1249
Alk P. nkat	375 ± 63	480 ± 156	786 ± 339
γGT nkat	277 ± 91	3162 ± 2427	4362 ± 4502
α_2 globulin g/l	6.9 ± 0.7	7.15 ± 1.35	6.6 ± 1.9
Albumin g/l	43 ± 2	43 ± 4	34 ± 7
Total protein g/l	73 ± 3	68 ± 5	71 ± 13
TOTAL BLOOD			
ESR mm 1st hour	6 ± 2	15 ± 14	37 ± 25
PT %	100	98 ± 7	65 ± 25
URINES			
Mn nmol/24 h	65.2 ± 34.4	34.4 ± 19.7	26.2 ± 30.4
Cu µmol/24 h	0.82 ± 0.45	0.75 ± 0.4	0.68 ± 0.4
Zn µmol/24 h	4.9 ± 2.1	6.9 ± 2.8	10.0 ± 6.6
Fe µmol/24 h	2.04 ± 1	2.55 ± 1.5	2.2 ± 3.1
Mn nmol/mmol creat.	8.55 ± 4.24	4.41 ± 2.09	4.41 ± 3.00
Cu µmol/mmol creat.	0.11 ± 0.33	0.19 ± 0.09	0.14 ± 0.11
Fe µmol/mmol creat.	0.18 ± 0.09	0.70 ± 0.50	0.45 ± 0.69
Zn µmol/mmol creat.	0.54 ± 0.34	0.9 ± 0.3	2.0 ± 1.2

TABLE II : CORRELATIONS BETWEEN SERUM MANGANESE AND SOME OTHER BIOLOGICAL PARAMETERS

	Chronic Alcoholic	Cirrhosis
S.Mn / Alk.P	$p > 0.9$ $r = 0.04$	$p < 0.2$ $r = 0.45$
S.Mn / SGPT	$p < 0.9$ $r = 0.06$	$p < 0.9$ $r = -0.18$
S.Mn / γGT	$p < 0.9$ $r = 0.06$	$p < 0.01$ $r = 0.77$
S.Mn / Alb	$p < 0.9$ $r = -0.10$	$p < 0.01$ $r = 0.65$
S.Mn / $\alpha_2$glob.	$p < 0.05$ $r = 0.66$	$p < 0.05$ $r = 0.67$
S.Mn / total prot.	$p < 0.5$ $r = 0.17$	$p < 0.5$ $r = 0.24$
S.Mn / ESR	$p < 0.5$ $r = 0.32$	$p < 0.9$ $r = 0.06$
S.Mn / PT	$p < 0.9$ $r = -0.06$	$p < 0.1$ $r = 0.47$
S.Mn / Bilirubin	$p < 0.1$ $r = 0.58$	$p < 0.02$ $r = -0.70$
S.Mn / U Mn	$p < 0.5$ $r = -0.24$	$p < 0.9$ $r = 0.11$
S.Mn / S Cu	$p < 0.3$ $r = 0.41$	$p < 0.05$ $r = 0.59$
S.Mn / U Cu	$p < 0.9$ $r = -0.21$	$p < 0.5$ $r = 0.09$
S.Mn / S Fe	$p < 0.10$ $r = -0.52$	$p > 0.9$ $r = -0.03$

S.Mn=serum manganese, U.Mn=urine manganese, S.Cu=serum copper, U.Cu=urine copper, S.Fe=serum iron, Alb=albumin, total prot.= total proteins, PT=prothrombin time, ESR=erythrocyte sedimentation rate, SGPT=glutamic pyruvic transaminase, Alk.P= alkaline phosphatase, γGT=glutamyl transferase, α_2 glob. =α_2 globuline.

Discussion

During chronic alcoholism, manganese status remains normal, but during cirrhosis we find an increase of serum manganese. The origin of this increase may be an accumulation of manganese in the body, a release from the liver storage, an increase of a manganoprotein or a mangano-enzyme without relationship with a disturbance of manganese metabolism (as ceruloplasmin increase in inflammation).
We try to explain this increase of serum manganese making correlation with some biological markors of different liver functions.
We found no relation with a disturbance of cholestasis that may have explained such an increase, as manganese is excreted by bile, (manganese is not correlated with Alkaline phosphatase), nor with hepatocellular damage that can make the manganese stored cross the cell membranes (no correlation with SGPT), nor with hepatocellular failure (no correlation with a decrease of PT).
We cannot explain the increase by a retention of manganese in the body even if urinary manganese is lowered. First, the increase in serum is not correlated to the urinary decrease, secondly in few patient we could practice a determination of manganese in a liver biopsy that don't showed any elevation.

So, the only explanation we found is an increase of manganese linked to the inflammatory process (We found a close correlation with copper and with α_2 globulin). In a previous study we found that a large part of manganese in serum is bound to an α_2 globulin certainly an α_2 macroglobulin (12). We can suppose that this protein may play a part in the inflammatory process and we are now trying to look at a close relationship between manganese and different inflammatory diseases.

References

1. UNDERWOOD E.J.: Trace elements in human and animal nutrition. Academic Press, New York, 4°ed. 1977.

2. WITZLEBEN C.L., PITLICK P., BERGMEYER J., BENORT R.: Am. J. Path. 53, 404 (1968).

3. LEACH R.M., LILBURN M.S.: Wld. Rev. Nutr. Diet. 32, 123-134 (1978).

4. DOISY E.A.: Environmental Health, Hemphill D Ed. University of Missouri, Press Columbia, U.S.A. 1973.

5. VERSIECK J., HOSTE J., BARBIER F.: Acta Gastro-enterol. Belg. 39, 340-349 (1976).

6. VERSIECK J., BARBIER F., SPEECKE A., HOSTE J.: Clin. Chem. 20, 9, 1141-1145 (1974).

7. SULLIVAN J.F., BLOTCKY A.J., JETTON M.M., HAHN K.J., BURCH R.E. J. Nutr. 109, 1432-1437 (1979).

8. KANABROCKI E.L., CASE L.F., GRAHAM L., FIELDS T., MILLER E.B., OESTER V.T., KAPLAN E.: Journal of Nuclear Medicine 8, 166-172 (1967).

9. FAVIER A., RUFFIEUX D.: Biomedicine & Pharmacotherapy 37,462-466 (1983).

10. FAVIER A., RUFFIEUX D.: Clin. Chim. Acta 124, 239-244 (1982).

11. FAVIER A., MALJOURNAL B., DECOUX G., RUFFIEUX D.: Ann. Biol. Clin. 41, 45-50 (1983).

12. LAFOND J.L., DURON C., FAVIER A.: International Symposium on Health and interactions of essential and toxic elements - LUND 1983 13-18 June.

D i s c u s s i o n Session XI - 2

Subjects of the discussion

- synthesis of metallothionein by Cd
- difference between generation and concentration
- metallothionein in connection to the metabolism of Zn
- Zn-supplementation during pregnancy
- toxic levels for Zn
- Zn deficient status
- bone abnormalities
- the general effect of the removal of the kidney

Contributors: Binnerts, Heilmaier, Hurley, Plantin, Weser, Westermark

LIST OF PARTICIPANTS

Aito, Mr. Antero
Institute of Occup. Health
Haartmaninkatu 1
SF-00290 Helsinki 29
Finland

Akesson, Mr. Björn
Dept. of Clin. Chem.
Univ. Hospital
S-22185 Lund
Sweden

Anke, Mr. M.
Sekt. Tierproduktion u. Vet. Med.
Karl-Marx-Univ., Leipzig
Dornburgeer Str. 24
DDR-69 Jena
German Dem. Rep.

Bartsch, Mr. Pierre
CRC Ulg
B-4000 Sant Tilman
Belgium

Bayer, Mrs. Gisela
Erich-Ollenhauer-Str. 32a
D-6200 Wiebaden
FR Germany

Beguin, Mr. Yves
10 Av. de l'Observatoire
B-4000 Liege
Belgium

Benzo, Mrs. Zully
Centro de Quimica, IVIC
Apartado 1827
YV-1010-A Caracas
Venezuela

Berg, Mr. Dieter
GSF, Nuklearbiologie
Ingolstädter Landstraße 1
D-8042 Neuherberg
FR Germany

Binnerts, Mr. Willem, T.
Dept. of Animal Physiol.
Agricult. Univ.
Haarweg 10
NL-6709 PJ Wageningen
The Netherlands

Blanusa, Mrs. Maja
Institute for Medical Res.
M. Pijade 158
YU-41000 Zagreb
Yugoslavia

Brätter, Mr. Peter
Hahn-Meitner-Institut
Glienicker Straße 100
D-1000 Berlin 39
FR Germany

Brockhaus, Mr. Arthur
Med. Inst. f. Umwelthygiene
Univ. Düsseldorf
Gurlittstraße 53
D-4000 Düsseldorf
FR Germany

Byrne, Mr. Anthony
Nuclear Chem. Section
J. Stefan Institute
YU-61000 Ljubljana
Yugoslavia

Chauchie, Mr. Philippe
Av. H. Hooven 9
B-1040 Bruxelles
Belgium

Clarkson, Mrs. Johanna
Department of Pathology
Univ. Hospital of Wales
Heath Park
GB-Cardiff, CF4 4 Xw
Great Britain

Cochran, Mr. Malcom
Flinders Medical Center
AUS-Adelaide 5042
Australia

Cornelis, Mrs. Rita
Lab. of Analyt. Chem.
Proefuinstraat 86
B-9000 Gent
Belgium

Cossack, Mr. Zafrallah
Interuniversitair Reactor Institute
Nuclear Biotechn.
Mekelweg 15
NL-2629 JB Delft
The Netherlands

Couzy, Mr. Francois
INSA, 20 Avenue Albert Einstein
F-69621 Villeurbonne, CEDEX
France

Crößmann, Mr. Gerd
Lanw. Untersuchungs- u. Forschungsanstalt
Nevinghoff 40
D-4400 Münster
FR Germany

D'Hollander, Mrs. Maria
Univ. Instelling, APEN
Universiteitsplein 1
B-2610 Wilryk
Belgium

Damsgaard, Mrs. Else
Research Establishment Riso
Isotope Devision
DK-4000 Roskilde
Denmark

Earle, Mrs. Kay Eliz
Imperial College Reactor Centre
Sillwood Park, Sunninghill
GB-Ascot, Berks
Great Britain

Etschenberg, Mr. Eugen
Nattermann & Co GmbH
Nattermann-Allee 1
D-5000 Köln 20
FR Germany

Everett, Mr. George
Cornell Univ. United States
Dept. of Agric. U.S. Plant, Soil a.
Nucl. Lab. Tower Road
Ithaka, New YOrk
U.S.A.

Fagioli, Mr. Francesco
Istituto Chimico
Via L. Borsari 46
I-44100 Ferrara
Italy

Faure, Mr. Henri
Hopital des S. Biochimie C
F-38700 Latroniche,
France

Favier, Mr. Alain
CHRG Grenoble la Trouche
F-Grenoble
France

Feinendegen, Mr. L.E.
Institute of Medicine
Nuclear Research Center Jülich GmbH
D-5170 Jülich
FR Germany

Frank, Mr. Adrian
National Veterinary Inst.
S-75007 Uppsala
Sweden

Galgian, Mrs. Vera
Forschungsinstitut f. Kinderernährg.
Weistück 11
D-4600 Dortmund 50
FR Germany

Gammelgaard, Mrs. Bente
Royal Danish School of Pharm.
Dept. of Pharmaceut. Chem.
2 Universitetsparken
DK-2100 Copenhagen
Denmark

Ganther, Mr. H.E.
Univ. of Wisconsin
Dept. of Nutr. Sciences
1415 Linden Drive
Madison, Wiscons. 53706
U.S.A.

Gardiner, Mr. Philip
Inst. f. Chemie, KFA Jülich
D-5170 Jülich
FR Germany

Galwik, Mr. Dieter
Hahn-Meitner-Institut f. Kernforschg.
Glienicker Str. 100
D-1000 Berlin 39
FR Germany

Gessner, Mrs. Hildegard
Hahn-Meitner-Inst. f. Kernforschung
Abt. Spurenelemente i.d. Biomedizin
Glienicker Str. 100
D-1000 Berlin 39
FR Germany

Gökmen, Mrs. Inci, G.
Middle East Techn. Univ., Dept. of Chem.
TR-Ankara
Turkey

Grobecker, Mr. K.H.
Univ. Gießen
Institut f. Pflanzenökologie
D-6300 Gießen
FR Germany

Gulba, Mr. Dietrich
Med. Hochschule Hannover, Abt. Kardiologie
Konstanty-Gutschow-Str. 8
D-3000 Hannover
FR Germany

Gundermann, Mr. Karl-Josef
Nattermann & Co GmbH, Clin. Res.
Nattermann-Allee 1
D-5000 Köln
FR Germany

Hau, Mr. Richard
Rheinbadstr. 6
D-7880 Bad Säckingen
FR Germany

Heilmaier, Mr. Herbert
GSF, Abt. f. Toxikologie
Ingolstäder Landstraße 1
D-8042 Neuherberg
FR Germany

Heller, Mr. Wolfgang
Chir. Univ. Klinik
Calwer Str. 7
D-7400 Tübingen
FR Germany

Herber, Mr. Robert F.M.
Coronel Lab.
11 Const. Huygerstraat 20
NL-1054 BW Amsterdam
The Netherland

Hudnik, Mr. Vida
Chemical Institute
Boris Kidric, Majdrihova 19
YU-61000 Ljubljana
Yugoslavia

Hurley, Mrs. Lucille
Dept. of Nutrition, Univ. of California
Davis, Ca 95616
U.S.A.

Johansson, Mr. Erland
The Gustaf Werner Inst., Dept. of Phys. Biol.
P.O. Box 531
S-75121 Uppsala
Sweden

Juguet, Mr. Bernard
Laboratoire d' Hygiène
1 rue des Hospitalieres
F-75004 Paris
France

Kaczmar, Mr. Bohdan
GSF, Abt. f. Nuklearbiologie
Ingolstädter Landstr. 1
D-8042 Neuherberg
FR Germany

Kantola, Mrs. Marina
Dept. of Chemistry
Univ. Kuopid
P.O. Box 6
SF-70211 Kuopid 21
Finland

Kauranen, Mr. Pentti
Dept. of Chem., Univ. Kuopid
P.O. Box 6
SF-70211 Kuopid 21
Finland

Kiem, Mr. Josef
IME-KFA
Postfach 19 13
D-5170 Jülich
FR Germany

Kirchgessner, Mr. M.
Techn. Univ. München
Inst. f. Ernährungswissenschaften
D-8050 Freising-Weihenstephan
FR Germany

Knapp, Mr. G.
Sorger Weg 16
A-8047 Graz
Austria

Koh, Mr. Tee-Siaw
Inst. of Med. and Vet. Sc.
P.O. Box 14
AUS-Adelaide, SA 5000
Australia

Kollmer, Mr. W.E.
GSF, Abt. f. Nuklearbiologie
Ingolstädter Landstr. 1
D-8042 Neuherberg,
FR Germany

Korunova, Mr. Vlasta
Inst. of Nucl. Biol. and Radiochem.
CSSR-Prag
Czecho-Slovakia

Kriegel, Mr. H.
GSF, Abt. f. Nuklearbiologie
Ingolstädter Landstr. 1
D-8042 Neuherberg
FR Germany

Kübler, Mr. Ulrich
Spectral Bioanal. Lab.
Knöbelstr. 28
D-8000 München
FR Germany

Kurfürst, Mr. U.
Grün-Optik
Solmser Str. 90
D-6330 Wetzlar
FR Germany

Lafond
Laboratoire de Biochimie C
F-38700 Latroniche
France

Lakomaa, Mrs. Eeva-Liisa
Techn. Res. Centre of Finl. Reactor Lab.
Otakaari 3 A
SF-02150 Espoo 15
Finland

Lieser, Mr. K.H.
Techn. Hochschule Darmstadt
D-6100 Darmstadt
FR Germany

Lietha, Mr. Roman
Züricher Str. 146/48
CH-B 640 Rapperswil
Switzerland

Lill, Mrs. Gisela
GSF, Phys.-Techn. Abteilung
Ingolstädter Landstraße 1
D-8042 Neuherberg
FR Germany

Llabador, Mr. Yvan
Lab. de Chimie Nucl. de Bordeaux
Le Haut Vigneau
F-33170 Gradignan
France

Locatelli, Mr. Clinio
Istituto Chimico
Via L. Borsari 46
I-44100 Ferrara
Italy

Luten, Mr. Joop
Institute for Fisheryproducts-TNO
Dokweg 37
NL-Ymuiden
The Netherlands

Machari, Mrs. Rajalakshmi
Nation. Inst. of Nutrion
IND-Tarnakha-Hyderabad
India

Makin, Mr. Paul
Hauptstr. 26 D/6/1
A-2351 Wiener Neudorf
Austria

Marczenko, Mr. Zygmunt
Techn. Univ. of Warsaw
ul. Noakowskiego 3
PL-Warsaw 00-664
Poland

Matsubara, Mr. Junko
Univ. of Tokyo
7-3-1 Hongo
J-Bunkyo Tokyo
Japan

Matsusaka, Mr. Nadndri
GSF, Forschungszentrum Neuherberg
Ingolstäder Landstr. 1
D-8042 Neuherberg
FR Germany

McWeeny
Univ. of Agriculture, Fisheries, Food
Haldin House
GB-Norvich NR2 4 SX
Great Britain

Merian, Mr. Ernst
Im Kirschgarten 22
CH-4106 Thervil

Morches, Mr. Bernd
Klinikum d. Univ. Mainz
Postf. 3960
D-6500 Mainz
FR Germany

Muntau, Mr. H.W.
Joint Res. Center ISPRA
I-Ispra/Varese
Italy

Muramatsu, Mr. Yasuyuki
Int. Atomic Energy Agency
Div. of Life Sciences
P.O. Box 100
A-1400 Vienna
Austria

Negretti, Mrs. Virginia
Han-Meitner-Institut
Bereich Kernchemie u. Reaktor
Glienicker Str. 100
D-100 Berlin 39
FR Germany

Neidhart, Mr. Bernd
Inst. f. Arb.-Physiologie
Ardeystr. 67
D-4600 Dortmund
FR Germany

Neve, Mr. Jean
Univ. Libre de Bruxelles
Inst. de Pharmacie
Campus Plaine 205-1
B-1050 Brussels
Belgium

Ohmori, Mr. Sayoko
Osaka Prefectural Inst. of Public Health
J-Osaka
Japan

Omang, Mr. Severre H.
Oslo Boad of Health
St. Olavs Plass 5
N-Oslo
Norwegen

Orvini, Mr. Edoardo
C.N.R. Centre for Radiochem. and
Activation Analysis
Via Taramelli 10
I-27100 Pavia
Italy

Oster, Mr. Oskar
Universitätsklinik Mainz, Ab. für
Klinische Chemie u. Labormedizin
Langenbeckstraße
D-6500 Mainz
FR Germany

Ottaway, Mr. J.M.
Dept. of Pure a. Appl. Chem.
295 Cathedral Street
GB-Glasgow
Great Britain

Pfeiffer, Mr. Carl Curt
Princeton Brain Bio Center
R862 Route 518
Skillmann, NJ 08558
U.S.A.

Piepponen, Mr. Sulo
Techn. Res. Centre Finland
Food Res. Laboratory
Biologinkuja 1
SF-02150 Espoo 15
Finland

Piwonka, Mr. Jorge
Katharinenstr. 17
D-7070 Schwäbisch-Gmünd
FR Germany

Plantin, Mr. Lars-Olof
Forskningscentrum, Huddinge, LSjukhus
S-14186 Huddinge
Sweden

Pszonicki, Mr. Leon
Int. Atomic Energy Agency
Laboratory Seibersdorf
Wagramer Str. 5, P.O. Box 100
A-1400 Vienna
Austria

Ribas, Mr. Bartolome
Dept. of Biochemistry
Faculty of Pharmacy
E-Madrid 3
Spain

Rillaerts, Mr. Eddi
Universitaire Instelling Antwerpen
Lab. Endocrinol.
Universiteitsplein 1
B-2610 Wilrijk
Belgium

Ringdal, Mr. Ole
Inst. of Nutrition, Direct. of Fisheries
P.O. Box 4285
N-5013 Bergen
Norway

Rösick, Mr. Ullrich
Hahn-Meitner-Institut
Glienicker Str. 100
D-1000 Berlin 39
FR Germany

Rueter, Mr. Jochen
Univ. Dortmund, Inst. f. Arbeitsmedizin
D-4600 Dortmund
FR Germany

Runkel, Mr. Karl-Heinz
GSF, Abt. f. Toxikologie
Ingolstädter Landstr. 1
D-8042 Neuherberg
FR Germany

Safaighi, Mr. Hasan
D-7400 Tübingen
FR Germany

Salgo, Mr. Lazlo
Paedriatric Dept. of Med. School
P.O. Box 471
H-6701 Szeged
Hungary

Sansoni, Mr. Bruno
Kernforschungsanlage Jülich GmbH
Zentralabteilung für Chemische Analysen
D-5170 Jülich
FR Germany

Schaupp, Mrs. Karin
Fa. Leopold & Co
Hafnerstr. 361
A-8055 Graz
Austria

Schelenz, Mr. R. Franz W.
Bundesforschungsanstalt f. Ernährung
Zentrallabor f. Isotopentechnik
Engesserstr. 20
D-7500 Karlsruhe
FR Germany

Schenkel, Mr. Hans
Univ. Hohenheim, Inst. f. Tierernährg.
Postfach 70 05 62
D-7000 Stuttgart 70
FR Germany

Schmidt, Mr. P.F.
Inst. f. Med. Physik
Hufferstr. 68
D-4400 Münster
FR Germany

Schotman, Mr. Anton J.H.
Klin. Veterinär Facult.
Yalelaan 16
NL-Utrecht
The Netherlands

Schramel, Mr. Peter
GSF, Phys.-Techn. Abt.
Ingolstädter Landstr. 1
D-8042 Neuherberg
FR Germany

Schroll, Mr. Armin
Deutsches Herzzentrum
Lothstr. 11
D-8000 München 2
FR Germany

Simonoff, Mr. Gabriel
Laboratoire de Chimie Nucleaire de Bordeaux
Le Haut Vigenau
F-33170 Gradignan
France

Simonoff, Mrs. Monique
Laboratoire de Chimie Nucleaire de Bordeaux
Le Haut Vigenau
F-33170 Gradignan
France

Stampfl, Mr. Andreas
GSF, Abt. f. Pharmakologie
Ingolstädter Landstr. 1
D-8042 Neuherberg
FR Germany

Stoeppler, Mr. M.
KFA Jülich, Inst. f. Chemie
D-5170 Jülich
FR Germany

Summer, Mr. K.H.
GSF, Abt. f. Toxikologie
Ingolstädter Landtsr. 1
D-8042 Neuherberg
FR Germany

Terai, Mr. Minoru
Dept. of Chem., Faculty of Science, Metropol. Univ.
2-1-1, Hukasawa, Setagaya
J-Tokyo, 158
Japan

Tjioe, Mr. P.S.
Interuniversity Reactor Inst.
15 Mekelweg
NL-2629 JB Delft
The Netherlands

Toelg, Mr. Günther
Max-Planck-Institut f. Metallf.
D-7070 Schwäbisch-Gmünd

Van den Hamer, Mr.
Interuniversity Reactor Inst.
15 Mekelweg
NL-2629 JB Delft
The Netherlands
FR Germany

Vandeteren, Mrs. Carine
INW-RUG
B-9000 Gent
Belgium

Verbuecken, Armand
Dept. Chem., Univ. of Antwerpen
B-2610 Wilryk
Belgium

Vernois, Mr. Vincent
Lab. "Pierre Süc", C.E.N. Saclay
F-91191 Fif s/Yvette
France

Wallaeys, Mr. Bart
Research Associate
INW Proeftuinstraat 86
B-9000 Gent
Belgium

Weber, Mr. G.
Univ. de Liège
B-4000 Liège
Belgium

Weigert, Mr. Peter
Bundesgesundheitsamt Berlin
D-1000 Berlin
FR Germany

Welz, Mr. Bernhard
Bodenseewerk Perkin-Elmer
Postfach 11 20
D-7770 Überlingen
FR Germany

Weser, Mr. U.
Anorg. Biochem., Phys.-Chem.
Institut, Univ. Tübingen
D-7400 Tübingen
FR Germany

Westermarck, Mr. T.
Central Inst. f. the Mentally Subnormal
SF-02400 Kirkkunumni
Finland

White, Mr. Colin
CSIRO, P. O. Wembley 6014
West Australia

Wick, Mr. Roland R.
GSF, Abt. f. Pathologie
Ingolstädter Landstr. 1
D-8042 Neuherberg
FR Germany

Woittiez, Mr. J.R.W.
ECN
NL-1755 ZG Petten
The Netherlands

Yukawa, Mr. Masae
Nat. Inst. of Radiol. Science
J-Anakawa, Chiba-Shi
Japan

Zawislak, Mr. Richard
Lab. d' Analyses
Inst. de Chimie Biolog.
11 rue Human
F-67085 Strasbourg, CEDEX
France

AUTHOR INDEX

SUBJECT INDEX

Note:
Page numbers refer only to the first page of each contribution